AF722755

Galois Theory and Modular Forms

Developments in Mathematics

VOLUME 11

Series Editor:

Krishnaswami Alladi, *University of Florida, U.S.A.*

Aims and Scope

Developments in Mathematics is a book series publishing

(i) Proceedings of Conferences dealing with the latest research advances,

(ii) Research Monographs, and

(iii) Contributed Volumes focussing on certain areas of special interest

Editors of conference proceedings are urged to include a few survey papers for wider appeal. Research monographs, which could be used as texts or references for graduate level courses, would also be suitable for the series. Contributed volumes are those where various authors either write papers or chapters in an organized volume devoted to a topic of special/current interest or importance. A contributed volume could deal with a classical topic, which is once again in the limelight owing to new developments.

Galois Theory and Modular Forms

Edited by

Ki-ichiro Hashimoto
Waseda University
Tokyo, Japan

Katsuya Miyake
Tokyo Metropolitan University
Tokyo, Japan

Hiroaki Nakamura
Okayama University
Okayama, Japan

Kluwer Academic Publishers
Boston/ Dordrecht/ London

Distributors for North, Central and South America:
Kluwer Academic Publishers
101 Philip Drive
Assinippi Park
Norwell, Massachusetts 02061 USA
Telephone (781) 871-6600
Fax (781) 871-6528
E-Mail <kluwer@wkap.com>

Distributors for all other countries:
Kluwer Academic Publishers Group
Post Office Box 322
3300 AH Dordrecht, THE NETHERLANDS
Telephone 31 78 6576 000
Fax 31 78 6576 474
E-Mail <orderdept@wkap.nl>

Electronic Services <http://www.wkap.nl>

Library of Congress Cataloging-in-Publication

Hashimoto, Ki-ichiro/ Miyake, Katsuya/ Nakamura, Hiroaki
Galois Theory and Modular Forms
ISBN 1-4020-7689-4

Copyright © 2004 by Kluwer Academic Publishers

All rights reserved. No part of this publication may be reproduced, stored in a retrieval system or transmitted in any form or by any means, electronic, mechanical, photo-copying, microfilming, recording, or otherwise, without the prior written permission of the publisher, with the exception of any material supplied specifically for the purpose of being entered and executed on a computer system, for exclusive use by the purchaser of the work.
Permissions for books published in the USA: permissions@wkap.com
Permissions for books published in Europe: permissions@wkap.nl
Printed on acid-free paper.

Printed in the United States of America

Contents

Preface

This volume is an outgrowth of the research project "The Inverse Galois Problem and its Application to Number Theory" which was carried out in three academic years from 1999 to 2001 with the support of the Grant-in-Aid for Scientific Research (B) (1) No.11440013. In September, 2001, an international conference "Galois Theory and Modular Forms" was held at Tokyo Metropolitan University after some preparatory workshops and symposia in previous years. The title of this book came from that of the conference, and the authors were participants of those meetings. All of the articles here were critically refereed by experts. Some of these articles give well prepared surveys on branches of research areas, and many articles aim to bear the latest research results accompanied with carefully written expository introductions.

When we started our research project, we picked up three areas to investigate under the key word "Galois groups"; namely, "generic polynomials" to be applied to number theory, "Galois coverings of algebraic curves" to study new type of representations of absolute Galois groups, and explicitly described "Shimura varieties" to understand well the Galois structures of some interesting polynomials including Brumer's sextic for the alternating group of degree 5. The topics of the articles in this volume are widely spread as a result. At a first glance, some readers may think this book somewhat unfocussed. The editors, however, believe it is of interest to present this collection of articles because they discuss those topics each of which could trace its source back to a spring of advanced ideas on Galois theory.

On the Inverse Galois Problem, Hilbert gave a remarkable result on the symmetric groups over the rational number field at the beginning of the last century. Then E. Noether proposed the so called "Noether's Problem" which became one of basic problems of modern algebra. If we go back to the 19th century, it was Abel who found new Abelian equations other than cyclotomic ones with moduli and values at division points of periods of elliptic functions. He was also interested in characterizing Abelian polynomials which were algebraically solvable. Then, for example, Kronecker formulated the Kronecker-Weber Theorem, and proposed 'the fondest dream in his youth' to determine all of the Abelian extensions of the rational number field, and those of imaginary quadratic fields by special values of the exponential function, and elliptic, and elliptic modular functions. The theory of complex multiplications of elliptic curves was certainly a starting point towards the later vast studies of automorphic forms and of Shimura varieties. The reader will find several articles in this book which concern with the beautiful theory and examples explaining/indicating mysterious phenomena arising in various arithmetic functions and varieties related to modular forms.

In the 1920's, algebraic number theory succeeded in establishing class field theory, and, as a result, opened the door to non-Abelian worlds. These were there behind the work of Hilbert and the abstractions of E. Noether stated above. Later in the latter half of the previous century, when the classification of finite simple groups was realized, we see a quite new development of the Inverse Galois Problem. For example, Thompson gave a stimulating plan for finite simple groups and their automorphism groups. Matzat is certainly one of the most active mathematicians who took this course. The editors are very happy to include his substantial article in the book, which establishes a foundation for the recent active studies of the differential Galois theory.

Investigation of structures of Galois groups is also an important stream in the above history of the Inverse Galois Problem. In 1953, Iwasawa showed that the commutator subgroup of the Galois group of the maximal solvable extension over the rational number field is a free pro-finite-solvable group with a countable number of generators. Then Shafarevich's conjecture claims that the commutator subgroup of the Galois group of the algebraic closure of the rational number field should be a free profinite group with a countable number of generators. Recently, our knowledge on structures of profinite fundamental groups has been vastly progressed under the name of "anabelian geometry". The reader will find an excellent paper by S. Mochizuki, where a new frontier has been pioneered for our understanding the arithmetic fundamental groups of curves over p-adic fields.

Classification of Galois extensions is another important branch of the Inverse Galois Problem. Here let us mention a problem which is concerned with "generic polynomials" for finite cyclic groups. When the local class field theory was established by the global class field theory, E. Noether immediately pointed out that the converse should be the right way; namely, the global class field theory should be constructed on the basis of the much simpler local class field theory. As a result of both theories, we know that every Abelian extension of a local completion field can be obtained by an Abelian extension of the base global field. This has not, however, been proved yet in any simple and direct manners. One way may be to utilize "generic polynomials" for finite cyclic groups over some small algebraic number fields. It should be noted that there exist no generic polynomials for cyclic group of order 8 over the rational number field.

The editors would like to express their hearty gratitude to all the participants and the speakers at our meetings, to the authors of articles here and the referees behind them, for their intimate cooperation at various stages of our modest project, all of which have been indispensable for collecting resulted fruits successfully in the present proceedings volume. Special thanks should also go to Dr. Yasuhiro Kishi and Mr. Masafumi Imaoka for their efforts in arranging the articles of the volume in the publisher's format.

Katsuya Miyake,
Ki-ichiro Hashimoto,
Hiroaki Nakamura

Part I

Arithmetic geometry

Galois Theory and Modular Forms
K. Hashimoto, K. Miyake, and H. Nakamura, Editors pp 3–12
© 2003 Kluwer Academic Publishers

THE ARITHMETIC OF WEIERSTRASS POINTS ON MODULAR CURVES $X_0(p)$

Scott AHLGREN*
Department of Mathematics,
University of Illinois,
Urbana, Illinois 61801
ahlgren@math.uiuc.edu

1. Introduction

The purpose of this paper is to describe some recent results regarding the arithmetic properties of Weierstrass points on modular curves $X_0(p)$ for primes p. We begin with some generalities; most of these can be found, for example, in the book of Farkas and Kra [F-K]. Suppose that X is a compact Riemann surface of genus $g \geq 2$. If r is a positive integer, then let $\mathcal{H}^r(X)$ denote the space of holomorphic r-differentials on X. Each $\mathcal{H}^r(X)$ is a finite-dimensional vector space over $\mathbb{C}$; we denote its dimension by $d_r(X)$. A point $Q \in X$ is called an *r-Weierstrass point* if there exists a non-zero differential $\omega \in \mathcal{H}^r(X)$ such that

$$\mathrm{ord}_Q\omega \geq d_r(X).$$

Suppose that $Q \in X$ and that $\{\omega_1, \dots, \omega_{d_r(X)}\}$ is a basis of $\mathcal{H}^r(X)$ which is adapted to the point Q; in other words we have

$$0 = \mathrm{ord}_Q\omega_1 < \mathrm{ord}_Q\omega_2 < \cdots < \mathrm{ord}_Q\omega_{d_r(X)}.$$

Then we define the *r-Weierstrass weight* of Q as

$$\mathrm{wt}_r(Q) := \sum_{j=1}^{d_r(X)} (\mathrm{ord}_Q\omega_j - j + 1)\,; \tag{1.1}$$

*The author thanks the National Science Foundation for its support through grants DMS 01-96443 and DMS 01-34577.
[received: June 13, 2002; accepted in revised form: October 16, 2002]

this definition does not depend on the particular basis chosen. Two basic facts are that

$$\mathrm{wt}_r(Q) > 0 \iff Q \text{ is an } r\text{-Weierstrass point}$$

and that

$$\sum_{Q\in X} \mathrm{wt}_r(Q) = d_r(X)(g-1)(2r-1+d_r(X)). \tag{1.2}$$

In this paper we will be exclusively concerned with modular curves; these are a class of Riemann surfaces which play an important role in number theory. If N is a positive integer, then we denote by $\Gamma_0(N)$ the usual congruence subgroup

$$\Gamma_0(N) := \left\{ \begin{pmatrix} a & b \\ c & d \end{pmatrix} \in \mathrm{SL}_2(\mathbb{Z}) \ : \ c \equiv 0 \pmod{N} \right\}.$$

Let $\mathbb{H}$ denote the upper half-plane of complex numbers, and define the *modular curve* $X_0(N)$ as the compactification of the quotient $Y_0(N) := \Gamma_0(N)\backslash\mathbb{H}$. Part of the importance of these curves arises from the fact that the curve $X_0(N)$ is the moduli space for pairs (E, C), where E is an elliptic curve and C is a cyclic subgroup of order N.

1-Weierstrass points (or, more simply, Weierstrass points) on modular curves have been studied by a number of authors. Works of Atkin [At] and Lehner-Newman [L-N], for example, determine some conditions under which the cusp at ∞ is or is not a Weierstrass point (it remains an open problem to classify those N for which ∞ is a Weierstrass point). Weierstrass points also play an important role in the work of Lehner-Newman [L-N] and Ogg [O1] which determines the complete list of N for which $X_0(N)$ is hyperelliptic.

For the curves $X_0(p)$ with p prime, it is known by work of Atkin and Ogg (see [O2]) that ∞ is not a Weierstrass point. Moreover, Ogg [O2] proved that if p is prime and if Q is a $\mathbb{Q}$-rational Weierstrass point on $X_0(p)$, then the reduction $\widetilde{Q}$ of Q modulo p is supersingular (i.e. the underlying elliptic curve is supersingular).

In the remainder of this paper (which is purely expository) our goal will be to describe recent results of the author with K. Ono and M. Papanikolas [A-O], [A-P] which develop the relationship between Weierstrass points on $X_0(p)$ and supersingular points in characteristic p. As we shall see, the result which is obtained for 1-Weierstrass points is quite satisfactory, while the situation for higher Weierstrass points remains somewhat more nebulous.

2. Weierstrass points on $X_0(p)$ and supersingular j-invariants

If $p \geq 5$ is prime, then the genus of the modular curve $X_0(p)$ is given by

$$g_p := \begin{cases} (p-13)/12, & \text{if } p \equiv 1 \pmod{12}, \\ (p-5)/12, & \text{if } p \equiv 5 \pmod{12}, \\ (p-7)/12, & \text{if } p \equiv 7 \pmod{12}, \\ (p+1)/12, & \text{if } p \equiv 11 \pmod{12}. \end{cases}$$

We assume throughout that $p \geq 23$; this ensures that $g_p \geq 2$ (if $g_p < 2$ then there are no r-Weierstrass points for any r). Since ∞ is not a Weierstrass point (from which it follows that the conjugate cusp at 0 is not a Weierstrass point) and $d_1(X_0(p)) = g_p$, we see from (1.2) that

$$\sum_{Q \in Y_0(p)} \mathrm{wt}_1(Q) = g_p^3 - g_p. \tag{2.1}$$

Throughout, we agree that $q := e^{2\pi iz}$, and we denote by $j(z)$ the usual elliptic modular function on $\mathrm{SL}_2(\mathbb{Z})$; we have

$$j(z) = q^{-1} + 744 + 196884q + \cdots .$$

To investigate the arithmetic of Weierstrass points on $X_0(p)$, we study the polynomials

$$F_p(x) := \prod_{Q \in Y_0(p)} (x - j(Q))^{\mathrm{wt}_1(Q)}. \tag{2.2}$$

Here $j(Q)$ means $j(\tau)$, where $\tau \in \mathbb{H}$ is any point which corresponds to Q under the standard identification. By (2.1), we see that the degree of $F_p(x)$ is $g_p^3 - g_p$. We define the supersingular polynomial in characteristic p by

$$S_p(x) := \prod_{\substack{E/\overline{\mathbb{F}}_p \\ \text{supersingular}}} (x - j(E))$$

(here the product runs over supersingular isomorphism classes of elliptic curves over the algebraic closure of the finite field $\mathbb{F}_p$). Then $S_p(x) \in \mathbb{F}_p[x]$, the degree of $S_p(x)$ is $g_p + 1$, and $S_p(x)$ splits completely over $\mathbb{F}_{p^2}$ (the paper of Kaneko and Zagier [K-Z] describes this polynomial from a number of perspectives).

The main result of [A-O] is the following.

Theorem 1. *If p is prime, then $F_p(x)$ has p-integral rational coefficients and satisfies*
$$F_p(x) \equiv S_p(x)^{g_p(g_p-1)} \pmod{p}.$$

We now give a brief description of the important elements of the proof of Theorem 1. The starting point is the classical fact that the Weierstrass weight of a point $Q \in X_0(p)$ is determined by the order of vanishing of a certain differential form at Q. Let $S_2(\Gamma_0(p))$ denote the space of weight two cusp forms on $\Gamma_0(p)$, and let $\{f_1, \dots, f_{g_p}\}$ be any basis for this space. We then define the Wronskian modular form

$$W_p(f_1, \dots, f_{g_p})(z) := \begin{vmatrix} f_1(z) & f_2(z) & \cdots & f_{g_p}(z) \\ f_1'(z) & f_2'(z) & \cdots & f_{g_p}'(z) \\ \vdots & \vdots & \vdots & \vdots \\ f_1^{(g_p-1)}(z) & f_2^{(g_p-1)}(z) & \cdots & f_{g_p}^{(g_p-1)}(z) \end{vmatrix},$$

and we define $\mathcal{W}_p(z)$ as that scalar multiple of $W_p(f_1, \dots, f_{g_p})(z)$ whose leading coefficient equals 1. In this way, $\mathcal{W}_p(z)$ is a cusp form of weight $g_p(g_p+1)$ on $\Gamma_0(p)$ which is independent of the choice of $f_1, \dots, f_{g_p}$; moreover, in view of the bijection between the spaces $\mathcal{H}^1(X_0(p))$ and $S_2(\Gamma_0(p))$, we have the following fundamental fact:

$$\operatorname{div}\left(\mathcal{W}_p(z)(dz)^{\frac{1}{2}g_p(g_p+1)}\right) = \sum_{Q \in Y_0(p)} \mathrm{wt}_1(Q) \cdot Q. \tag{2.3}$$

Rohrlich [R] gave a precise description of $\mathcal{W}_p(z) \pmod{p}$ as a modular form on $\mathrm{SL}_2(\mathbb{Z})$. To describe his result requires the introduction of some notation. For even integers $k \geq 4$, let B_k denote the usual Bernoulli number, and define the Eisenstein series $E_k(z)$ (a modular form of weight k for $\mathrm{SL}_2(\mathbb{Z})$) by

$$E_k(z) := 1 - \frac{2k}{B_k} \sum_{n=1}^{\infty} \sum_{d|n} d^{k-1} q^n.$$

Also, define the usual delta function by

$$\Delta(z) = \frac{E_4(z)^3 - E_6(z)^2}{1728} = q - 24q^2 + 252q^3 + \cdots;$$

$\Delta(z)$ is the unique normalized cusp form of weight 12 on $\mathrm{SL}_2(\mathbb{Z})$. Finally, if $p \geq 23$ is prime, then define $r_p \in \{0, 6, 8, 14\}$ by $r_p := p + 1 - 12g_p$. With this notation (and the agreement that $E_0 = 1$), Rohrlich obtains the following.

Theorem 2 (Rohrlich). *If $p \geq 23$ is prime then $\mathcal{W}_p(z)$ has p-integral rational coefficients and satisfies*

$$\mathcal{W}_p \equiv \Delta^{g_p(g_p+1)/2} E_{r_p}^{g_p} E_{14}^{g_p(g_p-1)/2} \pmod{p}.$$

In addition to Rohrlich's result, a certain multiplicative construction is crucial for the proof of Theorem 1. In particular, the following is proved in [A-O]. We recall that if $f(z)$ is a function of the upper half-plane, k is an integer, and $\gamma = \begin{pmatrix} a & b \\ c & d \end{pmatrix} \in \mathrm{GL}_2^+(\mathbb{R})$, then

$$(f|_k\gamma)(z) := (ad-bc)^{\frac{k}{2}}(cz+d)^{-k} f\left(\frac{az+b}{cz+d}\right).$$

Lemma 2.1 ([A-O, Lemma 3.3]). *Suppose that $p \geq 23$ is prime, and let $\widetilde{\mathcal{W}_p}(z)$ be defined by the product*

$$\prod_{A \in \Gamma_0(p) \backslash \mathrm{SL}_2(\mathbb{Z})} \mathcal{W}_p(z)\big|_{g_p(g_p+1)} A,$$

normalized to have leading coefficient equal to 1. *Then $\widetilde{\mathcal{W}_p}(z)$ is a cusp form of weight $g_p(g_p+1)(p+1)$ on $\mathrm{SL}_2(\mathbb{Z})$. Moreover, $\widetilde{\mathcal{W}_p}(z)$ has p-integral rational coefficients and satisfies*

$$\widetilde{\mathcal{W}_p}(z) \equiv \mathcal{W}_p(z)^2 \pmod{p}.$$

Remark 1. This result holds more generally for modular forms which are eigenforms of the Fricke involution; see Proposition 5.2 of [A-P] for a precise statement.

Remark 2. Since this is a multiplicative construction, the divisor of the modular form $\widetilde{\mathcal{W}_p}(z)$ retains much of the information present in the divisor of the original form $\mathcal{W}_p(z)$ (some care must of course be taken to recover this information, particularly at the elliptic fixed points of $\mathrm{SL}_2(\mathbb{Z})$). In view of (2.3), the construction is particularly useful for this reason.

It follows from the von Staudt-Claussen congruences that $E_{p-1} \equiv 1 \pmod{p}$. Using this fact together with Theorem 2 and Lemma 2.1, we conclude that

$$\widetilde{\mathcal{W}_p}(z) \equiv E_{p-1}(z)^{g_p^2-g_p} \Delta(z)^{g_p(g_p+1)} E_{r_p}(z)^{2g_p} E_{14}(z)^{g_p(g_p-1)} \pmod{p}. \tag{2.4}$$

Moreover, the modular forms on either side of (2.4) have the same weight.

It is a well-known result of Deligne (see, for example, [K-Z] or [Sw]) that

$$S_p(x) \equiv \prod_{\substack{\tau \in \mathrm{SL}_2(\mathbb{Z})\backslash\mathbb{H} \\ E_{p-1}(\tau)=0}} (x - j(\tau)) \pmod{p}. \tag{2.5}$$

By (2.2), (2.3), and Remark 2 above, we see that the values of the j-function at the zeros of $\widetilde{\mathcal{W}}_p(z)$ correspond roughly to the roots of the polynomial $F_p(x)$. However, by (2.4) we see that many of these j-values should also correspond roughly to the j-zeros of E_{p-1}; by (2.5) these are just the supersingular j-invariants in characteristic p. Theorem 1 follows from these considerations; of course it takes quite a bit of work to make all of this precise.

3. Higher Weierstrass points on $X_0(p)$

Many of the ingredients (Rohrlich's theorem, for example) which conspire to give such a clean result in the case of 1-Weierstrass points rely in turn on the isomorphism between $\mathcal{H}^1(X_0(p))$ and $S_2(\Gamma_0(p))$. The absence of such an isomorphism for the spaces $\mathcal{H}^r(X_0(p))$ when $r \geq 2$ causes much complication in these cases. For example, when $r \geq 2$, it is not clear whether or not ∞ is an r-Weierstrass point. Moreover, the p-integrality (or non p-integrality) of the analogue of the polynomial $F_p(x)$ becomes an issue. However, under a suitable assumption (which we shall discuss in detail below), we obtain in [A-P] an analogue of Theorem 1 for r-Weierstrass points on $X_0(p)$ when $2 \leq r \leq 5$.

Suppose that $p \geq 23$ is prime and that $r \geq 2$. Then, in analogy with (2.2), we define the polynomial

$$F_p^{(r)}(x) := \prod_{Q \in Y_0(p)} (x - j(Q))^{\mathrm{wt}_r(Q)}.$$

We note, using (1.2) and the fact that $d_r(X_0(p)) = (2r-1)(g_p - 1)$ for $r \geq 2$, that $F_p^{(r)}$ has degree $(2r-1)^2(g_p-1)^2 g_p - 2\mathrm{wt}_r(\infty)$. Also, we define

$$S_p^*(x) := \begin{cases} S_p(x), & \text{if } p \equiv 1 \pmod{12}, \\ S_p(x)/x, & \text{if } p \equiv 5 \pmod{12}, \\ S_p(x)/(x-1728), & \text{if } p \equiv 7 \pmod{12}, \\ S_p(x)/x(x-1728), & \text{if } p \equiv 11 \pmod{12}. \end{cases}$$

In every case, $S_p^*(x)$ is actually a polynomial. Further, we define the polynomial

$$f_{p,r}(x) := \begin{cases} 1, & \text{if } p \equiv 1 \pmod{12}, \\ x^{\lceil \frac{1}{3}(2r-1)^2(g_p-1)(g_p-2) \rceil}, & \text{if } p \equiv 5 \pmod{12}, \\ (x-1728)^{\frac{1}{2}(2r-1)^2(g_p-1)(g_p-2)}, & \text{if } p \equiv 7 \pmod{12}, \\ x^{\lceil \frac{1}{3}(2r-1)^2(g_p-1)(g_p-2) \rceil} & \\ \times (x-1728)^{\frac{1}{2}(2r-1)^2(g_p-1)(g_p-2)}, & \text{if } p \equiv 11 \pmod{12}. \end{cases}$$

We obtain an analogue of Theorem 1 when $2 \leq r \leq 5$ under the assumption that the space $\mathcal{H}^r(X_0(p))$ is *good* at p; the precise definition (as well as an explanation of the assumption that $r \leq 5$) will be given below.

Theorem 3. *Suppose that p is prime and that $2 \leq r \leq 5$. Suppose that $\mathcal{H}^r(X_0(p))$ is good at p. Then $F_p^{(r)}(x)$ has rational p-integral coefficients, and there exists a polynomial $H(x) \in \mathbb{F}_p[x]$ such that*

$$F_p^{(r)}(x) \equiv H(x) \cdot f_{p,r}(x) \cdot S_p^*(x)^{(2r-1)^2(g_p-1)(g_p-2)} \pmod{p}.$$

Remark. In view of the definitions of $S_p^*(x)$ and $f_{p,r}(x)$, Theorem 3 shows that $F_p^{(r)}(x) \pmod{p}$ is in every case divisible by a large power of the full supersingular polynomial $S_p(x)$. Computations suggest that the powers of x and $x-1728$ present in the definition of $f_{p,r}(x)$ may in fact be replaced by larger powers of these monomials.

To define what it means to be "good" requires some further notation. If $r \geq 2$ then let $S_{2r}^*(\Gamma_0(p))$ be the subspace of $S_{2r}(\Gamma_0(p))$ which corresponds to $\mathcal{H}^r(X_0(p))$ under the standard identification of meromorphic r-differentials on $X_0(p)$ with meromorphic modular forms of weight $2r$ on $\Gamma_0(p)$. Let $d := d_r(X_0(p))$. Then we see that $H^r(X_0(p))$ has a basis of the form $\{\omega_1, \omega_2, \ldots, \omega_d\}$, where for each i we have $\omega_i = f_i(z)(dz)^r$, and $f_i(z) \in S_{2r}^*(\Gamma_0(p))$ has rational Fourier coefficients. In order to determine such a basis uniquely, we stipulate further that the modular forms $f_i(z)$ have expansions at ∞ of the form

$$\begin{aligned} f_1(z) &= q^{r_1} + O(q^{r_1+1}), \\ f_2(z) &= q^{r_2} + O(q^{r_2+1}), \\ &\vdots \qquad \vdots \\ f_d(z) &= q^{r_d} + O(q^{r_d+1}), \end{aligned}$$

where $r = r_1 < r_2 < \cdots < r_d$, and the coefficient of q^{r_i} in f_j is zero if $i \neq j$. Then we call $\mathcal{H}^r(X_0(p))$ *good* at p if the modular forms $f_1, \ldots, f_d$ have p-integral Fourier coefficients.

The question of whether or not a given space $H^r(X_0(p))$ is good at p is an interesting one. Computations uncover a few "bad" primes, but suggest that most such spaces are indeed good. For example, the only prime $p < 800$ for which $\mathcal{H}^2(X_0(p))$ fails to be good at p is $p = 373$. Using W. Stein's modular forms package in MAGMA, we have carried out the following computations.

Proposition 3.1.

(1) $\mathcal{H}^2(X_0(p))$ *is good at* p *for* $23 \leq p < 800$ *with the exception of* $p = 373$.

(2) $\mathcal{H}^3(X_0(p))$ *is good at* p *for* $23 \leq p < 800$ *with the exception of* $p = 373, 643$.

(3) $\mathcal{H}^4(X_0(p))$ *is good at* p *for* $23 \leq p < 800$.

(4) $\mathcal{H}^5(X_0(p))$ *is good at* p *for* $23 \leq p < 400$ *with the exception of* $p = 379$.

The method of proof of Theorem 3 follows the rough outline of the proof of Theorem 1, although there are additional difficulties which are not present in the case $r = 1$. The assumption that $\mathcal{H}^r(X_0(p))$ is good at p ensures p-integrality of the relevant Wronskian in this case (and therefore the p-integrality of the polynomial $F_p^{(r)}$). We again use a multiplicative construction as in Lemma 2.1. However, we lack Rohrlich's Theorem 2 when $r \geq 2$; it is therefore necessary to develop some analogues of results of Serre regarding connections between the reductions modulo p of spaces of modular forms on $\Gamma_0(p)$ and those on $\mathrm{SL}_2(\mathbb{Z})$. These results are strong enough to obtain a statement as in Theorem 3 only when $r \leq 5$ (this assumption ensures that there are no "oldforms" of weight $2r$ on $\Gamma_0(p)$).

4. Examples

We give some examples of the phenomena described by Theorems 1 and 3. As an example of Theorem 1, we consider the case when $p = 31$. Here $g_p = 2$, and

$$S_{31}(x) = (x + 8)(x + 27)(x + 29).$$

Moreover, we find that

$$\begin{aligned} F_{31}(x) = x^6 &- 1559739496886640x^5 - 6159702699716996017077 3x^4 \\ &+ 570086541463862006632745 28x^3 \\ &- 24149765663444687311928192763 42x^2 \\ &- 133372401227986529665702756133 0208x \\ &- 93888205239878524150070589491402 2689 \\ \equiv (x+8)^2&(x+27)^2(x+29)^2 \pmod{31}. \end{aligned}$$

Next we consider the case of Theorem 3 corresponding to $p = 53$ and $r = 2$. Here $g_p = 4$ and $\mathrm{wt}_2(\infty) = 0$. We have

$$S_{53}(x) = x(x+3)(x+7)(x^2+50x+39) = x \cdot S_{53}^*(x),$$

and

$$F_{53}^{(2)}(x) \equiv S_{53}(x)^{54} \cdot x^{18} \cdot H_1(x) \pmod{53},$$

where $H_1(x)$ is the square of a polynomial of degree 18.

Finally we consider the case of Theorem 3 corresponding to $p = 73$ and $r = 2$. Here $g_p = 5$, $\mathrm{wt}_2(\infty) = 1$, and

$$S_{73}(x) = S_{73}^*(x) = (x+17)(x+64)(x^2+57x+8)(x^2+68x+9).$$

We find that

$$F_{73}^{(2)}(x) \equiv S_{73}(x)^{108} \cdot H_2(x) \pmod{73},$$

where $H_2(x)$ is the square of a polynomial of degree 35.

5. Some questions

In closing, we mention a few natural questions related to the results described above.

(1) Is it possible to find a proof of Theorem 1 (or of Theorem 3) which relies on the underlying geometry of the curve $X_0(p)$?

(2) Are there interesting Galois-theoretic properties of the polynomials $F_p(x)$ and $F_p^{(r)}(x)$?

(3) Is it possible to find a description of those p and r for which ∞ is an $r-$ Weierstrass point on $X_0(p)$?

(4) Is it possible to give a description of those p and r for which $\mathcal{H}^r(X_0(p))$ fails to be good at p? Are there infinitely many such p for any given r?

(5) Is there a more natural description of what it means to be "good"?

(6) Is it possible to obtain a more precise description of the polynomials $H(x)$ which appear in Theorem 3? Do the roots of these polynomials contain interesting arithmetic information?

(7) In the same vein, can the polynomial $S_p^*(x)$ in Theorem 3 always be replaced by the full supersingular polynomial $S_p(x)$?

Acknowledgments

The author wishes to thank K. Hashimoto, K. Miyake, and H. Nakamura, as well as others involved in the organization of the Saga and TMU conferences on Galois Theory and Modular Forms, for all of their efforts towards running two successful meetings.

References

[A-O] S. Ahlgren and K. Ono, *Weierstrass points on $X_0(p)$ and supersingular j-invariants*, Math. Ann. **325** (2003), 355–368.

[A-P] S. Ahlgren and M. Papanikolas, *Higher Weierstrass points on $X_0(p)$*, Trans. Amer. Math. Soc. **355** (2003), 1521–1535.

[At] A. O. L. Atkin, *Weierstrass points at cusps of $X_0(N)$*, Ann. of Math. **85** (1967), 42–45.

[F-K] H. M. Farkas and I. Kra, "Riemann surfaces," Springer-Verlag, New York, 1992.

[K-Z] M. Kaneko and D. Zagier, *Supersingular j-invariants, hypergeometric series, and Atkin's orthogonal polynomials*, in "Computational perspectives on number theory (Chicago, IL,1995)," Amer. Math. Soc., Providence, RI, 97–126, 1998.

[L-N] J. Lehner and M. Newman, *Weierstrass points on $\Gamma_0(N)$*, Ann. of Math. **79** (1964), 360–368.

[O1] A. Ogg, *Hyperelliptic modular curves*, Bull. Soc. Math. France **102** (1974), 449–462.

[O2] A. Ogg, *On the Weierstrass points of $X_0(N)$*, Illinois J. Math. **22** (1978), 31–35.

[R] D. Rohrlich, *Weierstrass points and modular forms*, Illinois J. Math. **29** (1985), 134–141.

[Sw] H. P. F. Swinnerton-Dyer, *On ℓ-adic representations and congruences for modular forms*, in "Modular functions of one variable, III," 1–55, Lecture Notes in Mathematics, 350. Springer-Verlag, Berlin.

Galois Theory and Modular Forms pp 13–38
K. Hashimoto, K. Miyake, and H. Nakamura, Editors
© 2003 Kluwer Academic Publishers

SEMISTABLE ABELIAN VARIETIES WITH SMALL DIVISION FIELDS

Armand BRUMER
Department of Mathematics
Fordham University,
Bronx, NY 10458
brumer@fordham.edu

Kenneth KRAMER*
Department of Mathematics
Queens College (CUNY)
Flushing, NY 11367
kramer@forbin.qc.edu

1. Introduction

The conjecture of Shimura-Taniyama-Weil, now proved through the work of Wiles and disciples, is only part of the Langlands program. Based on a comparison of the local factors ([And], [Ser1]), it also predicts that the L-series of an abelian surface defined over $\mathbb{Q}$ should be the L-series of a Hecke eigen cusp form of weight 2 on a suitable group commensurable with $\mathrm{Sp}_4(\mathbb{Z})$. The only decisive examples are related to lifts of automorphic representations of proper subgroups of Sp_4, for example the beautiful work of Yoshida ([Yos], [BSP]).

We do not know how to construct "non-trivial" weight 2 forms for groups such as the paramodular group or the Hecke subgroup $\Gamma_0(p)$ in genus 2. In fact, it seems that not a single example is known which is not a lift. Not even a conjectural dimension formula for weight 2 cusp forms has yet been proposed ([Ibu], [Has]).

It seems interesting to consider existence and non-existence on the arithmetic side. If the special fiber $\mathcal{A}_p$ of the Néron model of the abelian variety A has no unipotent part, we say that A has semistable reduction

*This author was supported in part by a grant from the City University of New York PSC-CUNY Research Award Program.
[received: June 7, 2002 ; accepted in revised form: October, 16, 2002]

at p. In that case, the exponent $\mathfrak{f}_p(A)$ of p in the conductor of A is the toroidal dimension of $\mathcal{A}_p$. For example, the modular variety $J_0(p)$ has conductor p^d, where d is its dimension. To ensure that the putative automorphic form not be a lift, we could study surfaces of prime conductor. This guarantees that we are not dealing, for instance, with a surface isogenous to a product of elliptic curves, the Weil restriction of scalars of an elliptic curve defined over a quadratic number field or a surface with non-trivial endomorphisms. Serre's conjecture [Ser4] implies that such a surface should be $\mathbb{Q}$-isogenous to the Jacobian of a curve also defined over $\mathbb{Q}$. This led us to try to understand curves C of genus 2 whose Jacobian variety $J = J(C)$ has prime conductor p. A minimal model for C has bad reduction at p, but also may have primes of bad reduction where J has good reduction. We call such primes, at which J reduces to the product of 2 elliptic curves, primes of *mild* reduction. Because of mild primes, we could not immediately rule out the possibility that all six Weierstrass points of C be rational, that is $\mathbb{Q}(J[2]) = \mathbb{Q}$. In the present investigation, we find that $\mathrm{Gal}(\mathbb{Q}(J[2])/\mathbb{Q})$ cannot even be nilpotent, as we shall explain below.

One expects that $\mathrm{Gal}(\mathbb{Q}(A[\ell])/\mathbb{Q})$ should be, in general, as large as is compatible with the endomorphisms of A and the Weil pairing. When $\mathrm{End}\, A = \mathbb{Z}$, this has been proved for ℓ sufficiently large by Serre [Ser5], in all odd dimensions as well as in dimensions 2 and 6. Note that $\mathbb{Q}(A[\ell])$ always contains the cyclotomic field $\mathbb{Q}(\boldsymbol{\mu}_\ell)$, whether or not A is principally polarized. As examples of the results obtained here, we have:

Proposition 2.4. *Let $A/\mathbb{Q}$ be a semistable abelian variety with good reduction at a prime $\ell \leq 19$ and set $H = \mathrm{Gal}(\mathbb{Q}(A[\ell])/\mathbb{Q}(\boldsymbol{\mu}_\ell))$. Then ℓ annihilates the maximal abelian quotient H/H'. So H is an ℓ-group if it is nilpotent.*

Theorem 1.1. *Suppose that $A/\mathbb{Q}$ is semistable, with bad reduction only at p, and that H is nilpotent. If $\ell = 2$ or 3, then $\mathfrak{f}_p(A) = \dim A$, so the reduction at p is totally toroidal. Moreover $p \equiv 1 \bmod 8$ if $\ell = 2$ and $p \equiv 1 \bmod 3$ if $\ell = 3$.*

Consider elliptic curves of prime conductor p with some rational ℓ-torsion. For $\ell = 2$, Neumann [Ne] and Setzer [Set] found that such elliptic curves exist if and only if $p = n^2 + 64$ or $p = 17$. For odd ℓ, Miyawaki [Mi] showed that the well-known examples with $(\ell, p) = (3, 19)$, (3,37) or (5,11) are the only ones. Under the stronger assumption that A has prime conductor, Theorem 1.1 implies:

Corollary 1.2. *Let $A/\mathbb{Q}$ be an abelian variety of prime conductor p such that H is nilpotent. If $\ell = 2$, then A is a Neumann-Setzer elliptic curve. If $\ell = 3$, then $p = 19$ or 37 and A is a Miyawaki elliptic curve.*

Let $\mathcal{C}$ be the $\mathbb{Q}$-isogeny class of an abelian variety A with semistable bad reduction at p. We denote by Φ_A the group of connected components of the special fiber of its Néron model at p. We say that A is ℓ-maximal if $\mathrm{ord}_\ell |\Phi_A(\bar{\mathbb{F}}_p)|$ is maximal among varieties in $\mathcal{C}$. A major role will be played by the collection

$$\mathcal{V}_\ell(\mathcal{C}) = \{A \in \mathcal{C} \mid \mathrm{ord}_\ell |\Phi_A(\bar{\mathbb{F}}_p)| \text{ is maximal}\} \tag{1}$$

of ℓ-maximal varieties. This notion depends on the choice of p, which will be clear from the context.

We show that $\mathbb{Q}(A[\ell]) = \mathbb{Q}(\boldsymbol{\mu}_{2\ell})$ for all $A \in \mathcal{V}_\ell(\mathcal{C})$ if $\mathcal{C}$ is the isogeny class of a product of elliptic curves of conductor p with some rational ℓ-torsion, cf. Proposition 5.3. The following is a converse when $\ell = 2$ or 3.

Theorem 1.3. *Let $\ell = 2$ or 3 and let $\mathcal{C}$ be a $\mathbb{Q}$-isogeny class of semistable abelian varieties with bad reduction only at p. If $\mathbb{Q}(A[\ell]) \subseteq \mathbb{Q}(\boldsymbol{\mu}_{2\ell})$ for all $A \in \mathcal{V}_\ell(\mathcal{C})$, then $\mathcal{C}$ is the isogeny class of a product of Neumann-Setzer ($\ell = 2$) or Miyawaki ($\ell = 3$) curves of conductor p.*

A brief outline of the paper follows. With the help of class field theory and the conductor bounds of Fontaine, we obtain in section 2 the information we require about the group $H = \mathrm{Gal}(\mathbb{Q}(A[\ell])/\mathbb{Q}(\boldsymbol{\mu}_\ell))$.

Denote by $\mathbb{T}_\ell(A)$ the Tate module of A and by $\mathcal{D}_v$ the decomposition group in $G_\mathbb{Q}$ of a place v over p. In section 3, we recall the definition of the $\mathcal{D}_v$-submodules of $\mathbb{T}_\ell(A)$ introduced by Grothendieck [Gro]. Now suppose A is semistable with bad reduction only at p and that H is an ℓ-group. Let a denote the dimension of the abelian variety in the special fiber of A at p. We use the Grothendieck modules together with ℓ-maximality and a strange lifting lemma to construct a pure $\mathbb{Z}_\ell$-submodule W_A of $\mathbb{T}_\ell(A)$ of rank $2a$. Although W_A is canonically only a $\mathcal{D}_v$-module, we obtain the rather surprising result that it is a $G_\mathbb{Q}$-module under suitable hypotheses. Then, in section 4, we use the formal group associated to A at ℓ and Raynaud's theory of $\mathbb{F}$-vector space schemes to prove that $W_A = 0$, so that $a = 0$ and A is totally toroidal at p. The proof of Theorem 1.1 may be found at the end of this section.

A generalization of Theorem 1.3 is proved in section 5, using the theorem of Faltings on Tate's conjecture. We then give two examples for which H is an ℓ-group, namely, $J_0(31)$ with $\ell = 2$ and $J_0(41)$ with $\ell = 5$. Since they clearly are not isogenous to a product of elliptic curves, any attempt to weaken the hypotheses of Theorem 1.3 will require some care.

Fontaine [Fo] proved that there are no abelian varieties of positive dimension with everywhere good reduction over $\mathbb{Q}$. We showed, in [BK2], that no semistable abelian variety exists with bad reduction exactly at one prime $p \leq 7$ and Schoof [Scho] improved this result by different methods.

In our next paper, using other techniques, we plan to study the existence of abelian surfaces $A/\mathbb{Q}$ of conductor p in the spirit of [BK1]. When A is principally polarized, $G = \mathrm{Gal}(\mathbb{Q}(A[2])/\mathbb{Q})$ is a subgroup of $\mathrm{GSp}_4(\mathbb{F}_2) \simeq \mathcal{S}_6$, the symmetric group on 6 letters. As an application of Theorem 1.1, we will show that G can only be $\mathcal{S}_n$ for $3 \leq n \leq 6$, $\mathcal{S}_n \times \mathbb{Z}/2$ for $n = 3, 4$, or the wreath product $\mathcal{S}_3 \wr \mathbb{Z}/2$, with each of these possibilities occurring infinitely often under the Schinzel hypothesis [SS].

Our criteria will show that $p = 277$ is the smallest prime for which there is a curve of genus 2 whose Jacobian has prime conductor. One such curve is given by

$$y^2 + y = x^5 - 2x^3 + 2x^2 - x.$$

We thus propose, as a challenge to our automorphic friends, a verification that all modular forms of weight 2 for smaller prime level are "lifts", but that at least one interesting eigenform exists for 277.

This paper concentrates on one of the results presented by the first author at this Conference. He is grateful to K. Hashimoto and K. Miyake for their kind invitation and for their support both moral and financial. Other results, more influenced by the work of Hashimoto and Ibukiyama, will be published later. Thanks to H. Nakamura for helping run the Conference so pleasantly and seamlessly.

2. Controlling the ℓ-division field

Let S be a set of rational primes not containing the prime ℓ nor the archimedean prime ∞. We shall say that a finite extension $L/\mathbb{Q}$ is (ℓ, S)-controlled if it has the following properties:

(L1) L is Galois over $\mathbb{Q}$ and contains $\boldsymbol{\mu}_\ell$;

(L2) $L/\mathbb{Q}$ is unramified outside each place over $S \cup \{\ell, \infty\}$;

(L3) the ramification degree for each place of L over S divides ℓ;

(L4) the higher ramification groups $\mathcal{D}_\lambda^{(u)}$ are trivial for $u > 1/(\ell - 1)$, where $\mathcal{D}_\lambda$ is the decomposition group at a place λ over ℓ in L.

If $S = \{p\}$ consists of one prime, we say $L/\mathbb{Q}$ is (ℓ, p)-controlled. Thanks to Grothendieck [Gro, §2.5] and Fontaine [Fo, Thm. A], the ℓ-division field of a semistable abelian variety $A/\mathbb{Q}$ with good reduction at ℓ is (ℓ, S)-controlled when S contains the primes of bad reduction. See also [BK2, §3].

Since the upper ramification numbering behaves well with respect to quotients, a subfield of L containing $\boldsymbol{\mu}_\ell$ and Galois over $\mathbb{Q}$ also is (ℓ, S)-controlled. Similarly, property (L4) is inherited by the compositum of finitely many Galois extensions that enjoy it. Properties (L1) and (L2) certainly are preserved under composition of (ℓ, S)-controlled extensions. By uniqueness of the tamely ramified extension of degree ℓ over the unramified closure of $\mathbb{Q}_p$, property (L3) is preserved under composition.

It is helpful to highlight the impact of (L4) on the following local situation, a variant of which was treated in [BK2, Lem. 6].

Lemma 2.1. *Let $E \supseteq F \supseteq \mathbb{Q}_\ell(\boldsymbol{\mu}_\ell) \supseteq \mathbb{Q}_\ell$ be a tower of local fields such that $E/\mathbb{Q}_\ell$ is Galois and $E/\mathbb{Q}_\ell(\boldsymbol{\mu}_\ell)$ is an abelian ℓ-extension. The higher ramification groups* $\mathrm{Gal}(E/\mathbb{Q}_\ell)^{(u)}$ *are trivial for all $u > 1/(\ell-1)$ if and only if each of the abelian conductor exponents $\mathfrak{f}(E/F)$ and $\mathfrak{f}(F/\mathbb{Q}_\ell(\boldsymbol{\mu}_\ell))$ is at most* 2.

Proof. First consider a general Galois extension of local fields E/K and put $H = \mathrm{Gal}(E/K)$. Assume only that F is the fixed field of a normal subgroup $N \triangleleft H$ and let $\bar{H} = H/N$. Write $\varphi_{E/K}$ for the Herbrand function [Ser2, IV, §3] from the lower to upper ramification numbering of H. Let us verify that $H_x = 1$ if and only if both N_x and $\bar{H}_{x'}$ are trivial, where $x' = \varphi_{E/F}(x)$. If $H_x = 1$, then $N_x = H_x \cap N = 1$ and by Herbrand's theorem [Ser2, IV, §3, Lem. 5], we also have $\bar{H}_{x'} = H_xN/N = 1$. Conversely, if $\bar{H}_{x'} = 1$, then $H_x \subseteq N$. It follows that $H_x = N_x$, so if N_x also is trivial, we may conclude that $H_x = 1$. This observation will be used below with $K = \mathbb{Q}_\ell(\boldsymbol{\mu}_\ell)$.

Now assume all the given hypotheses and let $D = \mathrm{Gal}(E/\mathbb{Q}_\ell)$. Since the tame ramification degree of the extension $E/\mathbb{Q}_\ell$ is $\ell - 1$, we have $\varphi_{E/\mathbb{Q}_\ell}(1) = |D_1|/|D_0| = 1/(\ell-1)$. Furthermore, $D_x = H_x$ for all $x \geq 1$ because H is the ℓ-Sylow subgroup of D. Then we have the following chain of equivalences:

$$\begin{aligned} D^{(u)} = 1 \text{ for all } u > 1/(\ell-1) &\Leftrightarrow D_{1+\epsilon} = 1 \text{ for all } \epsilon > 0 \\ &\Leftrightarrow H_{1+\epsilon} = 1 \text{ for all } \epsilon > 0 \\ &\Leftrightarrow N_{1+\epsilon} = 1 \text{ and } \bar{H}_{1+\epsilon} = 1 \text{ for all } \epsilon > 0, \end{aligned}$$

using our opening observation and the fact that $\varphi_{E/F}(1) = 1$ for the last equivalence.

The conductor exponent of an abelian extension E/K is given by $\mathfrak{f}(E/K) = \varphi_{E/K}(c) + 1$, where c is the largest integer such that $H_c \neq 1$. When H is an ℓ-group, we have $\varphi_{E/K}(1) = 1$. For an abelian ℓ-extension E/K we therefore find that $\mathfrak{f}(E/K) \leq 2$ if and only if $H_{1+\epsilon} = 1$ for all $\epsilon > 0$. Applying this fact to the abelian extensions E/F and $F/\mathbb{Q}_\ell(\boldsymbol{\mu}_\ell)$, we may conclude that $D^{(u)} = 1$ for all $u > 1/(\ell - 1)$ if and only if both $\mathfrak{f}(E/F) \leq 2$ and $\mathfrak{f}(F/\mathbb{Q}_\ell(\boldsymbol{\mu}_\ell)) \leq 2$. □

We use below the fact, proved in [MM], that $\mathbb{Q}(\boldsymbol{\mu}_\ell)$ has class number one exactly when $\ell \leq 19$.

Lemma 2.2. *Suppose M is (ℓ, S)-controlled and unramified outside ℓ, with $\ell \leq 19$. If M is abelian over $F = \mathbb{Q}(\boldsymbol{\mu}_{2\ell})$, then $M = F$.*

Proof. To analyze the extension M/F, we introduce the usual notation of class field theory. Let $\mathbb{A}_F^\times$ be the idele group of F and write $\mathbf{F}^\times$ for the image of $F^\times$ on the diagonal. Let F_v be the completion of F at a place v and put U_v for the connected component of the identity in $F_v^\times$. If v is non-archimedean, set $U_v^{(m)} = \{x \in U_v \mid v(x-1) \geq m\}$.

Observe that there is a unique prime λ over ℓ in F and let $\mathfrak{l}$ be a prime of M over λ. By (L4), we have $\mathrm{Gal}(M_\mathfrak{l}/\mathbb{Q}_\ell)^{(u)} = 1$ for all $u > 1/(\ell-1)$. For brevity, write $\mathfrak{f}_\lambda$ for the abelian conductor exponent of the local extension $M_\mathfrak{l}/F_\lambda$. Since M/F is unramified outside ℓ, there is a surjection $\mathbb{A}_F^\times/\mathcal{N} \twoheadrightarrow \mathrm{Gal}(M/F)$, where $\mathcal{N}$ is the norm subgroup

$$\mathcal{N} = (U_\lambda^{(\mathfrak{f}_\lambda)} \times \prod_{v \neq \lambda} U_v) \cdot \mathbf{F}^\times.$$

Let V_F be the image of the global units of F in $\Gamma = U_\lambda/U_\lambda^{(\mathfrak{f}_\lambda)}$. In our case $U_v \simeq \mathbb{C}^\times$ for archimedean places, so we have the exact sequence

$$0 \to \Gamma/V_F \to \mathbb{A}_F^\times/\mathcal{N} \to \mathbb{A}_F^\times/((\prod_v U_v) \cdot \mathbf{F}^\times) \to 0, \qquad (1)$$

in which the last quotient is isomorphic to the ideal class group of F and is trivial by assumption.

The group $U_\lambda/U_\lambda^{(1)}$ is generated by the images of global cyclotomic units when ℓ is odd and is trivial when $\ell = 2$. Since $U_\lambda^{(1)}/U_\lambda^{(f_\lambda)}$ is an ℓ-group, we find that Γ/V_F is an ℓ-group and so $\mathrm{Gal}(M/F)$ is an ℓ-group by exact sequence (1). Then $\mathfrak{f}_\lambda \leq 2$ by Lemma 2.1. The group $U_\lambda^{(1)}/U_\lambda^{(2)}$ is generated by the image of a global primitive ℓ^{th} root of unity when ℓ is odd or by the image of i when $\ell = 2$. Thus Γ/V_F and *a fortiori* $\mathrm{Gal}(M/F)$ is trivial. □

Remark. Under the hypothesis that $\mathrm{Gal}(M/F)$ is an abelian ℓ-group, similar reasoning shows that $M = F$ when ℓ is a regular prime.

Lemma 2.3. *Suppose that $\ell \leq 19$ and that L is an (ℓ, S)-controlled extension containing $F = \mathbb{Q}(\boldsymbol{\mu}_{2\ell})$. Let E be the maximal subfield* *of L abelian over F. Then* $\mathrm{Gal}(E/F)$ *is annihilated by ℓ and* $\dim_{\mathbb{F}_\ell} \mathrm{Gal}(E/F) \leq s$, *where s is the number of primes over S in F.*

Proof. For each prime $\mathfrak{p}_j$ over p in F, choose a prime $\mathfrak{P}_j$ over $\mathfrak{p}_j$ in L. Since p does not ramify in $F/\mathbb{Q}$, the inertia group $\mathcal{I}_j = \mathcal{I}_{\mathfrak{P}_j}$ is contained in $H = \mathrm{Gal}(L/F)$. Let N be the subgroup of H generated by $\mathcal{I}_j$ for $j = 1, \dots, s$ and by the commutator subgroup H'. The previous Lemma shows that $N = H$. By (L3), each $\mathcal{I}_j$ has order dividing ℓ, so $\mathrm{Gal}(E/F)$ is annihilated by ℓ and its $\mathbb{F}_\ell$-dimension is at most s. □

Remark. The bound on $\dim_{\mathbb{F}_\ell} \mathrm{Gal}(E/F)$ above can sometimes be sharpened by a class field theoretic analysis similar to that of Lemma 2.2. Define

$$\Gamma_S = \prod_{v \in S_F} U_v / U_v^\ell,$$

where S_F is the set of primes over S in F. Recall that λ denotes the prime over ℓ in F and let $\mathcal{U}$ be the image in Γ_S of those global units ϵ in F such that $\epsilon \equiv 1 \bmod \lambda^2$. Then $\mathrm{Gal}(E/F)$ is a quotient of $\Gamma_S/\mathcal{U}$.

Proposition 2.4. *Let $A/\mathbb{Q}$ be a semistable abelian variety with good reduction at a prime $\ell \leq 19$ and set* $H = \mathrm{Gal}(\mathbb{Q}(A[\ell])/\mathbb{Q}(\boldsymbol{\mu}_\ell))$. *Then ℓ annihilates the maximal abelian quotient H/H'. If H is nilpotent, then H is an ℓ-group.*

Proof. If ℓ is odd, or if $\ell = 2$ and $i \in \mathbb{Q}(A[\ell])$, let $L = \mathbb{Q}(A[\ell])$. Otherwise, let $L = \mathbb{Q}(i, A[2])$, in which case, $\mathrm{Gal}(L/\mathbb{Q}(i)) \simeq H$. Then L is an (ℓ, S)-controlled extension and our result follows from Lemma 2.3. Note that if H is nilpotent, it is the product of all its Sylow subgroups, so it must be an ℓ-group. □

Lemma 2.5. *Suppose L is an (ℓ,p)-controlled extension containing $F = \mathbb{Q}(\boldsymbol{\mu}_{2\ell})$ and ℓ is a regular prime. Put $G = \mathrm{Gal}(L/\mathbb{Q})$ and assume $H = \mathrm{Gal}(L/F)$ is an ℓ-group. Let $\tilde{\sigma}$ be a generator for the inertia group $\mathcal{I}_{\mathfrak{P}}$ of a prime $\mathfrak{P}$ over p in L and suppose $\tilde{\tau} \in G$ restricts to a generator of $\mathrm{Gal}(F/\mathbb{Q})$. Then the conjugates of $\tilde{\sigma}$ by powers of $\tilde{\tau}$ generate H, while $\tilde{\sigma}$ and $\tilde{\tau}$ generate G. If p is unramified in L, then $L = F$.*

Proof. Note that $\mathcal{I}_{\mathfrak{P}}$ is cyclic of order dividing ℓ by (L3). Let N be the subgroup of H generated by H', H^ℓ and the conjugates of $\tilde{\sigma}$ by powers of $\tilde{\tau}$. Then N is normal in G and corresponds to an (ℓ,p)-controlled extension M/F, unramified outside ℓ, such that $\mathrm{Gal}(M/F)$ is an elementary abelian ℓ-group. So $N = H$ by the Remark after Lemma 2.2. Our claim now follows from Burnside's lemma. □

Next, we give a description of the maximal $(2,p)$-controlled 2-extension of $\mathbb{Q}$. Since we do not need this result, we leave the proof to the reader as an exercise in class field theory. A general study of extensions of number fields with wild ramification of bounded depth can be found in recent work of Hajir and Maire [HM].

Proposition 2.6. *Consider $K = \mathbb{Q}(\sqrt{-p})$ and let n be the 2-part of the class number of K. If $p \equiv 3 \bmod 4$, define $M = K(i)$. If $p \equiv 1 \bmod 4$, let M be the 2-part of the ray class field of conductor 2 over K. Then $\mathrm{Gal}(M/K)$ is cyclic of order $2n$ and $\mathrm{Gal}(M/\mathbb{Q})$ is dihedral. Moreover, M is $(2,p)$-controlled and contains every $(2,p)$-controlled 2-extension of $\mathbb{Q}$.*

Finally, we summarize for later use some elementary facts about pure submodules unrelated to the rest of this section.

Lemma 2.7. *Suppose X and Y are pure submodules of a free $\mathbb{Z}_\ell$-module T of finite rank. Write $\bar{X} = (X + \ell T)/\ell T$ for the projection of X to $T/\ell T$. Then:*

(i) $\overline{X \cap Y} \subseteq \bar{X} \cap \bar{Y}$ *and* $\overline{X + Y} = \bar{X} + \bar{Y}$;

(ii) $X \cap Y$ *is pure;*

(iii) *if* $\bar{X} = 0$, *then* $X = 0$;

(iv) *if* $\bar{X} \cap \bar{Y} = 0$, *then* $X + Y$ *is pure and a direct sum.*

Let T' be a free $\mathbb{Z}_\ell$-module of finite rank and suppose there is a perfect pairing $e : T \times T' \to \mathbb{T}_\ell(\mu)$. Then $X^\perp + Y^\perp \subseteq (X \cap Y)^\perp$, with equality if $X^\perp + Y^\perp$ is pure.

Proof. Property (i) is clear and does not use purity. The natural injection

$$T/(X \cap Y) \hookrightarrow T/X \oplus T/Y$$

implies (ii). By Nakayama's Lemma and the isomorphism $\bar{X} \simeq X/(X \cap \ell T) = X/\ell X$ we have (iii). For (iv), suppose $\ell z = x + y$ for some $z \in T$, $x \in X$ and $y \in Y$. Then the coset $\bar{x} = -\bar{y}$ in $T/\ell T$ is an element of $\bar{X} \cap \bar{Y} = 0$. Hence $\bar{x} = \bar{y} = 0$. By purity of X and Y, we may write $x = \ell x_1$ and $y = \ell y_1$ for some $x_1 \in X$ and $y_1 \in Y$. Since T is torsion-free, we then have $z = x_1 + y_1 \in X + Y$. Hence $T/(X + Y)$ is torsion-free and (iv) is verified. Given the perfect pairing e, we clearly have $X^\perp + Y^\perp \subseteq (X \cap Y)^\perp$. Equality holds if $X^\perp + Y^\perp$ is pure because both sides have the same rank. □

3. Decomposition of the Tate module

Suppose $A/\mathbb{Q}$ is an abelian variety with good reduction at ℓ and semistable bad reduction at p. Grothendieck described certain submodules of the Tate module $\mathbb{T}_\ell(A)$ that are Galois modules for a decomposition group over p in $G_\mathbb{Q}$. A convenient summary of the information we need also appears in [Ed, §2]. In this section, we use our assumptions that the ℓ-division field of A is small and that A has good reduction outside p to create $G_\mathbb{Q}$-submodules of $\mathbb{T}_\ell(A)$ from these Grothendieck modules. First we establish the relevant notation.

Let $\mathcal{A}_p$ denote the special fiber of the Néron model of A at p. Its connected component $\mathcal{A}_p^0$ admits a decomposition

$$0 \to \mathcal{T} \to \mathcal{A}_p^0 \to \mathcal{B} \to 0$$

in which $\mathcal{T}$ is a torus and $\mathcal{B}$ an abelian variety defined over $\mathbb{F}_p$. Setting $\dim \mathcal{T} = t$ and $\dim \mathcal{B} = a$, we have $t + a = \dim A$. Write $\Phi_A = \mathcal{A}_p/\mathcal{A}_p^0$ for the group of connected components. Denote the perfect pairing on the Tate modules of A and its dual abelian variety $\hat{A}$ by

$$e_\infty : \mathbb{T}_\ell(A) \times \mathbb{T}_\ell(\hat{A}) \to \mathbb{T}_\ell(\boldsymbol{\mu}) = \varprojlim \boldsymbol{\mu}_{\ell^n}. \tag{1}$$

Let $L_\infty = \mathbb{Q}(A[\ell^\infty])$ be the ℓ-division tower of A and set $G_\infty = \mathrm{Gal}(L_\infty/\mathbb{Q})$. Since we may be moving among abelian varieties $\mathbb{Q}$-isogenous to A, it is important to note that L_∞ only depends on the $\mathbb{Q}$-isogeny class of A. Clearly $G_\mathbb{Q}$ acts on $\mathbb{T}_\ell(A)$ through G_∞. Fix an embedding $\iota : L_\infty \to \bar{\mathbb{Q}}_p$ and let v be the corresponding valuation. Write $\mathcal{D}_v \supseteq \mathcal{I}_v$ for its decomposition and inertia groups in G_∞. For $g \in G_\infty$, let gv denote the valuation corresponding to $\iota \circ g^{-1}$.

As in [Gro, §2.5], we define $\mathcal{M}_1 = \mathcal{M}_1(A, v)$ to be the submodule of $\mathbb{T}_\ell(A)$ fixed by $\mathcal{I}_v$ and $\mathcal{M}_2 = \mathcal{M}_2(A, v)$ to be the submodule of $\mathbb{T}_\ell(A)$ orthogonal to $\mathcal{M}_1(\hat{A}, v)$ under the e_∞-pairing. Clearly $g \in G_\infty$ acts by $g(\mathcal{M}_j(A, v)) = \mathcal{M}_j(A, gv)$ for $j = 1, 2$. Thus $\mathcal{M}_1$ and $\mathcal{M}_2$ are $\mathcal{D}_v$-modules.

The Igusa-Grothendieck theorem [Gro, Thm. 2.5] asserts that if A is semistable at p, then $\mathcal{M}_2 \subseteq \mathcal{M}_1$. Further, the successive quotients in the decomposition

$$\mathbb{T}_\ell(A) \supset \mathcal{M}_1 \supseteq \mathcal{M}_2 \supset 0, \tag{2}$$

are torsion-free $\mathbb{Z}_\ell$-modules. Both $\mathcal{M}_1$ and $\mathcal{M}_2$ are modules for $\mathcal{D}_v/\mathcal{I}_v \simeq G_{\mathbb{F}_p}$. We may identify $\mathcal{M}_2 \simeq \mathbb{T}_\ell(\mathcal{T})$ and $\mathcal{M}_1/\mathcal{M}_2 \simeq \mathbb{T}_\ell(\mathcal{B})$, so the $\mathbb{Z}_\ell$-ranks of $\mathcal{M}_2$ and $\mathcal{M}_1/\mathcal{M}_2$ are t and $2a$ respectively. Using the e_∞-pairing, one can see that $(g-1)(\mathbb{T}_\ell(A)) \subseteq \mathcal{M}_2$ for all $g \in \mathcal{I}_v$.

Let $\mathbb{T}_\ell(\varphi)$ denote the map of Tate modules induced by a $\mathbb{Q}$-isogeny $\varphi : A \to A'$ and set $\mathcal{M}'_j = \mathcal{M}_j(A', v)$ for $j = 1, 2$. Then $\mathbb{T}_\ell(\varphi)(\mathcal{M}_j) \subseteq$

$\mathcal{M}'_j$. We can also show that $\mathbb{T}_\ell(\varphi)^{-1}(\mathcal{M}'_j) \subseteq \mathcal{M}_j$ by using the existence of a quasi-inverse isogeny $\varphi' : A' \to A$ such that $\varphi' \circ \varphi$ (resp. $\varphi \circ \varphi'$) is multiplication by m on A (resp. A'), where m is the exponent of the kernel of φ. Indeed, if $\mathbb{T}_\ell(\varphi)(x) \in \mathcal{M}'_j$, then $mx \in \mathbb{T}_\ell(\varphi')(\mathcal{M}'_j) \subseteq \mathcal{M}_j$. But $\mathbb{T}_\ell(A)/\mathcal{M}_j$ is torsion-free, so $x \in \mathcal{M}_j$.

The following result, mildly strengthening [BK2, Lem. 3], describes a more subtle effect of isogeny. Write $\bar{\mathcal{M}}_1^{(n)}$ and $\bar{\mathcal{M}}_2^{(n)}$ for the projections of $\mathcal{M}_1$ and $\mathcal{M}_2$ to the nth layer $\mathbb{T}_\ell(A)/\ell^n\mathbb{T}_\ell(A) \simeq A[\ell^n]$. Write $|X|_\ell$ for the order of the ℓ-primary subgroup of a finite abelian group X. See (1) for our definition of ℓ-maximal abelian varieties.

Lemma 3.1. *Suppose $A/\mathbb{Q}$ has good reduction at ℓ and semistable bad reduction at p. Let $\varphi : A \to A'$ be a $\mathbb{Q}$-isogeny whose kernel κ has exponent ℓ^n. Then*

$$|\Phi_{A'}(\bar{\mathbb{F}}_p)|_\ell \cdot |\kappa/(\kappa \cap \bar{\mathcal{M}}_1^{(n)})| = |\Phi_A(\bar{\mathbb{F}}_p)|_\ell \cdot |\kappa \cap \bar{\mathcal{M}}_2^{(n)}|.$$

In particular, if A is ℓ-maximal and κ is a Galois submodule of $A[\ell^n]$ contained in $\bar{\mathcal{M}}_1^{(n)}$, then $\kappa \cap \bar{\mathcal{M}}_2^{(n)} = 0$.

Proof. The main argument is sketched in [BK2, Lem. 3] for $n = 1$ and carries over with obvious modifications. As a simplication to the formula given there, we may use the equality $|\Phi_A(\bar{\mathbb{F}}_p)| = |\Phi_{\hat{A}}(\bar{\mathbb{F}}_p)|$, arising from the existence of a pairing

$$\Phi_A(\bar{\mathbb{F}}_p) \times \Phi_{\hat{A}}(\bar{\mathbb{F}}_p) \to \mathbb{Q}/\mathbb{Z}, \tag{3}$$

defined by Grothendieck and shown to be non-degenerate for semistable abelian varieties by various authors. For this pairing and its history, see [Mc]. □

Corollary 3.2. *If A has semistable bad reduction at an odd prime p and is 2-maximal for p, then $\mathbb{Q}(A[2]) \neq \mathbb{Q}$. If, in addition, $\mathbb{Q}(A[2])$ is unramified outside 2 and ∞, then $\mathbb{Q}(A[2]) = \mathbb{Q}(i)$.*

Proof. If $\mathbb{Q}(A[2]) = \mathbb{Q}$, then $\bar{\mathcal{M}}_2^{(1)}$ is a non-trivial Galois module, in contradiction to the last assertion of the previous lemma. When $\mathbb{Q}(A[2])$ is unramified outside 2 and ∞, property (L4) shows that the discriminant $d_{L/\mathbb{Q}}$ of $L = \mathbb{Q}(A[2], i)$ satisfies $|d_{L/\mathbb{Q}}|^{1/[L:\mathbb{Q}]} < 4$. Odlyzko's lower bounds for root discriminants give $[L : \mathbb{Q}] \leq 5$. It follows from Lemma 2.5 that $L = \mathbb{Q}(i)$, whence $\mathbb{Q}(A[2]) = \mathbb{Q}(i)$. □

Recall that there is a dictionary (see [Sch, §2.5]) between G_∞-submodules X of $T = \mathbb{T}_\ell(A)$ and abelian varieties $\mathbb{Q}$-isogenous to A.

If $\ell^n T \subseteq X$ then $\kappa = X/\ell^n T \subseteq A[\ell^n]$ is the kernel of an isogeny $\varphi : A \to A' = A/\kappa$. Write φ' for the quasi-inverse isogeny, such that $\varphi' \circ \varphi$ is multiplication by ℓ^n on A and $\varphi \circ \varphi'$ is multiplication by ℓ^n on A'. Then $\mathbb{T}_\ell(\varphi')$ provides a Galois isomorphism of $\mathbb{T}_\ell(A')$ onto X. In particular, $X/\ell X \simeq \mathbb{T}_\ell(A')/\ell\mathbb{T}_\ell(A') \simeq A'[\ell]$.

Lemma 3.3. *Let $A/\mathbb{Q}$ be an abelian variety and let W be a $\mathbb{Z}_\ell$-submodule of $T = \mathbb{T}_\ell(A)$ stabilized by a subgroup $\mathcal{H}$ of G_∞. Define $W_n = W + \ell^n T$ and let $\mathcal{S}$ be the set of integers $n \geq 0$ such that W_n is stabilized by G_∞. Let $\mathcal{C}_0$ be a collection of abelian varieties containing at least those that correspond to the G_∞-submodules W_n for $n \in \mathcal{S}$. Define*

$$\mathcal{N} = \{g \in G_\infty \mid g \text{ acts by homothety on } A'[\ell] \text{ for all } A' \in \mathcal{C}_0\}.$$

If $\mathcal{H}$ and $\mathcal{N}$ generate G_∞, then G_∞ stabilizes W.

Proof. Since $W_0 = T$, we have $0 \in \mathcal{S}$. Suppose $n \in \mathcal{S}$. Then W_n is stabilized by G_∞, so W_n corresponds to $A' = A/\kappa_n$, where $\kappa_n = W_n/\ell^n T$. By assumption, A' is in $\mathcal{C}_0$, so the elements of $\mathcal{N}$ act as homotheties on $A'[\ell] \simeq W_n/\ell W_n$. For $w \in W$, consider a coset $w + \ell W_n$ in $W_n/\ell W_n$. If $g \in \mathcal{N}$ acts on this coset by multiplication by α, then $g(w) + \ell W_n = \alpha w + \ell W_n$ and therefore

$$g(w) \in \alpha w + \ell W_n = \alpha w + \ell(W + \ell^n T) \subseteq W + \ell^{n+1} T = W_{n+1}.$$

Hence $\mathcal{N}$ stabilizes W_{n+1}. Given that G_∞ is generated by $\mathcal{H}$ and $\mathcal{N}$ and that $\mathcal{H}$ already stabilizes W, we find that G_∞ stabilizes W_{n+1}. Therefore $n+1$ is in $\mathcal{S}$ and it follows that G_∞ stabilizes W_n for all $n \geq 0$. Hence G_∞ stabilizes W. □

We now explain the special assumptions on which our study of abelian varieties with small ℓ-division fields rests. Suppose $A/\mathbb{Q}$ is semistable, with bad reduction only at p and that $H = \mathrm{Gal}(\mathbb{Q}(A[\ell])/\mathbb{Q}(\boldsymbol{\mu}_\ell))$ is an ℓ-group. Let $\mathcal{C}$ denote the $\mathbb{Q}$-isogeny class of A. Thanks to the theorem of Faltings [Fa, Satz 6] and its extension to unpolarized abelian varieties by Zarhin [Zar, Thm. 1], the set of isomorphism classes in $\mathcal{C}$ is finite. Put L for the compositum of the ℓ-division fields of the varieties in $\mathcal{C}$ and, if necessary, adjoin $i = \sqrt{-1}$ when $\ell = 2$. Then L contains $F = \mathbb{Q}(\boldsymbol{\mu}_{2\ell})$ and L is an (ℓ, p)-controlled extension, as defined in section 2.

Once and for all, we choose a topological generator σ for the cyclic pro-ℓ group $\mathcal{I}_v \subset G_\infty$. There exists an element $\tau \in G_\infty$ whose restriction to F generates $\mathrm{Gal}(F/\mathbb{Q})$. When ℓ is odd, we choose τ to have order $\ell - 1$, using the fact that H is an ℓ-group. When $\ell = 2$, we choose τ to be the generator of some inertia group at an archimedean place of L_∞, *i.e.* a complex conjugation. In view of Lemma 2.5, the restrictions of σ

and τ to L generate $\mathrm{Gal}(L/\mathbb{Q})$ when ℓ is a regular prime. If $\ell = 2$ or 3 then $\tau^2 = 1$. However, our methods only require that the action of τ on $\mathbb{T}_\ell(A)$ satisfy a polynomial of degree at most 2, so they also apply, for example, if A is of GL_2-type.

Basic Assumptions.

(C1) $A/\mathbb{Q}$ is semistable, with bad reduction only at p;

(C2) ℓ is a regular prime and $H = \mathrm{Gal}(\mathbb{Q}(A[\ell])/\mathbb{Q}(\boldsymbol{\mu}_\ell))$ is an ℓ-group;

(C3) the action of τ on $\mathbb{T}_\ell(A)$ satisfies a polynomial of degree at most 2.

These conditions depend only on the $\mathbb{Q}$-isogeny class $\mathcal{C}$ of A. Indeed, $\mathbb{Q}(B[\ell^\infty]) = L_\infty$ is the same for all B in $\mathcal{C}$. Furthermore, H is an ℓ-group if and only if $\mathrm{Gal}(\mathbb{Q}(B[\ell])/\mathbb{Q}(\boldsymbol{\mu}_\ell))$ is an ℓ-group, since $\mathrm{Gal}(L_\infty/\mathbb{Q}(B[\ell]))$ is pro-ℓ. Finally, the minimal polynomial of τ can be read from $\mathbb{T}_\ell(A) \otimes \mathbb{Q}_\ell$.

We will show, under (C1) and (C2), that the following $\mathbb{Z}_\ell$-submodules of $\mathbb{T}_\ell(A)$ are in fact G_∞-modules:

$$W_A = \bigcap_{j\geq 0} \tau^j(\mathcal{M}_1(A,v)) \qquad \text{and} \qquad \mathrm{Y_A} = \sum_{\mathrm{j}\geq 0} \tau^{\mathrm{j}}(\mathcal{M}_2(\mathrm{A},\mathrm{v}))\,. \tag{4}$$

Proposition 3.4. *Suppose $A/\mathbb{Q}$ satisfies (C1) and (C2). Then W_A and Y_A are stabilized by G_∞. Furthermore W_A is a pure submodule of $\mathbb{T}_\ell(A)$.*

Proof. Let $\mathcal{H}$ be the closed subgroup of G_∞ generated by σ and τ. Clearly, τ acts on W_A and Y_A, in view of the definitions (4). Since σ is the identity map on $\mathcal{M}_1(A,v)$, it acts as the identity on W_A. As noted above, $(\sigma-1)(\mathbb{T}_\ell(A)) \subseteq \mathcal{M}_2(A,v)$, so σ acts on Y_A. Hence W_A and Y_A are modules for $\mathcal{H}$.

Let L be the compositum of the ℓ-division fields of all the abelian varieties $\mathbb{Q}$-isogenous to A and, if necessary when $\ell = 2$, also adjoin i. We wish to apply Lemma 4.6, with $T = \mathbb{T}_\ell(A)$, $\mathcal{G} = G_\infty$ and $W = W_A$ or Y_A. Then $\mathcal{N}$ certainly contains $\mathrm{Gal}(L_\infty/L)$. According to Lemma 2.5, the restrictions of σ and τ to L generate $\mathrm{Gal}(L/\mathbb{Q})$. Therefore $\mathcal{H}$ and $\mathcal{N}$ generate G_∞. It follows that W_A and Y_A are G_∞-modules.

The purity of $\mathcal{M}_1(A,v)$ implies the purity of W_A as a $\mathbb{Z}_\ell$-submodule of $\mathbb{T}_\ell(A)$ by Lemma 2.7. □

Recall from (1) the definition of an ℓ-maximal abelian variety. If X is a $\mathbb{Z}_\ell$-submodule of $\mathbb{T}_\ell(A)$, we write $\bar{X} = (X + \ell\mathbb{T}_\ell(A))/\ell\mathbb{T}_\ell(A)$ for the projection of X to $\mathbb{T}_\ell(A))/\ell\mathbb{T}_\ell(A) \simeq A[\ell]$.

Lemma 3.5. *Let A be an abelian variety satisfying (C1) and (C2). Put $\bar{\mathcal{M}}_1 = \bar{\mathcal{M}}_1(A, v)$ and define*

$$\kappa = \bigcap_{j \geq 0} \tau^j(\bar{\mathcal{M}}_1).$$

If A is ℓ-maximal, then $\kappa \cap \bar{\mathcal{M}}_2 = 0$.

Proof. By Lemma 2.5, the restrictions of σ and τ generate $\mathrm{Gal}(\mathbb{Q}(A[\ell])/\mathbb{Q})$. But σ acts trivially on $\mathcal{M}_1$ and therefore on κ, while τ acts on κ from the definition. Hence κ is a Galois invariant subspace of $A[\ell]$ contained in $\bar{\mathcal{M}}_1$ and we conclude by Lemma 3.1. □

Corollary 3.6. *If A satisfies (C1) and (C2), then there are at least 2 primes over p in $F = \mathbb{Q}(\boldsymbol{\mu}_{2\ell})$. In addition, if $\ell = 2$, then $p \equiv 1 \bmod 8$.*

Proof. We may assume that A is ℓ-maximal. Suppose ℓ is odd and there is only one prime over p in F. Then the decomposition group $\mathcal{D}_v$ projects onto the group $\mathrm{Gal}(F/\mathbb{Q})$ of order $\ell - 1$, with pro-ℓ kernel. Hence we could have chosen τ in $\mathcal{D}_v$ and so κ in the lemma is simply $\bar{\mathcal{M}}_1$. Therefore $\bar{\mathcal{M}}_2 = \kappa \cap \bar{\mathcal{M}}_2 = 0$. This implies that the toroidal dimension of the bad fiber of A is 0, contradicting bad reduction at p.

For $\ell = 2$, we showed in [BK2, Prop. 5] that $p \equiv 1 \bmod 4$, whence there are 2 primes over p in $F = \mathbb{Q}(i)$. Suppose $p \equiv 5 \bmod 8$ and let ϕ be a Frobenius element of $\mathcal{D}_v$, in the sense that ϕ induces the pth power map on the residue field. Then $\mathcal{D}_v = \langle \phi, \sigma \rangle$ and $\phi\sigma\phi^{-1} = \sigma^p$. Since $\phi(\sqrt{2}) = -\sqrt{2}$, the restrictions of τ, ϕ and σ to $E = \mathbb{Q}(i, \sqrt{2}, \sqrt{p})$ clearly generate $\mathrm{Gal}(E/\mathbb{Q})$. But E is the maximal elementary 2-extension of $\mathbb{Q}$ unramified outside $\{2, p, \infty\}$, so $G_\infty = \langle \tau, \phi, \sigma \rangle$ by Burnside's Lemma. Hence $\mathcal{D}_v$ has index 2, so is normal in G_∞.

Now we may write $\tau\sigma\tau^{-1} = \phi^x \sigma^y$ for some $x, y \in \mathbb{Z}_2$. Checking the action of both sides of this equation on a generator for $\mathbb{T}_2(\boldsymbol{\mu})$, we find that $x = 0$, so τ normalizes $\mathcal{I}_v$. Hence $\mathcal{M}_1$ is stabilized by τ and we get a contradiction as in the previous case. Therefore $p \equiv 1 \bmod 8$. □

From hereon in, we impose the additional assumption (C3). Since the dual abelian variety $\hat{A}$ is isogenous to A, it also satisfies (C3). Write $\bar{\mathcal{M}}'_j$ for the projection of $\mathcal{M}'_j = \mathcal{M}_j(\hat{A}, v)$ to $\hat{A}[\ell]$. In the perfect pairing

$$A[\ell] \times \hat{A}[\ell] \to \boldsymbol{\mu}_\ell \tag{5}$$

induced by the e_∞-pairing (1), we have $(\bar{\mathcal{M}}_1)^\perp = \bar{\mathcal{M}}'_2$. Recall that t and a respectively denote the toroidal and abelian dimension of the bad fiber of A at p.

Proposition 3.7. *Suppose A satisfies (C1), (C2), (C3) and is ℓ-maximal. Then W_A and Y_A are pure $\mathbb{Z}_\ell$-submodule of $\mathbb{T}_\ell(A)$ of rank $2a$ and $2t$ respectively, stabilized by G_∞. Under the e_∞-pairing, $W_A^\perp = Y_{\hat{A}}$ and $Y_A^\perp = W_{\hat{A}}$. We have a direct sum decomposition $\mathbb{T}_\ell(A) = W_A \oplus Y_A$. Furthermore, the natural map $W_A \to \mathcal{M}_1(A,v)/\mathcal{M}_2(A,v)$ is an isomorphism of $\mathbb{Z}_\ell$-modules.*

Proof. Hypothesis (C3) implies that $\tau^2(\mathcal{M}_2) \subseteq \mathcal{M}_2 + \tau(\mathcal{M}_2)$, so

$$Y_A = \sum_{j\geq 0} \tau^j(\mathcal{M}_2) = \mathcal{M}_2 + \tau(\mathcal{M}_2).$$

Furthermore, $\tau^2(\bar{\mathcal{M}}_2') \subseteq \bar{\mathcal{M}}_2' + \tau(\bar{\mathcal{M}}_2')$ and so $\tau^2(\bar{\mathcal{M}}_1) \supseteq \bar{\mathcal{M}}_1 \cap \tau(\bar{\mathcal{M}}_1)$ by the $\mathbb{F}_\ell$-vector space duality (5). Hence the Galois invariant subspace κ of Lemma 3.5 reduces to $\kappa = \bar{\mathcal{M}}_1 \cap \tau(\bar{\mathcal{M}}_1)$ and $\kappa \cap \bar{\mathcal{M}}_2 = 0$ by that lemma. Since $\mathcal{M}_2 \subseteq \mathcal{M}_1$, we have

$$\tau(\bar{\mathcal{M}}_2) \cap \bar{\mathcal{M}}_2 \subseteq \tau(\bar{\mathcal{M}}_1) \cap \bar{\mathcal{M}}_2 = \kappa \cap \bar{\mathcal{M}}_2 = 0. \tag{6}$$

Therefore, $Y_A = \mathcal{M}_2 + \tau(\mathcal{M}_2)$ is pure and a direct sum, of rank $2t$, by Lemma 2.7(iv). Similarly, $Y_{\hat{A}}$ is pure in $\mathbb{T}_\ell(\hat{A})$, so we have

$$Y_{\hat{A}} = \sum_{j\geq 0} \tau^j(\mathcal{M}_2') = \sum_{j\geq 0} \tau^j(\mathcal{M}_1^\perp) = (\bigcap_{j\geq 0} \tau^j(\mathcal{M}_1))^\perp = W_A^\perp.$$

Hence W_A is pure of rank $2a$. In view of Lemma 2.7(i) and (6), we have

$$\bar{W}_A \cap \bar{\mathcal{M}}_2 \subseteq (\bar{\mathcal{M}}_1 \cap \tau(\bar{\mathcal{M}}_1)) \cap \bar{\mathcal{M}}_2 = \tau(\bar{\mathcal{M}}_1) \cap \bar{\mathcal{M}}_2 = 0.$$

Hence $W_A + \mathcal{M}_2$ is pure of rank $t + 2a$ and a direct sum. By the obvious inclusion and equality of ranks, we have $W_A + \mathcal{M}_2 = \mathcal{M}_1$. It follows that

$$W_A \simeq (W_A + \mathcal{M}_2)/\mathcal{M}_2 = \mathcal{M}_1/\mathcal{M}_2.$$

By (6) and Lemma 2.7(iv), we also find that $\tau(\mathcal{M}_1) + \mathcal{M}_2$ is a pure submodule of $\mathbb{T}_\ell(A)$ and a direct sum. Equality of ranks implies that $\mathbb{T}_\ell(A) = \tau(\mathcal{M}_1) \oplus \mathcal{M}_2$ and so $\mathbb{T}_\ell(A) = \tau(W_A) \oplus \tau(\mathcal{M}_2) \oplus \mathcal{M}_2 = W_A \oplus Y_A$, as claimed. Finally, we have already observed in Proposition 3.4 that W_A and Y_A are G_∞-modules. □

4. The ℓ-adic formal group and the proof of theorem 1.1

Suppose that A is an abelian variety of dimension d defined over $\mathbb{Q}_\ell$, with good reduction modulo ℓ. The kernel of reduction has the structure of a formal group of height h and dimension d, with $d \leq h \leq 2d$. We put $\mathcal{F}_A$ for this formal group, suppressing the dependence on ℓ. The $G_{\mathbb{Q}_\ell}$-module $C = \mathcal{F}_A[\ell^\infty]$ is isomorphic to $(\mathbb{Q}_\ell/\mathbb{Z}_\ell)^h$ as an abelian group. Furthermore, C is a connected ℓ-divisible group over $\mathbb{Z}_\ell$ while $A[\ell^\infty]/C$ is étale.

Set $\mathbb{V}_\ell(A) = \mathbb{T}_\ell(A) \otimes_{\mathbb{Z}_\ell} \mathbb{Q}_\ell$ and $\mathbb{V}_\ell(\mathcal{F}_A) = \mathbb{T}_\ell(\mathcal{F}_A) \otimes_{\mathbb{Z}_\ell} \mathbb{Q}_\ell$ for the Tate vector spaces of A and $\mathcal{F}_A$. Clearly the inertia group inside $G_{\mathbb{Q}_\ell}$ acts trivially on $\mathbb{V}_\ell(A)/\mathbb{V}_\ell(\mathcal{F}_A)$, as this quotient corresponds to the Tate vector space of the reduction of A over $\mathbb{F}_\ell$. In the special case of *ordinary reduction* (*i.e.* $h = d$), we have

$$\mathbb{V}_\ell(\mathcal{F}_A)^\perp = \mathbb{V}_\ell(\mathcal{F}_{\hat{A}}) \tag{1}$$

under the Weil pairing $\mathbb{V}_\ell(A) \times \mathbb{V}_\ell(\hat{A}) \to \mathbb{Q}_\ell(1)$, where $\hat{A}$ denotes the dual abelian variety of A. (See for example [CG, p. 154].)

Lemma 4.1. *Let A be an abelian variety of dimension d defined over $\mathbb{Q}_\ell$ with good reduction modulo ℓ. If the tame ramification degree of the extension $\mathbb{Q}_\ell(A[\ell])/\mathbb{Q}_\ell$ divides $\ell - 1$, then A is ordinary.*

Proof. We extend the base to the ring of integers R in the unramified closure $K = \mathbb{Q}_\ell^{\mathrm{nr}}$ to obtain an algebraically closed residue field. According to Raynaud (see [Gru, Thm. 4.4]), the finite group scheme $A[\ell]$ admits a composition series whose simple constituents are $\mathbb{F}$-vector space schemes for varying finite fields $\mathbb{F}$ of characteristic ℓ. Such a constituent M is a cyclic $\mathbb{F}$-module, determined as follows. By simplicity, the wild ramification subgroup of G_K must act trivially on M, so the action of G_K factors through a quotient $\mathrm{Gal}(E/K)$, such that $E \subseteq K(A[\ell])$ and E/K is tamely ramified. Suppose $[E : K] = r$. Then $\mathbb{F} = \mathbb{F}_\ell(\boldsymbol{\mu}_r)$, with the action of a primitive r^{th} root of unity induced by the action of a generator for $\mathrm{Gal}(E/K)$.

In our case, r divides $\ell - 1$, whence $\mathbb{F} = \mathbb{F}_\ell$ and M therefore is a group scheme of order ℓ. These are classified by Oort-Tate. Thus each simple constituent M_j of $A[\ell]$ has the form $M_j \simeq G^\ell_{a_j,b_j}$, with $a_j b_j = \ell$ in the standard notation [Gru, Theorem 2.1]. Since the valuation v of R is unramified, we have $v(a_j) \in \{0, 1\}$. Then $G^\ell_{a_j,b_j}$ is étale (resp. connected) if $v(a_j) = 0$ (resp. 1). Let n_0 (resp. n_1) be the number of a_j such that $v(a_j) = 0$ (resp. 1).

The exponent of the different of an affine group scheme $\mathcal{G}$ may be defined as follows [Gru, p. 62]. Suppose $\mathcal{G} = \mathrm{spec}(\mathfrak{A})$, where $\mathfrak{A}$ is an R-algebra. Let $\Omega^1_{\mathfrak{A}/R}$ be the module of Kähler differentials of $\mathfrak{A}$ and define $\delta(\mathcal{G})$ to be the length of the R-module $s^*\Omega^1_{\mathfrak{A}/R} = \Omega^1_{\mathfrak{A}/R} \otimes_{\mathfrak{A}} R$, where the tensor product is formed over the zero-section $s : \mathfrak{A} \to R$.

Since the ℓ-divisible group arising from ℓ-power torsion on A has dimension d, we find that $\delta(A[\ell]) = dv(\ell) = d$ for example by [Gru, Prop. 3.4]. Furthermore, $\delta(G^{\ell}_{a_j,b_j}) = v(a_j)$ by [Gru, Prop. 2.3]. But δ behaves like an additive Euler characteristic on short exact sequences. Hence $d = \sum_j v(a_j) = n_1$. Note that the simple constituents that belong to the connected group scheme $\mathcal{F}_A[\ell]$ are precisely those for which $v(a_j) = 1$. By definition of the height h of $\mathcal{F}_A$ we have

$$\ell^h = |\mathcal{F}_A[\ell]| = \prod_{v(a_j)=1} |G^{\ell}_{a_j,b_j}| = \ell^{n_1}.$$

Hence $h = n_1 = d$ as claimed. □

We now return to the global situation: A is an abelian variety defined over $\mathbb{Q}$ and we impose the basic assumptions (C1) and (C2) of section 3. Preserving previous notation, we set $L_\infty = \mathbb{Q}(A[\ell^\infty])$ and $G_\infty = \mathrm{Gal}(L_\infty/\mathbb{Q})$. As usual, we have a topological generator σ for the inertia group $\mathcal{I}_v$ inside G_∞ at a place v over p and we have an element $\tau \in G_\infty$ of order $[F : \mathbb{Q}]$, where $F = \mathbb{Q}(\boldsymbol{\mu}_{2\ell})$, such that the restriction of τ generates $\mathrm{Gal}(F/\mathbb{Q})$. According to Lemma 2.5, the restrictions of σ and τ to the ℓ-division field of A generate its Galois group. By fixing an embedding of L_∞ to $\bar{\mathbb{Q}}_\ell$, we get a natural $\mathcal{D}_v$-isomorphism $\mathbb{T}_\ell(A) \to \mathbb{T}_\ell(A_{\mathbb{Q}_\ell})$, where $A_{\mathbb{Q}_\ell}$ is obtained by base extension to $\mathbb{Q}_\ell$. We are interested in elements of $\mathbb{T}_\ell(A)$ whose image lands in $\mathbb{T}_\ell(\mathcal{F}_A) \subseteq \mathbb{T}_\ell(A_{\mathbb{Q}_\ell})$.

Lemma 4.2. *Let $A/\mathbb{Q}$ be an abelian variety satisfying (C1) and (C2). Suppose X is a G_∞-submodule of $\mathbb{T}_\ell(A)$ with trivial action by the inertia group $\mathcal{I}_v$. Let K_∞ be the fixed field of the kernel of the representation of G_∞ afforded by X. Then K_∞ is unramified outside ℓ and totally ramified over ℓ. Furthermore $X \subseteq \mathbb{T}_\ell(\mathcal{F}_A)$.*

Proof. By assumption, K_∞ is unramified at v. But then $K_\infty/\mathbb{Q}$ is unramified at all places over p because it is Galois.

It is convenient to pass to the compositum $K_\infty F$. Let λ be the place of $K_\infty F$ over ℓ determined by our choice of embedding $L_\infty \to \bar{\mathbb{Q}}_\ell$. Since the prime over ℓ in F is totally ramified, we may verify that K_∞ is totally ramified over ℓ by showing that the inertia group $\mathcal{I}_{F,\lambda}$ inside $H_\infty = \mathrm{Gal}(K_\infty F/F)$ is equal to H_∞. Suppose on the contrary that $\mathcal{I}_{F,\lambda}$ is a proper subgroup of H_∞. Condition (C2) implies that H_∞ is pro-ℓ, so there exists a subgroup of index ℓ in H_∞, containing $\mathcal{I}_{F,\lambda}$. The corresponding fixed field is a cyclic extension of F of degee ℓ unramified everywhere, whose existence contradicts our assumption that ℓ is a regular prime.

Put $\bar{X}^{(n)} = (X + \ell^n\mathbb{T}_\ell(A))/\ell^n\mathbb{T}_\ell(A) \subseteq A[\ell^n]$ for the nth layer of X. Then we may represent elements of X in the form $x = \varprojlim x_n$ with $x_n \in \bar{X}^{(n)}$ compatible under the projection maps. Since ℓ is totally ramified in K_∞, the reduction $\bar{x}_n$ modulo λ is an element of $\bar{A}(\mathbb{F}_\ell)$. Suppose the ℓ-part of the order of $\bar{A}(\mathbb{F}_\ell)$ is ℓ^m. For all $n \geq m+1$, it follows that $\bar{x}_{n-m} = \ell^m\bar{x}_n = 0$. Hence $x_{n'}$ is in the kernel of reduction for all layers $n' = n - m \geq 1$ and we have $X \subseteq \mathbb{T}_\ell(\mathcal{F}_A)$, as claimed. □

Theorem 4.3. *Let ℓ be a regular prime and let $A/\mathbb{Q}$ be a semistable abelian variety with bad reduction only at p, such that $H = \mathrm{Gal}(\mathbb{Q}(A[\ell])/\mathbb{Q}(\boldsymbol{\mu}_\ell))$ is an ℓ-group. Assume that the action of τ on $\mathbb{T}_\ell(A)$ satisfies a polynomial of degree at most 2. Then A has ordinary reduction modulo ℓ and totally toroidal reduction modulo p.*

Proof. We may assume that among the members of its $\mathbb{Q}$-isogeny class $\mathcal{C}$, the variety A is ℓ-maximal; that is, $A \in \mathcal{V}_\ell(\mathcal{C})$ in the notation of (1). The dual abelian varieties A and $\hat{A}$ are $\mathbb{Q}$-isogenous and $\hat{A}$ also is ℓ-maximal because $|\Phi_{\hat{A}}(\bar{\mathbb{F}}_p)| = |\Phi_A(\bar{\mathbb{F}}_p)|$ by the perfect pairing (3). In view of Proposition 3.7, we have pure submodules W_A of $\mathbb{T}_\ell(A)$ and $W_{\hat{A}}$ of $\mathbb{T}_\ell(\hat{A})$, of rank $2a$ and stabilized by G_∞. Furthermore, the Weil pairing induces a perfect pairing

$$W_A \times W_{\hat{A}} \to \mathbb{T}_\ell(\boldsymbol{\mu}). \tag{2}$$

Because $\mathcal{I}_v$ acts trivially on W_A, we find that $W_A \subseteq \mathbb{T}_\ell(\mathcal{F}_A)$ by Lemma 4.2. Similarly, $W_{\hat{A}} \subseteq \mathbb{T}_\ell(\mathcal{F}_{\hat{A}})$. Under the assumption that H is an ℓ-group, Lemma 4.1 implies that A has ordinary reduction modulo ℓ, so the Weil pairing on $W_A \times W_{\hat{A}}$ is trivial by (1). This contradicts the perfect pairing (2) unless $W_A = W_{\hat{A}} = 0$. Hence $a = 0$ and A is totally toroidal at p. □

Proof of Theorem 1.1. Assume that $A/\mathbb{Q}$ is semistable, with bad reduction only at p, that $\ell = 2$ or 3 and that $H = \mathrm{Gal}(\mathbb{Q}(A[\ell])/\mathbb{Q}(\boldsymbol{\mu}_\ell))$ is nilpotent. Then H is in fact an ℓ-group by Proposition 2.4. We have $p \equiv 1 \bmod 8$ when $\ell = 2$ and $p \equiv 1 \bmod 3$ when $\ell = 3$ by Corollary 3.6. Since $\tau^2 = 1$ when $\ell = 2$ or 3, we may conclude by Theorem 4.3. □

5. Small ℓ-division fields

As building blocks for semistable abelian varieties with small ℓ-division fields and bad reduction at only one prime p, we consider elliptic curves $E/\mathbb{Q}$ of conductor p such that the mod-ℓ representation $\rho_{E,\ell} : G_\mathbb{Q} \to \mathrm{Aut}(E[\ell]) \simeq \mathrm{GL}_2(\mathbb{F}_\ell)$ is not surjective. Let Δ_E denote the minimal discriminant of E and recall that $\operatorname{ord}_p(\Delta_E) = -\operatorname{ord}_p(j_E)$, where j_E is the j-invariant. If $\ell \geq 7$ or $\operatorname{ord}_p(j_E) \not\equiv 0 \bmod \ell$, then E admits a $\mathbb{Q}$-isogeny of degree ℓ by [Ser3, Prop. 21]. However, if $\ell \leq 7$ and $|\Delta_E|$ is an ℓth power, E admits an isogeny of degree ℓ, for example by [BK1, Prop. 9.2]. It then follows, as in [Ser3, p. 307], that the isogeny class of E contains a curve with a rational point of order ℓ. The only such examples for odd ℓ occur in the well-known cases $(\ell, p) = (3, 19)$, $(3, 37)$ and $(5, 11)$, referred to below as Miyawaki curves [Mi]. For $\ell = 2$, according to Neumann [Ne] or Setzer [Set], we have $p = 17$ or $p = u^2+64$. In the latter case, there is a unique isogeny class consisting of the curves

$$y^2 + xy = \begin{cases} x^3 + \dfrac{u-1}{4}x^2 - x, & \Delta = \ p, \\ \\ x^3 + \dfrac{u-1}{4}x^2 + 4x + u, & \Delta = -p^2, \end{cases} \tag{1}$$

where the sign of u is chosen to guarantee that $u \equiv 1 \bmod 4$.

Since the order of the connected component group of a semistable elliptic curve $C/\mathbb{Q}_p$ is $|\Phi_C(\bar{\mathbb{F}}_p)| = \operatorname{ord}_p(\Delta_C)$, one easily verifies that each Neumann-Setzer or Miyawaki isogeny class contains a unique curve $E_\ddagger$ with maximal connected component group. Its discriminant satisfies

$$\operatorname{ord}_p(\Delta_\ddagger) = \begin{cases} 4 & \text{if } (\ell, p) = (2, 17), \\ \ell & \text{otherwise.} \end{cases} \tag{2}$$

From the Tate parametrization over $\mathbb{Q}_p$, we find that $\mathbb{Q}(E_\ddagger[\ell])$ is unramified over p and so $\mathbb{Q}(E_\ddagger[\ell]) = F = \mathbb{Q}(\boldsymbol{\mu}_{2\ell})$ by Lemma 2.5 and Corollary 3.2. According to [MO], $E_\ddagger$ is the strong Weil curve in its isogeny class.

Write $E = E_\ddagger$, suppressing the dependence on ℓ and p, which should be clear in context. Recall that the ℓ^∞-division tower L_∞ is the same for all curves in the isogeny class of E. Let $G_\infty = \operatorname{Gal}(L_\infty/\mathbb{Q})$ and observe that $\operatorname{Gal}(L_\infty/F)$ is pro-ℓ. As usual, $\sigma \in G_\infty$ denotes a topological generator of the inertia group at a place v over the bad prime p and $\tau \in G_\infty$ denotes an element of order $[F : \mathbb{Q}]$ whose restriction to F generates $\operatorname{Gal}(F/\mathbb{Q})$.

The Grothendieck module $\mathcal{M}_1(E, v) = \mathcal{M}_2(E, v)$ is a pure $\mathbb{Z}_\ell$-module of rank 1 in $\mathbb{T}_\ell(E)$ and we have $\mathbb{T}_\ell(E) = Y_E = \mathcal{M}_2(E, v) \oplus \tau(\mathcal{M}_2(E, v))$ by Proposition 3.7. Choose a generator P for $\mathcal{M}_2(E, v)$. With respect to the generating set $\{P, \tau(P)\}$ for $\mathbb{T}_\ell(E)$, we obtain a matrix representation

$$\rho_E : G_\infty \to \operatorname{Aut}(\mathbb{T}_\ell(E)) \simeq \operatorname{GL}_2(\mathbb{Z}_\ell).$$

The Tate parametrization of E over $\mathbb{Q}_p$ shows that

$$\rho_E(\sigma) = \begin{pmatrix} 1 & s \\ 0 & 1 \end{pmatrix}, \tag{3}$$

where s equals $\operatorname{ord}_p(\Delta_\ddagger)$ up to multiplication by a unit in $\mathbb{Z}_\ell$, so $s \equiv 0 \bmod \ell$. Since the eigenvalues of τ are 1 and $\omega = \chi(\tau)$, where χ is the cyclotomic character, we have

$$\rho_E(\tau) = \begin{pmatrix} 0 & -\omega \\ 1 & 1+\omega \end{pmatrix}. \tag{4}$$

The following lemma will be used to study abelian varieties $\mathbb{Q}$-isogenous to products of $E_\ddagger$ for fixed ℓ and p. It is sufficient for our purposes to state it for abelian varieties A, B defined over $\mathbb{Q}$. The representation ρ_B of $\operatorname{Gal}(\mathbb{Q}(B[\ell^\infty])/\mathbb{Q})$ afforded by $\mathbb{T}_\ell(B)$ naturally extends to the completed group ring

$$\Lambda_\ell(B) = \mathbb{Z}_\ell[[\operatorname{Gal}(\mathbb{Q}(B[\ell^\infty])/\mathbb{Q})]].$$

Lemma 5.1. *Let $\varphi : B \to A$ be a $\mathbb{Q}$-isogeny of abelian varieties. For given g_1, g_2 in $\operatorname{Gal}(\mathbb{Q}(B[\ell^\infty])/\mathbb{Q})$, assume there exists $\gamma \in \Lambda_\ell(B)$ such that $\rho_B(g_1) = \rho_B(g_2 + \ell\,\gamma)$. Then the restrictions of g_1 and g_2 to $A[\ell]$ are equal.*

Proof. Given $a \in A[\ell]$, we may find $b \in B$ of ℓ-power order, say ℓ^n, such that $\varphi(b) = a$. Since g_1 acts on b through the representation ρ_B modulo ℓ^n, we have $g_1(b) = g_2(b) + \ell\,\gamma(b)$. Then

$$g_1(a) = g_1(\varphi(b)) = \varphi(g_1(b)) = \varphi(g_2(b) + \ell\,\gamma(b)) = g_2(a) + \gamma(\ell\, a) = g_2(a),$$

because φ commutes with Galois and $\ell\, a = 0$. □

Proposition 5.2. *Suppose the abelian variety A is $\mathbb{Q}$-isogenous to a product of Neumann-Setzer or Miyawaki curves for fixed ℓ and p. Let $G = \operatorname{Gal}(\mathbb{Q}(A[\ell])/\mathbb{Q})$.*

(a) *If $\ell = 2$, then G is a quotient of $\mathbb{Z}/2 \times \mathbb{Z}/2$.*

(b) *If $\ell = 3$, then G is a quotient of S_3.*

(c) *If $\ell = 5$, then $H = \operatorname{Gal}(\mathbb{Q}(A[5])/\mathbb{Q}(\boldsymbol{\mu}_5))$ is abelian of exponent dividing 5 and rank at most 2. Furthermore, $\tilde{\tau}^{-2} h \tilde{\tau}^2 = h^{-1}$ for all $h \in H$.*

Proof. Since A is isogenous to E^d, where $d = \dim A$ and $E = E_\ddagger$, we have $\mathbb{Q}(A[\ell^\infty]) = \mathbb{Q}(E^d[\ell^\infty]) = L_\infty$. Furthermore, $\mathbb{T}_\ell(E^d) \simeq \oplus_1^d \mathbb{T}_\ell(E)$ affords the representation $\rho_{E^d} = \oplus_1^d \rho_E$.

Suppose $\ell = 2$ or 3, so that $\rho_E(\tau) = \left(\begin{smallmatrix} 0 & 1 \\ 1 & 0 \end{smallmatrix}\right)$. Then $\rho_E(\sigma\tau - \tau\sigma^{-1}) = s\,\mathbf{1}$, where $\mathbf{1}$ is the identity in $M_2(\mathbb{Z}_\ell)$. Therefore $\rho_{E^d}(\sigma\tau - \tau\sigma^{-1}) = s\,\mathbf{1}$, where $\mathbf{1}$ now denotes the identity in $M_{2d}(\mathbb{Z}_\ell)$. By Lemma 5.1, the restrictions $\tilde{\sigma}$ and $\tilde{\tau}$ of σ and τ to G therefore satisfy $\tilde{\sigma}\tilde{\tau} = \tilde{\tau}\tilde{\sigma}^{-1}$. But $\tilde{\sigma}$ and $\tilde{\tau}$ generate G by Lemma 2.5 and $\tilde{\sigma}^\ell = 1$ by property (L3) of (ℓ, S)-controlled extensions. Hence G is a quotient of $\mathbb{Z}/2 \times Z/2$ if $\ell = 2$ or a quotient of $\mathcal{S}_3$ if $\ell = 3$.

For $\ell = 5$, we find that $\rho_E(\sigma\tau^2 - \tau^2\sigma^{-1}) = (1+\omega)s\,\mathbf{1} \in M_2(\mathbb{Z}_5)$, so

$$\rho_{E^d}(\sigma\tau^2 - \tau^2\sigma^{-1}) = (1+\omega)s\,\mathbf{1} \in M_{2d}(\mathbb{Z}_5).$$

Now $\tilde{\sigma}\tilde{\tau}^2 = \tilde{\tau}^2\tilde{\sigma}^{-1}$ by Lemma 5.1. Acccording to Lemma 2.5, the subgroup H of G is generated by the conjugates of $\tilde{\sigma}$ under the action of $\tilde{\tau}$. Since $\tilde{\tau}^2$ normalizes $\tilde{\sigma}$, to prove that H is abelian, it suffices to show that $\tilde{\sigma}$ commutes with $\tilde{\sigma}^\tau = \tilde{\tau}^{-1}\tilde{\sigma}\tilde{\tau}$. The reader may verify by judicious use of $\rho_E(\sigma - 1)$ and $\rho_E(\tau)$ that the image of ρ_E on the group ring $\mathbb{Z}_5[[G_\infty]]$ contains all 2×2 matrices congruent to 0 modulo s. One can also check that

$$\rho_E(\sigma^\tau\,\sigma - \sigma\,\sigma^\tau) = -s^2\,\omega \begin{pmatrix} 1 & 2(1+\omega) \\ 0 & -1 \end{pmatrix}.$$

Hence $\rho_E(\sigma^\tau\,\sigma - \sigma\,\sigma^\tau)$ belongs to $5\,\rho_E(\mathbb{Z}_5[[G_\infty]])$ and so $\rho_{E^d}(\sigma^\tau\,\sigma - \sigma\,\sigma^\tau)$ belongs to $5\,\rho_{E^d}(\mathbb{Z}_5[[G_\infty]])$. It follows from Lemma 5.1 that $\tilde{\sigma}^\tau$ and $\tilde{\sigma}$ do commute. We may conclude that H is abelian of exponent dividing 5 and rank at most 2, generated by $\tilde{\sigma}$ and $\tilde{\sigma}^\tau$. Furthermore, the action of conjugation by $\tilde{\tau}^2$ on H is given by inversion. □

Proposition 5.3. *Let $\mathcal{C}$ be the $\mathbb{Q}$-isogeny class of E^d, where E is an elliptic curve of conductor p with some rational ℓ-torsion. Then $\mathbb{Q}(A[\ell]) = \mathbb{Q}(\boldsymbol{\mu}_{2\ell})$ for all abelian varieties A in $\mathcal{V}_\ell(\mathcal{C})$.*

Proof. We may assume that $E = E_\ddagger$ is ℓ-maximal. Put

$$\mathcal{M}_2 = \mathcal{M}_2(E^d, v) = \oplus_1^d \mathcal{M}_2(E, v).$$

In view of (3), we have $\sigma\tau(z) = \tau(z) + s\,z$ for all $z \in \mathcal{M}_2$.

Suppose $A \in \mathcal{C}$ and let $\varphi : E^d \to A$ be the corresponding isogeny. The basic assumptions (C1), (C2) of section 3 clearly are satisfied by E^d. Condition (C3) holds because $\rho_{E^d} = \oplus \rho_E$ is a sum of d copies of a fixed representation of GL_2-type. Since these conditions are isogeny invariant, they also hold for A. Suppose, in addition, that A is ℓ-maximal. Then Proposition 3.7 implies that $\mathbb{T}_\ell(A) = \mathcal{M}_2' \oplus \tau(\mathcal{M}_2')$, where $\mathcal{M}_2' = \mathcal{M}_2(A, v) = \mathcal{M}_1(A, v)$.

By construction, σ acts trivially on $\mathcal{M}_2'$. To examine the action of σ on $\tau(\mathcal{M}_2')$, recall from the start of section 3 that $\mathcal{M}_2'$ contains $\varphi(\mathcal{M}_2)$

with finite index, say n. Given $z' \in \mathcal{M}_2'$, we may therefore find $z \in \mathcal{M}_2$ such that $n\,z' = \varphi(z)$. It follows that

$$n\sigma\tau(z') = \varphi(\sigma\tau(z)) = \varphi(\tau(z) + s\,z) = n\tau(z') + ns\,z'.$$

But $\mathbb{T}_\ell(A)$ is torsion-free, so $\sigma\tau(z') = \tau(z') + s\,z'$. Since $s \equiv 0 \bmod \ell$, we find that σ acts trivially on the first layer $\tau(\bar{\mathcal{M}}_2')$ of $\tau(\mathcal{M}_2')$ and therefore on all of $A[\ell] = \bar{\mathcal{M}}_2' \oplus \tau(\bar{\mathcal{M}}_2')$. Hence $\mathbb{Q}(A[\ell]) = \mathbb{Q}(\boldsymbol{\mu}_{2\ell})$ by Lemma 2.5 and Corollary 3.2. □

Lemma 5.4. *Suppose that $\mathcal{C}$ is a $\mathbb{Q}$-isogeny class of abelian varieties of dimension d satisfying (C1), (C2), (C3) and that $\mathbb{Q}(A[\ell]) \subseteq \mathbb{Q}(\boldsymbol{\mu}_{2\ell})$ for each A in $\mathcal{V}_\ell(\mathcal{C})$. Then there exists a $\mathbb{Z}_\ell$-submodule X of $\mathbb{T}_\ell(A)$ of rank 2 and stabilized by G_∞ such that $\mathbb{T}_\ell(A) \simeq X^d$ as G_∞-modules.*

Proof. Theorem 4.3 implies that A has totally toroidal reduction modulo p, so $\mathcal{M}_2 = \mathcal{M}_2(A,v) = \mathcal{M}_1(A,v)$. For all $A \in \mathcal{V}_\ell(\mathcal{C})$, we have $\bar{\mathcal{M}}_2 \cap \tau(\bar{\mathcal{M}}_2) = 0$ by (6) and $\mathbb{T}_\ell(A) = Y_A = \mathcal{M}_2 \oplus \tau(\mathcal{M}_2)$ by Proposition 3.7.

Choose a set of free generators $m_1, \ldots, m_d$ for $\mathcal{M}_2$. For $j = 1, \ldots, d$, define $X^{(j)}$ to be $\mathbb{Z}_\ell$-span of m_j and $\tau(m_j)$. Each $X^{(j)}$ is a pure submodule of $\mathbb{T}_\ell(A)$ of rank 2 and $\mathbb{T}_\ell(A) = \oplus_{j=1}^d X^{(j)}$.

We wish to use Lemma 3.3 to show that $X = X^{(j)}$ is a G_∞-module. Take $\mathcal{H}$ to be the group generated by τ, so X certainly is an $\mathcal{H}$-module by (C3). Suppose $X_n = X + \ell^n\mathbb{T}_\ell(A)$ is a G_∞-module and let $\varphi : A \to A'$ be the $\mathbb{Q}$-isogeny whose kernel is $\kappa = (X + \ell^n\mathbb{T}_\ell(A))/\ell^n\mathbb{T}_\ell(A) \simeq \mathbb{Z}/\ell^n \oplus \mathbb{Z}/\ell^n$.

Write $\bar{\mathcal{M}}_2^{(n)}$ for the projection of $\mathcal{M}_2$ to $\mathbb{T}_\ell(A)/\ell^n\mathbb{T}_\ell(A) \simeq A[\ell^n]$. It is clear that $\kappa \cap \bar{\mathcal{M}}_2^{(n)}$ is isomorphic to one copy of $\mathbb{Z}/\ell^n$, generated by the coset of m_j. Hence $|\kappa \cap \bar{\mathcal{M}}_2^{(n)}| = \ell^n$ and we have $|\Phi_{A'}(\bar{\mathbb{F}}_p)|_\ell = |\Phi_A(\bar{\mathbb{F}}_p)|_\ell$ by Lemma 3.1. Therefore A' also is ℓ-maximal. Then by assumption, $\mathbb{Q}(A'[\ell]) \subseteq F = \mathbb{Q}(\boldsymbol{\mu}_{2\ell})$. Hence the subgroup $\mathcal{N}$ defined in Lemma 3.3 contains $\mathrm{Gal}(L_\infty/F)$. But $\mathrm{Gal}(L_\infty/F)$ and $\mathcal{H}$ certainly generate G_∞. We may conclude from Lemma 3.3 that X is a G_∞-module.

The following standard argument now shows that the $X^{(j)}$'s are isomorphic as G_∞-modules. Reasoning as above, we find that for each $j \neq 1$, the $\mathbb{Z}_\ell$-submodule of $\mathbb{T}_\ell(A)$ of rank 2 spanned by $m_1 + m_j$ and $\tau(m_1 + m_j)$ also is stabilized by G_∞. But then the matrix representation of G_∞ afforded by the $\mathbb{Z}_\ell$-span of m_1 and $\tau(m_1)$ must be identical to the representation afforded by the $\mathbb{Z}_\ell$-span of m_j and $\tau(m_j)$. It now follows that there is a G_∞-isomorphism $\mathbb{T}_\ell(A) \simeq X^d$, where X is the $\mathbb{Z}_\ell$-module spanned by m and $\tau(m)$ for any choice of $m \in \mathcal{M}_2$ such that $m \notin \ell\mathcal{M}_2$. □

For the proof of our final result, we need to observe that ℓ-maximality is preserved for certain products of abelian varieties. Note that $A =$

$B \times C$ satifies (C1), (C2), (C3) of section 3 if and only if each factor does.

Lemma 5.5. *Suppose A satisfies (C1), (C2), (C3). If $A = B \times C$ is the product of ℓ-maximal abelian varieties, then A also is ℓ-maximal.*

Proof. To show A is ℓ-maximal, it suffices to consider $\mathbb{Q}$-isogenies $\varphi : A \to A'$ of ℓ-power degree. Set $\kappa = \operatorname{Ker}\varphi$ and assume κ has exponent ℓ^n. Theorem 5.4 implies that A has totally toroidal reduction at p, so $\mathcal{M}_1(A,v) = \mathcal{M}_2(A,v)$ and similarly for B and C. Moreover, inside $\mathbb{T}_\ell(A) = \mathbb{T}_\ell(B) \times \mathbb{T}_\ell(C)$, we have $\mathcal{M}_1(A,v) = \mathcal{M}_1(B,v) \times \mathcal{M}_1(C,v)$. Write $\bar{\mathcal{M}}_2^{(n)}$ for projection to $\mathbb{T}_\ell/\ell^n\mathbb{T}_\ell$ and let

$$\begin{aligned}\kappa_0 &= \kappa \cap \bar{\mathcal{M}}_2^{(n)}(A,v)\\ &= \{ (\bar{b},\bar{c}) \in \kappa \mid b \in \mathcal{M}_2(B,v),\, c \in \mathcal{M}_2(C,v) \} \subset A[\ell^n].\end{aligned}$$

In view of (6), we have

$$\bar{\mathcal{M}}_2^{(n)}(B,v) \cap \tau(\bar{\mathcal{M}}_2^{(n)}(B,v)) = \bar{\mathcal{M}}_2^{(n)}(C,v) \cap \tau(\bar{\mathcal{M}}_2^{(n)}(C,v)) = 0,$$

whence $\kappa_0 \cap \tau(\kappa_0) = 0$. But κ is a Galois module, so it contains $\kappa_0 + \tau(\kappa_0)$ and therefore $|\kappa| \geq |\kappa_0|\,|\tau(\kappa_0)| = |\kappa_0|^2$. It follows from Lemma 3.1 that $|\Phi_A(\bar{\mathbb{F}}_p)|_\ell \geq |\Phi_{A'}(\bar{\mathbb{F}}_p)|_\ell$. Hence A is ℓ-maximal. □

Theorem 5.6. *Let $\mathcal{C}$ be a $\mathbb{Q}$-isogeny class of semistable abelian varieties of dimension d with bad reduction only at p. Assume ℓ is a regular prime and $\mathbb{Q}(A[\ell]) \subseteq \mathbb{Q}(\mu_{2\ell})$ for every A in $\mathcal{V}_\ell(\mathcal{C})$. Suppose the action of τ on $\mathbb{T}_\ell(A)$ satisfies a polynomial of degree at most 2. Then $\ell = 2$, 3 or 5 and there exists an elliptic curve E of conductor p with some rational ℓ-torsion such that $\mathcal{C}$ is the $\mathbb{Q}$-isogeny class of E^d.*

Proof. We may apply Lemma 5.4 to $A \in \mathcal{V}_\ell(\mathcal{C})$ to obtain a $\mathbb{Z}_\ell$-submodule X of $\mathbb{T}_\ell(A)$ of rank 2 stabilized by G_∞, such that $\mathbb{T}_\ell(A) \simeq X^d$ is a G_∞-isomorphism. According to the Tate conjecture, proved by Faltings ([Sch]), we have

$$\operatorname{End}_{\mathbb{Q}}(A) \otimes \mathbb{Z}_\ell \simeq \operatorname{End}_{G_\infty} \mathbb{T}_\ell(A).$$

But the commutant $\operatorname{End}_{G_\infty} \mathbb{T}_\ell(A)$ clearly contains $M_d(\mathbb{Z}_\ell)$. It follows that the $\mathbb{Z}_\ell$-rank of $\operatorname{End}_{\mathbb{Q}}(A) \otimes \mathbb{Z}_\ell$ is at least d^2.

Suppose A is simple, so that $D = \operatorname{End}_{\mathbb{Q}}(A) \otimes \mathbb{Q}$ is a division algebra. We have shown above that $[D : \mathbb{Q}] \geq d^2$. But D acts on the space of invariant holomorphic differentials $\Omega^1(A)$, so

$$d = \dim_{\mathbb{Q}} \Omega^1(A) = [D : \mathbb{Q}] \dim_D \Omega^1(A) \geq d^2.$$

Hence $d = 1$ and A is a Neumann-Setzer ($\ell = 2$) or Miyawaki ($\ell = 3$ or 5) elliptic curve.

If $A' \in \mathcal{V}_\ell(\mathcal{C})$ is not $\mathbb{Q}$-simple, we may find proper abelian subvarieties B' and C', such that A' is $\mathbb{Q}$-isogenous to $B' \times C'$. Certainly B' and C' inherit properties (C1), (C2), (C3) from A'. Choose B and C to be ℓ-maximal in the isogeny class of B' and C' respectively. Then A' is isogenous to $A = B \times C$ and Lemma 5.5 shows that A is in $\mathcal{V}_\ell(\mathcal{C})$. By assumption, the ℓ-division field of A therefore is contained in $\mathbb{Q}(\mu_{2\ell})$ and so *a fortiori* the same is true of the ℓ-division fields of B and C. Thus we may complete the proof by induction. □

Corollary 5.7. *Let $\mathcal{C}$ be a $\mathbb{Q}$-isogeny class of semistable abelian varieties of dimesion d with bad reduction only at p and suppose* $\mathrm{Gal}(\mathbb{Q}(A[2])/\mathbb{Q})$ *is abelian for all $A \in \mathcal{V}_2(\mathcal{C})$. Then $\mathcal{C}$ is the isogeny class of E^d, where E is a Neumann-Setzer curve of conductor p.*

Proof. Theorem 1.1 tells us that A is totally toroidal, so for $A \in \mathcal{V}_\ell(\mathcal{C})$, we have $A[2] = \bar{\mathcal{M}}_2 \oplus \tau(\bar{\mathcal{M}}_2)$ by (6) and equality of dimension. Since σ acts trivially on $\mathcal{M}_2$ and the restrictions of σ and τ to $\mathbb{Q}(A[2])$ commute by assumption, we find that σ acts trivially on $A[2]$. It follows that p is unramified in $\mathbb{Q}(A[2])$, whence $\mathbb{Q}(A[2]) = \mathbb{Q}(i)$ by Corollary 3.2. Our claim is now implied by Theorem 5.6. □

The following two examples show that the ℓ-division field of a *simple* semistable abelian variety of dimension greater than 1 with one place of bad reduction may be an ℓ-group. Compare with Theorem 5.6 and Theorem 1.3.

From a model for the modular curve $X_0(41)$ given by Weber [Web], one finds the minimal model

$$y^2 + (x^4 + 2x^3 - x)y = -3x^6 - 16x^5 - 29x^4 - 14x^3 + 13x^2 + 9x - 4.$$

Its Jacobian $J_0(41)$ has real multiplications by the maximal order in the totally real cubic field of discriminant 148 and bad reduction only at 41, where it is totally toroidal. The 2-division field L_0 of $J_0(41)$ can be seen to be cyclic over $K = \mathbb{Q}(\sqrt{-41})$ and dihedral over $\mathbb{Q}$, as we expect from Proposition 2.6. Compare with Proposition 5.2(a). Since the class number of K is 8, the maximal $(2, 41)$-controlled extension has degree 32 over $\mathbb{Q}$. However $[L_0 : \mathbb{Q}] = 16$.

A minimal model for $X_0(31)$ is given by

$$y^2 + (x^3 + x^2 + 2x + 1)y = -x^4 - 2x^3 - 3x^2 - 2x - 1.$$

Its Jacobian variety $A = J_0(31)$ admits real multiplications by the ring of integers of $\mathbb{Q}(\sqrt{5})$ and is the full Eisenstein quotient at 5, cf. [Ma].

For the endomorphism $\pi = \sqrt{5}$, we have $A[\pi] = \mathbb{Z}/5\mathbb{Z} \oplus \boldsymbol{\mu}_5$ and so $H = \mathrm{Gal}(\mathbb{Q}(A[5])/\mathbb{Q}(\boldsymbol{\mu}_5))$ is an elementary abelian 5-group. One can check that its $\mathbb{F}_5$-rank is 3, in contrast to Proposition 5.2(c). Using the Remark after Lemma 2.3, we find that $\mathbb{Q}(A[5])$ is the maximal $(5, 31)$-controlled extension abelian over $\mathbb{Q}(\boldsymbol{\mu}_5)$.

Acknowledgments

The first author wishes to express his indebtedness for the generous hospitality, patience, friendship and stimulation afforded him by Professor Hashimoto and his students at Waseda University. He also thanks Professors K. Miyake of Tokyo Metropolitan University, T. Ibukiyama of Osaka University and H. Saito of Kyoto University, their institutions and colleagues. Together, they made it possible to spend a wonderful month in Japan, beyond the week of this Conference.

References

[And] A. N. Andrianov, "Quadratic forms and Hecke operators," Grundlehren der Math., 286. Springer-Verlag, 1987.

[BSP] S. Böcherer and R. Schulze-Pillot, *Siegel modular forms and theta series attached to quaternion algebras*, Nagoya Math. J. **121** (1991), 35–96.

[BK1] A. Brumer and K. Kramer, *The rank of elliptic curves*, Duke Math. J., **44** (1977), 715–743.

[BK2] A. Brumer and K. Kramer, *Non-existence of certain semistable abelian varieties*, Manuscripta Math. **106** (2001), 291–304.

[CG] J. Coates and R. Greenberg, *Kummer theory for abelian varieties over local fields*, Invent. Math. **124** (1996), 129–174.

[Ed] B. Edixhoven, *On the prime to p-part of the group of connected components of Néron models*, Compositio Math. **97** (1995), 29–49.

[Fa] G. Faltings, *Endlichkeitssätze für abelsche Varietäten über Zahlkörpern*, Invent. Math. **73** (1983), 349–366.

[Fo] J.-M. Fontaine, *Il n'y a pas de variété abélienne sur* $\mathbb{Z}$, Invent. Math. **81** (1985), 515–538.

[Gro] A. Grothendieck, *Modèles de Néron et monodromie*, in "Groupes de monodromie en géométrie algébrique. I," Lecture Notes in Mathematics, 288. Springer-Verlag, New York, 1973.

[Gru] F. Grunewald, *Some facts from the theory of group schemes*, in "Rational Points (G. Faltings and G. Wüstholz, ed.)," Aspects of Mathematics, E6. F. Vieweg & Sohn, Braunschweig, 1984.

[HM] F. Hajir and C. Maire, *Extensions of number fields with wild ramification of bounded depth*, Inter. Math. Res. Notices (2002), 667–696.

[Has] K.-I. Hashimoto, *The dimension of the spaces of cusp forms on Siegel upper half-plane of degree two*, I, J. Fac. Sci. Univ. Tokyo Sect. IA Math. **30** (1983), 403–488.

[Ibu] T. Ibukiyama, *On relations of dimension of automorphic forms of* $\mathrm{Sp}(2,\mathbb{R})$ *and its compact twist* $\mathrm{Sp}(2)$, I, Adv. Studies in Pure Math. **7** (1985), 7–29.

[MM] J. M. Masley and H. L. Montgomery, *Cyclotomic fields with unique factorization*, J. Reine Ang. Math. **286/287** (1976), 248–256.

[Ma] B. Mazur, *Modular curves and the Eisenstein ideal*, Publ. Math. I.H.E.S. **47** (1976), 33–186.

[Mc] W. McCallum, *Duality theorems for Néron models*, Duke Math. J. **53** (1986), 1093–1124.

[MO] J.-F. Mestre and J. Oesterlé, *Courbes de Weil semi-stables de discriminant une puissance m-ième*, J. Reine Ang. Math. **400** (1989), 171–184.

[Mi] I. Miyawaki, *Elliptic curves of prime power conductor with* $\mathbb{Q}$*-rational points of finite order*, Osaka Math J., **10** (1973), 309–323.

[Ne] O. Neumann, *Elliptische Kurven mit vorgeschriebenem Reduktionsverhalten. II*, Math. Nach. **56** (1973), 269–280.

[Sch] N. Schappacher, *Tate's conjecture on the endomorphisms of abelian varieties*, in "Rational Points (G. Faltings and G. Wüstholz, ed.)," Aspects of Mathematics, E6. F. Vieweg & Sohn, Braunschweig, 1984.

[Scho] R. Schoof, *Semistable abelian varieties over* $\mathbb{Q}$, (2001), preprint.

[Ser1] J.-P. Serre, *Facteurs locaux des fonctions zêta des variétés algébriques*, Sém. DPP, éxposé **19** (1969–1970).

[Ser2] J.-P. Serre, "Local Fields," Lecture Notes in Mathematics, 67. Springer-Verlag, New York, 1979.

[Ser3] J.-P. Serre, *Propriétés galoisiennes des points d'ordre fini des courbes elliptiques*, Invent. Math. **15** (1972), 259–331.

[Ser4] J.-P. Serre, *Sur les représentations modulaires de degré* 2 *de* $\mathrm{Gal}(\overline{\mathbb{Q}}/\mathbb{Q})$, Duke Math. J. **54** (1987), 179–230.

[Ser5] J.-P. Serre, Oeuvres. Collected papers. IV (1986), 1–55, Springer-Verlag, 2000.

[Set] C. B. Setzer, *Elliptic curves of prime conductor*, J. Lond. Math. Soc. **10** (1975), 367–378.

[SS] Schinzel, A. and Sierpiński, W., *Sur certaines hypothèses concernant les nombres premiers*, Acta Arithmetica **4** (1958), 185–208.

[Web] H.-J. Weber, *Hyperelliptic simple factors of* $J_0(N)$ *with dimension at least* 3, Exper. Math. J. **6** (1997), 273–287.

[Yos] H. Yoshida, *On Siegel modular forms obtained from theta series*, J. Reine Ang. Math. **352** (1984), 184–219.

[Zar] Yu. G. Zarhin, *A finiteness theorem for unpolarized abelian varieties over number fields with prescribed places of bad reduction*, Invent. Math. **79** (1985), 309–321.

Galois Theory and Modular Forms pp 39–64
K. Hashimoto, K. Miyake, and H. Nakamura, Editors
© 2003 Kluwer Academic Publishers

Q-CURVES WITH RATIONAL j-INVARIANTS AND JACOBIAN SURFACES OF GL_2-TYPE

Ki-ichiro HASHIMOTO
Department of Mathematical Sciences
Waseda University,
3-4-1, Okubo Shinjuku-ku, Tokyo,
169-8555 Japan
khasimot@mse.waseda.ac.jp

Abstract We study elliptic curves over quadratic fields with rational j-invariants regarded as $\mathbf{Q}$-curves, in connection with jacobian surfaces of genus two curves over $\mathbf{Q}$. We discuss their minimality as $\mathbf{Q}$-curves, and the classification, as well as the Neben type characters of the associated modular forms. This can be described as the sign change phenomenon by quartic twists of curves over $\mathbf{Q}$. We also study their 2-fold covers by genus two curves. Among others, we construct a parametric family $\{C(j)\}$ of genus two curves over $\mathbf{Q}$ covering minimal $\mathbf{Q}$-curves over $\mathbf{Q}(\sqrt{j-12^3})$ with j-invariant j. We find in the twists of $\{C(j)\}$ concrete equations of curves over $\mathbf{Q}$ whose jacobians are isogenous over an extension of $\mathbf{Q}$ to Shimura's abelian surfaces A_f attached to normalized eigen forms $f \in S_2(N, (\frac{N}{*}))$, whose Fourier coefficients belong to $\mathbf{Q}(\sqrt{-1})$, in all known non CM cases i.e., $N = 37$, 65, 104, 157, 397, and 877.

0. Introduction

Let $\Gamma_1(N)$ be the congruence subgroup

$$\Gamma_1(N) = \left\{ \begin{pmatrix} a & b \\ c & d \end{pmatrix} \in \mathrm{SL}_2(\mathbf{Z}) \mid c \equiv 0,\, a, d \equiv 1 \pmod{N} \right\},$$

and let $S_2(\Gamma_1(N))$ be the space of cusp forms of weight 2 with respect to $\Gamma_1(N)$. To each normalized Hecke eigen form $f \in S_2(\Gamma_1(N))$, one can

[received: July 1, 2002; accepted in revised form: December 10, 2002]

attach as in Theorem 7.14, [Sh1] an abelian variety A_f over $\mathbf{Q}$ obtained as a $\mathbf{Q}$-simple factor of $J_1(N)$, the jacobian variety of the modular curve $X_1(N)$. A_f is called the Shimura's abelian variety associated with f. Let $f(z) = \sum_{n=1}^{\infty} a_f(n)q^n$ ($q = e^{2\pi\sqrt{-1}z}$) be its Fourier expansion and let $K_f = \mathbf{Q}(\{a_f(n)\})$ be the field generated by the Fourier coefficients of f. Put $g := [K_f : \mathbf{Q}]$. Then A_f is of dimension g, and is characterized, up to $\mathbf{Q}$-isogeny, by the property $\mathrm{End}_{\mathbf{Q}}(A_f) \otimes \mathbf{Q} = K_f$ where $\mathrm{End}_{\mathbf{Q}}(A_f)$ denotes the ring of endomorphisms of A_f which are defined over $\mathbf{Q}$, and the action of $a_f(p) \in K_f$ corresponds to that of the Hecke operator $T(p)$. It then follows that the L-function of $A_f/\mathbf{Q}$ is given by

$$L(A_f/\mathbf{Q}, s) = \prod_{i=1}^{g} L({}^{\sigma_i}f, s),$$

where $\{\sigma_1, \dots, \sigma_d\}$ denote the set of embeddings of K_f into $\mathbf{C}$, and

$${}^{\sigma}f = \sum_{n=1}^{\infty} {}^{\sigma}a_f(n)q^n.$$

In general an abelian variety A over $\mathbf{Q}$ is called of GL_2-type if $\mathrm{End}_{\mathbf{Q}}(A) \otimes \mathbf{Q}$ is a commutative field of degree equal to $\dim A$. The so called modularity conjecture claims that any such $A/\mathbf{Q}$ should be isogenous over $\mathbf{Q}$ to A_f for some N and $f \in S_2(\Gamma_1(N))$ (c.f. [Ri3], [HHM]).

In this paper we are concerned, not to the proof of this conjecture, but to study the following closely related problems in the case $g = 2$:

- How many, for a given quadratic field K, abelian surfaces are there over $\mathbf{Q}$ which are of GL_2-type with $\mathrm{End}_{\mathbf{Q}}(A) \otimes \mathbf{Q} \cong K$?
- Constuct as many curves of genus 2 over $\mathbf{Q}$ as possible, whose jacobian surfaces are of GL_2-type, with $\mathrm{End}_{\mathbf{Q}}(\mathrm{Jac}(C)) \otimes \mathbf{Q} \cong K$.

Of special interest to us is the following problem:

- To each eigen form $f \in S_2(\Gamma_1(N))$ such that $[K_f : \mathbf{Q}] = 2$, find a curve $C/\mathbf{Q}$ such that $\mathrm{Jac}(C)$ is isogenous over $\overline{\mathbf{Q}}$ to A_f.

Here of course we require to find an isogeny to be defined over a field as small as possible. Also note that the existence of such a curve C over $\mathbf{Q}$ is not a priori clear. Moreover, if f has nontrivial Neben type character, one cannot in general expect that the isogeny $\mathrm{Jac}(C) \to A_f$ is defined over $\mathbf{Q}$, since the Euler factors of L-functions of these have different shapes.

These problems have been studied in our previous work [Ha] in the case $K_f = \mathbf{Q}(\sqrt{-5})$ and f is an eigen form with quadratic Neben type character. We discussed with an explicitly constructed family, several arithmetic properties of *minimal* elliptic **Q**-curves of degree 5 over real quadratic fields and jacobian surfaces of GL_2-type over **Q** associated to each other. And we found in this family the solutions to the third problem for all known cases such that $K_f = \mathbf{Q}(\sqrt{-5})$.

The present paper treats the case $K_f = \mathbf{Q}(\sqrt{-1})$, and obtains the similar results as [Ha]. Thus we are interested in those A_f which decompose over $\mathbf{Q}(\sqrt{N})$ into the product of elliptic curve B_f with itself which has rational j-invariant. One may call B_f a **Q**-curve of "*degree one*", so that the study of such objects, as presented here, could have been done before [Ha]. We should note that the modularlity for such **Q**-curves has been completely settled (see [BCDT]).

As for the first of the above problems, we shall establish the complete classification of ε-minimal **Q**-curves ($\varepsilon = \pm 1$) of degree one over quadratic fields K, in the sense that the isogeny $\mu_\sigma : E \to {}^\sigma E$ is also defined over K and satisfies

$$^\sigma\mu_\sigma \circ \mu_\sigma = \varepsilon \times id. \qquad (\varepsilon = \pm 1).$$

In the case $\varepsilon = +1$, the question is easily answered: All (+1)-minimal **Q**-curves of degree one over K are obtained as the twist $E^{Tate}(j)_\delta$ by some $\delta \in \mathbf{Q}^\times$ of the Tate's model

$$E^{\mathrm{Tate}}(j) : \ Y^2 + XY = X^3 - \frac{36}{j-1728}X - \frac{1}{j-1728}.$$

We call an abelian surface $A/\mathbf{Q}$ to be of *fake* GL_2-type, if $\mathrm{End}_{\mathbf{Q}}(A) \otimes \mathbf{Q} \cong \mathbf{Q} \oplus \mathbf{Q}$, while $\mathrm{End}_K(A) \otimes \mathbf{Q} \cong \mathrm{M}_2(\mathbf{Q})$ for a quadratic field K. Then we shall prove, for a quadratic field $K = \mathbf{Q}(\sqrt{m})$, the following

Theorem 0.1. *For each* $j \in \mathbf{Q} (j \neq 0, 1728)$*, the set of* K*-isomorphism classes of* (+1)*-minimal* **Q***-curves of degree one over* K *with* j*-invariant* j *is in one to one correspondence with* $\mathbf{Q}^\times / S_K$ *where* $S_K := (K^\times)^2 \cap \mathbf{Q}^\times$*. Moreover, to each class of* $\delta \in \mathbf{Q}^\times / S_K$ *one can assign a* **Q***-isogeny class of abelian surfaces of fake* GL_2*-type over* **Q** *represented by* $E^{\mathrm{Tate}}(j)_\delta \times E^{\mathrm{Tate}}(j)_{m\delta}$.

We shall show that abelian surfaces of fake GL_2-type over **Q** occur as jacobian surfaces of algebraic curves, and they play imortant roles in studying abelian surfaces $A/\mathbf{Q}$ of GL_2-type over **Q** such that $\mathrm{End}_{\mathbf{Q}}(A) \otimes \mathbf{Q} \cong \mathbf{Q}(\sqrt{-1})$.

The problem of (−1)-minimal **Q**-curves over K is more difficult, and to classify them one needs to study the quartic twists of elliptic curves over **Q**. We note that if E is a (−1)-minimal **Q**-curve of degree 1 over

K then $\mathrm{Res}_{K/\mathbf{Q}}(E)$ is an abelian surface over $\mathbf{Q}$ of GL_2-type such that

$$\mathrm{End}_{\mathbf{Q}}(\mathrm{Res}_{K/\mathbf{Q}}(E)) \otimes \mathbf{Q} \cong \mathbf{Q}(\sqrt{-1}) \qquad \text{(c.f. [Ri3])}.$$

Let $\mathcal{L}(K)$ be the set of quartic cyclic extensions $L/\mathbf{Q}$ which contain K. We shall prove

Theorem 0.2. *Suppose that $j \in \mathbf{Q}$ and that $K := \mathbf{Q}(\sqrt{|j-1728|})$ is a real quadratic field with $\mathcal{L}(K) \neq \emptyset$. Then for each $\delta \in K^\times$ satisfying $\mathrm{N}_{K/\mathbf{Q}}(\delta) = c^2|j-1728|$ $(c \in \mathbf{Q}^\times)$, the twisted curve $E^{Tate}(j)_\delta$ is a (-1)-minimal $\mathbf{Q}$-curve over K, so that*

$$\mathrm{End}_{\mathbf{Q}}(\mathrm{Res}_{K/\mathbf{Q}}(E^{Tate}(j)_\delta)) \otimes \mathbf{Q} \cong \mathbf{Q}(\sqrt{-1}).$$

In particular, $\mathrm{Res}_{K/\mathbf{Q}}(E^{Tate}(j)_\delta)$ is an abelian surface over $\mathbf{Q}$ of GL_2-type. Moreover, the map

$$E^{Tate}(j)_\delta \mapsto \mathrm{N}_{K/\mathbf{Q}}(\delta)\ \mathcal{S}_K$$

induces a bijection from the set of K-isomorphism classes of (-1)-minimal $\mathbf{Q}$-curves of degree 1 over K to $\mathbf{Q}^\times/\mathcal{S}_K$.

As for the second and third questions listed above, our strategy is, as in [Ha], to construct a generic family of genus two curves which are double covers of elliptic curves with j-invariants j. For this purpose, we study in full detail the action of involutive automorphisms on the space of sextic polynomials. By this result we obtain the most general family of split curves of genus two, from which we are able to find the one parameter family $C(j)$ which is defined by the equation $Y^2 = F(X;j)$,

$$\begin{aligned} F(X;j) := X^6 &+ 144X^5 - (j-1728)X^4 + 224(j-1728)X^3 \\ &- (j-1728)^2X^2 + 144(j-1728)^2X + (j-1728)^3. \end{aligned}$$

One of our main results, which has several interesting applications, is stated as:

Theorem 0.3. *There exists a morphism $\psi : C(j) \to E(j)$ of degree two which is defined over $\mathbf{Q}(\sqrt{j-1728})$, where $E(j)$ is a twist of the Tate's generic family of elliptic curves and is defined by*

$$E(j):\ 3W^2 = 4(j-1728)U^3 - 3jU + j, \qquad j(E(j)) = j.$$

We can show in particular that for all known cases of $f \in S_2(\Gamma_1(N))$ such that $K_f = \mathbf{Q}(\sqrt{-1})$, the solutions to the above problem are in our family $C(j)$.

This paper is organized as follows: In Sect.1, we introduce some basic definitions such as $\mathbf{Q}$-curves of degree one over quadratic fields K, their minimality, and the group $\mathcal{S}_K$ which plays an important role in our theory. In Sect.2 we discuss some general theory of quadratic extensions $K(\sqrt{\delta})$ of K which are cyclic quartic fields over $\mathbf{Q}$. To each $K(\sqrt{\delta})$ we associate a quartic twists of $E(j)$, which will be shown to be a (-1)-minimal $\mathbf{Q}$-curves over K. We also discuss the other case where $K(\sqrt{\delta})$ is a $(2,2)$-extension of $\mathbf{Q}$, and associate to them the twists of $E(j)/K$, to obtain abelian surfaces over $\mathbf{Q}$ *of fake* GL_2-*type*. In Sect.3, we describe one of our main results which gives a family $C(j)$ of genus two curves that are the double covers of elliptic curves $E(j)$, and their jacobian varieties are of fake GL_2-type. The construction of such family is described in the final Sect.6. In Sect.4 we shall give concrete examples of $C(j)$ with quadratic twists whose jacobians are isogenous over cyclic quartic fields to Shimura's abelian surfaces A_f attached to cusp forms f of quadratic Neben type character of level $N = 37, 65, 104, 157, 397$, and 877. In Sect.5, we shall give another example of curve of genus two over $\mathbf{Q}$ whose jacobian is isogenous over $\mathbf{Q}$ to A_f for $N = 65$.

1. Minimal Q-curves of degree 1 over quadratic fields

Let $K = \mathbf{Q}(\sqrt{m})$ be a quadratic field and let $j \in K$ be such that $j \neq 0, 12^3$. As in [Ha] we denote by $\mathcal{E}_K(j)$ the set of K-isomorphism classes of elliptic curves E over K with $j(E) = j$. Then for $j \neq 0, 12^3$, the set $\mathcal{E}_K(j)$ is seen to be a homogeneous space over $K^\times$, where the action of $K^\times$ is given by the quadratic twist. Namely each $\delta \in K^\times$ defines a bijection $E \mapsto E \otimes \sqrt{\delta} := E_\delta$ from $\mathcal{E}_K(j)$ to itself. Here if E is defined by a Weierstrass equation

$$E: \; Y^2 = X^3 + AX + B \qquad (A, B \in K),$$

then E_δ can be defined by

$$E_\delta: \; \delta Y^2 = X^3 + AX + B. \tag{1}$$

In what follows, we always assume that our elliptic curves are defined over a fixed quadratic field $K = \mathbf{Q}(\sqrt{m})$, and have no complex multiplication. In such a case we have

$$\mathcal{E}_K(j) \cong K^\times/(K^\times)^2, \quad E_\delta \mapsto \delta(K^\times)^2.$$

Recall that a $\mathbf{Q}$-curve E over K is an elliptic curve defined over K such that there exists an isogeny $\mu_\sigma : E \to {}^\sigma E$, wher σ denotes the nontrivial automorphism of $K/\mathbf{Q}$. It is called of degree d, if d is the minimum of $\deg(\mu_\sigma)$. Then d is a positive square free integer.

Definition 1. *A* $\mathbf{Q}$*-curve* E *over a quadratic field* K *is called minimal if* μ_σ *is also defined over* K*. It is called* ε*-minimal if* ${}^\sigma\mu_\sigma \circ \mu_\sigma = \varepsilon d \times id$. ($\varepsilon = \pm 1$).

Our aim is to study and classify all ε-minimal $\mathbf{Q}$-curves over K of degree 1. This means that μ_σ is an isomorphism, so that E and ${}^\sigma E$ have the same j-invariant, i.e. $j(E) \in \mathbf{Q}$. For such j, we choose as a standard model of each class in $\mathcal{E}_K(j)$ the following curve $E^{\mathrm{Tate}}(j)$ which is defined over $\mathbf{Q}(j) = \mathbf{Q}$:

$$E^{\mathrm{Tate}}(j) : \; Y^2 + XY = X^3 - \frac{36}{j-1728}X - \frac{1}{j-1728}. \tag{2}$$

One has

$$j(E^{\mathrm{Tate}}(j)) = j, \qquad \Delta(E^{\mathrm{Tate}}(j)) = \frac{j^2}{(j-1728)^3}$$

for the j-invariant and the discriminant of $E^{\mathrm{Tate}}(j)$. By a simple change of coordinates X, Y, one sees that $E^{\mathrm{Tate}}(j)$ is isomorphic over $\mathbf{Q}$ to the following model:

$$6(j-1728)Y^2 = 4(j-1728)X^3 - 3jX + j. \tag{3}$$

$E^{\mathrm{Tate}}(j)$ is obviously a $(+1)$-minimal $\mathbf{Q}$-curve over K, with $\mu_\sigma = id$. For $\delta \in K^\times$ put $E_\delta := E^{\mathrm{Tate}}(j) \otimes \sqrt{\delta}$. We have an isomorphism

$$\nu : E^{\mathrm{Tate}}(j) \longrightarrow E_\delta, \qquad (x, y) \mapsto (x, y/\sqrt{\delta}),$$

which is defined over $K(\sqrt{\delta})$. By our assumption that E_δ has no complex multiplication, we see that the K-isomorphisms $E_\delta \to {}^\sigma E_\delta$ are given by

$$\mu_\sigma : E_\delta \longrightarrow {}^\sigma E_\delta, \qquad (x, y) \mapsto \left(x, \pm(\sqrt{\delta}/\sqrt{{}^\sigma\delta})y\right).$$

Hence E_δ/K is a minimal $\mathbf{Q}$-curve if and only if $\sqrt{\delta}/\sqrt{{}^\sigma\delta} \in K^\times$, or equivalently

$$\mathrm{N}_{K/\mathbf{Q}}(\delta) = \frac{\delta}{{}^\sigma\delta} \cdot ({}^\sigma\delta)^2 \in (K^\times)^2.$$

We introduce a subgroup $\mathcal{S}_K$ of $\mathbf{Q}^\times$ defined as

$$\mathcal{S}_K := (K^\times)^2 \cap \mathbf{Q}^\times. \tag{4}$$

From $K = \mathbf{Q}(\sqrt{m})$ we have $\mathcal{S}_K = \langle (\mathbf{Q}^\times)^2, m \rangle = (\mathbf{Q}^\times)^2 \cup m(\mathbf{Q}^\times)^2$. The above argument gives the following

Lemma 2. *For $\delta \in K^\times$, the curve $E_\delta := E^{\mathrm{Tate}}(j) \otimes \sqrt{\delta}$ is a minimal $\mathbf{Q}$-curve over K if and only if $\mathrm{N}_{K/\mathbf{Q}}(\delta) \in \mathcal{S}_K$.*

To study the sign of the minimal $\mathbf{Q}$-curve E_δ, and the structure of the associated abelian surface $\mathrm{Res}_{K/\mathbf{Q}}(E(j)_\delta)$, we distinguish the two cases according as $\delta \in K^\times \backslash \mathbf{Q}^\times$ or $\delta \in \mathbf{Q}^\times$, each of which will be studied in the followings sections.

2. Sign change of Tate curves by quartic twists

Here we study the case $\delta \in K^\times \backslash \mathbf{Q}^\times$. We first recall some elementary facts in Galois theory. For a quadratic field K, we denote by $\mathcal{L}(K)$ the set of quartic cyclic extensions $L/\mathbf{Q}$ containing K.

Lemma 3. (Serre [Ser, Theorem(1.2.1)]) *Let $K = \mathbf{Q}(\sqrt{m})$ be a quadratic field. Then the quadratic extension $K(\sqrt{\delta})$, $(\delta \in K^\times)$ of K belongs to $\mathcal{L}(K)$ if and only if $\mathrm{N}_{K/\mathbf{Q}}(\delta) = c^2 m$ for some $c \in \mathbf{Q}^\times$.*

In particular, we have $\mathcal{L}(K) \neq \emptyset$ if and only if $m \in \mathrm{N}_{K/\mathbf{Q}}(K^\times)$, or equivalently $-1 \in \mathrm{N}_{K/\mathbf{Q}}(K^\times)$. It follows that K must be *real* quadratic. Moreover, if $L = K(\sqrt{\delta}) \in \mathcal{L}(K)$ then $\tilde{\sigma} : \sqrt{\delta} \mapsto \sqrt{m}/\sqrt{\delta}$ gives a generator of the Galois group $\mathrm{Gal}(L/\mathbf{Q})$. Note also that one has $\tilde{\sigma}^2(\sqrt{\delta}) = -\sqrt{\delta}$.

Suppose that $K = \mathbf{Q}(\sqrt{m})$ satisfies the above condition, and choose $\delta_0 \in K^\times$ with $\mathrm{N}_{K/\mathbf{Q}}(\delta_0) = m$. Then the set $\mathcal{L}(K)$ is parametrized as

Corollary 4. *For any $c \in \mathbf{Q}^\times$ the field $L_c := K(\sqrt{c\delta_0})$ belongs to $\mathcal{L}(K)$, and $c \mapsto L_c$ induces a bijection*

$$\mathbf{Q}^\times/\mathcal{S}_K \cong \mathcal{L}(K).$$

Proof. Since $\mathrm{N}_{K/\mathbf{Q}}(c\delta_0) = c^2 m$ we see from Lemma 3 that $L_c := K(\sqrt{c\delta_0})$ belongs to $\mathcal{L}(K)$. Conversely, for any member $L \in \mathcal{L}(K)$ we can find as above $\delta \in K^\times$ so that $L = K(\sqrt{\delta})$ and $\mathrm{N}_{K/\mathbf{Q}}(\delta) = c^2 m$, with $c \in \mathbf{Q}^\times$. Then we see that $\alpha := \delta c {\delta_0}^{-1}$ satisfies $\mathrm{N}_{K/\mathbf{Q}}(\alpha) = 1$, hence by Hilbert's theorem 90, $\alpha = \sigma(\beta)/\beta = \mathrm{N}_{K/\mathbf{Q}}(\beta)/\beta^2$, so that $\delta = c\delta_0 \mathrm{N}_{K/\mathbf{Q}}(\beta)\beta^{-2}$. It follows that

$$L = K(\sqrt{\delta}) = K(\sqrt{c_1\delta_0}), \qquad (c_1 := c\mathrm{N}_{K/\mathbf{Q}}(\beta) \in \mathbf{Q}^\times).$$

Finally we observe that

$$L_{c_1} = L_{c_2} \iff c_1 {c_2}^{-1} \in \mathbf{Q}^\times \cap (K^\times)^2 = \mathcal{S}_K.$$

This completes the proof. □

We now recall an argument of [Ha] on the sign change of minimal $\mathbf{Q}$-curves. Let E be a minimal $\mathbf{Q}$-curve over a *real* quadratic field $K = \mathbf{Q}(\sqrt{m})$, with an isogeny $\mu: E \to {}^{\sigma}E$ satisfying ${}^{\sigma}\mu_{\sigma}\circ\mu_{\sigma} = \varepsilon d$, where $\varepsilon = \pm 1$; $\deg(\mu_{\sigma}) = d$ $(d > 0)$ is a positive square free integer, and σ is the non-trivial automorphism of $K/\mathbf{Q}$.

We assume that $\mathcal{L}(K) \neq \emptyset$, and consider the isomorphism

$$\nu: E \to E_{\delta} \qquad (x, y) \mapsto (x, \frac{y}{\sqrt{\delta}}),$$

over L, where $E_{\delta} = E \otimes \sqrt{\delta}/K$ denotes a quadratic twist of E/K. Define $\mu_{\sigma}^{*} := {}^{\sigma}\nu\circ\mu_{\sigma}\circ\nu^{-1}$. Then μ_{σ}^{*} is an isogeny $E_{\delta} \to {}^{\sigma}E_{\delta}$ which is defined over K. We have the following commutative diagram

$$\begin{array}{ccc} (x,y) \in E = E^{\text{Tate}}(j) & \xrightarrow{\nu} & E_{\delta} \ni (x, y/\sqrt{\delta}) = (x^{*}, y^{*}) \\ {\scriptstyle \mu_{\sigma}=id.}\Big\downarrow & & \Big\downarrow{\scriptstyle \mu_{\sigma}^{*}} \\ (x,y) \in E = E^{\text{Tate}}(j) & \xrightarrow{{}^{\tilde{\sigma}}\nu} & E_{\delta} \ni (x, y/\sqrt{{}^{\sigma}\delta}) = (x^{*}, \frac{\sqrt{\delta}}{{}^{\tilde{\sigma}}\sqrt{\delta}}y^{*}). \end{array}$$

It then follows that

$$\begin{aligned} {}^{\sigma}\mu_{\sigma}^{*}\circ\mu_{\sigma}^{*} &= \left({}^{\tilde{\sigma}^2}\nu\circ{}^{\sigma}\mu_{\sigma}\circ{}^{\tilde{\sigma}}\nu^{-1}\right)\circ\left({}^{\tilde{\sigma}}\nu\circ\mu_{\sigma}\circ\nu^{-1}\right) \\ &= {}^{\tilde{\sigma}^2}\nu\circ\left({}^{\sigma}\mu_{\sigma}\circ\mu_{\sigma}\right)\circ\nu^{-1} \\ &= \varepsilon d\left({}^{\tilde{\sigma}^2}\nu\circ\nu^{-1}\right) \\ &= -\varepsilon d \times id. \end{aligned}$$

This proves the following

Proposition 5. ([Ha]) *Assumption being as above, E_{δ}/K is again a minimal $\mathbf{Q}$-curve over K with an isogeny $\mu_{\sigma}^{*}: E_{\delta} \to {}^{\sigma}E_{\delta}$ defined over K satisfying ${}^{\sigma}\mu_{\sigma}^{*}\circ\mu_{\sigma}^{*} = -\varepsilon d \times id$.*

We shall now apply this to $E(j)$ with $d = 1$.

Theorem 2.1. *Suppose that $j \in \mathbf{Q}$ and that $K := \mathbf{Q}(\sqrt{|j - 1728|})$ is a real quadratic field with $\mathcal{L}(K) \neq \emptyset$. Then for each $\delta \in K^{\times}$ satisfying $\mathrm{N}_{K/\mathbf{Q}}(\delta) = c^2|j - 1728|$ $(c \in \mathbf{Q}^{\times})$, the twisted curve $E(j)_{\delta}$ is a (-1)-minimal $\mathbf{Q}$-curve over K, so that*

$$\mathrm{End}_{\mathbf{Q}}(\mathrm{Res}_{K/\mathbf{Q}}(E(j)_{\delta})) \otimes \mathbf{Q} \cong \mathbf{Q}(\sqrt{-1}). \tag{5}$$

In particular, $\mathrm{Res}_{K/\mathbf{Q}}(E(j)_{\delta})$ is an abelian surface over $\mathbf{Q}$ of GL_2-type.

The last assertion (5) follows from a result of Ribet [Ri3].

Next we study the case $\delta \in \mathbf{Q}^\times$. If $\delta \in \mathbf{Q}^\times \cap (K^\times)^2 = \mathcal{S}_K$ then E_δ is K-isomorphic to $E = E^{\mathrm{Tate}}(j)$. So assume that $\delta \in \mathbf{Q}^\times \backslash \mathcal{S}_K$. In such a case E_δ is again defined over $\mathbf{Q}$, and the isomorphism $\nu : E = E^{\mathrm{Tate}}(j) \to E_\delta$ is defined over a quadratic field $\mathbf{Q}(\sqrt{\delta})$ different from K. It follows that $K(\sqrt{\delta})/\mathbf{Q}$ is a $(2,2)$-extension and one can take $\tilde{\sigma} \in \mathrm{Gal}(K(\sqrt{\delta})/\mathbf{Q}$ such that $\tilde{\sigma}|K = \sigma$. Then one sees from the diagram below that E_δ is a $(+1)$-minimal $\mathbf{Q}$-curve over K.

$$\begin{array}{ccc} E = E^{\mathrm{Tate}}(j) & \xrightarrow{\nu} & E_\delta \\ {\scriptstyle \mu_\sigma = id.} \downarrow & & \downarrow {\scriptstyle \mu_{\tilde{\sigma}}^* = id.} \\ E = E^{\mathrm{Tate}}(j) & \xrightarrow{\tilde{\sigma}\nu} & E_\delta. \end{array}$$

We note here that the two curves $E_\delta, E_{m\delta}$, which are not isomorphic over $\mathbf{Q}$, become isomorphic over K. Therefore it is meaningful to consider an abelian surface $E_\delta \times E_{m\delta}$ as a model over $\mathbf{Q}$ of $\mathrm{Res}_{K/\mathbf{Q}}(E(j)_\delta)$.

Definition 6. *An abelian surface* $A/\mathbf{Q}$ *is called of fake* GL_2-*type over* $\mathbf{Q}$, *if the* $\mathbf{Q}$-*algebra of endomorphisms of* A *defined over* $\mathbf{Q}$ *is*

$$\mathrm{End}_{\mathbf{Q}}(A) \otimes \mathbf{Q} \cong \mathbf{Q} \oplus \mathbf{Q},$$

while $\mathrm{End}_K(A) \otimes \mathbf{Q} \cong \mathrm{M}_2(\mathbf{Q})$ *for a quadratic field* K.

Thus an abelian surface $A/\mathbf{Q}$ is of fake GL_2-type over $\mathbf{Q}$ if it is $\mathbf{Q}$-isogenous to a product $E_1 \times E_2$ of elliptic curves over $\mathbf{Q}$ which are K-isogenous but not $\mathbf{Q}$-isogenous to each other. We shall show in the next paragraph that abelian surfaces of fake GL_2-type over $\mathbf{Q}$ occur as jacobian sufaces of algebraic curves, and they play imortant roles in studying abelian surfaces $A/\mathbf{Q}$ of GL_2-type over $\mathbf{Q}$ such that $\mathrm{End}_{\mathbf{Q}}(A) \otimes \mathbf{Q} \cong \mathbf{Q}(\sqrt{-1})$.

3. Genus two curves with jacobian of fake GL_2-type

Let $C(j)/\mathbf{Q}(j)$ be a family of curves of genus two defined by the equation $Y^2 = F(X; j)$,

$$\begin{aligned} F(X;j) := X^6 + 144X^5 - (j-1728)X^4 + 224(j-1728)X^3 \\ - (j-1728)^2X^2 + 144(j-1728)^2X + (j-1728)^3. \end{aligned} \tag{6}$$

We shall describe the construction of such family in the final section. Our first result is stated as follows:

Theorem 3.1. *There exists a morphism* $\psi : C(j) \to E(j)$ *of degree two which is defined over* $\mathbf{Q}(\sqrt{j-1728})$, *where* $E(j)$ *is a twist of* $E^{\mathrm{Tate}}(j)$

corresponding to the quadratic extension $\mathbf{Q}(\sqrt{2(j-1728)})$ *of* $\mathbf{Q}(j)$ *and is defined by*

$$E(j):\ 3W^2 = 4(j-1728)U^3 - 3jU + j. \tag{7}$$

Proof. We put, for simplicity, $m := j - 1728$. The curve $E(j)$ is easily seen to be $\mathbf{Q}(j)$-isomorphic to a quadratic twist $E^{\mathrm{Tate}}(j)_{2m}$ of $E^{\mathrm{Tate}}(j)$. We observe that a covering map $\psi : C(j) \to E(j)$, which is defined over $\mathbf{Q}(\sqrt{m})$, is given by

$$\psi^*(U) = \frac{(\sqrt{m}-72)X^2 + 2\sqrt{m}(\sqrt{m}+24)X + m(\sqrt{m}-72)}{2\sqrt{m}(X+\sqrt{m})^2},$$

$$\psi^*(W) = \frac{-48Y}{(X+\sqrt{m})^3}.$$

□

One can easily check the following identity for the pullback of invariant 1-forms:

$$\psi^*\Big(\frac{dU}{W}\Big) = -2\sqrt{m}\frac{dX}{Y} + 2\frac{XdX}{Y}. \tag{8}$$

Let σ be the generator of the Galois group of the quadratic extension $\mathbf{Q}(\sqrt{m})/\mathbf{Q}$. Then ${}^\sigma\psi$ is also a morphism of degree 2 from $C(j)$ to $E(j)$, and we have

$${}^\sigma\psi^*\Big(\frac{dU}{W}\Big) = 2\sqrt{m}\frac{dX}{Y} + 2\frac{XdX}{Y}. \tag{9}$$

Since $\psi^*(\frac{dU}{W})$ and ${}^\sigma\psi^*(\frac{dU}{W})$ are linearly independent, the morphism

$$\Phi := (\psi, {}^\sigma\psi) : C(j) \to E(j) \times {}^\sigma E(j), \quad P \mapsto (\psi(P), {}^\sigma\psi(P))$$

induces an isogeny

$$\Phi : \operatorname{Jac} C(j) \to E(j) \times {}^\sigma E(j) = E(j) \times E(j),$$

which is defined over $\mathbf{Q}(\sqrt{m})$. Thus we have seen that $E(j)$ is a $(+)$-minimal $\mathbf{Q}$-curve of degree one over $\mathbf{Q}(\sqrt{m})$, the isogeny $\mu_\sigma : E(j) \to {}^\sigma E(j) = E(j)$ being the identity map. It follows that all endomorphisms of $\operatorname{Jac} C(j)$ are defined over $\mathbf{Q}(\sqrt{m})$, and in fact the $\mathbf{Q}$-algebra of endomorphisms of $\operatorname{Jac} C(j)$ defined over $\mathbf{Q}(\sqrt{m})$ is

$$\operatorname{End}_{\mathbf{Q}(\sqrt{m})}(\operatorname{Jac} C(j)) \otimes \mathbf{Q} \cong \mathrm{M}_2(\mathbf{Q}). \tag{10}$$

Theorem 3.2. *The jacobian variety* $\operatorname{Jac} C(j)$ *is of fake* GL_2*-type over* $\mathbf{Q}(j)$. *Namely the* $\mathbf{Q}$*-algebra of endomorphisms of* $\operatorname{Jac} C(j)$ *defined over* $\mathbf{Q}(j)$ *is described as*

$$\operatorname{End}_{\mathbf{Q}(j)}(\operatorname{Jac} C(j)) \otimes \mathbf{Q} \cong \mathbf{Q} \oplus \mathbf{Q}. \tag{11}$$

Proof. The assertion means that $\operatorname{Jac} C(j)$ is isogenous over $\mathbf{Q}(j)$ to a product of two elliptic curves which are not $\mathbf{Q}(j)$-isogenous but are $\mathbf{Q}(\sqrt{m})$-isogenous to each other. This indicates the existence of non-trivial morphisms

$$f : C(j) \to E(j), \qquad f_m : C(j) \to E(j)_m,$$

which are defined over $\mathbf{Q}(j)$, where $E(j)_m$ denotes the quadratic twists of $E(j)$ by $\sqrt{m}$. In fact f is given by $f(P) = \psi(P) \oplus_{E(j)} {}^{\sigma}\psi(P)$. A computation shows that f is written explicitly as $f((X,Y)) = (U,W)$, with

$$f^*(U) = \frac{3m^2 + 6mX^2 - 192X^3 - X^4}{(3m + X^2)^2}$$
$$f^*(W) = \frac{6(m^2 - 72mX - mX^2 + 8X^3)Y}{(3m + X^2)^3}.$$

Similarly the morphism $f_m : C(j) \to E(j)_m$ is given by

$$f_m(P) = \psi_m(P) \oplus_{E(j)_m} {}^{\sigma}\psi_m(P),$$
$$\psi_m(P) := (U, W/\sqrt{m}), \qquad \psi(P) = (U,W).$$

Then after some computation one obtains $f_m((X,Y)) = (U,W)$, with

$$f_m^*(U) = \frac{-m^2 - 192mX + 6mX^2 + 3X^4}{(m + 3X^2)^2}$$
$$f_m^*(W) = \frac{6(8m - mX - 72X^2 + X^3)Y}{(m + 3X^2)^3}.$$

One can now easily check the equalities

$$f^*(\frac{dU}{W}) = 4\frac{XdX}{Y}$$
$$f_m^*(\frac{dU}{W}) = -4m\frac{dX}{Y}.$$

Since $f^*(\frac{dU}{W})$ and $f_m^*(\frac{dU}{W})$ are linearly independent, it follows that the morphism (f, f_m), which is defined over $\mathbf{Q}$, induces an isogeny from $\operatorname{Jac} C(j)$ to $E(j) \times E(j)_m$. □

Theorem 3.2 can be proved also by the following argument. Using (10) we identify $\operatorname{End}_{\mathbf{Q}(\sqrt{m})}(\operatorname{Jac} C(j)) \otimes \mathbf{Q}$ with $\mathrm{M}_2(\mathbf{Q})$. Then, from ${}^{\sigma}\Phi = ({}^{\sigma}\psi, \sigma)$ we see that the action of σ on $\operatorname{End}_{\mathbf{Q}(\sqrt{m})}(\operatorname{Jac} C(j)) \otimes \mathbf{Q}$ corresponds to the inner automorphism $\operatorname{Int}(J)$, $J = \begin{pmatrix} 0 & 1 \\ 1 & 0 \end{pmatrix}$ on $\mathrm{M}_2(\mathbf{Q})$.

It follows that $\mathrm{End}_{\mathbf{Q}(j)}(\mathrm{Jac}\, C(j)) \otimes \mathbf{Q}$ corresponds to the subalgebra consisting of $M \in \mathrm{M}_2(\mathbf{Q})$ such that $MJ = JM$. Hence we have

$$\mathrm{End}_{\mathbf{Q}(j)}(\mathrm{Jac}\, C(j)) \otimes \mathbf{Q} \cong \{\begin{pmatrix} a & b \\ b & a \end{pmatrix} \mid a, b \in \mathbf{Q}\} \cong \mathbf{Q} \oplus \mathbf{Q}.$$

Remark 7. *It is interesting to observe that the sextic polynomial $F(X; j)$ in (6) is irreducible over $\mathbf{Q}(j)$, and its splitting field is a regular extension of $\mathbf{Q}$ whose Galois group over $\mathbf{Q}(j)$ is isomorphic to S_4, the symmetric group of degree 4. This fact can be proved by computing several resolvent polynomials in the roots of $F(X; j)$.*

4. Shimura's abelian surfaces

Let N the discriminant of a real quadratic field $K = \mathbf{Q}(\sqrt{N})$, and let $\chi(*) = (\frac{N}{*})$ be the Dirichlet character associated to $K/\mathbf{Q}$. Let $\Gamma_0(N)$ be the modular group

$$\Gamma_0(N) := \left\{\begin{pmatrix} a & b \\ c & d \end{pmatrix} \in \mathrm{SL}_2(\mathbf{Z}) \mid c \equiv 0 \pmod{N}\right\},$$

and denote by $S_2(N, \chi)$ the space of cusp forms of weight 2 and Neben type character $\chi(*)$. We assume that there exists a Hecke eigen form $f \in S_2(N, \chi)$ such that the field $K_f = \mathbf{Q}(\{a_n\})$ generated by the Fourier coefficients of f is an (imaginary) quadratic field. Then A_f is decomposed over K into the product

$$A_f \approx B_f \times {}^{\sigma}B_f \qquad \text{(isogenous over } K = \mathbf{Q}(\sqrt{N})\text{)},$$

where B_f/K is called the Shimura's elliptic curve. It has the following remarkable properties:

(i) B_f/K has good reduction at all finite primes of K.

(ii) B_f/K is a minimal $\mathbf{Q}$-curve, and its degree ($=$ Min $\deg(\mu_\sigma)$) is the square free part of $D(K_f/\mathbf{Q})$:

$$\mu_\sigma/K \ : B_f \to {}^{\sigma}B_f.$$

We assume that $K_f = \mathbf{Q}(\sqrt{-1})$. This condition is equivalent to $j(B_f) \in \mathbf{Q}$. Up to isogeny, six such B_f of non CM type are known; their levels are $N = 37, 65, 104, 157, 397, 877$, and the dimensions of the corresponding spaces $S_2(N, \chi)$ are $2, 4, 12, 12, 32, 72$, respectively. The defining equations for B_f/K were studied by Shiota [Shi], Pinch [Pi], and Cremona [Cr]. On the other hand, elliptic curves over real quadratic fileds with everywhere good reduction has been studied by Comalada [Co], Pinch [Pi], Cremona [Cr], Kagawa [Ka], Kida [Ki] and others. The K-isomorphism classes of known such curves over our $K = \mathbf{Q}(\sqrt{N})$ are as in the following table (c.f. [Co], [Cr], [Ki]).

Table 1a

N	Name	j	$E(j)_\delta$	K-isogeny
37	37A	2^{12}	$\delta_{37A} = 37 + 6\sqrt{37}$	$5: A \to B$
37	37B	$2^{12}{\cdot}211^3$	$\delta_{37B} = -29{\cdot}139{\cdot}211{\cdot}(37 + 6\sqrt{37})$	$5: B \to A$
65	65A	17^3	$\delta_{65A} = -7{\cdot}17(65 + \sqrt{65})$	$2: A \to C$
65	65B	17^3	$\delta_{65B} = -7{\cdot}17(65 + 7\sqrt{65})$	$2: B \to D$
65	65C	257^3	$\delta_{65C} = -2{\cdot}7{\cdot}73{\cdot}257(65 + 8\sqrt{65})$	$2: C \to A$
65	65D	257^3	$\delta_{65D} = -7{\cdot}73{\cdot}257(1105 + 137\sqrt{65})$	$2: D \to B$
877	877A	$2^{12}{\cdot}7^3$	$\delta_{877A} = 5{\cdot}7{\cdot}(8149{\cdot}877 + 241326\sqrt{877})$	
104	104A	2^6	$\delta_{104A} = -2(26 + 5\sqrt{26})$	$5: A \to C$
104	104B	2^6	$\delta_{104B} = -(26 + 5\sqrt{26})$	$5: B \to D$
104	104C	$-2^6{\cdot}719^3$	$\delta_{104C} = 2{\cdot}19{\cdot}199{\cdot}719(26 + 5\sqrt{26})$	$5: C \to A$
104	104D	$-2^6{\cdot}719^3$	$\delta_{104D} = 19{\cdot}199{\cdot}719(26 + 5\sqrt{26})$	$5: D \to B$
157	157A	-13^3	$\delta_{157A} = 5{\cdot}13{\cdot}(17{\cdot}157 + 213\sqrt{157})$	
397	397A	11^3	$\delta_{397A} = -11{\cdot}(173{\cdot}397 + 3447\sqrt{397})$	

The columns after N and the names are their j-invariants and the models over K which are obtained by twisting $E(j)$ with $\sqrt{\delta}$. It turns out that each of them is isomorphic to Shimura's elliptic curve B_f for some $f \in S_2(N, \chi)$. Also it is important to observe that their j-invariants are all perfect cubes, and that the real quadratic fields $\mathbf{Q}(\sqrt{N})$ coincide with $\mathbf{Q}(\sqrt{|j - 12^3|})$, as indicated in the following table:

Table 1b

Name	j	$j - 1728$	$\Delta(E(j)) = 2^{\pm 6} j^2/(j - 12^3)^3$
37A	2^{12}	$2^6{\cdot}37$	37^{-3}
37B	$2^{12}{\cdot}211^3$	$2^6{\cdot}29^2{\cdot}139^2{\cdot}37$	$29^{-6}{\cdot}139^{-6}{\cdot}211^6{\cdot}37^{-3}$
65A	17^3	$5{\cdot}7^2{\cdot}13$	$2^6{\cdot}5^{-3}{\cdot}7^{-6}{\cdot}13^{-3}{\cdot}17^6$
65B	17^3	$5{\cdot}7^2{\cdot}13$	$2^6{\cdot}5^{-3}{\cdot}7^{-6}{\cdot}13^{-3}{\cdot}17^6$
65C	257^3	$5{\cdot}7^2{\cdot}13{\cdot}73^2$	$2^6{\cdot}5^{-3}{\cdot}7^{-6}{\cdot}13^{-3}{\cdot}73^{-6}{\cdot}257^6$
65D	257^3	$5{\cdot}7^2{\cdot}13{\cdot}73^2$	$2^6{\cdot}5^{-3}{\cdot}7^{-6}{\cdot}13^{-3}{\cdot}73^{-6}{\cdot}257^6$
877A	$2^{12}{\cdot}7^3$	$2^6{\cdot}5^2{\cdot}877$	$5^{-6}{\cdot}7^6{\cdot}877^{-3}$
104A	2^6	$-2^7{\cdot}13$	$-2^{-3}{\cdot}13^{-3}$
104B	2^6	$-2^7{\cdot}13$	$-2^{-3}{\cdot}13^{-3}$
104C	$-2^6{\cdot}719^3$	$-2^7{\cdot}13{\cdot}19^2{\cdot}199^2$	$-2^{-3}{\cdot}13^{-3}{\cdot}19^{-6}{\cdot}199^{-6}{\cdot}719^6$
104D	$-2^6{\cdot}719^3$	$-2^7{\cdot}13{\cdot}19^2{\cdot}199^2$	$-2^{-3}{\cdot}13^{-3}{\cdot}19^{-6}{\cdot}199^{-6}{\cdot}719^6$
157A	-13^3	$-5^2{\cdot}157$	$-2^6{\cdot}5^{-6}{\cdot}13^6{\cdot}157^{-3}$
397A	11^3	-397	$-2^6{\cdot}11^6{\cdot}397^{-3}$

We shall now state our second main results Theorem 4.1, and Theorem 4.2, which describe Shimura's abelian surfaces as jacobian of curves belonging to the family $C(j)$ with quadratic twists.

Theorem 4.1. *Let* $N = 37, 65, 877$, *and* $f \in S_2(N, \chi)$ *be an eigenform with* $K_f = \mathbf{Q}(\sqrt{-1})$. *Then* A_f *is isogenous over* L_N *to the jacobian of the following curve* $C(j) \otimes \sqrt{r}$ *defined over* $\mathbf{Q}$, *where* L_N *is a cyclic quartic field given as* $L_N = \mathbf{Q}\left(\sqrt{\delta_N}\right)$ *with*

$$\delta_N = 2(-37 + \sqrt{37}), \quad 2(65 + \sqrt{65}), \quad -2(877 + 29\sqrt{877})$$

respectively, which satisfies $K = \mathbf{Q}(\sqrt{N}) \subseteq L_N \subseteq \mathbf{Q}(\zeta_N)$.

Table 2a

N	$C(j) \otimes \sqrt{r}$	$C:\ rY^2 = f(X)$
37	$C(2^{12}) \otimes \sqrt{-1}$	$-Y^2 = X^6 + 18X^5 - 37X^4$ $+1036X^3 - 1369X^2 + 24642X + 50653$
37	$C(2^{12}211^3) \otimes \sqrt{29{\cdot}139{\cdot}211}$	$29{\cdot}139{\cdot}211Y^2 = 4031X^6 + 18X^5 - 149147X^4$ $+1036X^3 - 5518439X^2 + 24642X + 204182243$

Table 2b

N	$C(j) \otimes \sqrt{r}$	$C:\ rY^2 = f(X)$
65	$C(17^3) \otimes \sqrt{-2{\cdot}7{\cdot}17}$	$-2{\cdot}7{\cdot}17Y^2 = (X^2 + 2X + 65)$ $\times(7X^4 + 130X^3 - 1170X^2 + 8450X + 29575)$
65	$C(257^3) \otimes \sqrt{-2{\cdot}7{\cdot}73{\cdot}257}$	$-2{\cdot}7{\cdot}73{\cdot}257Y^2 = (X^2 - 16X + 65)$ $\times(511X^4 + 8320X^3 + 66690X^2$ $+ 540800X + 2158975)$
877	$C(2^{12}7^3) \otimes \sqrt{-5{\cdot}7}$	$-5{\cdot}7Y^2 = 5X^6 + 18X^5 - 4385X^4 + 24556X^3$ $-3845645X^2 + 13844322X + 3372630665$

Proof. The polynomials $f(X)$ listed above are obtained from $F(X; j)$ by a minor change of variable X over $\mathbf{Q}$. Therefore the assertions are immediate consequences of Theorem 3.1, except for the fields over which the isogenies are defined. Denote the pure generator of L_N over $\mathbf{Q}(\sqrt{N})$ by $\sqrt{\delta_N}$. Then we see from the argument of Sect.2 that, to prove the last assertion it suffices to check that, up to a square factor, the quotient δ_{NA}/δ_N coincides with the twisting factor r of $C(j)$. The quartic Dirichlet character ψ, which corresponds to the extension $L_N/\mathbf{Q}$ is given as follows. Note that 5 (resp. 2) is a primitive root modulo 37 (resp. 5, 13 and 877). Thus for $N = 37, 877$ the pair $\{\psi^{\pm 1}\}$ of Dirichlet characters of order 4 defined mod N is unique, and we have

$$N = 37: \ \psi(p) = (\sqrt{-1})^k \quad \text{for} \quad p \equiv 5^k \pmod{37}$$
$$N = 877: \ \psi(p) = (\sqrt{-1})^k \quad \text{for} \quad p \equiv 2^k \pmod{877}.$$

L_N is then generated over $\mathbf{Q}$ by the so called Gaussian period. Namely we have $L_N = \mathbf{Q}(\eta_j)$ for

$$\eta_j := \sum_{i=0}^{(N-1)/4-1} \zeta^{g^{4i+j}} \qquad (0 \le j \le 3) \tag{12}$$

where $g \in \mathbf{Z}/N\mathbf{Z}$ is a fixed primitive root modulo N. The irreducible equation $G_4(X; N)$ over $\mathbf{Q}$ for η_j is known (c.f. [H, pp.489-494]; see also [BE]), and is easily computed for each N. Indeed we have

$$G_4(X; 37) = X^4 + X^3 + 5X^2 + 7X + 49$$
$$G_4(X; 877) = X^4 + X^3 + 110X^2 + 3234X + 16317,$$

from which we obtain

$$L_{37} = \mathbf{Q}\left(\sqrt{2(-37+\sqrt{37})}\right), \quad L_{877} = \mathbf{Q}\left(\sqrt{-2(877+29\sqrt{877})}\right).$$

It follows that for $N = 37$ (resp. $N = 877$), $L_N = \mathbf{Q}(\sqrt{\delta_N})$ with $\delta_{37} = 2\sqrt{37}(1-\sqrt{37})$ (resp $\delta_{877} = -2\sqrt{877}(29+\sqrt{877})$), so that we observe

$$\begin{aligned} \delta_{37} &= -(7-\sqrt{37})^2 \cdot (37 + 6\sqrt{37}) \\ &= -(7-\sqrt{37})^2 \times \delta_{37A}, \\ \delta_{37B} &= -29 \cdot 139 \cdot 211 \times \delta_{37A}, \end{aligned}$$

and

$$\delta_{877A} = -5 \cdot 7 \cdot \left(\frac{385+\sqrt{877}}{12}\right)^2 \times \delta_{877}.$$

Thus we see that the factors -1, $29 \cdot 139 \cdot 211$, and $-5 \cdot 7$ are exactly the twisting factors r of $C(2^{12})$, $C(2^{12} \cdot 211^3)$, $C(2^{12} \cdot 7^3)$, respectively. Hence we can conclude that A_f is isogenous to $\mathrm{Jac}C(j) \otimes \sqrt{r}/\mathbf{Q}$ over L_N.

On the other hand, since $(\mathbf{Z}/65\mathbf{Z})^\times \cong (\mathbf{Z}/5\mathbf{Z})^\times \times (\mathbf{Z}/13\mathbf{Z})^\times$ we see that there are six cyclic subgroups of order 4 of the Dirichlet characters defined mod 65. They are divided into three classes according as the quadratic subfields corresponding to their subgroups of order 2, which are $\mathbf{Q}(\sqrt{5})$, $\mathbf{Q}(\sqrt{13})$ and $\mathbf{Q}(\sqrt{65})$. Among the two which correspond to $\mathbf{Q}(\sqrt{65})$, we choose the following one generated by :

$$N = 65: \ \psi(p) = (\sqrt{-1})^{j-k} \quad \text{for} \quad p \equiv 2^j \pmod 5, \quad p \equiv 2^k \pmod{13}.$$

We have then $\mathrm{Ker}(\psi) = \langle 2 \rangle$ ($\subseteq (\mathbf{Z}/65\mathbf{Z})^\times$) so that the corresponding quartic field L_{65} is generated over $\mathbf{Q}$ by

$$\eta := \sum_{i=1}^{12} \zeta_{65}^{2^i}.$$

The equation for η over $\mathbf{Q}$ is easily computed to be $X^4 - X^3 - 24X^2 + 4X + 16$, and solving this we obtain

$$\eta = \frac{1}{4}\left((1+\sqrt{65}) + \sqrt{130 + 2\sqrt{65}}\right).$$

It follows that $L_{65} = \mathbf{Q}(\sqrt{\delta_{65}})$ with $\delta_{65} = 2\sqrt{65}(1 + 2\sqrt{65})$, and we have

$$\delta_{65A} = -7{\cdot}17(65 + \sqrt{65}) = -2{\cdot}7{\cdot}17{\cdot}2^{-2} \times \delta_{65}.$$

Thus we see that the factor $-2{\cdot}7{\cdot}17$ coincides with the twisting factor r of $C(17^3)$. It follows that A_f is isogenous to $\mathrm{Jac}C(j) \otimes \sqrt{r}/\mathbf{Q}$ over L_N. This completes the proof of Theorem 4.1. □

The L-function of $C(j)/\mathbf{Q}$ or $\mathrm{Jac}C(j)/\mathbf{Q}$ is defined by

$$L(C(j)/\mathbf{Q}, s) = L(\mathrm{Jac}C(j)/\mathbf{Q}, s) := \prod_p L_p(C(j)/\mathbf{Q}, p^{-s})^{-1},$$

where for good primes p the Euler p-factor is given by

$$L_p(C(j)/\mathbf{Q}, u) = (1 - A_p u + pu^2)(1 - B_p u + pu^2).$$

Here A_p, B_p are real numbers which are determined by the congruence zeta function of $\overline{C(j)}_p := C(j) \otimes \mathbf{F}_p$ as

$$\begin{aligned} \mathrm{Z}(\overline{C(j)}_p/\mathbf{F}_p, u) &:= \exp\left(\sum_{m=1}^{\infty} \frac{N_m}{m} u^m\right) \qquad N_m := \#\left(\overline{C(j)}_p(\mathbf{F}_{p^m})\right) \\ &= \frac{(1-\alpha_p u)(1-\overline{\alpha}_p u)(1-\beta_p u)(1-\overline{\beta}_p u)}{(1-u)(1-pu)} \\ &= \frac{(1 - A_p u + pu^2)(1 - B_p u + pu^2)}{(1-u)(1-pu)}, \end{aligned} \tag{13}$$

so that $A_p = \alpha_p + \overline{\alpha}_p$, $B_p = \beta_p + \overline{\beta}_p$. We note that α_p, β_p are complex numbers satisfying

$$\alpha_p\overline{\alpha}_p = \beta_p\overline{\beta}_p = p, \quad |\alpha_p| = |\beta_p| = p^{1/2}.$$

By the isogeny theorem of Faltings [Fa], the above results imply the following equality for $C := C(j) \otimes \sqrt{r}$

$$\begin{aligned} L(C/\mathbf{Q}, s) &= L(f \otimes \psi, s)L(\bar{f} \otimes \psi, s) \\ &= L((B_f/k) \otimes \psi, s) \\ &= L((A_f/\mathbf{Q}) \otimes \psi, s), \end{aligned}$$

which can also be expressed as

$$A_p\psi(p) = a_f(p), \qquad B_p\psi(p) = a_{\bar{f}}(p). \tag{14}$$

We tabulate some of these quantities for small p.

Table 3a

	$N = 37$ $C(2^{12})$			$N = 65$ $C(17^3)$		
p	$a_f(p)$	$A_p,\ B_p$	$\psi(p)(\frac{-1}{p})$	$a_f(p)$	$A_p,\ B_p$	$\psi(p)(\frac{-238}{p})$
3	-1	$-1, -1$	$+1$	$-2\sqrt{-1}$	$2, -2$	$\sqrt{-1}$
11	-3	$-3, -3$	$+1$	$-2\sqrt{-1}$	$2, -2$	$\sqrt{-1}$
17	$-2\sqrt{-1}$	$2, -2$	$\sqrt{-1}$	0	$*, *$	0
19	$-6\sqrt{-1}$	$6, -6$	$-\sqrt{-1}$	$6\sqrt{-1}$	$6, -6$	$\sqrt{-1}$
23	$-4\sqrt{-1}$	$4, -4$	$-\sqrt{-1}$	$-6\sqrt{-1}$	$6, -6$	$-\sqrt{-1}$
29	$4\sqrt{-1}$	$4, -4$	$-\sqrt{-1}$	6	$-6, -6$	-1
31	0	$0, 0$	$\sqrt{-1}$	$-6\sqrt{-1}$	$6, -6$	$-\sqrt{-1}$
37	$-1 - 6\sqrt{-1}$	$*, *$	$*$	6	$-6, -6$	-1
41	-3	$3, 3$	-1	$8\sqrt{-1}$	$8, -8$	$-\sqrt{-1}$
43	$6\sqrt{-1}$	$6, -6$	$-\sqrt{-1}$	$6\sqrt{-1}$	$6, -6$	$-\sqrt{-1}$
47	3	$-3, -3$	-1	-8	$-8, -8$	$+1$

Table 3b

	$N = 877$ $C(2^{12}{\cdot}7^3)$				$N = 877$ $C(2^{12}{\cdot}7^3)$		
p	$a_f(p)$	$A_p,\ B_p$	$\psi(p)(\frac{-35}{p})$	p	$a_f(p)$	$A_p,\ B_p$	$\psi(p)(\frac{-35}{p})$
3	1	$-1, -1$	-1	11	$-4\sqrt{-1}$	$4, -4$	$\sqrt{-1}$
13	0	$0, 0$	$\sqrt{-1}$	17	$-2\sqrt{-1}$	$2, -2$	$\sqrt{-1}$
19	$-6\sqrt{-1}$	$6, -6$	$\sqrt{-1}$	23	0	$0, 0$	-1
29	6	$-6, -6$	-1	31	-3	$3, 3$	-1
37	3	$-3, -3$	-1	41	6	$-6, -6$	-1
43	1	$-1, -1$	-1	47	$4\sqrt{-1}$	$4, -4$	$\sqrt{-1}$

Table 4

N	$C(j)\otimes\sqrt{r}$	$C(j)\otimes\sqrt{r}:\ rY^2=f(X)$
104	$C(2^6)\otimes\sqrt{\pm 13}$	$\pm 13Y^2 = X^6 - 109\,X^4 + 40\,X^3$ $+\,3635\,X^2 - 408\,X - 42047$
104	$C(-2^6 719^3)\otimes\sqrt{\pm 13\cdot 19\cdot 199\cdot 719}$	$\pm 13\cdot 19\cdot 199\cdot 719Y^2 = 3781\,X^6 + 18\,X^5$ $+98306\,X^4 - 728\,X^3 - 2555956\,X^2$ $+\,12168\,X - 66454856$
157	$C(-2197)\otimes\sqrt{\pm 2\cdot 5\cdot 13}$	$\pm 2\cdot 5\cdot 13Y^2 = 5\,X^6 + 144\,X^5$ $+785\,X^4 - 35168\,X^3 - 123245\,X^2$ $+\,3549456\,X - 19349465$
397	$C(11^3)\otimes\sqrt{\pm 2\cdot 11}$	$\pm 2\cdot 11Y^2 = X^6 + 144\,X^5$ $+397\,X^4 - 88928\,X^3 - 157609\,X^2$ $+\,22695696\,X - 62570773$

Theorem 4.2. *Let $N = 104, 157, 397$, and $f \in S_2(N,\chi)$ be an eigenform with $K_f = \mathbf{Q}(\sqrt{-1})$. Then A_f is isogenous over $\mathbf{Q}(\sqrt{-1})L_N$ to the jacobian of the following curve $C(j)\otimes\sqrt{r}$ defined over $\mathbf{Q}$, where L_N is a cyclic quartic field given as $L_N = \mathbf{Q}\left(\sqrt{\delta_N}\right)$ with*

$$\delta_N = -(26+5\sqrt{26}),\quad 2(-157+11\sqrt{157}),\quad -2(397+19\sqrt{397})$$

respectively, which satisfies $K = \mathbf{Q}(\sqrt{N}) \subseteq L_N \subseteq \mathbf{Q}(\zeta_{2N})$.

Proof. As in Theorem 4.1, the polynomials $f(X)$ listed above are obtained from $F(X;j)$ by changing variable X linearly over $\mathbf{Q}$. Therefore it suffices to check the assertions for the fields over which the isogenies are defined. This is again equivalent to show that up to a square factor, the quotient of two δ's δ_{NA}/δ_N coincides with the twisting factor r of $C(j)$.

First note that the quartic Dirichlet character ψ, which corresponds to the extension $L_N/\mathbf{Q}$ is given as follows. For $N = 157, 397$ the pair $\{\psi^{\pm 1}\}$ of Dirichlet characters of order 4 defined mod N is unique, and using 20 as a primitive root modulo N we have

$$N = 157,\ 397:\ \psi(p) = (\sqrt{-1})^k \quad \text{for} \quad p \equiv 20^k \pmod{N}.$$

L_N is generated over $\mathbf{Q}$ by the Gaussian period η_j (see (12)) whose irreducible equation $G_4(X;N)$ over $\mathbf{Q}$ is

$$G_4(X;157) = X^4 + X^3 + 20X^2 - 206X + 517$$
$$G_4(X;397) = X^4 + X^3 + 50X^2 - 918X + 3069.$$

From this we obtain $L_N = \mathbf{Q}(\sqrt{\delta_N})$ with

$$\delta_{157} = 2(-157 + 11\sqrt{157}), \quad \delta_{397} = -2(397 + 19\sqrt{397}).$$

It follows that for $N = 157$ (resp. $N = 397$), we have

$$\delta_{157A}\left(2(25 - 2\sqrt{157})\right)^2 = -2{\cdot}5{\cdot}13 \times \delta_{157},$$

and

$$\delta_{397A} = 2{\cdot}11{\cdot}\left(\frac{20 + \sqrt{397}}{6}\right)^2 \times \delta_{397A}.$$

This proves the assertion for $N = 157, 397$.

Finally suppose that $N = 104 = 8{\cdot}13$. Then there is no Dirichlet character ψ defined modulo N such that ψ^2 corresponds to $\mathbf{Q}(\sqrt{N})$. Hence we have to raise the level to $2N = 208$:

$$(\mathbf{Z}/208\mathbf{Z})^\times \cong (\mathbf{Z}/16\mathbf{Z})^\times \times (\mathbf{Z}/13\mathbf{Z})^\times \cong \{\pm 1\} \times \langle 5 \rangle \times \langle 2 \rangle,$$

on which there are 8 candidates for ψ satisfying the above condition. They are determined by the values at -1,5, 2:

$$(\psi(-1), \psi(5), \psi(2)) = (\varepsilon_1, (\sqrt{-1})^{\varepsilon_2}, (\sqrt{-1})^{\varepsilon_3}), \quad (\varepsilon_i = \pm 1).$$

We choose ψ for which $(\varepsilon_1, \varepsilon_2, \varepsilon_3) = (1, 1, -1)$. Then we have $\mathrm{Ker}(\psi) = \langle 5, 23 \rangle \subset (\mathbf{Z}/208\mathbf{Z})^\times$ so that the corresponding quartic field L_{208} is generated over $\mathbf{Q}$ by

$$\eta := \sum_{i=0}^{3} \sum_{j=0}^{5} \zeta^{5^i 23^j} \qquad (\zeta = e^{2\pi\sqrt{-1}/208}).$$

It is not difficult to see that the equation for η over $\mathbf{Q}$ is $X^4 + 52X^2 + 26$, and solving this we obtain $\eta = \sqrt{-26 - 5\sqrt{26}}$. It follows that $L_{104} = \mathbf{Q}(\sqrt{\delta_{104}})$ with $\delta_{104} = -26 - 5\sqrt{26}$, and we have

$$\begin{aligned} \delta_{104A}{\cdot}(\sqrt{26})^2 &= -4{\cdot}13{\cdot}(26 + 5\sqrt{26}) = -13{\cdot}2^2 \times \delta_{26}, \\ \delta_{104C}{\cdot}(\sqrt{26})^2 &= -13{\cdot}19{\cdot}199{\cdot}719{\cdot}2^2 \times \delta_{26}. \end{aligned}$$

Thus we see that the factor -13, $-13{\cdot}19{\cdot}199{\cdot}719$ coincide with the twisting factors r of $C(2^6)$ $C(-2^6 719^3)$. It follows that A_f is isogenous to $\mathrm{Jac}C(j) \otimes \sqrt{r}/\mathbf{Q}$ over L_N. This completes the proof of Theorem 4.1. $\square$

The assertion of Theorem 4.2 is slightly weaker then Theorem 4.1; in fact it is equivalent to the following equality ($C = C(j) \otimes \sqrt{r}$)

$$\begin{aligned} L(C/\mathbf{Q}(\sqrt{-1}), s) &= L(f \otimes \psi, s) L(\bar{f} \otimes \psi, s) \\ &\quad \times L(f \otimes \psi \otimes (\frac{-1}{*}), s) L(\bar{f} \otimes \psi \otimes (\frac{-1}{*}), s) \\ &= L((B_f/k) \otimes \psi, s) L((B_f/k) \otimes \psi \otimes (\frac{-1}{*}), s) \\ &= L((A_f/\mathbf{Q}) \otimes \psi, s) L((A_f/\mathbf{Q}) \otimes \psi \otimes (\frac{-1}{*}), s) \end{aligned}$$

which can also be expressed as

$$\begin{aligned} &\{A_p\psi(p), A_p\psi(p)(\frac{-1}{p}), B_p\psi(p), B_p\psi(p)(\frac{-1}{p})\} \\ &\qquad = \{a_f(p), a_f(p)(\frac{-1}{p}), a_{\bar{f}}(p), a_{\bar{f}}(p)(\frac{-1}{p})\}. \end{aligned} \tag{15}$$

We tabulate the quantities $\{A_p, B_p, a_f(p)\}$ for small p.

Table 5a

$N = 397$ $C(11^3)$				$N = 397$ $C(11^3)$			
p	$a_f(p)$	A_p, B_p	$\psi(p)(\frac{22}{p})$	p	$a_f(p)$	A_p, B_p	$\psi(p)(\frac{22}{p})$
3	2	$2, -2$	-1	5	$-2\sqrt{-1}$	$2, -2$	$\sqrt{-1}$
7	$-3\sqrt{-1}$	$-3, -3$	$-\sqrt{-1}$	11	0	$*, *$	0
13	0	$0, 0$	$-\sqrt{-1}$	17	$4\sqrt{-1}$	$4, -4$	$\sqrt{-1}$
19	-2	$2, -2$	-1	23	6	$6, -6$	1
29	-3	$-3, -3$	1	31	-4	$4, -4$	-1
37	-9	$9, 9$	-1	41	$8\sqrt{-1}$	$8, -8$	$\sqrt{-1}$
43	0	$0, 0$	1	47	6	$6, -6$	-1

Table 5b

	$N=104$ $C(2^6)$			$N=157$ $C(-13^3)$		
p	$a_f(p)$	$A_p,\ B_p$	$\psi(p)(\frac{2}{p})$	$a_f(p)$	$A_p,\ B_p$	$\psi(p)(\frac{130}{p})$
3	$-\sqrt{-1}$	$-1,-1$	$+1$	-2	$2,-2$	-1
7	$-3\sqrt{-1}$	$-3,-3$	$\sqrt{-1}$	$-3\sqrt{-1}$	$-3,-3$	$\sqrt{-1}$
11	2	$2,-2$	-1	6	$6,-6$	$+1$
17	3	$-3,-3$	-1	3	$-3,-3$	-1
19	0	$0,0$	-1	0	$0,0$	$+1$
23	-6	$6,-6$	$+1$	$-5\sqrt{-1}$	$5,5$	$-\sqrt{-1}$
29	$6\sqrt{-1}$	$6,-6$	$\sqrt{-1}$	$4\sqrt{-1}$	$4,-4$	$\sqrt{-1}$
31	0	$0,0$	$\sqrt{-1}$	0	$0,0$	$+1$
37	3	$-3,-3$	-1	-3	$3,3$	-1
41	$10\sqrt{-1}$	$10,-10$	$-\sqrt{-1}$	$2\sqrt{-1}$	$2,-2$	$\sqrt{-1}$
43	$9\sqrt{-1}$	$9,9$	$\sqrt{-1}$	$9\sqrt{-1}$	$-9,-9$	$-\sqrt{-1}$
47	$7\sqrt{-1}$	$-7,-7$	$-\sqrt{-1}$	6	$6,-6$	$+1$

5. Another example for $N = 65$

In general the field over which the isogeny $\mathrm{Jac}C(j) \to A_f$ is defined is bigger than $\mathbf{Q}$ when the Neben type character χ of f is nontrivial. Indeed the assertion is equivalent from the isogeny theorem of Faltings [Fa] to the equality of their L-functions, i.e.

$$(1 - A_p u + pu^2)(1 - B_p u + pu^2) = (1 - a_f(p)u + \chi(p)pu^2)(1 - b_{\bar{f}}(p)u + \chi(p)pu^2)$$

for almost all primes p, where A_p, B_p are real numbers as defined in (13). In very rare cases, however, it happens that the isogeny can be defined over smaller field. In this paragraph we shall present an example of a curve $C/\mathbf{Q}$ such that $\mathrm{Jac}C$ is isogenous to A_f over $\mathbf{Q}$, for $N = 65$.

Theorem 5.1. *Let $C_{65}/\mathbf{Q}$ be a curve of genus two defined over $\mathbf{Q}$ by the equation*

$$C_{65}: \ Y^2 = -5X(X+13)(5X+1)(5X^2+2X+13). \qquad (16)$$

Then C_{65} has an involution defined over the quadratic field $K := \mathbf{Q}(\sqrt{65})$

$$\rho: \ (X,Y) \mapsto \left(X + \frac{13}{5X}, (1 + \frac{13\sqrt{65}}{25X^3})Y \right)$$

and the quotient curve $C_{65}/\langle\rho\rangle$ is an elliptic curve over K defined by

$$E_{65}: \ W^2 = U(U-65)(U+2-2\sqrt{65}). \qquad (17)$$

Furthermore, E_{65}/k is a (-1)-minimal **Q***-curve with $j(E_{65}) = 17^3$ and has a trivial conductor over K. Finally* Jac$C$ *is isogenous to A_f, $f \in S_2(65, \chi)$ over* **Q**.

Proof. Putting

$$U := X + \frac{13}{5X}, \qquad W := \left(1 + \frac{13\sqrt{65}}{25X^3}\right) Y,$$

we see in the same way as in Theorem 3.1 that the equation (16) becomes (17). Then we transform (17) into Weierstrass equation $W^2 = 4U^3 - g_2U - g_3$, with

$$g_2 = \frac{544(33 - \sqrt{65})}{3}, \qquad g_3 = \frac{1792(1105 - 49\sqrt{65})}{27} \tag{18}$$

Now we observe that for $\lambda := (1 + \sqrt{65})/8$ we have

$$\frac{{}^\sigma g_2}{g_2} = \frac{33 + \sqrt{65}}{33 - \sqrt{65}} = \lambda^4, \qquad \frac{{}^\sigma g_3}{g_3} = \frac{1105 + 49\sqrt{65}}{1105 - 49\sqrt{65}} = \lambda^6.$$

Hence we obtain an isomorphism over K

$$\mu_\sigma/K : E_{65} \to {}^\sigma E_{65}, \quad (U, W) \mapsto (\lambda^2 U, \lambda^3 W).$$

Moreover, we have

$$N_{k/\mathbf{Q}}(\lambda) = -1 \quad \Rightarrow \quad {}^\sigma\mu \circ \mu = (-1)id.$$

□

To our best knowledge, the curve $C_{65}/\mathbf{Q}$ is the unique example for which there exists an isogeny over **Q** from Jac$C$ to A_f with $f \in S_2(\Gamma_0(N), \chi)$ an eigenform with nontrivial Neben type character χ.

6. Construction of $C(j)/\mathbf{Q}(j)$

Here we shall describe the construction of the family $\{C(j)\}$ given in (6). We start from the following well known

Lemma 8. *For a curve C/K of genus two to form a double cover of some elliptic curve, it is necessary and sufficient that C has non hyperelliptic involution.*

The curve C can be defined by an equation $Y^2 = F(X)$ with a separable polynomial $F(X) \in K[X]$ of degree 5 or 6. Then a non hyperelliptic involution ρ of C induces an involutive automorphism $\overline{\rho}$ of the rational function field $K(X)$. Since

$$\mathrm{Aut}_K(K(X)) = \mathrm{PGL}_2(K), \qquad \overline{\rho} = \begin{pmatrix} a & b \\ c & d \end{pmatrix} : X \mapsto \overline{\rho}(X) = \frac{aX+b}{cX+d},$$

we can write

$$\overline{\rho} : K(X) \longrightarrow K(X), \qquad \overline{\rho}(X) = \frac{aX+b}{cX-a}.$$

with $a, b, c \in K$, $m := a^2 + bc \neq 0$. It is easy to see that $\overline{\rho}$ can be extended to an automorphism of C of order 2 (resp. order 4) if and only if

$$(cX-a)^6 F\left(\frac{aX+b}{cX-a}\right) = \varepsilon m^3 F(X) \qquad (m = a^2 + bc). \tag{19}$$

with $\varepsilon = +1$ (resp. $\varepsilon = -1$). To obtain such $F(X)$ is nothing but to obtain the eigen vectors of $\Psi_6(\overline{\rho})$, where Ψ_6 is the symmetric tensor representation of degree 6. on the space of binary sextic forms.

Lemma 9. *The characteristic polynomial of* $\Psi_6(\overline{\rho})$ *is* $(X - m^3)^4(X + m^3)$. *Moreover, the polynomials* $\{F_j^{(+)}(X) \mid 1 \leq j \leq 4\}$ (*resp.* $\{F_j^{(-)}(X) \mid 1 \leq j \leq 3\}$) *tabulated below in Table* 6 (*resp. Table* 7), *form a basis of the eigenspace corresponding to the eigenvalue* m^3 (*resp.* $-m^3$).

Table 6

$F_1^{(+)}(X)$	$(b^2 + 2abX + 2a^2X^2 + bcX^2)\times(b^4 + 4ab^3X + 5a^2b^2X^2 - b^3cX^2$ $+2a^3bX^3 - 2ab^2cX^3 + a^4X^4 + a^2bcX^4 + b^2c^2X^4)$
$F_2^{(+)}(X)$	$-ab^5 - 5a^2b^4X + b^5cX - 10a^3b^3X^2 + 5ab^4cX^2 - 10a^4b^2X^3$ $+10a^2b^3cX^3 - 5a^5bX^4 + 10a^3b^2cX^4 + 8a^4bcX^5 + 3a^2b^2c^2X^5$ $+ b^3c^3X^5 + a^5cx^6$
$F_3^{(+)}(X)$	$a^2b^4 + 4a^3b^3X - 2ab^4cX + 6a^4b^2X^2 - 8a^2b^3cX^2 + b^4c^2X^2$ $+4a^5bX^3 - 12a^3b^2cX^3 + 4ab^3c^2X^3 + 2a^6X^4 - 5a^4bcX^4$ $+9a^2b^2c^2X^4 + b^3c^3X^4 - 2a^5cX^5 + 4a^3bc^2X^5 + a^4c^2X^6$
$F_4^{(+)}(X)$	$(-ab + 2bcX + acX^2)\times(a^2b^2 + 3a^3bX - ab^2cX$ $+3a^4X^2 - 2a^2bcX^2 + b^2c^2X^2 - 3a^3cX^3 + abc^2X^3 + a^2c^2X^4)$

Table 7

$F_1^{(-)}(X)$	$(-b-2aX+cX^2)\times(b^4+4ab^3X+7a^2b^2X^2+b^3cX^2$ $+6a^3bX^3+2ab^2cX^3+3a^4X^4+3a^2bcX^4+b^2c^2X^4)$
$F_2^{(-)}(X)$	$(-b-2aX+cX^2)\times(-ab^4)-3a^2b^3X+b^4cX-4a^3b^2X^2$ $+2ab^3cX^2-2a^4bX^3+3a^2b^2cX^3+b^3c^2X^3-a^5X^4)$
$F_3^{(-)}(X)$	$(-b-2aX+cX^2)\times(a^2b^3+2a^3b^2X-2ab^3cX$ $+2a^4bX^2-3a^2b^2cX^2+b^3c^2X^2-4a^3bcX^3-a^4cX^4)$

From the above result, we obtain a family of genus two curves with 7 free parameters a, b, c, p, q, r, s, which covers elliptic curves by degree two.

$$C:\ Y^2=pF_1^{(+)}(X)+qF_2^{(+)}(X)+rF_3^{(+)}(X)+sF_4^{(+)}(X). \tag{20}$$

Furthermore, we see from (19) that the non hyperelliptic involution ρ of C is given by

$$\iota:(X,Y)\mapsto\left(\frac{aX+b}{cX-a},\frac{m\sqrt{m}Y}{(cX-a)^3}\right),\qquad (m=a^2+bc).$$

It follows that the quotient curve $C/\langle\iota\rangle$ is of genus one, and has a model over the quadratic extension $\mathbf{Q}(a,\ b,\ c,\ p,\ q,\ r,\ s,\ \sqrt{m})$ of $\mathbf{Q}(a,\ b,\ c,\ p,\ q,\ r,\ s)$, which is defined by the following equation:

$$C/\langle\iota\rangle:\ W^2=(-2a+2\sqrt{m}+cU)\times G(U), \tag{21}$$

$$\begin{aligned}G(U)=\;&(2a^3b^3p-6ab^4cp+6a^2b^3cq-2b^4c^2q-2a^3b^2cr+6ab^3c^2r\\&-6a^2b^2c^2s+2b^3c^3s)+(6a^4b^2p-9a^2b^3cp-3b^4c^2p+13a^3b^2cq\\&+ab^3c^2q-4a^4bcr+9a^2b^2c^2r+b^3c^3r-9a^3bc^2s+3ab^2c^3s)U\\&+(6a^5bp+8a^4bcq+3a^2b^2c^2q+b^3c^3q-2a^5cr+4a^3bc^2r\\&-3a^4c^2s+3a^2bc^3s)U^2+(2a^6p+3a^4bcp+3a^2b^2c^2p\\&+b^3c^3p+a^5cq+a^4c^2r+a^3c^3s)U^3\end{aligned}$$

We note that the RHS of the above equation is linear in p,q,r,s. The j-invariant of $C/\langle\iota\rangle$ is written in the form

$$j(C/\langle\iota\rangle)=j_0(a,b,c,p,q,r,s)+\sqrt{m}j_1(a,b,c,p,q,r,s),$$

with $j_0,j_1\in\mathbf{Q}(a,b,c,p,q,r,s)$. Now we look for a specialization of our free parameters a, b, c, p, q, r, s which satisfies $j_1(a,b,c,p,q,r,s)=$

0. Although to solve this question in its full generality is difficult , it is rather easy to find a sufficient condition, since the number of the parameter is big enough.

Proposition 10. *The equation* $j_1(a,b,c,p,q,r,s) = 0$ *is satisfied by* $a = 0$, $r = -bp/c$, $s = 7bq/(9c)$. *Then* $F(X)$ *in* (20) *becomes*

$$F(X) = 9b^3p + 9b^2cqX - 9b^2cpX^2 + 14bc^2qX^3 - 9bc^2pX^4 + 9c^3qX^5 + 9c^3pX^6 \tag{22}$$

and the j*-invariant of* $C/\langle\iota\rangle$ *is given by*

$$j(C/\langle\iota\rangle) = \frac{1728(12bp^2 + cq^2)}{cq^2} \in \mathbf{Q}(b,c,p,q). \tag{23}$$

Making the transformation

$$b \mapsto \frac{b}{p^2}, \quad c \mapsto \frac{20736b}{(j-1728)q^2}, \quad X \mapsto \frac{qX}{144p}$$

we now obtain the family $C(j)$ and covering map of degree 2: $C(j) \to E(j)$.

References

[BCDT] C. Breuil, B. Conrad, F. Diamond and R. Taylor, *On the modularlity of elliptic curves over* **Q**, J. Amer. Math. Soc. **14** (2001), 843–939.

[BE] B. C. Berndt and R. J. Evans, *The Determination of Gauss Sums*, Bulletin Amer. Math. Soc. **5** (1981), 107–129.

[Co] S. Comalada, Elliptic curves with trivial conductor over quadratic fields, Pacific J. Math. **144** (1990), 233–258.

[Cr] J. E. Cremona, *Modular symbols for* $\Gamma_1(N)$ *and elliptic curves with everywhere good reduction*, Math. Proc. Camb. Phil. Soc. **111** (1992), 199–218.

[Fa] G. Faltings, *Endlichkeitssätze für abelsche Varietäten über Zahlkörpern*, Invent. Math. **73** (1983), 349–366; Erratum: **75** (1984), 381.

[Hs] Y. Hasegawa, **Q**-*curves over quadratic fields*, Manuscripta Math. **94** (1997), 347–364.

[HHM] Y. Hasegawa, K.Hashimoto and F.Momose, *Modularity conjecture for* **Q**-*curves and QM-curves*, International J. Math. **10** (1999), 1011–1036.

[Ha] K. Hashimoto, **Q**-*curves of degree* 5 *and jacobian surfaces of* GL_2-*type*, Manuscripta Math. **98** (1999), 165–182.

[H] H. Hasse, "Vorlesungen über Zahlentheorie, zweite Auf.," Springer-Verlag, Berlin, 1964.

[Ig] J. Igusa, *Arithmetic variety of moduli for gunus two*, Ann. of Math. **72** (1960), 612–649.

[Ka] T. Kagawa, *Elliptic curves with everywhere good reduction over quadratic fields*, dissertation, Waseda Univ., 1998.

[Ki] M. Kida, *Computing Elliptic curves having good reduction everywhere over quadratic fields*, preprint (1999).

[Pi] R. G. E. Pinch, *Elliptic curves with good reduction away from* 3, Math. Proc. Camb. Phil. Soc. **101** (1987), 451–459.

[Py] E. Pyle, *Abelian varieties over* $\mathbf{Q}$ *with large endomorphism algebras and their simple components over* $\bar{\mathbf{Q}}$, dissertation, Univ. of California at Berkeley, 1995.

[Ri1] K. Ribet, *Galois representations attached to eigenforms with Nebentypus*, in "Modular Functions of One Variable V," 17–52, Lecture Notes in Mathematics, 601. Springer-Verlag, Berlin/Heidelberg/New York, 1977.

[Ri2] K. Ribet, *Endomorphism algebras of abelian varieties attached to newforms of weight* 2, in "Seminar on Number Theory, Paris 1979–80," 263–276, Progress in Math., 12. Birkhäuser, Boston, Mass., 1981.

[Ri3] K. Ribet, *Abelian varieties over* $\mathbf{Q}$ *and modular forms*, in "1992 Proceedings of KAIST Mathematics Workshop," Korea Advanced Institute of Science and Technology, Taejon, 1992, 53–79.

[Ser] J.-P. Serre, "Topics in Galois Theory," Research Notes in Mathematics, 1. Jones and Bartlett Publ., 1992.

[Se] B. Setzer, *Elliptic curves with good reduction everywhere over quadratic fields and having rational* j*-invariant*, Illinois J. Math. **25** (1981), no 2, 233–245.

[Sh1] G. Shimura, "Introduction to the arithmetic theory of automorphic functions," Publications of the Mathematical Society of Japan, 11. Iwanami Shoten, Publishers, Tokyo, 1971.

[Si] J. H. Silverman, "The Arithmetic of Elliptic Curves," Graduate Texts in Mathematics, 106. Springer-Verlag, 1986.

[Shi] K. Shiota, *On explicit models of Shimura's elliptic curves*, J. Math. Soc. Japan **38** (1986), 649–659.

[Wi] A. Wiles, *Modular elliptic curves and Fermat's Last Theorem*, Ann. of Math. **141** (1995), 443–551.

Galois Theory and Modular Forms
K. Hashimoto, K. Miyake, and H. Nakamura, Editors pp 65–76
© 2003 Kluwer Academic Publishers

POINTS DEFINED OVER CYCLIC QUARTIC EXTENSIONS ON AN ELLIPTIC CURVE AND GENERALIZED KUMMER SURFACES

Masato KUWATA
Kanagawa Institute of Technology
1030 Shimo-Ogino
Atsugi-shi, Kanagawa,
243-0292 Japan
kuwata@gen.kanagawa-it.ac.jp

1. Introduction

Let E be an elliptic curve over a number field k. By the Mordell-Weil theorem the group $E(K)$ of K-rational points on E, where K/k is a finite extension of k, is a finitely generated abelian group. We fix E/k once and for all, and we study the behavior of the rank of the group $E(K)$ as K varies through a certain family. We are particularly interested in the family $\mathcal{F}_k(G)$ of all Galois extensions K/k whose Galois group $\mathrm{Gal}(K/k)$ is isomorphic to a prescribed finite group G. In this article we focus on the case $G = \mathbb{Z}/4\mathbb{Z}$.

One case that has been well studied is the case where $G = \mathbb{Z}/2\mathbb{Z}$. If an elliptic curve E/k is given by the Weierstrass equation $y^2 = x^3 + Ax + B$, and d is a nonzero element of k, the quadratic twist of E by d, denoted by E_d, is given by the equation $dy^2 = x^3 + Ax + B$. Since we have the relation $\operatorname{rank} E(k(\sqrt{d})) = \operatorname{rank} E(k) + \operatorname{rank} E_d(k)$, studying the behavior of the rank of $E(k(\sqrt{d}))$, as $k(\sqrt{d})$ varies through $\mathcal{F}_k(\mathbb{Z}/2\mathbb{Z})$, is equivalent to studying the family $\{E_d(k) \mid d \in k^\times/(k^\times)^2\}$. In this case it is very easy to find values of d such that $\operatorname{rank} E_d(k)$ is positive. Indeed, take any integer m in k, and write $m^3 + Am + B = dl^2$, then for almost all m, the point $(m, l) \in E_d(k)$ is of infinite order.

[received: July, 12; accepted in revised form: August, 15, 2002]

If k is the field of rational numbers $\mathbb{Q}$, our problem is conjecturally equivalent to the vanishing of the quadratic twists of the L-function of E (see [5] for more detail). Let $L(E,s) = \sum_{n=1}^{\infty} a_n n^{-s}$ be the L-function of $E/\mathbb{Q}$, and let $\chi : \mathrm{Gal}(\bar{\mathbb{Q}}/\mathbb{Q}) \to \mathbb{C}^\times$ be a Dirichlet character of order 2. The twist of $L(E,s)$ by χ is given by $L(E,s,\chi) = \sum_{n=1}^{\infty} \chi(n) a_n n^{-s}$. In this case we have a functional equation that relates $L(E, 2-s, \chi)$ and $L(E,s,\chi)$, and this functional equation forces $L(E,1,\chi)$ to vanish for at least one half of quadratic characters.

When χ runs through Dirichlet characters of order n, there is no a priori reason that the twist $L(E,s,\chi)$ vanishes for infinitely many χ's. However, numerical experiments seem to suggest that $L(E,s,\chi)$ vanishes quite often when the order of the character is small (see [2]).

If n is greater than 2, we do not dispose of an obvious method to find a point on E defined over some cyclic extension of degree n. It turns out that we have a higher dimensional analogue of E_d that is a variety over k whose k-rational point corresponds to a point on E defined over some cyclic extension K/k of degree n. When $G = \mathbb{Z}/3\mathbb{Z}$, $\mathbb{Z}/4\mathbb{Z}$ or $\mathbb{Z}/6\mathbb{Z}$, this variety is a surface which belongs to the class of surfaces called *generalized Kummer surfaces*. Earlier, we obtained some results for the case $G = \mathbb{Z}/3\mathbb{Z}$ by studying this surface (see [4]). When $G = \mathbb{Z}/4\mathbb{Z}$, it turns out that we have a stronger result:

Theorem 1.1. *Let E be an elliptic curve over a number field k. Then there exist infinitely many cyclic quartic extensions K/k such that* $\operatorname{rank} E(K)$ *is strictly greater than* $\operatorname{rank} E(K_2)$*, where K_2 is the unique intermediate quadratic extension of k in K.*

Corollary 1.2. *Let $F_{k,G} = \bigcup_{K \in \mathcal{F}_k(G)} K$ be the compositum of all the Galois extensions of k whose Galois group is isomorphic to G. Then the quotient group $E(F_{k,\mathbb{Z}/4\mathbb{Z}})/E(F_{k,\mathbb{Z}/2\mathbb{Z}})$ is not finitely generated.*

To prove our theorem we find a k-rational curve contained in the generalized Kummer surface in question, and write down its equation explicitly.

The author would like to thank Professor Hershy Kisilevsky and Dr. Jack Fearnley for useful conversations. He would also thank Professors Katsuya Miyake and Ki-ichiro Hashimoto for giving me an opportunity to present an earlier version of this paper at the conference "Galois theory and Modular forms" at the Tokyo Metropolitan University.

2. A generalized Kummer surface

Let E be an elliptic curve over k. Consider the automorphism ρ of $E \times E$ given by

$$\begin{array}{rccc} \rho : & E \times E & \longrightarrow & E \times E \\ & (P, Q) & \longmapsto & (Q, -P). \end{array}$$

It is easy to see that ρ is of order 4, and ρ^2 is the multiplication-by-(-1) map. Let X be the quotient of $E \times E$ by the group $\langle \rho \rangle$ generated by ρ, and let $\widetilde{X}$ be its minimal non-singular model obtained by blowing-ups. $\widetilde{X}$ is a generalized Kummer surface (cf. Katsura [3], Bertin [1]). In particular $\widetilde{X}$ is a $K3$ surface.

The set of fixed points of ρ consists of the four points of the form (T, T), where T is a 2-torsion point. The set of fixed points of ρ^2 consists of the sixteen points of the form (T, S), where T and S are 2-torsion points. Thus ρ acts freely away from these sixteen points. We denote by X° the open set of X obtained by removing the image of these sixteen points in X.

Lemma 2.1. *If the equivalence classes $[P, Q]$ is in the set of k-rational points $X^\circ(k)$, then (P, Q) satisfies one of the following:*

(1) *P and Q are defined over k.*

(2) *P and Q are both defined over some quadratic extension $k(\sqrt{d})/k$. If $\tau \in \mathrm{Gal}(k(\sqrt{d})/k)$ is the generator, then we have $\tau(P) = -P$ and $\tau(Q) = -Q$.*

(3) *P and Q are both defined over some cyclic quartic extension K/k. If we choose a suitable generator $\sigma \in \mathrm{Gal}(K/k)$, then we have $\sigma(P) = Q$ and $\sigma(Q) = -P$.*

Proof. Let P and Q be points on $E(\bar{k})$, and suppose that $[P, Q] \in X^\circ(k)$. If σ is an element of $\mathrm{Gal}(\bar{k}/k)$, then $(\sigma(P), \sigma(Q))$ is one of the following four pairs: (P, Q), $\rho(P, Q) = (Q, -P)$, $\rho^2(P, Q) = (-P, -Q)$ or $\rho^3(P, Q) = (-Q, P)$. Note that these four pairs are distinct since $[P, Q]$ is in X°. We can therefore define a map $\psi : \mathrm{Gal}(\bar{k}/k) \to \langle \rho \rangle$. Since the automorphism ρ is defined over k, it commutes with any element of $\mathrm{Gal}(\bar{k}/k)$. Thus, if $\sigma_1(P, Q) = \rho^i(P, Q)$ and $\sigma_2(P, Q) = \rho^j(P, Q)$, then $\sigma_1(\sigma_2(P, Q)) = \sigma_1(\rho^j(P, Q)) = \rho^i(\rho^j(P, Q)) = \rho^{i+j}(P, Q)$. This shows that the map ψ is a homomorphism.

Let K be the Galois extension of k corresponding to $\ker \psi$ via Galois theory. Then $\mathrm{Gal}(K/k)$ is isomorphism to a subgroup of $\langle \rho \rangle$. If $\mathrm{Gal}(K/k) = \{\mathrm{id}\}$, then $K = k$, and both P and Q are defined over k. This is the case (1).

If $\mathrm{Gal}(K/k) \simeq \langle \rho^2 \rangle$, then K is a quadratic extension of k. Let $\tau \in \mathrm{Gal}(\bar{k}/k)$ be an element whose image in $\mathrm{Gal}(K/k)$ generates $\mathrm{Gal}(K/k)$. Then $\psi(\tau) = \langle \rho^2 \rangle$. This shows that $\tau(P,Q) = \rho^2(P,Q) = (-P,-Q)$. This is the case (2).

If $\mathrm{Gal}(K/k) \simeq \langle \rho \rangle$, then K is a cyclic quartic extension. Let $\sigma \in \mathrm{Gal}(\bar{k}/k)$ be an element whose image in $\mathrm{Gal}(K/k)$ maps to ρ by ψ. Then we have $\sigma(P,Q) = (Q,-P)$. This implies that $Q = \sigma(P)$ and $\sigma^2(P) = \sigma(Q) = -P$. This is the case (3). □

Conversely, if we have a point P defined over some cyclic quartic extension of k, we have the following.

Lemma 2.2. *Let P be a point whose field of definition is a cyclic quartic extension K/k. Let σ be a generator of $\mathrm{Gal}(K/k)$. Define $P' = P - \sigma^2(P)$. Then $[P', \sigma(P')]$ is in $X(k)$. Furthermore, if E does not have a k-rational 2-torsion point, then $[P', \sigma(P')]$ is in $X^\circ(k)$.*

Proof. First, note that P' is not O. Otherwise, we have $P = \sigma^2(P)$, which is a contradiction to the fact that the field of definition of P' is K. We see easily that $\sigma^2(P') = -P'$, which implies that $\sigma(P', \sigma(P')) = (\sigma(P'), \sigma^2(P')) = (\sigma(P'), -P') = \rho(P', \sigma(P'))$. This shows that $[P', \sigma(P')]$ is in $X(k)$.

Suppose that $[P', \sigma(P')]$ is not in $X^\circ(k)$. Then P' is a 2-torsion point of E. Moreover, P' is defined over the intermediate quadratic extension K_2 between K and k, since we have $\sigma^2(P') = -P' = P'$. Then we see that either P' itself or $P' + \sigma(P')$ is a k-rational 2-torsion point of E. □

3. Equation of the surface X

In this section we will write down an equation of X in order to study X in detail. We first fix an equation of E:

$$E : y^2 = x^3 + Ax + B. \tag{1}$$

For simplicity of exposition, we assume $AB \neq 0$ throughout. Let $k(E \times E)$ be the function field of $E \times E$. Then $k(E \times E)$ may be written as $k(x_1, x_2, y_1, y_2)$, where (x_1, y_1) and (x_2, y_2) both satisfy the equation (1). The automorphism induced by ρ on $k(E \times E)$, also denoted by ρ, satisfies

$$\rho(x_1) = x_2, \quad \rho(x_2) = x_1, \quad \rho(y_1) = y_2, \quad \rho(y_2) = -y_1.$$

Let Y be the quotient surface $E \times E/\langle \rho^2 \rangle = E \times E/\{\pm 1\}$, which is a singular model of the Kummer surface associated with $E \times E$.

Lemma 3.1. *The function field $k(Y)$ of the quotient surface Y is the subfield of $k(E \times E)$ generated by x_1, x_2 and $y = y_1/y_2$, which satisfy the equation*

$$(x_2^3 + Ax_2 + B)y^2 = x_1^3 + Ax_1 + B. \tag{2}$$

Furthermore, the automorphism ρ acts on $k(Y)$ by

$$\rho(x_1) = x_2, \quad \rho(x_2) = x_1, \quad \rho(y) = -\frac{1}{y}.$$

Proof. It is easy to see that the elements x_1, x_2 and $y = y_1/y_2$ are fixed by the automorphism ρ^2 and thus belong to $k(Y)$. Also, it is easy to see that x_1, x_2 and y satisfy the equation (2). We thus need to show that these three elements generate $k(Y)$. To see this it suffices to show that the degree of extension $[k(E \times E) : k(Y)]$ is 2. The element y_1 is a root of the quadratic equation in T with coefficients in $k(Y)$:

$$T^2 - x_1^3 - Ax_1 - B = 0.$$

Thus, we have $[k(E \times E) : k(Y)] \leq 2$. The fact that y_1 and y_2 are independent in $k(E \times E)$ implies that $k(E \times E) \neq k(Y)$. Thus, we conclude $[k(E \times E) : k(Y)] = 2$. The action of ρ on y is given by

$$\rho(y) = \frac{\rho(y_1)}{\rho(y_2)} = \frac{y_2}{-y_1} = -\frac{1}{y}.$$

□

Proposition 3.2. *The function field $k(X)$ of the quotient surface $X = E \times E/\langle\rho\rangle$ is the subfield of $k(E \times E)$ generated by*

$$\xi = x_1 + x_2, \quad \eta = x_2 y - \frac{x_1}{y}, \quad t = y - \frac{1}{y},$$

which satisfy the equation

$$-t\xi^3 + 3\xi^2\eta + \eta^3 + A\left(t^2 + 4\right)\eta + Bt\left(t^2 + 4\right) = 0. \tag{3}$$

Proof. First, we can easily verify that the elements ξ, η and t are fixed by the automorphism ρ, and thus belong to $k(X)$. It is easy to see that $k(\xi, \eta, y) = k(x_1, x_2, y) = k(Y)$. This shows that $k(Y)$ is a quadratic extension of $k(\xi, \eta, t)$ obtained by adding y satisfying $y^2 - ty - 1 = 0$.

To obtain a relation among ξ, η and t, we first express the equation (2) in terms of ξ, η and y, and then we eliminate y using the relation $y^2 - ty - 1 = 0$. □

4. Rational curves on the surface X

In this section we show that there exist infinitely many parametrized curves defined over k on the surface X. This allows us to show that there are infinitely many different cyclic quartic extensions such that the Mordell-Weil group over it increases from that over k.

Looking at the equation (2), we notice that the surface $Y = E \times E/\{\pm 1\}$ can be regarded as a family of cubic curves in the x_1x_2-plane parametrized by y. The projective model of this plane curve defined over $k(y)$ intersects with the line at infinity at three points. These are not defined over $k(y)$, but one of them is defined over the extension $k(\sqrt[3]{y})$. Writing $u = \sqrt[3]{y}$, we obtain a plane cubic curve defined over $k(u)$ given by

$$(x_2^3 + Ax_2 + B)u^6 = x_1^3 + Ax_1 + B \tag{4}$$

with a $k(u)$-rational point at infinity. The tangent line at this point intersects with the third point P (see Figure 1), which is again $k(u)$-rational.

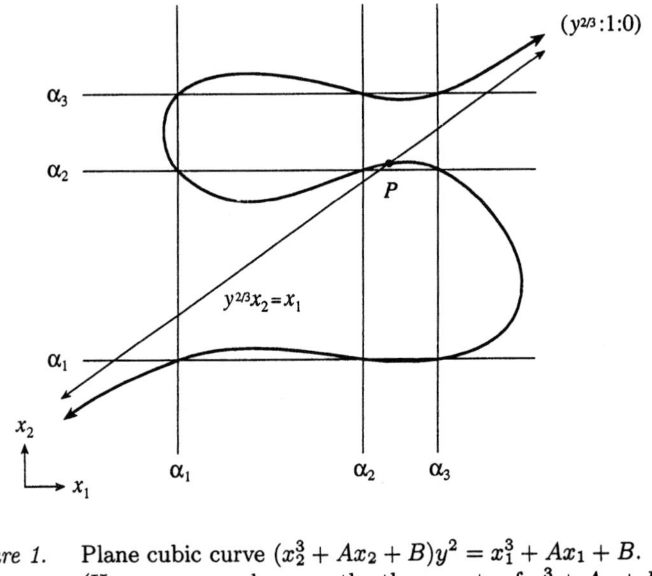

Figure 1. Plane cubic curve $(x_2^3 + Ax_2 + B)y^2 = x_1^3 + Ax_1 + B$. (Here, α_1,α_2 and α_3 are the three roots of $x^3 + Ax + B = 0$.)

A simple calculation shows that the coordinates of P are given by

$$(x_1, x_2) = \left(\frac{-B(u^4 + u^2 + 1)}{A(u^2 + 1)}, \frac{-B(u^4 + u^2 + 1)}{Au^2(u^2 + 1)} \right). \tag{5}$$

In other words the triple cover Y' of Y defined by the equation (4) has a k-rational parametrized curve. The automorphism of ρ on Y can be lifted to the automorphism of Y' by defining

$$\rho' : (x_1, x_2, u) \mapsto (x_2, x_1, -1/u).$$

It is easy to see that the parametrized curve obtained above is stable under the action of ρ. We have a commutative diagram:

$$\begin{array}{ccc} Y' & \longrightarrow & Y \\ {\scriptstyle \rho'}\downarrow & & \downarrow{\scriptstyle \rho} \\ Y' & \longrightarrow & Y \end{array}$$

Let X' be the quotient $Y'/\langle\rho\rangle$. The diagram above induce the diagram:

$$\begin{array}{ccc} Y' & \longrightarrow & Y \\ \downarrow & & \downarrow \\ X' & \longrightarrow & X \end{array}$$

Proposition 4.1. *The function field of the quotient* $X' = Y'/\langle\rho\rangle$ *is the subfield of* $k(E \times E)$ *generated by*

$$\xi = x_1 + x_2, \quad \eta = x_2 u^3 - \frac{x_1}{u^3}, \quad s = u - \frac{1}{u},$$

which satisfy the equation

$$\begin{aligned} &- s(s^2+3)\xi^3 + 3\xi^2\eta + \eta^3 \\ &\qquad + A(s^2+1)^2(s^2+4)\eta + Bs(s^2+1)^2(s^2+3)(s^2+4) = 0. \quad (6) \end{aligned}$$

The covering map $X' \to X$ *is given by* $(\xi, \eta, s) \mapsto (\xi, \eta, s^3 + 3s)$.

Proof. The proof goes similarly to that of Proposition 3.2. The last part is a consequence of the simple calculation

$$s^3 = \left(u - \frac{1}{u}\right)^3 = u^3 - \frac{1}{u^3} - 3\left(u - \frac{1}{u}\right) = y - \frac{1}{y} - 3s = t - 3s.$$

□

We think of the equation (6) as that of a cubic curve in $\xi\eta$-plane over the function field $k(s)$. This curve has two rational points, one at infinity and the other coming from P on Y. Calculations show that the latter has coordinates

$$(\xi, \eta) = \left(\frac{-B(s^2+3)}{A}, \frac{-Bs(s^2+3)}{A}\right).$$

Choosing the point at infinity $(1 : s : 0)$ as the origin, we regard X' as the elliptic curve over $k(s)$.

Proposition 4.2. *The equation* (6) *is transformed to the Weierstrass form*

$$\begin{aligned} Y^2 = X^3 &- 48A^2(s^2+4)^2X \\ &- 16(s^2+4)^3\big((4A^3+27B^2)s^2(s^2+3)^2+8A^3\big) \end{aligned} \tag{7}$$

by the transformation

$$X = \frac{4(s^2+4)}{(s\xi-\eta)}\Big(As(s^2+2)\xi + A(2s^2+1)\eta + 3Bs(s^2+1)(s^2+3)\Big),$$

$$\begin{aligned} Y = \frac{12s(s^2+4)}{(s\xi-\eta)^2}\Big(&3B(s^2+3)\big(s(s^2+2)\xi^2 - 2\xi\eta - s\eta^2\big) \\ &+ 2A(s^2+1)^2(s^2+4)\big(A\eta + Bs(s^2+3)\big)\Big). \end{aligned}$$

The point $(\xi,\eta) = \big(-B(s^2+3)/A, -Bs(s^2+3)/A\big)$ *is transformed to*

$$\begin{aligned} (X,Y) = \bigg(&\frac{4}{A^2}\left((A^3+9B^2)(s^2+3)^2 - A^3\right), \\ &\frac{36B(s^2+3)}{A^3}\left((A^3+6B^2)(s^2+3)^2 - A^3\right),\bigg) \end{aligned} \tag{8}$$

which is of infinite order.

Proof. Converting to the Weierstrass form is standard and straight forward. The elliptic surface given by (7) has the following types of singular fibers under the condition $AB \neq 0$.

- I_0^* at $s^2+4=0$,
- I_2 at $s(s^2+3)=0$,
- I_1 at $(4A^3+27B^2)s^2(s^2+3)^2+16A^3=0$.

If the point given by (8) is a torsion point, its specialization at each fiber is a torsion point of the same order. Since a fiber of type I_0^* is isomorphic to $\mathbb{G}_a \times (\mathbb{Z}/2\mathbb{Z})^2$ as an algebraic group, its torsion points are of order 2. Since the Y-coordinate of the point (8) is not 0, it is not a point of order 2, and thus it is of infinite order. □

A $k(s)$-rational point on the elliptic curve (7) is one-to-one correspondence with a $k(s)$-rational point $(\xi(s),\eta(s))$ on the plane curve (6), which in turn gives a parametrized curve $(\xi(s),\eta(s),s^3+3s)$ on the surface X. Since Proposition 4.2 gives us infinitely many $k(s)$-rational points on (7), we have infinitely many k-rational parametrized curve on X.

5. Points on E defined over cyclic quartic extensions

We have seen that the surface X has a k-rationally parametrized curve

$$(\xi, \eta, t) = \left(\frac{-B(s^2+3)}{A}, \frac{-Bs(s^2+3)}{A}, s^3+3s \right) \tag{9}$$

with parameter s. This gives a point on E defined over a quartic extension of $k(s)$.

Lemma 5.1. *Let v be a root of the quadratic equation in T with $k(u)$-coefficients:*

$$A(u^2+1)T^2 + A^3Bu^4(u^2+1)^2 + B^3(u^4+u^2+1)^3 = 0.$$

Then $k(v)$ is a cyclic quartic extension of $k(s)$, and the elliptic curve E has four $k(v)$-rational points

$$(x, y) = \left(\frac{-B(u^4+u^2+1)}{A(u^2+1)}, \pm\frac{v}{A(u^2+1)} \right),$$
$$\left(\frac{-B(u^4+u^2+1)}{Au^2(u^2+1)}, \pm\frac{v}{Au^3(u^2+1)} \right).$$

Proof. We have seen that the point on X given by (9) comes from the point P on the Kummer surface Y. We obtain four points in the lemma by taking the preimage of P in $E \times E$. The arguments in §2 show that the field of definition of these points, $k(v)$, is a cyclic quartic extension of $k(s)$. This may be seen directly as v satisfies the following quartic equation in T with $k[s]$-coefficients

$$\begin{aligned} &A^2(s^2+4)T^4 \\ &\quad + AB(s^2+4)(s^4+3s^2+1)\big(A^3(s^2+4) + B^2(s^2+3)^3\big)T^2 \\ &\qquad + B^2\big(A^3(s^2+4) + B^2(s^2+3)^3\big)^2 = 0. \end{aligned} \tag{10}$$

□

Proof of Theorem 1.1. First we show that the points obtained in Lemma 5.1 are of infinite order. If we let

$$d(u) = -\frac{A^3Bu^4(u^2+1)^2 + B^3(u^4+u^2+1)^3}{A(u^2+1)},$$

then the points in Lemma 5.1 may be considered as $k(u)$-rational point on the twisted curve

$$E_{d(u)} : d(u)y^2 = x^3 + Ax + B.$$

If we regard this elliptic curve over $k(u)$ as an elliptic surface over k, then the only singular fibers it has are of Kodaira type I_0^*. This implies that the only possible torsion $k(u)$-rational points are points of order 2. However, our points are clearly not of order 2 since the y-coordinates are not 0.

Since $k(v)/k(u)$-trace of the $k(v)$-points in Lemma 5.1 are O, we see that the intersection between the subgroup generated by these $k(v)$-points and the group $E(k(u))$ is $\{O\}$. Thus we have $\operatorname{rank} E(k(v)) > \operatorname{rank} E(k(u))$.

By specializing s to different values of k, we obtain infinitely many points on E defined over the splitting field of (10). By Hilbert's irreducibility theorem these are cyclic quartic extensions for infinitely many values of s. Also, by a theorem of Silverman [6] on the specialization of a family of elliptic curves, points on an elliptic curve obtained by specializing a point of infinite order is once again of infinite order except for a finite number of values. This completes the proof. □

Proof of Corollary 1.2. Since points on E whose field of definition are different quartic extensions are clearly independent, the assertion follows immediate from the theorem. □

Remark 5.2. The discriminant of the equation (10) is given by

$$16A^6B^6s^4(s^2+4)^3(s^4+5s^2+5)^4\left(A^3(s^2+4)+B^2(s^2+3)^3\right)^6.$$

6. Elliptic curve with 2-torsion

If the elliptic curve E in question has a 2-torsion point, then the surface X has more k-rational curves on it. In this section we show explicit results for the case where E has three k-rational 2-torsion points. Other cases can be worked out similarly.

Let us fix the equation of E as

$$E : y^2 = (x-c)(x-d)(x+c+d).$$

Then the equation (3) of X becomes

$$-t\xi^3 + 3\xi^2\eta + \eta^3 - (c^2+cd+d^2)(t^2+4)\eta + cd(c+d)t(t^2+4) = 0.$$

We regard this as a cubic curve in the $\xi\eta$-plane over $k(t)$. We have several obvious $k(t)$-rational points, for example,

$$P_1 = (2c, ct), \quad P_2 = (2d, dt), \quad P_3 = (-2c-2d, -ct-dt).$$

These correspond to 2-torsion points on $E \times E$. The tangent line at P_1 intersects with the curve at $Q = (-c, ct)$. This gives four points on E given by

$$\left(-\frac{c}{2} + \frac{3ct}{2\sqrt{t^2+4}}, \sqrt{\frac{3c}{8}\left((c+2d)^2 - \frac{9c^2t^2}{t^2+4}\right)\left(1 - \frac{t}{\sqrt{t^2+4}}\right)}\right),$$
$$\left(-\frac{c}{2} - \frac{3ct}{2\sqrt{t^2+4}}, \sqrt{\frac{3c}{8}\left((c+2d)^2 - \frac{9c^2t^2}{t^2+4}\right)\left(1 + \frac{t}{\sqrt{t^2+4}}\right)}\right).$$

These are defined over the cyclic quartic extension that is the splitting field of the equation

$$T^4 + 3c(t^2+4)\big((2c+d)(c-d)t^2 - (c+2d)^2\big)T^2 + 9c^2(t^2+4)\big((2c+d)(c-d)t^2 - (c+2d)^2\big)^2 = 0.$$

By considering tangent lines at other points, we obtain more points defined over similar cyclic quartic extensions.

Remark 6.1. Choosing $P_1 = (2c, ct)$ as the origin, the equation of X may be converted to the Weierstrass form.

$$Y^2 = X^3 - 48(c^2+cd+d^2)(t^2+4)X + 16(t^2+4)^3\big((c-d)^2(c+2d)^2(2c+d)^2t^2 + 8(c^2+cd+d^2)\big)^3.$$

The rank of the Mordell-Weil group of this elliptic curve over $k(t)$ is 2 or 3 depending on whether or not E has complex multiplication. If E does not have complex multiplication, then the Mordell-Weil group is generated by the images of P_2 and P_3.

References

[1] J. Bertin, *Réseaux de Kummer et surfaces de K3*, Invent. Math. **93** (1988), 267–284.

[2] J. Fearnley, *Vanishing and non-vanishing of L-series of elliptic curves twisted by Dirichlet characters*, Ph.D. thesis, Concordia University, 2001.

[3] T. Katsura, *Generalized Kummer surfaces and their unirationality in characteristic p*, J. Fac. Sci., Univ. Tokyo, Sect. IA **34** (1987), 1–41.

[4] M. Kuwata, *Points defined over cyclic cubic extensions on an elliptic curve and generalized Kummer surfaces*, preprint.

[5] D. E. Rohrlich, *The vanishing of certain Rankin-Selberg convolutions*, in "Automorphic forms and analytic number theory (Montreal, PQ, 1989)," Univ. Montréal, Montreal, QC, 123–133, 1990.

[6] J. H. Silverman, *Heights and the specialization map for families of abelian varieties*, J. Reine Angew. Math. **342** (1983), 197–251.

Galois Theory and Modular Forms pp 77–122
K. Hashimoto, K. Miyake, and H. Nakamura, Editors
© 2003 Kluwer Academic Publishers

THE ABSOLUTE ANABELIAN GEOMETRY OF HYPERBOLIC CURVES

Shinichi MOCHIZUKI
Research Institute for Mathematical Sciences,
Kyoto University,
Kyoto 606-8502 Japan
motizuki@kurims.kyoto-u.ac.jp

Contents:

Introduction

Let X_K be a *hyperbolic curve* (cf. §0 below) over a *field K of characteristic* 0. Denote its *algebraic fundamental group* by Π_{X_K}. Thus, we have a *natural surjection*

$$\Pi_{X_K} \twoheadrightarrow G_K$$

of Π_{X_K} onto the *absolute Galois group* G_K of K.

When K is a *finite extension of* $\mathbb{Q}$ *or* $\mathbb{Q}_p$, and one *holds G_K fixed*, then it is known (cf. [Tama1], [Mzk6]; Theorem 1.3.4 of the present manuscript) that one may *recover the curve X_K in a functorial fashion* from Π_{X_K}. This sort of result may be thought of as a *"relative result"* (i.e., over G_K). Then the question naturally arises:

[received: May 26, 2002; accepted in revised form: February 20, 2003]

> *To what extent are the* **"absolute analogues"** *of this result valid — i.e., what if one does not hold* G_K *fixed?*

If K is a *number field*, then it is *still possible to recover* X_K from Π_{X_K} (cf. Corollary 1.3.5), by applying the *theorem of Neukirch-Uchida* (cf. Theorem 1.1.3). On the other hand, when K is a *p-adic local field* (i.e., a finite extension of $\mathbb{Q}_p$), the analogue of the theorem of Neukirch-Uchida *fails to hold*, and indeed, it is the opinion of the author at the time of writing that it is *unlikely* (in the p-adic local case) that one can recover X_K in general (i.e., in the fashion of Corollary 1.3.5) from Π_{X_K}.

In the present manuscript, we begin by *reviewing/surveying* in §1 the anabelian geometry of number fields, p-adic local fields, and hyperbolic curves from the point of view of the *goal of understanding to what extent the anabelian geometry of hyperbolic curves over p-adic local fields can be made* **"absolute"**. Our main result (Theorem 2.7), given in §2, states that when K is a *p-adic local field*, (although we may be unable to recover X_K itself) *one may recover (in a functorial fashion) the* **special fiber** *of* X_K, *together with its natural* **log structure**, *in an* **absolute fashion**, i.e., solely from the isomorphism class of the profinite group Π_{X_K}.

0. Notations and Conventions

Numbers:

We will denote by $\mathbb{N}$ the set of *natural numbers*, by which we mean the set of integers $n \geq 0$. A *number field* is defined to be a finite extension of the field of rational numbers $\mathbb{Q}$.

Topological Groups:

Let G be a *Hausdorff topological group*, and $H \subseteq G$ a *closed subgroup*. Let us write

$$Z_G(H) \stackrel{\text{def}}{=} \{g \in G \mid g \cdot h = h \cdot g,\ \forall\ h \in H\}$$

for the *centralizer* of H in G;

$$N_G(H) \stackrel{\text{def}}{=} \{g \in G \mid g \cdot H \cdot g^{-1} = H\}$$

for the *normalizer* of H in G; and

$$C_G(H) \stackrel{\text{def}}{=} \{g \in G \mid (g \cdot H \cdot g^{-1}) \bigcap H \text{ has finite index in } H,\ g \cdot H \cdot g^{-1}\}$$

for the *commensurator* of H in G. Note that: (i) $Z_G(H)$, $N_G(H)$ and $C_G(H)$ are *subgroups of* G; (ii) we have *inclusions*

$$H,\ Z_G(H) \subseteq N_G(H) \subseteq C_G(H)$$

and (iii) H is *normal* in $N_G(H)$.

Note that $Z_G(H)$, $N_G(H)$ are *always closed in* G, while $C_G(H)$ is *not necessarily closed in* G. Indeed, one may construct such an example as follows: Let

$$M \stackrel{\text{def}}{=} \prod_{\mathbb{N}} \mathbb{Z}_p$$

endowed with the *product topology* (of the various copies of $\mathbb{Z}_p$ equipped with their usual topology). Thus, M is a *Hausdorff topological group.* For $n \in \mathbb{N}$, write $F^n(M) \subseteq M$ for the sub-topological group given by the product of the copies of $\mathbb{Z}_p$ indexed by $m \geq n$. Write $\text{Aut}_F(M)$ for the *set of automorphisms of the topological group M that preserve the filtration $F^*(M)$ on M.* If $\alpha \in \text{Aut}_F(M)$, then for every $n \in \mathbb{N}$, α induces a continuous homomorphism $\alpha_n : M/F^n(M) \to M/F^n(M)$ which is clearly *surjective*, hence an *isomorphism* (since $M/F^n(M)$ is profinite and topologically finitely generated — cf. [FJ], Proposition 15.3). It thus follows that α induces an isomorphism $F^n(M) \xrightarrow{\sim} F^n(M)$, hence that the inverse of α also lies in $\text{Aut}_F(M)$. In particular, we conclude that $\text{Aut}_F(M)$ is a *group.* Equip $\text{Aut}_F(M)$ with the *coarsest topology* for which all of the homomorphisms $\text{Aut}_F(M) \to \text{Aut}(M/F^n(M))$ (where $\text{Aut}(M/F^n(M)) \cong GL_n(\mathbb{Z}_p)$ is equipped with its usual topology) are continuous. Note that relative to this topology, $\text{Aut}_F(M)$ forms a *Hausdorff topological group.* Now define G to be the *semi-direct product* of M with $\text{Aut}_F(M)$ (so G is a Hausdorff topological group), and H to be

$$\prod_{n \in \mathbb{N}} p^n \cdot \mathbb{Z}_p \subseteq \prod_{\mathbb{N}} \mathbb{Z}_p = M$$

(so $H \subseteq G$ is a closed subgroup). Then $C_G(H)$ is *not closed in* G. For instance, if one denotes by $e_n \in \prod_{\mathbb{N}} \mathbb{Z}_p$ the vector with a 1 in the n-th place and zeroes elsewhere, then the *limit* A_∞ (where

$$A_\infty(e_n) \stackrel{\text{def}}{=} e_n + e_{n+1}$$

for all $n \in \mathbb{N}$) *of the automorphisms* $A_m \in C_G(H)$ (where $A_m(e_n) \stackrel{\text{def}}{=} e_n + e_{n+1}$ if $n \leq m$, $A_m(e_n) \stackrel{\text{def}}{=} e_n$ if $n > m$) *is not contained in* $C_G(H)$.

Definition 0.1. (i) Let G be a profinite group. Then we shall say that G is *slim* if the centralizer $Z_G(H)$ of any open subgroup $H \subseteq G$ in G is trivial.
(ii) We shall say that a continuous homomorphism of profinite groups $G \to H$ is *relatively slim* if the centralizer in H of the image of every open subgroup of G is trivial.

(iii) We shall say that a closed subgroup $H \subseteq G$ of a profinite group G is *commensurably* (respectively, *normally*) *terminal* if the commensurator $C_G(H)$ (respectively, normalizer $N_G(H)$) is equal to H.

Remark 0.1.1. Thus, a profinite group G is *slim* if and only if the *identity morphism* $G \rightarrow G$ is *relatively slim*. Moreover, if $H \subseteq G$ is a closed subgroup such that the inclusion $H \hookrightarrow G$ is *relatively slim*, then both H and G are *slim*.

Remark 0.1.2. It is a formal consequence of the definitions that:

$$\text{commensurably terminal} \implies \text{normally terminal}$$

and that (if $H \subseteq G$ is a closed subgroup of a profinite group G, then):

$$H \subseteq G \text{ commensurably terminal, } H \text{ slim} \implies \text{the inclusion } H \hookrightarrow G \text{ is relatively slim.}$$

Remark 0.1.3. It was pointed out to the author by F. Oort that a profinite group is *slim* if and only if every open subgroup of G has trivial center. (Indeed, the necessity of this condition is clear. Its sufficiency may be shown as follows: If $H \subseteq G$ is an open subgroup, then for any $h \in Z_G(H)$, let $H' \subseteq G$ by the (necessarily open) subgroup generated by H and h. Thus, h lies in the center of H', which is trivial by assumption.) This property of a profinite group — i.e., that every open subgroup be center-free — was investigated in detail in [Naka1] (cf., e.g., Corollaries 1.3.3, 1.3.4).

Curves:

Suppose that $g \geq 0$ is an *integer*. Then a *family of curves of genus* g

$$X \rightarrow S$$

is defined to be a smooth, proper, geometrically connected morphism $X \rightarrow S$ whose geometric fibers are curves of genus g.

Suppose that $g, r \geq 0$ are *integers* such that $2g - 2 + r > 0$. We shall denote the *moduli stack of* r*-pointed stable curves of genus* g (where we assume the points to be *unordered*) by $\overline{\mathcal{M}}_{g,r}$ (cf. [DM], [Knud] for an exposition of the theory of such curves; strictly speaking, [Knud] treats the finite étale covering of $\overline{\mathcal{M}}_{g,r}$ determined by *ordering* the marked points). The open substack $\mathcal{M}_{g,r} \subseteq \overline{\mathcal{M}}_{g,r}$ of smooth curves will be referred to as the *moduli stack of smooth* r*-pointed stable curves of genus* g or, alternatively, as the *moduli stack of hyperbolic curves of type* (g, r).

A *family of hyperbolic curves of type* (g, r)

$$X \rightarrow S$$

is defined to be a morphism which factors $X \hookrightarrow Y \to S$ as the composite of an open immersion $X \hookrightarrow Y$ onto the complement $Y \backslash D$ of a relative divisor $D \subseteq Y$ which is finite étale over S of relative degree r, and a family $Y \to S$ of curves of genus g. One checks easily that, if S is *normal*, then the pair (Y, D) is *unique up to canonical isomorphism.* (Indeed, when S is the spectrum of a field, this fact is well-known from the elementary theory of algebraic curves. Next, we consider an arbitrary *connected normal* S on which a prime l is *invertible* (which, by Zariski localization, we may assume without loss of generality). Denote by $S' \to S$ the finite étale covering parametrizing *orderings of the marked points* and *trivializations of the l-torsion points of the Jacobian of* Y. Note that $S' \to S$ is *independent* of the choice of (Y, D), since (by the normality of S), S' may be constructed as the *normalization* of S in the function field of S' (which is independent of the choice of (Y, D) since the restriction of (Y, D) to the generic point of S has already been shown to be unique). Thus, the uniqueness of (Y, D) follows by considering the classifying morphism (associated to (Y, D)) from S' to the finite étale covering of $(\mathcal{M}_{g,r})_{\mathbb{Z}[\frac{1}{l}]}$ parametrizing orderings of the marked points and trivializations of the l-torsion points of the Jacobian [since this covering is well-known to be a scheme, for l sufficiently large].) We shall refer to Y (respectively, D; D; D) as the *compactification* (respectively, *divisor at infinity*; *divisor of cusps*; *divisor of marked points*) of X. A *family of hyperbolic curves* $X \to S$ is defined to be a morphism $X \to S$ such that the restriction of this morphism to each connected component of S is a *family of hyperbolic curves of type* (g, r) for some integers (g, r) as above.

1. Review of Anabelian Geometry

1.1 The Anabelian Geometry of Number Fields

In this §, we review well-known *anabelian* (and related) properties of the *Galois groups of number fields* and (mainly p-adic) *local fields.*

Let F be a *number field.* Fix an *algebraic closure* $\overline{F}$ of F and denote the resulting absolute *Galois group* of F by G_F. Let $\mathfrak{p}$ be a (not necessarily nonarchimedean!) *prime* of F. Write $G_{\mathfrak{p}} \subseteq G_F$ for the *decomposition group* (well-defined up to conjugacy) associated to $\mathfrak{p}$ and $F_{\mathfrak{p}}$ for the *completion* of F at $\mathfrak{p}$. Then we have the following result (cf. also Corollary 1.3.3 below for a generalization of the slimness of $G_{\mathfrak{p}}$, G_F):

Theorem 1.1.1 (Slimness and Commensurable Terminality). *Suppose that* $\mathfrak{p}$ *is* **nonarchimedean**. *Then:*
(i) The closed subgroup $G_{\mathfrak{p}} \subseteq G_F$ *is* **commensurably terminal.**

(ii) The inclusion $G_{\mathfrak{p}} \hookrightarrow G_F$ *is* **relatively slim**. *In particular,* $G_{\mathfrak{p}}$, G_F *are* **slim**.

Proof. Assertion (i) is a formal consequence of [NSW], Corollary 12.1.3. As for assertion (ii), the slimness of $G_{\mathfrak{p}}$ follows from *local class field theory* (cf., e.g., [Serre2]). (That is, if $\sigma \in G_{\mathfrak{p}}$ commutes with an open subgroup $H \subseteq G_{\mathfrak{p}}$, then σ induces the trivial action on the abelianization H^{ab}. But, by local class field theory, H^{ab} may be identified with the profinite completion of $K^{\times}$, where K is the finite extension of $F_{\mathfrak{p}}$ determined by H. Thus, σ acts trivially on all sufficiently large finite extensions K of $F_{\mathfrak{p}}$, so $\sigma = 1$, as desired.) Relative slimness thus follows formally from the slimness of $G_{\mathfrak{p}}$ and (i) (cf. Remark 0.1.2). □

Theorem 1.1.2 (Topologically Finitely Generated Closed Normal Subgroups). *Every topologically finitely generated closed normal subgroup of* G_F *is trivial.*

Proof. This follows from [FJ], Theorem 15.10. □

Theorem 1.1.3 (The Neukirch-Uchida Theorem on the Anabelian Nature of Number Fields). *Let* F_1, F_2 *be number fields. Let* $\overline{F}_1$ *(respectively,* $\overline{F}_2$*) be an algebraic closure of* F_1 *(respectively,* F_2*). Write* $\mathrm{Isom}(\overline{F}_2/F_2, \overline{F}_1/F_1)$ *for the set of field isomorphisms* $\overline{F}_2 \xrightarrow{\sim} \overline{F}_1$ *that map* F_2 *onto* F_1*. Then the natural map*

$$\mathrm{Isom}(\overline{F}_2/F_2, \overline{F}_1/F_1) \to \mathrm{Isom}(\mathrm{Gal}(\overline{F}_1/F_1), \mathrm{Gal}(\overline{F}_2/F_2))$$

is bijective.

Proof. This is the content of [NSW], Theorem 12.2.1. □

Remark 1.1.3.1. It is important to note, however, that the analogue of Theorem 1.1.3 for finite extensions of $\mathbb{Q}_p$ is *false* (cf. [NSW], p. 674). Nevertheless, by considering isomorphisms of Galois groups that *preserve the higher ramification filtration*, one may obtain *a partial analogue of this theorem for p-adic local fields* (cf. [Mzk5]).

Next, we would like to consider a situation that arises frequently in anabelian geometry. Suppose that G is *equal to* G_F or $G_{\mathfrak{p}}$ (where we assume now that $\mathfrak{p}$ is *nonarchimedean*!), and that we are given an *exact sequence of profinite groups*:

$$1 \to \Delta \to \Pi \to G \to 1$$

Suppose, moreover, that this sequence *splits over some open subgroup of* G, and that Δ is *topologically finitely generated.* The following result was related to the author by *A. Tamagawa*:

Lemma 1.1.4 (Intrinsic Characterization of Arithmetic Quotients). *(i) Suppose that $G = G_F$. Let $\Pi' \subseteq \Pi$ be an open subgroup. Then the kernel of the homomorphism $\Pi' \to G$ may be characterized as the unique maximal closed normal subgroup of Π' which is topologically finitely generated.*
(ii) Suppose that $G = G_{\mathfrak{p}}$. Assume further that for every open subgroup $\Pi'' \subseteq \Pi$, the abelianization $(\Delta'')^{\mathrm{ab}}$ of Δ'' (where $\Delta'' \stackrel{\mathrm{def}}{=} \Pi'' \bigcap \Delta$) satisfies the following property:

> (∗) *The maximal torsion-free quotient $(\Delta'')^{\mathrm{ab}} \twoheadrightarrow Q''$ of $(\Delta'')^{\mathrm{ab}}$ on which the action of $G'' \stackrel{\mathrm{def}}{=} \Pi''/\Delta''$ (by conjugation) is trivial is a finitely generated* **free** *$\widehat{\mathbb{Z}}$-module.*

Let $\Pi' \subseteq \Pi$ be an arbitrary open subgroup. Then:

$$[G : G'] \cdot [F_{\mathfrak{p}} : \mathbb{Q}_p] = \dim_{\mathbb{Q}_p}((\Pi')^{\mathrm{ab}} \otimes_{\widehat{\mathbb{Z}}} \mathbb{Q}_p) - \dim_{\mathbb{Q}_l}((\Pi')^{\mathrm{ab}} \otimes_{\widehat{\mathbb{Z}}} \mathbb{Q}_l)$$

(where $\Delta' \stackrel{\mathrm{def}}{=} \Delta \bigcap \Pi'$; $G' \stackrel{\mathrm{def}}{=} \Pi'/\Delta'$; p is the rational prime that $\mathfrak{p}$ divides; and l is any prime number distinct from p). (In fact, p may also be characterized as the unique prime number for which the difference on the right is nonzero for infinitely many prime numbers l.) In particular, the subgroup $\Delta \subseteq \Pi$ may be characterized as the intersection of those open subgroups $\Pi' \subseteq \Pi$ such that:

$$[G : G'] = [\Pi : \Pi']$$

(i.e., such that $[G : G'] \cdot [F_{\mathfrak{p}} : \mathbb{Q}_p] = [\Pi : \Pi'] \cdot ([G : G] \cdot [F_{\mathfrak{p}} : \mathbb{Q}_p])$).

Proof. Assertion (i) is a formal consequence of Theorem 1.1.2.

Now we turn to assertion (ii). Denote by K' the finite extension of $F_{\mathfrak{p}}$ determined by G'. Then:

$$[G : G'] \cdot [F_{\mathfrak{p}} : \mathbb{Q}_p] = [K' : \mathbb{Q}_p]$$

Next, let us observe that we have an *exact sequence*

$$0 \to \mathrm{Im}(\Delta') \to (\Pi')^{\mathrm{ab}} \to (G')^{\mathrm{ab}} \to 0$$

(where $\mathrm{Im}(\Delta')$ is the image of Δ' in $(\Pi')^{\mathrm{ab}}$) of finitely generated $\widehat{\mathbb{Z}}$-modules. Note, moreover, that we have a *natural surjection* $Q' \twoheadrightarrow \mathrm{Im}(\Delta')/(\text{torsion})$ (where Q' is as in (∗)). On the other hand, it follows formally from our assumption that $\Pi' \twoheadrightarrow G'$ *splits over an open subgroup of* G' (together with the well-known fact that the group cohomology of a finite group is always annihilated by the order to the group) that this natural surjection is, in fact, an *isomorphism*. In particular, it follows formally from $(*)$ that:

$$\dim_{\mathbb{Q}_p}((\Pi')^{\mathrm{ab}} \otimes_{\widehat{\mathbb{Z}}} \mathbb{Q}_p) - \dim_{\mathbb{Q}_l}((\Pi')^{\mathrm{ab}} \otimes_{\widehat{\mathbb{Z}}} \mathbb{Q}_l)$$
$$= \dim_{\mathbb{Q}_p}((G')^{\mathrm{ab}} \otimes_{\widehat{\mathbb{Z}}} \mathbb{Q}_p) - \dim_{\mathbb{Q}_l}((G')^{\mathrm{ab}} \otimes_{\widehat{\mathbb{Z}}} \mathbb{Q}_l)$$

Thus, to complete the proof of Lemma 1.1.4, it suffices to prove that:

$$[K' : \mathbb{Q}_p] = \dim_{\mathbb{Q}_p}((G')^{\mathrm{ab}} \otimes_{\widehat{\mathbb{Z}}} \mathbb{Q}_p) - \dim_{\mathbb{Q}_l}((G')^{\mathrm{ab}} \otimes_{\widehat{\mathbb{Z}}} \mathbb{Q}_l)$$

But this is a formal consequence of *local class field theory* (cf., e.g., [Serre2]; §1.2 below), i.e., the fact that $(G')^{\mathrm{ab}}$ is isomorphic to the profinite completion of $(K')^{\times}$. □

Typically, in applications involving hyperbolic curves, one shows that the condition $(*)$ of Lemma 1.1.4 is satisfied by applying the following:

Lemma 1.1.5 (Tate Modules of Semi-abelian Varieties). *Let K be a finite extension of $\mathbb{Q}_p$. Fix an algebraic closure $\overline{K}$ of K; write $G_K \stackrel{\mathrm{def}}{=} \mathrm{Gal}(\overline{K}/K)$. Let A be a semi-abelian variety over K. Denote the resulting (profinite)* **Tate module** *of A by:*

$$T(A) \stackrel{\mathrm{def}}{=} \mathrm{Hom}(\mathbb{Q}/\mathbb{Z}, A(\overline{K}))$$

Then the maximal torsion-free quotient $T(A) \twoheadrightarrow Q$ on which G_K acts trivially is a finitely generated free $\widehat{\mathbb{Z}}$-module.

Proof. A semi-abelian variety is an extension of an abelian variety by a torus. Thus, $T(A)$ is the extension of the Tate module of an abelian variety by the Tate module of a torus. Moreover, since (after restricting to some open subgroup of G_K) the Tate module of a torus is isomorphic to the direct sum of a finite number of copies of $\widehat{\mathbb{Z}}(1)$, we thus conclude that the image of the Tate module of the torus in Q is necessarily zero. In particular, we may assume for the remainder of the proof without loss of generality that A is an *abelian variety*.

Now it follows from the theory of [FC] (cf., in particular, [FC], Chapter III, Corollary 7.3), that $T(A)$ fits into *exact sequences* (of G_K-modules)

$$0 \to T_{\mathrm{good}} \to T(A) \to T_{\mathrm{com}} \to 0$$
$$0 \to T_{\mathrm{trc}} \to T_{\mathrm{good}} \to T(B) \to 0$$

where $T(B)$ is the Tate module of an abelian variety B over K with *potentially good reduction*; and $T_{\mathrm{com}} = M_{\mathrm{com}} \otimes_{\mathbb{Z}} \widehat{\mathbb{Z}}$, $T_{\mathrm{trc}} = M_{\mathrm{trc}} \otimes_{\mathbb{Z}} \widehat{\mathbb{Z}}(1)$ for finitely generated free $\mathbb{Z}$-modules M_{com}, M_{trc} on which G_K

acts via a *finite quotient.* [Here, "com" (respectively, "trc") stands for "combinatorial" (respectively, "toric").] It is thus evident that T_{trc} maps to 0 in Q. Moreover, by [Mzk4], Lemma 8.1 (the proof of which is valid for arbitrary B, even though in *loc. cit.*, this result is only stated in the case of a Jacobian), and the *Riemann Hypothesis for abelian varieties over finite fields* (cf., e.g., [Mumf], p. 206), it follows that $T(B)$ also maps to 0 in Q. Thus, we conclude that Q is equal to the maximal torsion-free quotient of T_{com} on which G_K acts trivially. Since $\widehat{\mathbb{Z}}$ is $\mathbb{Z}$-flat, however, this implies that Q is equal to the result of applying $\otimes_{\mathbb{Z}}\widehat{\mathbb{Z}}$ to the maximal torsion-free quotient of M_{com} on which G_K acts trivially. But this last quotient is manifestly finitely generated and free over $\mathbb{Z}$. This completes the proof. □

1.2 The Anabelian Geometry of *p*-adic Local Fields

In this §, we review certain well-known *"group-theoretic" properties of Galois groups of p-adic local fields*, i.e., properties preserved by *arbitrary isomorphisms* between such Galois groups.

For $i = 1, 2$, let p_i be a *prime number.* Let K_i be a finite extension of $\mathbb{Q}_{p_i}$. We denote the *ring of integers* (respectively, *maximal ideal; residue field*) of K_i by $\mathcal{O}_{K_i}$ (respectively, $\mathfrak{m}_{K_i}$; k_i). Also, we assume that we have chosen an *algebraic closures* $\overline{K}_i$ of K_i and write

$$G_{K_i} \stackrel{\mathrm{def}}{=} \mathrm{Gal}(\overline{K}_i/K_i)$$

for the corresponding absolute Galois group of K_i. Thus, by *local class field theory* (cf., e.g., [Serre2]), we have a natural isomorphism

$$(K_i^{\times})^{\wedge} \xrightarrow{\sim} G_{K_i}^{\mathrm{ab}}$$

(where the "$\wedge$" denotes the profinite completion of an abelian group; "$\times$" denotes the group of units of a ring; and "ab" denotes the maximal abelian quotient of a topological group). In particular, $G_{K_i}^{\mathrm{ab}}$ fits into an *exact sequence*

$$0 \to \mathcal{O}_{K_i}^{\times} \to G_{K_i}^{\mathrm{ab}} \to \widehat{\mathbb{Z}} \to 0$$

(arising from a similar exact sequence for $(K_i^{\times})^{\wedge}$). Moreover, we obtain natural inclusions

$$k_i^{\times} \hookrightarrow \mathcal{O}_{K_i}^{\times} \subseteq K_i^{\times} \hookrightarrow G_{K_i}^{\mathrm{ab}}$$
$$K_i^{\times}/\mathcal{O}_{K_i}^{\times} \xrightarrow{\sim} \mathbb{Z} \hookrightarrow G_{K_i}^{\mathrm{ab}}/\mathrm{Im}(\mathcal{O}_{K_i}^{\times})$$

(where "$\xrightarrow{\sim}$" denotes the morphism induced by the valuation on $K_i^{\times}$) by considering the Teichmüller representatives of elements of $k_i^{\times}$ and the Frobenius element, respectively. Also, in the following discussion we shall write:

$$\boldsymbol{\mu}_{\widehat{\mathbb{Z}}}(\overline{K}_i) \stackrel{\text{def}}{=} \text{Hom}(\mathbb{Q}/\mathbb{Z}, \overline{K}_i^{\times}); \quad \boldsymbol{\mu}_{\widehat{\mathbb{Z}}'}(\overline{K}_i) \stackrel{\text{def}}{=} \boldsymbol{\mu}_{\widehat{\mathbb{Z}}}(\overline{K}_i) \otimes_{\widehat{\mathbb{Z}}} \widehat{\mathbb{Z}}';$$
$$\boldsymbol{\mu}_{\mathbb{Q}/\mathbb{Z}}(\overline{K}_i) \stackrel{\text{def}}{=} \boldsymbol{\mu}_{\widehat{\mathbb{Z}}}(\overline{K}_i) \otimes_{\widehat{\mathbb{Z}}} \mathbb{Q}/\mathbb{Z}$$

(where $\widehat{\mathbb{Z}}' \stackrel{\text{def}}{=} \widehat{\mathbb{Z}}/\mathbb{Z}_p$). Finally, we denote the *cyclotomic character* of G_{K_i} by:

$$\chi_i : G_{K_i} \to \widehat{\mathbb{Z}}^{\times}$$

Proposition 1.2.1 (Invariants of Arbitrary Isomorphisms of Galois Groups of Local Fields). *Suppose that we are given an isomorphism of profinite groups:*

$$\alpha : G_{K_1} \xrightarrow{\sim} G_{K_2}$$

Then:
(i) We have: $p_1 = p_2$. Thus, (in the remainder of this proposition and its proof) we shall write $p \stackrel{\text{def}}{=} p_1 = p_2$.
(ii) α induces an isomorphism $I_{K_1} \xrightarrow{\sim} I_{K_2}$ between the respective **inertia subgroups** *of G_{K_1}, G_{K_2}.*
(iii) The isomorphism $\alpha^{\text{ab}} : G_{K_1}^{\text{ab}} \xrightarrow{\sim} G_{K_2}^{\text{ab}}$ induced by α preserves the images $\text{Im}(\mathcal{O}_{K_i}^{\times})$, $\text{Im}(k_i^{\times})$, $\text{Im}(K_i^{\times})$ of the natural morphisms discussed above.
(iv) The morphism induced by α between the respective quotients $G_{K_i}^{\text{ab}}/\text{Im}(\mathcal{O}_{K_i}^{\times})$ preserves the respective **Frobenius elements.**
(v) $[K_1 : \mathbb{Q}_p] = [K_2 : \mathbb{Q}_p]$; $[k_1 : \mathbb{F}_p] = [k_2 : \mathbb{F}_p]$. In particular, the ramification indices of K_1, K_2 over $\mathbb{Q}_p$ coincide.
(vi) The morphisms induced by α on the abelianizations of the various open subgroups of the G_{K_i} induce an isomorphism

$$\boldsymbol{\mu}_{\mathbb{Q}/\mathbb{Z}}(\overline{K}_1) \xrightarrow{\sim} \boldsymbol{\mu}_{\mathbb{Q}/\mathbb{Z}}(\overline{K}_2)$$

which is Galois-equivariant with respect to α. In particular, α preserves the **cyclotomic characters** *χ_i.*
(vii) The morphism $H^2(K_1, \boldsymbol{\mu}_{\mathbb{Q}/\mathbb{Z}}(\overline{K}_1)) \xrightarrow{\sim} H^2(K_2, \boldsymbol{\mu}_{\mathbb{Q}/\mathbb{Z}}(\overline{K}_2))$ induced by α (cf. (vi)) preserves the **"residue map"**

$$H^2(K_i, \boldsymbol{\mu}_{\mathbb{Q}/\mathbb{Z}}(\overline{K}_i)) \xrightarrow{\sim} \mathbb{Q}/\mathbb{Z}$$

of local class field theory (cf. [Serre2], §1.1).

Proof. Property (i) follows by considering the *ranks* of $G_{K_i}^{\text{ab}}$ over various $\mathbb{Z}_l$ (cf. Lemma 1.1.4, (ii)). Property (iii) for $\text{Im}(k_i^{\times})$ follows from the fact that $\mathrm{Im}(k_i^\times)$ may be recovered as the prime-to-p torsion subgroup of $G_{K_i}^{\mathrm{ab}}$. Property (v) follows for $[K_i : \mathbb{Q}_p]$ by considering the $\mathbb{Z}_p$-rank of $G_{K_i}^{\mathrm{ab}}$ (minus 1) and for $[k_i : \mathbb{F}_p]$ by considering the cardinality of $\mathrm{Im}(k_i^\times)$ (plus 1) — cf. (i), (iii). Property (ii) follows from the fact that whether or not a finite extension is unramified may be determined group-theoretically by considering the variation of the ramification index over $\mathbb{Q}_p$ (cf. (v)). Property (iii) for $\mathrm{Im}(\mathcal{O}_{K_i}^\times)$ follows formally from (ii) (since this image is equal to the image in $G_{K_i}^{\mathrm{ab}}$ of I_{K_i}). Property (iv) follows by applying (iii) for $\mathrm{Im}(k_i^\times)$ to the various open subgroups of G_{K_i} that correspond to *unramified* extensions of K_i and using the fact the Frobenius element is the unique element that acts on $\overline{k}_i^\times$ (where $\overline{k}_i$ denotes the algebraic closure of k_i induced by $\overline{K}_i$) as *multiplication by* $|k_1| = |k_2|$ [where, relative to our use of the term "multiplication", we think of the abelian group $\overline{k}_i^\times$ *additively*]. Here, we note that if L_i is a finite extension of K_i, then the inclusion

$$G_{K_i}^{\mathrm{ab}} \xrightarrow{\sim} (K_i^\times)^\wedge \hookrightarrow (L_i^\times)^\wedge \xrightarrow{\sim} G_{L_i}^{\mathrm{ab}}$$

may be reconstructed group-theoretically by considering the *Verlagerung*, or transfer, map (cf. [Serre2], §2.4). Property (iii) for $\mathrm{Im}(K_i^\times)$ follows formally from (iv). Property (vi) follows formally from (iii). Finally, property (vii) follows (cf. the theory of the Brauer group of a local field, as exposed, for instance, in [Serre2], §1) from the fact that the morphism $H^2(K_i, \boldsymbol{\mu}_{\mathbb{Q}/\mathbb{Z}}(\overline{K}_i)) \xrightarrow{\sim} \mathbb{Q}/\mathbb{Z}$ may be constructed as the composite of the natural isomorphism

$$H^2(K_i, \boldsymbol{\mu}_{\mathbb{Q}/\mathbb{Z}}(\overline{K}_i)) = H^2(G_{K_i}, \boldsymbol{\mu}_{\mathbb{Q}/\mathbb{Z}}(\overline{K}_i)) \xrightarrow{\sim} H^2(G_{K_i}, \overline{K}_i^\times)$$

— which is group-theoretic, by (iii) — with the inverse of the isomorphism

$$H^2(\mathrm{Gal}(K_i^{\mathrm{unr}}/K_i), (K_i^{\mathrm{unr}})^\times) \xrightarrow{\sim} H^2(G_{K_i}, \overline{K}_i^\times)$$

(where K_i^{unr} denotes the maximal unramified extension of K_i) — which is group-theoretic, by (ii), (iii) — followed by the natural isomorphism

$$\begin{aligned} H^2(\mathrm{Gal}(K_i^{\mathrm{unr}}/K_i), (K_i^{\mathrm{unr}})^\times) \xrightarrow{\sim} H^2(\mathrm{Gal}(K_i^{\mathrm{unr}}/K_i), \mathbb{Z}) &= H^2(\widehat{\mathbb{Z}}, \mathbb{Z}) \\ &= \mathbb{Q}/\mathbb{Z} \end{aligned}$$

— which is group-theoretic, by (ii), (iii), (iv). $\square$

Before proceeding, we observe that Proposition 1.2.1, (i), may be extended as follows: Write

$$\mathfrak{A}_{\mathbb{Q}} \stackrel{\mathrm{def}}{=} |\mathrm{Spec}(\mathbb{Z})| \bigcup \{\infty\}$$

(where "$|-|$" denotes the underlying set of a scheme) for the set of *"all arithmetic primes of* $\mathbb{Q}$*"*. If $v \in \mathfrak{A}_{\mathbb{Q}}$ is equal to $(0) \in |\mathrm{Spec}(\mathbb{Z})|$ (respectively, ∞), set $G_v \stackrel{\mathrm{def}}{=} G_{\mathbb{Q}}$ (respectively, $G_v \stackrel{\mathrm{def}}{=} \mathrm{Gal}(\mathbb{C}/\mathbb{R})$). If $v \in |\mathrm{Spec}(\mathbb{Z})| \subseteq \mathfrak{A}_{\mathbb{Q}}$ is equal to the prime determined by a prime number p, set $G_v \stackrel{\mathrm{def}}{=} G_{\mathbb{Q}_p}$.

Proposition 1.2.2 (Intrinsicity of Arithmetic Types). *For* $i = 1, 2$, *let* $v_i \in \mathfrak{A}_{\mathbb{Q}}$. *Suppose that* H_i *is an open subgroup of* G_{v_i}. *Then* $H_1 \cong H_2$ *implies* $v_1 = v_2$.

Proof. Indeed, open subgroups of $G_{\mathbb{Q}}$ may be distinguished by the fact that their abelianizations *fail to be topologically finitely generated.* (Indeed, consider the abelian extensions arising from adjoining roots of unity.) By contrast, abelianizations of open subgroups of $G_{\mathbb{R}}$ or $G_{\mathbb{Q}_p}$ (cf. the above discussion) are topologically finitely generated. Next, open subgroups of $G_{\mathbb{R}}$ may be distinguished from those of any $G_{\mathbb{Q}_p}$ by the fact they are *finite.* The remainder of Proposition 1.2.2 follows from Proposition 1.2.1, (i). $\square$

Next, let us write $\mathrm{Spec}(\mathcal{O}_{K_i})^{\log}$ for the log scheme obtained by equipping the scheme $\mathrm{Spec}(\mathcal{O}_{K_i})$ with the *log structure* defined by the divisor $V(\mathfrak{m}_{K_i})$. Thus, by pulling back this log structure via the natural morphism $\mathrm{Spec}(k_i) \hookrightarrow \mathrm{Spec}(\mathcal{O}_{K_i})$, we obtain a log scheme $\mathrm{Spec}(k_i)^{\log}$, which we denote by

$$k_i^{\log}$$

for short. Note that the "étale monoid" that defines the log structure on $k_i^{\log}$ "admits a global chart" in the sense that it is defined by a *single constant monoid* (in the Zariski topology of $\mathrm{Spec}(k_i)$) $M_{k_i^{\log}}$, which fits into a natural *exact sequence* (of monoids):

$$1 \to k_i^{\times} \to M_{k_i^{\log}} \to \mathbb{N} \to 0$$

Thus, the $k_i^{\times}$-torsor U_i determined by considering the inverse image of $1 \in \mathbb{N}$ in this sequence may be identified with the $k_i^{\times}$-torsor of *uniformizers* $\in \mathfrak{m}_{K_i}$ *considered modulo* $\mathfrak{m}_{K_i}^2$.

Next, let us write

$$G_{K_i} \twoheadrightarrow G_{k_i}^{\log}$$

for the quotient defined by the *maximal tamely ramified extension* K_i^{tame} of K_i. Thus, $G_{k_i}^{\log}$ may also be thought of as the *"logarithmic fundamental group"* $\pi_1(k_i^{\log})$ of the log scheme $k_i^{\log}$. Moreover, $G_{k_i}^{\log}$ fits into a natural *exact sequence:*

$$1 \to \boldsymbol{\mu}_{\widehat{\mathbb{Z}}}(\overline{k}_i) \to G_{k_i}^{\log} \to \widehat{\mathbb{Z}} \to 1$$

where, just as in the case of K_i, we write:

$$\boldsymbol{\mu}_{\widehat{\mathbb{Z}}}(\overline{k}_i) = \boldsymbol{\mu}_{\widehat{\mathbb{Z}}'}(\overline{k}_i) \stackrel{\text{def}}{=} \text{Hom}(\mathbb{Q}/\mathbb{Z}, \overline{k}_i^\times);$$
$$\boldsymbol{\mu}_{\mathbb{Q}/\mathbb{Z}}(\overline{k}_i) \stackrel{\text{def}}{=} \boldsymbol{\mu}_{\widehat{\mathbb{Z}}}(\overline{k}_i) \otimes_{\widehat{\mathbb{Z}}} \mathbb{Q}/\mathbb{Z}$$

The *"abelianization"* of this exact sequence yields an exact sequence:

$$1 \to k_i^\times \to (G_{k_i}^{\log})^{\text{ab}} \to \widehat{\mathbb{Z}} \to 1$$

Now we have the following:

Proposition 1.2.3. *(i) Any isomorphism $\alpha : G_{K_1} \xrightarrow{\sim} G_{K_2}$ (as in Proposition 1.2.1) induces an isomorphism $G_{k_1}^{\log} \xrightarrow{\sim} G_{k_2}^{\log}$ between the respective quotients.*
(ii) There is a natural isomorphism of $k_i^\times$-torsors between the **torsor of uniformizers** *U_i discussed above and the $H^1(\widehat{\mathbb{Z}}, \boldsymbol{\mu}_{\widehat{\mathbb{Z}}}(\overline{k}_i)) = k_i^\times$*-**torsor** *$H^1(G_{k_i}^{\log}, \boldsymbol{\mu}_{\widehat{\mathbb{Z}}}(\overline{k}_i))^{[1]}$ of elements of $H^1(G_{k_i}^{\log}, \boldsymbol{\mu}_{\widehat{\mathbb{Z}}}(\overline{k}_i))$ that map to the identity element in $H^1(\boldsymbol{\mu}_{\widehat{\mathbb{Z}}}(\overline{k}_i), \boldsymbol{\mu}_{\widehat{\mathbb{Z}}}(\overline{k}_i)) = \text{Hom}_{\widehat{\mathbb{Z}}'}(\boldsymbol{\mu}_{\widehat{\mathbb{Z}}}(\overline{k}_i), \boldsymbol{\mu}_{\widehat{\mathbb{Z}}}(\overline{k}_i))$. This isomorphism is defined by associating to a uniformizer $\pi \in U_i$ the $\boldsymbol{\mu}_{\widehat{\mathbb{Z}}}(\overline{k}_i)$-torsor over $k_i^{\log}$ determined by the roots $\pi^{1/N}$, as N ranges (multiplicatively) over the positive integers prime to p_i.*
(iii) The profinite group $G_{k_i}^{\log}$ is **slim**.

Proof. Property (i) follows from Proposition 1.2.1, (ii), together with the fact that the quotient $G_{K_i} \twoheadrightarrow G_{k_i}^{\log}$ may be identified with the quotient of G_{K_i} by the (unique) maximal pro-p subgroup of I_{K_i}. Next, since *any* morphism of $k_i^\times$-torsors is *necessarily* an isomorphism, property (ii) follows by observing that the stated association of coverings to uniformizers is indeed a morphism of $k_i^\times$-torsors — a *tautology*, which may by verified by considering the case $N = q_i - 1$ (where q_i is the cardinality of k_i), in which case this tautology amounts to the *computation*: $(\zeta^{1/N})^{q_i} = \zeta \cdot (\zeta^{1/N})$ (for $\zeta \in k_i^\times$). Finally, property (iii) follows formally from the fact that the quotient $G_{k_i}^{\log}/\text{Im}(\boldsymbol{\mu}_{\widehat{\mathbb{Z}}}(\overline{k}_i))$ is *infinite* and acts *faithfully* on all open subgroups of the closed subgroup $\text{Im}(\boldsymbol{\mu}_{\widehat{\mathbb{Z}}}(\overline{k}_i)) \subseteq G_{k_i}^{\log}$. □

In the following, let us denote by $(k_i^{\log})^\sim \to k_i^{\log}$ the *"universal covering"* of $k_i^{\log}$ defined by the extension K_i^{tame} of K_i. Thus, $G_{k_i}^{\log}$ acts naturally as the *group of covering transformations* of $(k_i^{\log})^\sim \to k_i^{\log}$.

Proposition 1.2.4 (The "Grothendieck Conjecture" for the Logarithmic Point). *Suppose that we are given an isomorphism of profinite groups:*

$$\lambda : G_{k_1}^{\log} \xrightarrow{\sim} G_{k_2}^{\log}$$

Then:
(i) We have: $|k_1| = |k_2|$; $p_1 = p_2$. *Thus, (in the remainder of this proposition and its proof) we shall write* $p \stackrel{\text{def}}{=} p_1 = p_2$.
(ii) λ *preserves the subgroups* $\text{Im}(\boldsymbol{\mu}_{\widehat{\mathbb{Z}}}(\overline{k}_i)) \subseteq G_{k_i}^{\log}$ *as well as the Frobenius elements in the quotients* $G_{k_i}^{\log}/\text{Im}(\boldsymbol{\mu}_{\widehat{\mathbb{Z}}}(\overline{k}_i))$.
(iii) Assume further that the morphism

$$\overline{k}_1^{\times} \xrightarrow{\sim} \overline{k}_2^{\times}$$

induced by λ *(by thinking of* $\overline{k}_i^{\times}$ *as* $\boldsymbol{\mu}_{\mathbb{Q}/\mathbb{Z}}(\overline{k}_i)$*) arises from an* **isomorphism of fields** $\overline{\sigma} : \overline{k}_1 \xrightarrow{\sim} \overline{k}_2$. *Then there exists a unique commutative diagram*

$$\begin{array}{ccc} (k_1^{\log})^{\sim} & \stackrel{(\sigma^{\log})^{\sim}}{\longrightarrow} & (k_2^{\log})^{\sim} \\ \downarrow & & \downarrow \\ k_1^{\log} & \stackrel{\sigma^{\log}}{\longrightarrow} & k_2^{\log} \end{array}$$

of log schemes, compatible with the natural action of $G_{k_i}^{\log}$ *on* $(k_i^{\log})^{\sim}$ *(for* $i = 1, 2$*), in which the vertical morphisms are the natural morphisms, and the horizontal morphisms are* **isomorphisms** *for which the morphisms on the underlying schemes are those induced by* $\overline{\sigma}$.

Proof. Property (i) follows by observing that p_i is the unique prime number such that 1 plus the cardinality of the torsion subgroup of $(G_{k_i}^{\log})^{\text{ab}}$ — i.e., the cardinality of k_i — is equal to a power of p_i. Property (ii) follows by thinking of the quotients $G_{k_i}^{\log}/\text{Im}(\boldsymbol{\mu}_{\widehat{\mathbb{Z}}}(\overline{k}_i))$ as the quotients of $G_{k_i}^{\log}$ obtained by forming the quotient of $(G_{k_i}^{\log})^{\text{ab}}$ by its torsion subgroup, and then using that the Frobenius element is the unique element that acts on the abelian group $\text{Im}(\boldsymbol{\mu}_{\widehat{\mathbb{Z}}}(\overline{k}_i))$ via multiplication by $|k_1| = |k_2|$. As for (iii), the morphism $\sigma^{\log}$ is the unique logarithmic extension of σ whose induced morphism $U_1 \xrightarrow{\sim} U_2$ is the morphism obtained (cf. Proposition 1.2.3, (ii)) by considering the morphism induced by λ between the $H^1(\widehat{\mathbb{Z}}, \boldsymbol{\mu}_{\widehat{\mathbb{Z}}}(\overline{k}_i)) = k_i^{\times}$-**torsors** $H^1(G_{k_i}^{\log}, \boldsymbol{\mu}_{\widehat{\mathbb{Z}}}(\overline{k}_i))^{[1]}$ (for $i = 1, 2$) — which are preserved by λ, by (ii). Note that here we also use (cf. (ii)) that the *Frobenius element* $\in \widehat{\mathbb{Z}}$ is preserved, since this element is necessary to ensure the compatibility of the identifications

$$H^1(\widehat{\mathbb{Z}}, \boldsymbol{\mu}_{\widehat{\mathbb{Z}}}(\overline{k}_i)) = k_i^{\times}$$

(cf. Proposition 1.2.3, (ii)). The morphism $(\sigma^{\log})^{\sim}$ is obtained by applying this construction of "$\sigma^{\log}$" to the various finite log étale coverings of $k_i^{\log}$ obtained by considering various open subgroups of $G_{k_i}^{\log}$. Here, the transition morphisms among coverings are induced by the *Verlagerung*, as in the proof of Proposition 1.2.1. Finally, the uniqueness of the lifting $(\sigma^{\log})^{\sim}$ of $\sigma^{\log}$ is a formal consequence of the fact that $G_{k_i}^{\log}$ is *center-free* (cf. Proposition 1.2.3, (iii)). □

1.3 The Anabelian Geometry of Hyperbolic Curves

Characteristic Zero:

Let K be a *field of characteristic* 0 whose absolute Galois group is *slim*. Let X be a *hyperbolic curve* of type (g, r) over $S \stackrel{\text{def}}{=} \text{Spec}(K)$. Fix an *algebraic closure* $\overline{K}$ of K and write $\overline{s} : \text{Spec}(\overline{K}) \to S$; $G_K \stackrel{\text{def}}{=} \text{Gal}(\overline{K}/K)$. Let $\overline{x} \in X(\overline{K})$ be a $\overline{K}$-valued point of X lying over $\overline{s}$. Then, setting $\Pi_X \stackrel{\text{def}}{=} \pi_1(X, \overline{x})$, we obtain an *exact sequence*

$$1 \to \Delta_X \to \Pi_X \to G_K \to 1$$

which determines a well-defined *continous homomorphism*

$$G_K \to \text{Out}(\Delta_X)$$

to the *group of outer automorphisms* $\text{Out}(\Delta_X)$ *of* Δ_X.

Lemma 1.3.1 (Slimness of Geometric and Arithmetic Fundamental Groups). *The profinite groups Δ_X, Π_X are* **slim**.

Proof. The slimness of Π_X is a formal consequence of the slimness of Δ_X and our *assumption that G_K is slim.* Thus, it remains to prove that Δ_X is slim. One approach to proving this fact is given in [Naka1], Corollary 1.3.4. Another approach is the following: Let $H \subseteq \Delta_X$ be an *open normal subgroup* for which the associated covering $X_H \to X_{\overline{K}} \stackrel{\text{def}}{=} X \times_K \overline{K}$ is such that X_H is a *curve of genus* ≥ 2. Thus, H^{ab} may be thought of as the profinite Tate module associated to the *generalized Jacobian* of the singular curve obtained from the unique smooth compactification of X_H by identifying the various *cusps* (i.e., points of the compactification not lying in X_H) to a *single point.* In particular, if conjugation by an element $\delta \in \Delta_X$ induces the *trivial action* on H^{ab}, then we conclude that the image of δ in Δ_X/H induces the trivial action on the generalized Jacobian just discussed, hence on X_H itself. But this implies that $\delta \in H$. By taking H to be sufficiently small, we thus conclude that $\delta = 1$. □

In particular, it follows formally from Lemma 1.3.1 that:

Corollary 1.3.2 (A Natural Exact Sequence). *We have a natural exact sequence of profinite groups:*

$$1 \to \Delta_X \to \text{Aut}(\Delta_X) \to \text{Out}(\Delta_X) \to 1$$

In particular, by pulling back this exact sequence, one may recover the exact sequence $1 \to \Delta_X \to \Pi_X \to G_K \to 1$ *entirely group-theoretically from the outer Galois representation* $G_K \to \mathrm{Out}(\Delta_X)$.

One example of the sort of "K" under consideration is the case of a *"sub-p-adic field"*:

Corollary 1.3.3 (Slimness of Sub-p-adic Fields). *The absolute Galois group of a* **sub-p-adic field** *(i.e., a field isomorphic to a subfield of a finitely generated field extension of* $\mathbb{Q}_p$*, for some prime number* p*) is* **slim**.

Proof. This fact is implied by the argument of the proof of [Mzk6], Lemma 15.8. □

In [Mzk6], the author (essentially) proved the following result (cf. [Mzk6], Theorem A, for a stronger version that treats arbitrary *dominant morphisms* and *open group homomorphisms* [i.e., which are not necessarily isomorphisms]):

Theorem 1.3.4 ("Sub-p-adic Profinite Grothendieck Conjecture"). *Suppose that* K *is a* **sub-p-adic field**, *and that* X *and* Y *are hyperbolic curves over* K. *Denote by* $\mathrm{Isom}_K(X, Y)$ *the set of* K*-isomorphisms* $X \xrightarrow{\sim} Y$*; by* $\mathrm{Isom}^{\mathrm{Out}}_{G_K}(\Delta_X, \Delta_Y)$ *the set of outer isomorphisms between the two profinite groups in parentheses that are compatible with the respective outer actions of* G_K. *Then the natural map*

$$\mathrm{Isom}_K(X, Y) \to \mathrm{Isom}^{\mathrm{Out}}_{G_K}(\Delta_X, \Delta_Y)$$

is bijective.

Thus, by combining Theorems 1.1.3; 1.3.4; Lemma 1.1.4, (i), we obtain the following:

Corollary 1.3.5 (Absolute Grothendieck Conjecture over Number Fields). *Let* K, L *be* **number fields**; X *(respectively,* Y*) a hyperbolic curve over* K *(respectively,* L*). Denote by* $\mathrm{Isom}(X, Y)$ *the set of scheme isomorphisms* $X \xrightarrow{\sim} Y$*; by* $\mathrm{Isom}^{\mathrm{Out}}(\Pi_X, \Pi_Y)$ *the set of outer isomorphisms between the two profinite groups in parentheses. Then the natural map*

$$\mathrm{Isom}(X, Y) \to \mathrm{Isom}^{\mathrm{Out}}(\Pi_X, \Pi_Y)$$

is bijective.

Remark 1.3.5.1. Since the analogue of Theorem 1.1.3 in the p-adic local case is false, it seems *unlikely* to the author at the time of writing that the analogue of Corollary 1.3.5 should hold over p-adic local fields.

Remark 1.3.5.2. In fact, the statement of Corollary 1.3.5 holds when K is taken to be an *arbitrary finitely generated field extension of* $\mathbb{Q}$. This follows by applying a theorem of F. Pop (in place of the Neukirch-Uchida theorem) — cf. the discussion on [NSW], p. 677, for more details.

One interesting result in the present context is the following, due to *M. Matsumoto* (cf. [Mtmo], Theorems 2.1, 2.2):

Theorem 1.3.6 (Kernels of Outer Galois Representations). *Let X be an* **affine hyperbolic curve** *over a* **sub-complex field** K *— i.e., a field isomorphic to a subfield of the field of complex numbers. Then the kernel of the resulting outer Galois representation*

$$G_K \to \mathrm{Out}(\Delta_X)$$

is contained in the kernel of the natural homomorphism $G_K \to G_{\mathbb{Q}}$.

Remark 1.3.6.1. Thus, in particular, when K is a *number field*, the homomorphism $G_K \to \mathrm{Out}(\Delta_X)$ is *injective*. This injectivity was first proven by Belyi in the case of hyperbolic curves of type $(g, r) = (0, 3)$. It was then conjectured by Voevodskii to be true for all (hyperbolic) (g, r) and proven by Voevodskii to be true for $g = 1$. Finally, it was proven by Matsumoto to hold for all (g, r) such that $r > 0$. To the knowledge of the author, the proper case remains open at the time of writing. We refer to the discussion surrounding [Mtmo], Theorem 2.1, for more details on this history.

Remark 1.3.6.2. One interesting aspect of the homomorphism appearing in Theorem 1.3.6 is that it allows one to interpret Theorem 1.3.4 (when $X = Y$) as a *computation of the centralizer of the image of this homomorphism* $G_K \to \mathrm{Out}(\Delta_X)$. This point of view is surveyed in detail in [Naka2].

Next, we would like to discuss various properties of the *inertia groups of the cusps* of a hyperbolic curve. For every *cusp* x of $X_{\overline{K}} \stackrel{\mathrm{def}}{=} X \times_K \overline{K}$ — i.e., point of the unique smooth compactification of $X_{\overline{K}}$ over $\overline{K}$ that does not lie in $X_{\overline{K}}$ — we have an associated *inertia group* (abstractly isomorphic to $\widehat{\mathbb{Z}}$)

$$I_x \subseteq \Delta_X$$

(well-defined, up to conjugation). If l is any *prime number*, then let us denote the *maximal pro-l quotient* of a profinite group by means of a superscript "(l)". Thus, we also obtain an inertia group $I_x^{(l)} \subseteq \Delta_X^{(l)}$ (abstractly isomorphic to $\mathbb{Z}_l$).

Lemma 1.3.7 (Commensurable Terminality of Inertia Groups). *The subgroups $I_x^{(l)} \subseteq \Delta_X^{(l)}$, $I_x \subseteq \Delta_X$ are commensurably terminal.*

Proof. Indeed, let σ be an element of the commensurator. If the asserted commensurable terminality is *false*, then by projecting to a finite quotient, we may assume that we have a *finite Galois covering*

$$Z \to X_{\overline{K}}$$

(of degree a power of l in the pro-l case), together with a *cusp* z of Z such that: (i) z maps to x; (ii) $z \neq z^\sigma$; (iii) z, z^σ have conjugate inertia groups in Δ_Z. We may also assume (by taking $\Delta_Z \subseteq \Delta_X$ to be sufficiently small) that Z has *genus* ≥ 2 and admits a cusp $z' \neq z, z^\sigma$. Then it is easy to see that Z admits an infinite abelian (pro-l, in the pro-l case) covering which is totally ramified at z, z', but not at z^σ. But this contradicts property (iii). □

Remark 1.3.7.1. As was pointed out to the author by the referee, in the case $r \geq 2$, the assertion of Lemma 1.3.7 is a formal consequence of Theorem B′ of [HR].

Now, let us assume that we are given *two hyperbolic curves* $(X_i)_{K_i}$ (for $i = 1, 2$), each defined over a finite extension K_i of $\mathbb{Q}_{p_i}$. Let us write q_i for the *cardinality of the residue field of* K_i. Choose basepoints for the $(X_i)_{K_i}$ and denote the resulting fundamental groups by $\Pi_{(X_i)_{K_i}}$. Also, let us denote the unique proper smooth curve over K_i that *compactifies* $(X_i)_{K_i}$ by $(Y_i)_{K_i}$. Suppose, moreover, that we are given an *isomorphism*

$$\alpha_X : \Pi_{(X_1)_{K_1}} \xrightarrow{\sim} \Pi_{(X_2)_{K_2}}$$

of profinite groups.

Lemma 1.3.8 (Group-Theoreticity of Arithmetic Quotients). *The isomorphism α_X is necessarily compatible with the quotients $\Pi_{(X_i)_{K_i}} \twoheadrightarrow G_{K_i}$.*

Proof. This follows formally from Lemmas 1.1.4, 1.1.5. □

Thus, Lemma 1.3.8, Proposition 1.2.1, (v), imply that $q_1 = q_2$.

Lemma 1.3.9 (Group-Theoreticity of the Cusps). *The types (g_i, r_i) of the hyperbolic curves $(X_i)_{K_i}$ coincide. In particular, for any prime number l, α_X maps inertia groups of cusps in Δ_{X_1} (respectively, $\Delta_{X_1}^{(l)}$) to inertia groups of cusps in Δ_{X_2} (respectively, $\Delta_{X_2}^{(l)}$).*

Proof. Whether or not $r_i = 0$ may be determined by considering whether or not Δ_{X_i} is *free* as a profinite group. When $r_i > 0$, one may compute r_i

by considering the *weight* — i.e., the number w such that the eigenvalues of the action are algebraic numbers of archimedean absolute value q_i^w — of the action of the Frobenius element $\in G_{k_i}$ (cf. Proposition 1.2.1, (iv)) as follows: First, we observe that (as is well-known) the weights of the action of Frobenius on $\Delta_{X_i}^{\mathrm{ab}} \otimes \mathbb{Q}_l$ (where l is a prime number distinct from p_1, p_2) belong to the set $\{0, 1, \frac{1}{2}\}$. (Here, we compute weights by choosing a *lifting* of the Frobenius element $\in G_{k_i}$ to G_{K_i}. Since (as is well-known) the action of the *inertia subgroup* of G_{K_i} on $\Delta_{X_i}^{\mathrm{ab}} \otimes \mathbb{Q}_l$ is *quasi-unipotent* (i.e., unipotent on an open subgroup of this inertia subgroup), it follows immediately that the weights are *independent* of the choice of lifting.) Now if M is a $\mathbb{Q}_l$-vector space on which Frobenius acts, let us write

$$M^{\mathrm{wt}\ w}$$

for the $\mathbb{Q}_l$-subspace of M on which Frobenius acts with eigenvalues of weight w. Then, setting $I_i \stackrel{\mathrm{def}}{=} \mathrm{Ker}(\Delta_{X_i}^{\mathrm{ab}} \otimes \mathbb{Q}_l \twoheadrightarrow \Delta_{Y_i}^{\mathrm{ab}} \otimes \mathbb{Q}_l)$, we have:

$$\begin{aligned} r_i - 1 &= \dim_{\mathbb{Q}_l}(I_i) = \dim_{\mathbb{Q}_l}(I_i^{\mathrm{wt}\ 1}) \\ &= \dim_{\mathbb{Q}_l}(\Delta_{X_i}^{\mathrm{ab}} \otimes \mathbb{Q}_l)^{\mathrm{wt}\ 1} - \dim_{\mathbb{Q}_l}(\Delta_{Y_i}^{\mathrm{ab}} \otimes \mathbb{Q}_l)^{\mathrm{wt}\ 1} \\ &= \dim_{\mathbb{Q}_l}(\Delta_{X_i}^{\mathrm{ab}} \otimes \mathbb{Q}_l)^{\mathrm{wt}\ 1} - \dim_{\mathbb{Q}_l}(\Delta_{Y_i}^{\mathrm{ab}} \otimes \mathbb{Q}_l)^{\mathrm{wt}\ 0} \\ &= \dim_{\mathbb{Q}_l}(\Delta_{X_i}^{\mathrm{ab}} \otimes \mathbb{Q}_l)^{\mathrm{wt}\ 1} - \dim_{\mathbb{Q}_l}(\Delta_{X_i}^{\mathrm{ab}} \otimes \mathbb{Q}_l)^{\mathrm{wt}\ 0} \end{aligned}$$

(where the fourth equality follows from the auto-duality (up to a Tate twist) of $\Delta_{Y_i}^{\mathrm{ab}} \otimes \mathbb{Q}_l$; and the second and fifth equalities follow from the fact that Frobenius acts on I_i with weight 1). On the other hand, the quantities appearing in the final line of this sequence of equalities are all "group-theoretic". Thus, we conclude that $r_1 = r_2$. Since $\dim_{\mathbb{Q}_l}(\Delta_{X_i}^{\mathrm{ab}} \otimes \mathbb{Q}_l) = 2g_i - 1 + r_i$ (respectively, $= 2g_i$) when $r_i > 0$ (respectively, when $r_i = 0$), this implies that $g_1 = g_2$, as desired.

Finally, the statement concerning preservation of inertia groups *follows formally* from the fact that "r_i is group-theoretic" (by applying this fact to coverings of X_i). Indeed, let l be a *prime number* (possibly equal to p_1 or p_2). Since r_i may be recovered group-theoretically, given any finite étale coverings

$$Z_i \to V_i \to X_i$$

such that Z_i is Galois, of degree a *power of* l, over V_i, one may determine group-theoretically whether or not $Z_i \to V_i$ is "*totally ramified* at a single point of Z_i and unramified elsewhere", since this condition is easily verified to be equivalent to the equality:

$$r_{Z_i} = \deg(Z_i/V_i) \cdot (r_{V_i} - 1) + 1$$

Moreover, the group-theoreticity of this condition extends immediately to the case of *pro-l* coverings $Z_i \to V_i$. Thus, by Lemma 1.3.7, we conclude that *the inertia groups of cusps in* $(\Delta_{X_i})^{(l)}$ (i.e., the maximal pro-l quotient of Δ_{X_i}) *may be characterized (group-theoretically!)* as the maximal subgroups of $(\Delta_{X_i})^{(l)}$ that correspond to (profinite) coverings satisfying this condition.

In particular, (by Lemma 1.3.7) the *set of cusps of* X_i may be reconstructed (group-theoretically!) as the set of $(\Delta_{X_i})^{(l)}$-orbits (relative to the action via conjugation) of such inertia groups in $(\Delta_{X_i})^{(l)}$. Thus, by applying this observation to arbitrary finite étale coverings of X_i, we recover the inertia subgroups of the cusps of Δ_{X_i} as the subgroups that fix *some* cusp of the universal covering $\widetilde{X}_i \to X_i$ of X_i determined by the basepoint in question. This completes the proof. □

Remark 1.3.9.1. As was pointed out to the author by the referee, the argument given in the second paragraph of the proof of Lemma 1.3.9 may be replaced by the more group-theoretic argument of [Tama2], Proposition 2.4.

Positive Characteristic:

For $i = 1, 2$, let k_i be a *finite field* of characteristic p; X_i a *hyperbolic curve* over k_i. Choose a *universal tamely ramified* (i.e., at the punctures of X_i) *covering* $\widetilde{X}_i \to X_i$ of X_i; write

$$\Pi_{X_i}^{\text{tame}} \stackrel{\text{def}}{=} \text{Gal}(\widetilde{X}_i/X_i)$$

for the corresponding *fundamental groups*. Thus, we obtain *exact sequences*:

$$1 \to \Delta_{X_i}^{\text{tame}} \to \Pi_{X_i}^{\text{tame}} \to G_{k_i} \to 1$$

(where G_{k_i} is the absolute Galois group of k_i determined by $\widetilde{X}_i$). As is well-known, the *Frobenius element* determines a natural isomorphism $\widehat{\mathbb{Z}} \cong G_{k_i}$.

Lemma 1.3.10 (Slimness of Fundamental Groups). *For* $i = 1, 2$, *the profinite groups* $\Delta_{X_i}^{\text{tame}}$, $\Pi_{X_i}^{\text{tame}}$ *are* **slim**.

Proof. The slimness of $\Delta_{X_i}^{\text{tame}}$ follows by exactly the same argument — i.e., by considering the action of $\Delta_{X_i}^{\text{tame}}$ on *abelianizations of open subgroups* — as that given in the proof of Lemma 1.3.1. [Alternatively, as was pointed out to the author by the referee, the slimness of $\Delta_{X_i}^{\text{tame}}$ follows from [Tama1], Proposition 1.11 (cf. Remark 0.1.3).]

By a similar argument, the slimness of $\Pi_{X_i}^{\text{tame}}$ follows formally from:

(i) the *slimness* of $\Delta_{X_i}^{\text{tame}}$;

(ii) some positive power of the character of G_{k_i} arising from the determinant of the *prime-to-p portion* of $(\Delta_{X_i}^{\text{tame}})^{\text{ab}}$ coincides with some positive power of the *cyclotomic character.*

(Here, we note (ii) is sufficient to deal with *both* the "l-primary portion" of $\widehat{\mathbb{Z}} \cong G_{k_i}$ (for $l \neq p$) *and* the "p-primary portion" of $\widehat{\mathbb{Z}} \cong G_{k_i}$.) □

Remark 1.3.10.1. Property (ii) in the proof of Lemma 1.3.10 was substantially simplified by a suggestion made to the author by the referee.

The following fundamental result is due to *A. Tamagawa* (cf. [Tama1], Theorem 4.3):

Theorem 1.3.11 (The Grothendieck Conjecture for Affine Hyperbolic Curves over Finite Fields). *Assume, for $i = 1, 2$, that X_i is* **affine**. *Then the natural map*

$$\text{Isom}(\widetilde{X}_1/X_1, \widetilde{X}_2/X_2) \to \text{Isom}(\Pi_{X_1}^{\text{tame}}, \Pi_{X_2}^{\text{tame}})$$

(from scheme-theoretic isomorphisms $\widetilde{X}_1 \xrightarrow{\sim} \widetilde{X}_2$ lying over an isomorphism $X_1 \xrightarrow{\sim} X_2$ to isomorphisms of profinite groups $\Pi_{X_1}^{\text{tame}} \xrightarrow{\sim} \Pi_{X_2}^{\text{tame}}$) is bijective.

Finally, we observe that, just as in the characteristic zero case, *inertia groups of cusps are commensurably terminal*: If x_i is a *cusp* of $(X_i)_{\overline{k}_i} \stackrel{\text{def}}{=} X_i \times_{k_i} \overline{k}_i$, then we have an associated *inertia group* (abstractly isomorphic to $\widehat{\mathbb{Z}}'$)

$$I_{x_i} \subseteq \Delta_{X_i}^{\text{tame}}$$

(well-defined, up to conjugation). If l is any *prime number distinct from* p, then we also obtain an inertia group $I_{x_i}^{(l)} \subseteq (\Delta_{X_i}^{\text{tame}})^{(l)}$ (abstractly isomorphic to $\mathbb{Z}_l$).

Lemma 1.3.12 (Commensurable Terminality of Inertia Groups). *The subgroups $I_{x_i}^{(l)} \subseteq (\Delta_{X_i}^{\text{tame}})^{(l)}$, $I_{x_i} \subseteq \Delta_{X_i}^{\text{tame}}$ are commensurably terminal.*

Proof. The proof is entirely similar to that of Lemma 1.3.7. [Alternatively, as was pointed out to the author by the referee, the assertion concerning $I_{x_i} \subseteq \Delta_{X_i}^{\text{tame}}$ follows from [Tama2], Lemma 2.1, while the assertion concerning $I_{x_i}^{(l)} \subseteq (\Delta_{X_i}^{\text{tame}})^{(l)}$ follows, at least in the case $r \geq 2$, formally from Theorem B′ of [HR].] □

2. Reconstruction of the Logarithmic Special Fiber

For $i = 1, 2$, let K_i be a *finite extension* of $\mathbb{Q}_{p_i}$ (cf. §1.2), and suppose that we are given a *hyperbolic curve* $(X_i)_{K_i}$ over K_i. Let us fix a $\overline{K}_i$-valued basepoint for $(X_i)_{K_i}$ and denote the resulting fundamental group $\pi_1((X_i)_{K_i})$ by $\Pi_{(X_i)_{K_i}}$. Suppose, moreover, that we are given an *isomorphism* $\alpha_X : \Pi_{(X_1)_{K_1}} \xrightarrow{\sim} \Pi_{(X_2)_{K_2}}$, which, by Lemma 1.3.8, necessarily fits into a *commutative diagram*

$$\begin{array}{ccc} \Pi_{(X_1)_{K_1}} & \xrightarrow{\alpha_X} & \Pi_{(X_2)_{K_2}} \\ \downarrow & & \downarrow \\ G_{K_1} & \xrightarrow{\alpha_K} & G_{K_2} \end{array}$$

where the vertical morphisms are the natural ones, and the horizontal morphisms are assumed to be *isomorphisms*. Note that by Proposition 1.2.1, (i), this already implies that $p_1 = p_2$; set $p \stackrel{\text{def}}{=} p_1 = p_2$. That such a diagram *necessarily arises "geometrically"* follows from the main theorem of [Mzk6] (cf. Theorem 1.3.4) — *if one assumes that* α_K *arises geometrically* (i.e., from an isomorphism of fields $K_1 \xrightarrow{\sim} K_2$). In this §, we would like to investigate what one can say *in general* (i.e., without assuming that α_K arises geometrically) concerning this sort of commutative diagram. In some sense, all the key arguments that we use here are *already present in [Mzk4]*, except that there, these arguments were applied to prove *different theorems*. Thus, in the following discussion, we explain how the *same arguments* may be used to prove Theorem 2.7 below.

Let us denote the *type* of the hyperbolic curve $(X_i)_{K_i}$ by (g_i, r_i). Also, we shall denote the *geometric fundamental group* by

$$\Delta_{X_i} \stackrel{\text{def}}{=} \text{Ker}(\Pi_{(X_i)_{K_i}} \to G_{K_i})$$

and the unique proper smooth curve over K_i that *compactifies* $(X_i)_{K_i}$ by $(Y_i)_{K_i}$.

Lemma 2.1 (Group-Theoreticity of Stability). *$(X_1)_{K_1}$ has stable reduction if and only if $(X_2)_{K_2}$ does.*

Proof. This follows (essentially) from the well-known *stable reduction criterion*: That is to say, $(X_i)_{K_i}$ has stable reduction if and only if the inertia subgroup of G_{K_i} acts *unipotently* on $\Delta_{Y_i}^{\text{ab}} \otimes \widehat{\mathbb{Z}}'$ and *trivially* on the (finite) *set of conjugacy classes of inertia groups of cusps in* Δ_{X_i} (i.e., the set of cusps of $(X_i)_{K_i} \otimes_{K_i} \overline{K}_i$ — cf. Lemma 1.3.9). [Note that this condition is *group-theoretic*, by Proposition 1.2.1, (ii).] □

Now let us assume that $(X_i)_{K_i}$ has *stable reduction* over $\mathcal{O}_{K_i}$. Denote the *stable model* of $(X_i)_{K_i}$ over $\mathcal{O}_{K_i}$ by:

$$(\mathcal{X}_i)_{\mathcal{O}_{K_i}} \to \mathrm{Spec}(\mathcal{O}_{K_i})$$

Here, in the case where $r_i > 0$, we mean by the term *"stable model"* the complement of the marked points in the unique stable pointed curve over $\mathcal{O}_{K_i}$ that extends the pointed curve over K_i determined by $(X_i)_{K_i}$. Then, by the theory of [Mzk4], §2, 8, there exists a well-defined *quotient*

$$\Pi_{(X_i)_{K_i}} \twoheadrightarrow \Pi^{\mathrm{adm}}_{(X_i)_{K_i}}$$

whose finite quotients correspond to (subcoverings of) admissible coverings of the result of base-changing $(\mathcal{X}_i)_{\mathcal{O}_{K_i}}$ to rings of integers of tamely ramified extensions of K_i. In particular, we have a natural exact sequence:

$$1 \to \Delta^{\mathrm{adm}}_{X_i} \to \Pi^{\mathrm{adm}}_{(X_i)_{K_i}} \to G^{\log}_{k_i} \to 1$$

(where $\Delta^{\mathrm{adm}}_{X_i}$ is defined so as to make the sequence exact). Moreover, $\Pi^{\mathrm{adm}}_{(X_i)_{K_i}}$ itself admits a natural quotient (cf. [Mzk4], §3)

$$\Pi_{(X_i)_{K_i}} \twoheadrightarrow \Pi^{\mathrm{adm}}_{(X_i)_{K_i}} \twoheadrightarrow \Pi^{\mathrm{et}}_{(X_i)_{K_i}}$$

whose finite quotients correspond to coverings of $(X_i)_{K_i}$ that extend to *finite étale coverings of* $(\mathcal{X}_i)_{\mathcal{O}_{K_i}}$ *which are tamely ramified at the cusps.* In particular, we have a natural exact sequence:

$$1 \to \Delta^{\mathrm{et}}_{X_i} \to \Pi^{\mathrm{et}}_{(X_i)_{K_i}} \to G_{k_i} \to 1$$

(where $\Delta^{\mathrm{et}}_{X_i}$ is defined so as to make the sequence exact).

Lemma 2.2 (Admissible and Étale Quotients). *(i) The profinite groups* $\Pi_{(X_i)_{K_i}}$*,* $\Pi^{\mathrm{adm}}_{(X_i)_{K_i}}$*, and* $\Delta^{\mathrm{adm}}_{X_i}$ *are all* **slim***.*
(ii) The morphism α_X *is* **compatible** *with the quotients*

$$\Pi_{(X_i)_{K_i}} \twoheadrightarrow \Pi^{\mathrm{adm}}_{(X_i)_{K_i}} \twoheadrightarrow \Pi^{\mathrm{et}}_{(X_i)_{K_i}}$$

of $\Pi_{(X_i)_{K_i}}$.

Proof. We begin with *assertion (i).* The slimness of $\Pi_{(X_i)_{K_i}}$ follows formally from Theorem 1.1.1, (ii); Lemma 1.3.1. Next, let us assume that $\Delta^{\mathrm{adm}}_{X_i}$ *has already been shown to be slim.* Then it follows formally from Proposition 1.2.3, (iii), that $\Pi^{\mathrm{adm}}_{(X_i)_{K_i}}$ is slim.

Thus, to complete the proof of assertion (i), it suffices to show that $\Delta^{\text{adm}}_{X_i}$ is *slim*. We argue as in Lemma 1.3.1, i.e., we consider the covering associated to an *open normal subgroup* $H \subseteq \Delta^{\text{adm}}_{X_i}$. To simplify notation, we assume that $(\mathcal{X}_i)_{k_i}$ *is* "that covering"; then it suffices to show (cf. the proof of Lemma 1.3.1) that any automorphism (k_i-linear) σ of $(\mathcal{X}_i)_{k_i}$ which acts *trivially* on $(\Delta^{\text{adm}}_{X_i})^{\text{ab}}$ is itself trivial. We may assume without loss of generality [i.e., by taking H to be sufficiently small] that $(\mathcal{X}_i)_{k_i}$ is *sturdy* (cf. [Mzk4], Definition 1.1) — i.e., that the normalizations of all the geometric irreducible components of $(\mathcal{X}_i)_{k_i}$ have genus ≥ 2. Then by [Mzk4], Proposition 1.3, it follows that σ acts trivially on the set of geometric irreducible components of $(\mathcal{X}_i)_{k_i}$. In particular, σ acts on each such geometric irreducible component individually. Thus, by considering the action of σ on the torsion points of the Jacobians of each these geometric irreducible components of $(\mathcal{X}_i)_{k_i}$ (cf. the proof of Lemma 1.3.1), we conclude that σ acts as the *identity* on each of these geometric irreducible components, as desired.

Next, we turn to assertion (ii). For $\Pi^{\text{adm}}_{(X_i)_{K_i}}$, this follows (essentially) from Proposition 8.4 of [Mzk4] (together with Lemma 1.3.9, which shows that α_X preserves the pro-p inertia groups associated to the cusps in $\Pi_{(X_i)_{K_i}}$). Of course, in [Mzk4], $K_1 = K_2$ and α_K is the identity, but in fact, the only property of α_K necessary for the proof of [Mzk4], Proposition 8.4 — which is, in essence, a formal consequence of [Mzk4], Lemma 8.1 (concerning *unramified quotients* of the p-adic Tate module of a semi-abelian variety over a p-adic local field) — is that α_K *preserve the inertia and wild inertia groups* (which we know, by Proposition 1.2.1, (ii); Proposition 1.2.3, (i), of the present paper).

Similarly, the portion of assertion (ii) concerning $\Pi^{\text{et}}_{(X_i)_{K_i}}$ follows (essentially) from [Mzk4], Proposition 3.2, at least in the case $r_i = 0$. That is to say, even though α_K is not necessarily the identity in the present discussion, the only properties of α_K that are necessary for the proof of [Mzk4], Proposition 3.2, are Proposition 1.2.3, (i); Proposition 1.2.4, (ii) (of the present paper).

Finally, to treat the case of arbitrary $r_i > 0$, we argue as follows: Consider an *open normal subgroup* $H \subseteq \Pi^{\text{adm}}_{(X_i)_{K_i}}$. Then observe that there exists an *open normal subgroup* $H' \subseteq \Pi^{\text{adm}}_{(X_i)_{K_i}}$ such that the orders of the finite groups $\Pi^{\text{adm}}_{(X_i)_{K_i}}/H$, $\Pi^{\text{adm}}_{(X_i)_{K_i}}/H'$ are *relatively prime* and such that the covering of $(\mathcal{X}_i)_{k_i}$ determined by H' is *sturdy*. Moreover, [by the assumption of relative primeness] it follows easily that the covering determined by H is *"of étale type"* (i.e., arises from a quotient of $\Pi^{\text{et}}_{(X_i)_{K_i}}$) if and only if it becomes a covering "of étale type" after base-change via the covering determined by H'. Thus, we conclude that we may assume without loss of generality that $(\mathcal{X}_i)_{k_i}$ is *sturdy*.

Now, let us observe that since $(\mathcal{X}_i)_{k_i}$ is sturdy, it admits admissible coverings "of étale type" with *arbitrarily large prime-to-p ramification at the cusps*. Thus, [since the condition that an admissible covering be "of étale type" amounts to the condition that there be *no ramification at the nodes* or *at the prime of* $\mathcal{O}_{K_i}$] we conclude that the covering of $(\mathcal{X}_i)_{k_i}$ determined by H is of étale type if and only if, for *every* open normal $H'' \subseteq \Pi^{\mathrm{adm}}_{(X_i)_{K_i}}$ which has *ramification index divisible by the prime-to-p portion of the order of the finite group* $\Pi^{\mathrm{adm}}_{(X_i)_{K_i}}/H$ *at all of the cusps*, it becomes a covering of étale type **after base-change** by the covering determined by H''. [Indeed, the necessity of this condition is clear; the sufficiency of this condition follows from the existence of coverings of étale type with arbitrarily large prime-to-p ramification at the cusps. Also, we note that by Lemma 1.3.9, this condition is *preserved* by α_X.] On the other hand, such *base-changed* coverings are [by construction] *necessarily unramified at the cusps*. Thus, the issue of whether or not this base-changed covering is of étale type *reduces* — by Lemma 1.3.9, which shows that α_X preserves the quotient $\Pi_{(X_i)_{K_i}} \twoheadrightarrow \Pi_{(Y_i)_{K_i}}$ — to the "$r_i = 0$" case, which has already been resolved. This completes the proof of assertion (ii) in the case $r_i > 0$. $\square$

Lemma 2.3 (Group-Theoreticity of Dual Semi-Graphs of the Special Fiber). *The morphism* α_X *induces an isomorphism*

$$\alpha_{X,\Gamma^c} : \Gamma^c_{(\mathcal{X}_1)_{k_1}} \xrightarrow{\sim} \Gamma^c_{(\mathcal{X}_2)_{k_2}}$$

between the "dual semi-graphs with compact structure" (i.e., the usual dual graphs $\Gamma_{(\mathcal{X}_i)_{k_i}}$*, together with extra edges corresponding to the cusps — cf. the Appendix) of the special fibers* $(\mathcal{X}_i)_{k_i}$ *of* $(\mathcal{X}_i)_{\mathcal{O}_{K_i}}$*. Moreover,* α_{X,Γ^c} *is functorial with respect to passage to finite étale coverings of the* $(X_i)_{K_i}$.

Proof. Indeed, if one *forgets about the "compact structure"*, then this is a formal consequence of Lemma 1.3.9 [which shows that α_X preserves the quotient $\Pi_{(X_i)_{K_i}} \twoheadrightarrow \Pi_{(Y_i)_{K_i}}$]; Lemma 2.2, (ii), and the theory [concerning the $r_i = 0$ case] of [Mzk4], §1 – 5, summarized in [Mzk4], Corollary 5.3. Even though α_K is not necessarily the identity in the present discussion, the only properties of α_K that are necessary for the proof of [Mzk4], Corollary 5.3 are Proposition 1.2.3, (i); Proposition 1.2.4, (ii) (of the present paper). That is to say, the point is that the *Frobenius element is preserved*, which means that the *weight filtrations* on l-adic cohomology (where l is a prime distinct from p) are, as well.

The *compatibility with the "compact structure"* follows from the pro-l (where $l \neq p$) portion of Lemma 1.3.9, together with the easily verified fact (cf. the proof of Lemma 1.3.7) that the inertia group of a cusp in $\Pi^{\mathrm{adm}}_{(X_i)_{K_i}}$ is contained (up to conjugacy) in the decomposition group of a *unique* irreducible component of $(\mathcal{X}_i)_{k_i}$. □

Next, we would like to show that α_X is necessarily *"of degree 1"*. This is essentially the argument of [Mzk4], Lemma 9.1, but we present this argument in detail below since we are working here under the assumption that α_K is *arbitrary*. For simplicity, *we assume until further notice is given* that:

> ($*^{\mathrm{CSSN}}$) $r_i = 0$ [i.e., the curves are *compact*] and, moreover, the special fiber $(\mathcal{X}_i)_{k_i}$ of $(\mathcal{X}_i)_{\mathcal{O}_{K_i}}$ is *singular* and *sturdy* (cf. [Mzk4], Definition 1.1) — i.e., the normalizations of all the geometric irreducible components of $(\mathcal{X}_i)_{k_i}$ have genus ≥ 2 — and has a *noncontractible dual semi-graph* $\Gamma^{\mathrm{c}}_{(\mathcal{X}_i)_{k_i}}$ — i.e., this semi-graph is *not a tree*.

(These conditions may always be achieved by replacing $(X_i)_{K_i}$ by a finite étale covering of $(X_i)_{K_i}$ — cf. [Mzk4], Lemma 2.9; [Mzk4], the first two paragraphs of the proof of Theorem 9.2.)

We begin by introducing some notation. Write:

$$V_i \stackrel{\mathrm{def}}{=} \Delta^{\mathrm{ab}}_{X_i};$$
$$H_i \stackrel{\mathrm{def}}{=} H_1^{\mathrm{sing}}(\Gamma^{\mathrm{c}}_{(\mathcal{X}_i)_{k_i}}, \mathbb{Z}) = H_1^{\mathrm{sing}}(\Gamma_{(\mathcal{X}_i)_{k_i}}, \mathbb{Z})$$

(where "H_1^{sing}" denotes the first singular homology group). Thus, by considering the coverings of $(\mathcal{X}_i)_{\mathcal{O}_{K_i}}$ induced by unramified coverings of the graph $\Gamma_{(\mathcal{X}_i)_{k_i}}$, we obtain natural (*group-theoretic!*) *"combinatorial quotients"*:

$$V_i \twoheadrightarrow (H_i)_{\widehat{\mathbb{Z}}} \stackrel{\mathrm{def}}{=} H_i \otimes \widehat{\mathbb{Z}}$$

Lemma 2.4 (Ordinary New Parts, after Raynaud). *For a "sufficiently large prime number l" (where "sufficiently large" depends only on p, g_i), and after possibly replacing K_i by a finite unramified extension of K_i, there exists a cyclic étale covering $(\mathcal{Z}_i)_{\mathcal{O}_{K_i}} \to (\mathcal{X}_i)_{\mathcal{O}_{K_i}}$ of $(\mathcal{X}_i)_{\mathcal{O}_{K_i}}$ of degree l such that the "new part" $V_i^{\mathrm{new}} \stackrel{\mathrm{def}}{=} \Delta^{\mathrm{ab}}_{(Z_i)_{K_i}}/\Delta^{\mathrm{ab}}_{(X_i)_{K_i}}$ of the abelianized geometric fundamental group of $(Z_i)_{K_i}$ satisfies the following:*

(i) *We have an exact sequence:*

$$0 \to V_i^{\mathrm{mlt}} \to (V_i^{\mathrm{new}})_{\mathbb{Z}_p} \stackrel{\mathrm{def}}{=} V_i^{\mathrm{new}} \otimes_{\widehat{\mathbb{Z}}} \mathbb{Z}_p \to V_i^{\mathrm{etl}} \to 0$$

— *where* V_i^{etl} *is an* **unramified** G_{K_i}*-module, and* V_i^{mlt} *is the* **Cartier dual** *of an unramified* G_{K_i}*-module.*

(ii) *The "combinatorial quotient" of* $\Delta^{\text{ab}}_{(Z_i)_{K_i}}$ *(arising from the coverings of* $(\mathcal{Z}_i)_{\mathcal{O}_{K_i}}$ *induced by unramified coverings of the dual semi-graph of the special fiber of* $(\mathcal{Z}_i)_{\mathcal{O}_{K_i}}$*) induces a* **nonzero** *quotient* $V_i^{\text{new}} \twoheadrightarrow (H_i^{\text{new}})_{\widehat{\mathbb{Z}}}$ *of* V_i^{new}.

Here, the injection $\Delta^{\text{ab}}_{(X_i)_{K_i}} \hookrightarrow \Delta^{\text{ab}}_{(Z_i)_{K_i}}$ *is the injection induced by pull-back via* $(Z_i)_{K_i} \to (X_i)_{K_i}$ *and Poincaré duality (or, alternatively, by the "Verlagerung").*

Proof. Note that since both conditions (i), (ii) are *group-theoretic*, they may be realized *simultaneously* for $i = 1, 2$. Now to satisfy *condition (i)*, it suffices — cf., e.g., the discussion in [Mzk4], §8, of "V_G", "$V_{G^{\text{ord}}}$" — to choose the covering so that the "new parts" of the Jacobians of the irreducible components of the special fiber of $(\mathcal{Z}_i)_{\mathcal{O}_{K_i}}$ are all *ordinary*. That this is possible for l sufficiently large is a consequence of a *theorem of Raynaud* (as formulated, for instance, in [Tama1], Lemma 1.9). Then let us observe that, so long as we choose the étale covering $(\mathcal{Z}_i)_{\mathcal{O}_{K_i}} \to (\mathcal{X}_i)_{\mathcal{O}_{K_i}}$ so that it is *nontrivial* over every irreducible component of $(\mathcal{X}_i)_{k_i}$, *condition (ii) is automatically satisfied*: Indeed, if we write $h_i \stackrel{\text{def}}{=} \text{rk}_{\mathbb{Z}}(H_i)$ — so $h_i > 0$ since $\Gamma^{\text{c}}_{(\mathcal{X}_i)_{k_i}}$ is assumed to be *noncontractible* — then to assert that condition (ii) fails to hold — i.e., that there are *"no new cycles in the dual graph"* — is to assert that we have an *equality of Euler characteristics*:

$$(\sum_j g_{Z,j}) + h_i - 1 = l\{(\sum_j g_{X,j}) + h_i - 1\}$$

(where the first (respectively, second) sum is the sum of the genera of the irreducible components of the geometric special fiber of $(\mathcal{Z}_i)_{\mathcal{O}_{K_i}}$ (respectively, $(\mathcal{X}_i)_{\mathcal{O}_{K_i}}$)). But, since

$$\sum_j (g_{Z,j} - 1) = \sum_j l(g_{X,j} - 1)$$

we thus conclude that $(l-1) = \{\sum_j (l-1)\} + h_i(l-1)$, hence that $1 = (\sum_j 1) + h_i$ — which is *absurd*, since both the sum and h_i are ≥ 1. This completes the proof. $\square$

Remark 2.4.1. The author would like to thank A. Tamagawa for explaining to him the utility of Raynaud's theorem in this sort of situation.

In the following discussion, to keep the notation simple, *we shall replace* $(X_i)_{K_i}$ by some $(Z_i)_{K_i}$ as in Lemma 2.4. Thus, V_i^{new} is a G_{K_i}-*quotient module* of V_i. Moreover, we have a *surjection*

$$V_i^{\text{new}} \twoheadrightarrow (H_i^{\text{new}})_{\widehat{\mathbb{Z}}}$$

such that the quotient $(H_i)_{\widehat{\mathbb{Z}}} \twoheadrightarrow (H_i^{\text{new}})_{\widehat{\mathbb{Z}}}$ is *defined over* $\mathbb{Z}$, i.e., arises from a quotient $H_i \twoheadrightarrow H_i^{\text{new}}$. (Indeed, this last assertion follows from the fact that the quotient $H_i \twoheadrightarrow H_i^{\text{new}}$ arises as the cokernel (modulo torsion) of the morphism induced on first singular cohomology modules by a finite (ramified) covering of graphs — i.e., the covering induced on dual graphs by the covering $(\mathcal{Z}_i)_{\mathcal{O}_{K_i}} \to (\mathcal{X}_i)_{\mathcal{O}_{K_i}}$ of Lemma 2.4.)

On the other hand, the *cup product* on group cohomology gives rise to a *nondegenerate (group-theoretic!) pairing*

$$V_i^\vee \otimes_{\widehat{\mathbb{Z}}} V_i^\vee \otimes_{\widehat{\mathbb{Z}}} \boldsymbol{\mu}_{\widehat{\mathbb{Z}}}(\overline{K}_i) \to M_i \stackrel{\text{def}}{=} H^2(\Delta_{X_i}, \boldsymbol{\mu}_{\widehat{\mathbb{Z}}}(\overline{K}_i)) \ (\cong \widehat{\mathbb{Z}})$$

(where we think of $V_i^\vee \stackrel{\text{def}}{=} \text{Hom}(V_i, \widehat{\mathbb{Z}})$ as $H^1(\Delta_{X_i}, \widehat{\mathbb{Z}})$), hence, by restriction to $(V_i^{\text{new}})^\vee \hookrightarrow V_i^\vee$, a *pairing*

$$(V_i^{\text{new}})^\vee \otimes_{\widehat{\mathbb{Z}}} (V_i^{\text{new}})^\vee \otimes_{\widehat{\mathbb{Z}}} \boldsymbol{\mu}_{\widehat{\mathbb{Z}}}(\overline{K}_i) \to M_i \stackrel{\text{def}}{=} H^2(\Delta_{X_i}, \boldsymbol{\mu}_{\widehat{\mathbb{Z}}}(\overline{K}_i)) \ (\cong \widehat{\mathbb{Z}})$$

which is still *nondegenerate* (over $\mathbb{Q}$), since it arises from an *ample line bundle* — namely, the restriction of the polarization determined by the theta divisor on the Jacobian of $(X_i)_{K_i}$ to the "new part" of $(X_i)_{K_i}$. This pairing determines an *"isogeny"* (i.e., a morphism which is an isomorphism over $\mathbb{Q}$):

$$(V_i^{\text{new}})^\vee \otimes_{\widehat{\mathbb{Z}}} \boldsymbol{\mu}_{\widehat{\mathbb{Z}}}(\overline{K}_i) \otimes_{\widehat{\mathbb{Z}}} M_i^\vee \hookrightarrow V_i^{\text{new}}$$

Thus, if we take the dual of the surjection discussed in the preceding paragraph, then we obtain an inclusion

$$(H_i^{\text{new}})_{\widehat{\mathbb{Z}}}^\vee \otimes \boldsymbol{\mu}_{\widehat{\mathbb{Z}}}(\overline{K}_i) \otimes M_i^\vee \hookrightarrow (V_i^{\text{new}})^\vee \otimes \boldsymbol{\mu}_{\widehat{\mathbb{Z}}}(\overline{K}_i) \otimes M_i^\vee \hookrightarrow V_i^{\text{new}}$$

which (as one sees, for instance, by applying the fact that $\boldsymbol{\mu}_{\widehat{\mathbb{Z}}}(\overline{K}_i)^{G_{K_i}} = 0$) maps into the kernel of the surjection $V_i^{\text{new}} \twoheadrightarrow (H_i^{\text{new}})_{\widehat{\mathbb{Z}}}$.

Next, let us observe that the *kernel* N_i of the *surjection of unramified* G_{K_i}*-modules* (i.e., G_{k_i}-modules)

$$V_i^{\text{etl}} \twoheadrightarrow (H_i^{\text{new}})_{\mathbb{Z}_p}$$

satisfies:

$$H^0(G_{k_i}, N_i \otimes \mathbb{Q}_p) = H^1(G_{k_i}, N_i \otimes \mathbb{Q}_p) = 0$$

(Indeed, N_i arises as a submodule of the module of p-power torsion points of an abelian variety over k_i, so the vanishing of these cohomology groups follows from the *Riemann Hypothesis for abelian varieties over finite fields* (cf., e.g., [Mumf], p. 206), i.e., the fact that (some power of) the Frobenius element of G_{k_i} acts on N_i with eigenvalues which are algebraic numbers with complex absolute values equal to a *nonzero* rational power of p.) In particular, we conclude that the above surjection admits a *unique* G_{K_i}*-equivariant splitting* $(H_i^{\text{new}})_{\mathbb{Z}_p} \hookrightarrow (V_i^{\text{etl}})_{\mathbb{Q}_p}$. Similarly, (by taking Cartier duals) the injection $(H_i^{\text{new}})^{\vee}_{\mathbb{Z}_p} \otimes \boldsymbol{\mu}_{\mathbb{Z}_p}(\overline{K}_i) \otimes M_i^{\vee} \hookrightarrow V_i^{\text{mlt}}$ also admits a unique G_{K_i}-equivariant splitting over $\mathbb{Q}_p$. Thus, *by applying these splittings*, we see that the G_{K_i}-action on $(V_i^{\text{new}})_{\mathbb{Z}_p}$ determines a *p-adic extension class*

$$\begin{aligned}(\eta_i)_{\mathbb{Z}_p} &\in \{(H_i^{\text{new}})^{\vee}_{\mathbb{Q}_p}\}^{\otimes 2} \otimes M_i^{\vee} \otimes (H^1(K_i, \boldsymbol{\mu}_{\widehat{\mathbb{Z}}}(\overline{K}_i))/H^1_f(K_i, \boldsymbol{\mu}_{\widehat{\mathbb{Z}}}(\overline{K}_i))) \\ &= \{(H_i^{\text{new}})^{\vee}_{\mathbb{Q}_p}\}^{\otimes 2} \otimes M_i^{\vee}\end{aligned}$$

where (by Proposition 1.2.1, (vii)) $H^1(K_i, \boldsymbol{\mu}_{\widehat{\mathbb{Z}}}(\overline{K}_i))$ may be identified with $(K_i^{\times})^{\wedge}$, and we define

$$H^1_f(K_i, \boldsymbol{\mu}_{\widehat{\mathbb{Z}}}(\overline{K}_i)) \stackrel{\text{def}}{=} \mathcal{O}^{\times}_{K_i} \subseteq (K_i^{\times})^{\wedge} \xrightarrow{\sim} H^1(K_i, \boldsymbol{\mu}_{\widehat{\mathbb{Z}}}(\overline{K}_i))$$

so the quotient group $(H^1(K_i, \boldsymbol{\mu}_{\widehat{\mathbb{Z}}}(\overline{K}_i))/H^1_f(K_i, \boldsymbol{\mu}_{\widehat{\mathbb{Z}}}(\overline{K}_i)))$ may be identified with $\widehat{\mathbb{Z}}$.

Next, let us observe that the kernel N_i' of $(V_i^{\text{new}})_{\widehat{\mathbb{Z}}'} \twoheadrightarrow (H_i^{\text{new}})_{\widehat{\mathbb{Z}}'}$ is an *unramified* representation of G_{K_i} (since it arises from the module of prime-to-p-power torsion points of a semi-abelian variety over k_i). Moreover, the injection of *unramified* G_{K_i}*-modules*

$$(H_i^{\text{new}})^{\vee}_{\widehat{\mathbb{Z}}'} \otimes \boldsymbol{\mu}_{\widehat{\mathbb{Z}}'}(\overline{K}_i) \otimes M_i^{\vee} \hookrightarrow N_i'$$

splits uniquely over $\mathbb{Q}$, since (by the Riemann Hypothesis for abelian varieties over finite fields — cf., e.g., [Mumf], p. 206) the Frobenius element of G_{k_i} acts on the smaller module (respectively, quotient by this smaller module) with *weight* 1 (respectively, $\frac{1}{2}$). Thus, just as in the p-adic case, we may construct a *prime-to-p-adic extension class* $(\eta_i)_{\widehat{\mathbb{Z}}'}$ from the G_{K_i}-action on $(V_i^{\text{new}})_{\widehat{\mathbb{Z}}'}$, which, together with $(\eta_i)_{\mathbb{Z}_p}$, yields an *extension class* (cf. [FC], Chapter III, Corollary 7.3):

$$\begin{aligned}\eta_i &\in \{(H_i^{\text{new}})^{\vee}_{\widehat{\mathbb{Z}}}\}^{\otimes 2} \otimes M_i^{\vee} \otimes \{H^1(K_i, \boldsymbol{\mu}_{\widehat{\mathbb{Z}}}(\overline{K}_i))/H^1_f(K_i, \boldsymbol{\mu}_{\widehat{\mathbb{Z}}}(\overline{K}_i))\} \otimes \mathbb{Q} \\ &= \{(H_i^{\text{new}})^{\vee}_{\widehat{\mathbb{Z}}}\}^{\otimes 2} \otimes M_i^{\vee} \otimes \mathbb{Q}\end{aligned}$$

That is to say, η_i may be thought of as a (*group-theoretically reconstructible!*) bilinear form:

$$\langle -, - \rangle_i : (H_i^{\text{new}})^{\otimes 2}_{\widehat{\mathbb{Z}}} \to (M_i^{\vee})_{\mathbb{Q}} \stackrel{\text{def}}{=} M_i^{\vee} \otimes \mathbb{Q}$$

Moreover:

Lemma 2.5. *Assume that* $(\mathcal{X}_i)_{\mathcal{O}_{K_i}}$ *arises as some "*$(\mathcal{Z}_i)_{\mathcal{O}_{K_i}}$*" as in Lemma 2.4. Then:*
(i) **(Positive Rational Structures)** *The image of* $(H_i^{\text{new}})^{\otimes 2}$ *under the morphism* $(H_i^{\text{new}})^{\otimes 2}_{\widehat{\mathbb{Z}}} \to (M_i^{\vee})_{\mathbb{Q}}$ *forms a rank one* $\mathbb{Z}$*-submodule of* $(M_i^{\vee})_{\mathbb{Q}}$*. Moreover, for any two nonzero elements* $a, b \in H_i$*,* $\langle a, a\rangle_i$ *differs from* $\langle b, b\rangle_i$ *by a factor in* $\mathbb{Q}_{>0}$ *(i.e., a positive rational number). In particular, this image determines a "*$\mathbb{Q}_{>0}$*-structure" on* $(M_i^{\vee})_{\mathbb{Q}}$*, i.e., a* $\mathbb{Q}$*-rational structure on* $(M_i^{\vee})_{\mathbb{Q}}$*, together with a collection of generators of this* $\mathbb{Q}$*-rational structure that differ from one another by factors in* $\mathbb{Q}_{>0}$*. Finally, this* $\mathbb{Q}_{>0}$*-structure is the same as the* $\mathbb{Q}_{>0}$*-structure on* $M_i^{\vee}$ *determined by the first Chern class of an ample line bundle on* $(X_i)_{K_i}$ *in* $M_i = H^2(\Delta_{X_i}, \boldsymbol{\mu}_{\widehat{\mathbb{Z}}}(\overline{K}_i))$*.*
(ii) **(Preservation of Degree)** *The isomorphism*

$$M_1 = H^2(\Delta_{X_1}, \boldsymbol{\mu}_{\widehat{\mathbb{Z}}}(\overline{K}_1)) \xrightarrow{\sim} H^2(\Delta_{X_2}, \boldsymbol{\mu}_{\widehat{\mathbb{Z}}}(\overline{K}_2)) = M_2$$

induced by α_X *preserves the elements on both sides determined by the first Chern class of a line bundle on* $(X_i)_{K_i}$ *of degree* 1*.*

Proof. Indeed, assertion (i), follows formally from [FC], Chapter III, Corollary 7.3, and Theorem 10.1, (iii) (by considering "new part" of the Jacobian of $(X_i)_{K_i}$ equipped with the polarization induced by the theta polarization on the Jacobian).

As for assertion (ii), the elements in question are the *unique* elements that, on the one hand, are *rational* and *positive* with respect to the structures discussed in assertion (i), and, on the other hand, *generate* M_i as a $\widehat{\mathbb{Z}}$-module. □

Remark 2.5.1. Note that *the conclusion of Lemma 2.5, (ii), is valid not just for* $(X_i)_{K_i}$*, but for* **any** *finite étale cover of the original* $(X_i)_{K_i}$, i.e., even if this cover does not arise as some "$(\mathcal{Z}_i)_{\mathcal{O}_{K_i}}$" as in Lemma 2.4. Indeed, this follows from the fact that the *crucial "*$\mathbb{Q}_{>0}$*-structure"* of Lemma 2.5, (i), is preserved by pull-back to such a cover, which just multiplies the Chern class at issue in Lemma 2.5, (ii), by the degree of the cover (an element of $\mathbb{Q}_{>0}$!).

Remark 2.5.2. In the discussion of [Mzk4], §9, it was not necessary to be as careful as we were in the discussion above in constructing the p-adic class $(\eta_i)_{\mathbb{Z}_p}$ (i.e., "μ_p" in the notation of *loc. cit.*). This is because in *loc. cit.*, we were working over a *single p-adic base-field "K"*. In this more restricted context, the extension class $(\eta_i)_{\mathbb{Z}_p}$ may be *extracted much more easily* from V_i by simply forming the quotient by the submodule of $H^1(K_i, \mathrm{Ker}((V_i)_{\mathbb{Z}_p} \twoheadrightarrow (H_i)_{\mathbb{Z}_p}))$ generated by the elements which are *"crystalline"*, or, more simply, of *"geometric origin"* (i.e., arise from $\mathcal{O}_K$-rational points of the formal group associated to the p-divisible group determined — via "Tate's theorem" (cf. Theorem 4 of [Tate]) — by the G_{K_i}-module $\mathrm{Ker}((V_i)_{\mathbb{Z}_p} \twoheadrightarrow (H_i)_{\mathbb{Z}_p}))$. Unfortunately, the author omitted a detailed discussion of this aspect of the argument in the discussion of [Mzk4], §9.

Remark 2.5.3. Relative to Remark 2.5.2, we note nevertheless that even in the discussion of [Mzk4], §9, it is still necessary to work (at least until one recovers the "$\mathbb{Q}_{>0}$-structure" — cf. Remark 2.5.1) with $(X_i)_{K_i}$ such that the dual graph of the special fiber $(\mathcal{X}_i)_{k_i}$ is *noncontractible.* This minor technical point was omitted in the discussion of [Mzk4], §9.

Next, let us write $(\mathcal{X}_i^{\log})_{\mathcal{O}_{K_i}}$ for the log scheme obtained by equipping $(\mathcal{X}_i)_{\mathcal{O}_{K_i}}$ with the log structure determined by the monoid of regular functions $\in \mathcal{O}_{(\mathcal{X}_i)_{\mathcal{O}_{K_i}}}$ which are invertible on the open subscheme $(X_i)_{K_i} \subseteq (\mathcal{X}_i)_{\mathcal{O}_{K_i}}$. Thus, in the terminology of [Kato2], $(\mathcal{X}_i^{\log})_{\mathcal{O}_{K_i}}$ is *log regular.* Also, let us write $(\mathcal{X}_i^{\log})_{k_i}$ for the log scheme obtained by equipping $(\mathcal{X}_i)_{k_i}$ with the log structure determined by restricting the log structure of $(\mathcal{X}_i^{\log})_{\mathcal{O}_{K_i}}$. Thus, the quotient $\Pi_{(X_i)_{K_i}} \twoheadrightarrow \Pi^{\mathrm{adm}}_{(X_i)_{K_i}}$ determines a *"universal admissible covering"*

$$(\widetilde{\mathcal{X}}_i^{\log})_{k_i} \to (\mathcal{X}_i^{\log})_{k_i}$$

of $(\mathcal{X}_i^{\log})_{k_i}$.

Now let us choose a *connected component* $\widetilde{\mathcal{I}}_i$ of the k_i-smooth locus (i.e., the complement of the nodes) of $(\widetilde{\mathcal{X}}_i^{\log})_{k_i}$. Write $\mathcal{I}_i \subseteq (\mathcal{X}_i)_{k_i}$ for the image of $\widetilde{\mathcal{I}}_i$ in $(\mathcal{X}_i)_{k_i}$. Thus,

$$\widetilde{\mathcal{I}}_i \to \mathcal{I}_i$$

is a *"tame universal covering"* of $\mathcal{I}_i$ (i.e., a universal covering of the hyperbolic curve $\mathcal{I}_i$ among those finite étale coverings that are tamely ramified at the "cusps" of this hyperbolic curve). In the following discussion, we shall also assume, for simplicity, that $\mathcal{I}_i$ is *geometrically connected* over k_i (a condition that may always be achieved by replacing K_i by a finite unramified extension of K_i).

Now the Galois group $\Pi_{\mathcal{I}_i}$ of this covering may also be thought of as the *quotient of the decomposition group in* $\Pi^{\mathrm{adm}}_{(X_i)_{K_i}}$ *of* $\widetilde{\mathcal{I}}_i$ *by its inertia group.* In particular, since $\Pi_{\mathcal{I}_i}$ is formed by taking the *quotient* by this inertia group, it follows that the surjection $\Pi^{\mathrm{adm}}_{(X_i)_{K_i}} \twoheadrightarrow G^{\log}_{k_i}$ induces a natural surjection

$$\Pi_{\mathcal{I}_i} \twoheadrightarrow G_{k_i}$$

whose kernel is the *geometric (tame) fundamental group* $\pi_1^{\text{tame}}((\mathcal{I}_i)_{\overline{k}_i})$ of $\mathcal{I}_i$.

Finally, we observe that *it makes sense to speak of* $\widetilde{\mathcal{I}}_1$ *and* $\widetilde{\mathcal{I}}_2$ *as corresponding via* α_X. Indeed, by Lemma 2.3, α_X induces an isomorphism between the *pro-graphs* determined by the $(\widetilde{\mathcal{X}}_i^{\log})_{k_i}$. Thus, the $\widetilde{\mathcal{I}}_i$ may be said to *correspond via* α_X when the vertices that they determine in these pro-graphs correspond. Moreover, when the $\widetilde{\mathcal{I}}_i$ correspond via α_X, it follows (by considering the stabilizer of the vertex determined by $\widetilde{\mathcal{I}}_i$) that α_X induces a bijection between the respective *decomposition groups* $\mathcal{D}_i$ in $\Pi^{\text{adm}}_{(X_i)_{K_i}}$ of $\widetilde{\mathcal{I}}_i$, as well as between the respective *inertia subgroups* of these decomposition groups $\mathcal{D}_i$ (which may be characterized group-theoretically as the *centers* of the subgroups $\mathcal{D}_i \bigcap \text{Ker}(\Pi^{\text{adm}}_{(X_i)_{K_i}} \twoheadrightarrow G_{k_i})$, since $\pi_1^{\text{tame}}((\mathcal{I}_i)_{\overline{k}_i})$ is *center-free* — cf. Lemma 1.3.10). Thus, in summary, α_X induces a *commutative diagram:*

$$\begin{array}{ccc} \Pi_{\mathcal{I}_1} & \xrightarrow{\sim} & \Pi_{\mathcal{I}_2} \\ \downarrow & & \downarrow \\ G_{k_1} & \xrightarrow{\sim} & G_{k_2} \end{array}$$

We are now ready (cf. [Mzk4], §7) to apply the *main result* of [Tama1]. This result states that commutative diagrams as above are in natural bijective correspondence with commutative diagrams

$$\begin{array}{ccc} \widetilde{\mathcal{I}}_1 & \xrightarrow{\sim} & \widetilde{\mathcal{I}}_2 \\ \downarrow & & \downarrow \\ \mathcal{I}_1 & \xrightarrow{\sim} & \mathcal{I}_2 \end{array}$$

lying over commutative diagrams

$$\begin{array}{ccc} \overline{k}_1 & \xrightarrow{\sim} & \overline{k}_2 \\ \downarrow & & \downarrow \\ k_1 & \xrightarrow{\sim} & k_2 \end{array}$$

(cf. Theorem 1.3.11). In particular, these commutative diagrams induce an isomorphism

$$(\widehat{\mathbb{Z}}' \cong)\ H^2_{\text{c}}((\mathcal{I}_1)_{\overline{k}_1}, \mu_{\widehat{\mathbb{Z}}'}(\overline{k}_1)) \xrightarrow{\sim} H^2_{\text{c}}((\mathcal{I}_2)_{\overline{k}_2}, \mu_{\widehat{\mathbb{Z}}'}(\overline{k}_2))\ (\cong \widehat{\mathbb{Z}}')$$

(where "H^2_{c}" denotes étale cohomology with compact supports — cf. [Milne], Chapter III, Proposition 1.29; Remark 1.30) which maps the element "1" (i.e., the element determined by the first Chern class of a line bundle of degree 1) on the left to the element "1" on the right. (Indeed, this follows from the fact that the morphism $\mathcal{I}_1 \xrightarrow{\sim} \mathcal{I}_2$ appearing in the above commutative diagram is an isomorphism, hence of degree 1.) Note that the isomorphism $\mu_{\widehat{\mathbb{Z}}'}(\overline{k}_1) \xrightarrow{\sim} \mu_{\widehat{\mathbb{Z}}'}(\overline{k}_2)$ that we use here is that obtained from the commutative diagram above, i.e., that provided by Theorem 1.3.11.

Lemma 2.6 (Compatibility of Isomorphisms Between Roots of Unity). *Assume that $(\mathcal{X}_i)_{\mathcal{O}_{K_i}}$ arises as some "$(\mathcal{Z}_i)_{\mathcal{O}_{K_i}}$" as in Lemma 2.4. Then the following diagram*

$$\begin{array}{ccc} \mu_{\widehat{\mathbb{Z}}'}(\overline{k}_1) & \xrightarrow{\sim} & \mu_{\widehat{\mathbb{Z}}'}(\overline{k}_2) \\ \downarrow & & \downarrow \\ \mu_{\widehat{\mathbb{Z}}'}(\overline{K}_1) & \xrightarrow{\sim} & \mu_{\widehat{\mathbb{Z}}'}(\overline{K}_2) \end{array}$$

— in which the vertical morphisms are the natural ones (obtained by considering Teichmüller representatives); the upper horizontal morphism is the morphism determined by Theorem 1.3.11; and the lower horizontal morphism is the morphism determined by Proposition 1.2.1, (vi) — **commutes**.

Proof. Indeed, the diagram in the statement of Lemma 2.6 induces a diagram:

$$\begin{array}{ccc} H^2_c((\mathcal{I}_1)_{\overline{k}_1}, \mu_{\widehat{\mathbb{Z}}'}(\overline{k}_1)) & \xrightarrow{\sim} & H^2_c((\mathcal{I}_2)_{\overline{k}_2}, \mu_{\widehat{\mathbb{Z}}'}(\overline{k}_2)) \\ \downarrow & & \downarrow \\ H^2_c((\mathcal{I}_1)_{\overline{k}_1}, \mu_{\widehat{\mathbb{Z}}'}(\overline{K}_1)) & \xrightarrow{\sim} & H^2_c((\mathcal{I}_2)_{\overline{k}_2}, \mu_{\widehat{\mathbb{Z}}'}(\overline{K}_2)) \end{array} \quad \ldots (*_1)$$

Moreover, we have a diagram

$$\begin{array}{ccc} H^2_c((\mathcal{I}_1)_{\overline{k}_1}, \mu_{\widehat{\mathbb{Z}}'}(\overline{K}_1)) & \xrightarrow{\sim} & H^2_c((\mathcal{I}_2)_{\overline{k}_2}, \mu_{\widehat{\mathbb{Z}}'}(\overline{K}_2)) \\ \downarrow & & \downarrow \\ H^2((X_1)_{\overline{K}_1}, \mu_{\widehat{\mathbb{Z}}'}(\overline{K}_1)) & \xrightarrow{\sim} & H^2((X_2)_{\overline{K}_2}, \mu_{\widehat{\mathbb{Z}}'}(\overline{K}_2)) \end{array} \quad \ldots (*_2)$$

where the *horizontal morphisms* are induced by α_K (cf. Proposition 1.2.1, (vi)), and the *vertical morphisms* are induced "group-theoretically" as follows: First, observe that [since we continue to *operate under the assumption* $(*^{CSSN})$]

$$H^2((X_i)_{\overline{K}_i}, \mu_{\widehat{\mathbb{Z}}'}(\overline{K}_i)) \cong H^2(\Delta_{X_i}, \mu_{\widehat{\mathbb{Z}}'}(\overline{K}_i)) \cong H^2(\Delta^{\mathrm{adm}}_{X_i}, \mu_{\widehat{\mathbb{Z}}'}(\overline{K}_i))$$

while

$$H^2_c((\mathcal{I}_i)_{\overline{k}_i}, \mu_{\widehat{\mathbb{Z}}'}(\overline{K}_i)) \cong H^2((\overline{\mathcal{I}}_i)_{\overline{k}_i}, \mu_{\widehat{\mathbb{Z}}'}(\overline{K}_i)) \cong H^2(\pi_1((\overline{\mathcal{I}}_i)_{\overline{k}_i}), \mu_{\widehat{\mathbb{Z}}'}(\overline{K}_i))$$

(where we write $\overline{\mathcal{I}}_i$ for the unique nonsingular compactification of $\mathcal{I}_i$). Moreover, since we continue to *operate under the assumption* $(*^{CSSN})$, it follows (cf. the discussion of "Second Cohomology Groups" in the Appendix) that the natural *"push-forward" morphism in étale cohomology*

$$(H^2(\pi_1((\overline{\mathcal{I}}_i)_{\overline{k}_i}), \mu_{\widehat{\mathbb{Z}}'}(\overline{K}_i)) \cong) \ H^2_c((\mathcal{I}_i)_{\overline{k}_i}, \mu_{\widehat{\mathbb{Z}}'}(\overline{K}_i)) \\ \to H^2((\mathcal{X}_i)_{k_i} \otimes_{k_i} \overline{k}_i, \mu_{\widehat{\mathbb{Z}}'}(\overline{K}_i)) \ (\cong H^2(\Delta^{\text{et}}_{X_i}, \mu_{\widehat{\mathbb{Z}}'}(\overline{K}_i)))$$

may be *reconstructed group-theoretically* (by using the various natural homomorphisms $\pi_1((\overline{\mathcal{I}}_i)_{\overline{k}_i}) \to \Delta^{\text{et}}_{X_i}$ [well-defined up to composite with an inner automorphism]). Thus, the *desired vertical morphisms* of diagram $(*_2)$ may be obtained by composing these "push-forward" morphisms with the morphisms

$$H^2(\Delta^{\text{et}}_{X_i}, \mu_{\widehat{\mathbb{Z}}'}(\overline{K}_i)) \to H^2(\Delta^{\text{adm}}_{X_i}, \mu_{\widehat{\mathbb{Z}}'}(\overline{K}_i))$$

induced by the surjections $\Delta^{\text{adm}}_{X_i} \twoheadrightarrow \Delta^{\text{et}}_{X_i}$ [which are group-theoretic by Lemma 2.2, (ii)]. In particular, we thus see that the vertical morphisms of diagram $(*_2)$ are *group-theoretic*, i.e., (in other words) *diagram* $(*_2)$ *commutes*.

Now let us compose the above two diagrams $(*_1)$, $(*_2)$ to form a single diagram:

$$\begin{array}{ccc} H^2_c((\mathcal{I}_1)_{\overline{k}_1}, \mu_{\widehat{\mathbb{Z}}'}(\overline{k}_1)) & \stackrel{\sim}{\to} & H^2_c((\mathcal{I}_2)_{\overline{k}_2}, \mu_{\widehat{\mathbb{Z}}'}(\overline{k}_2)) \\ \downarrow & & \downarrow \\ H^2((X_1)_{\overline{K}_1}, \mu_{\widehat{\mathbb{Z}}'}(\overline{K}_1)) & \stackrel{\sim}{\to} & H^2((X_2)_{\overline{K}_2}, \mu_{\widehat{\mathbb{Z}}'}(\overline{K}_2)) \end{array} \qquad \ldots (*_3)$$

Note that *this diagram* $(*_3)$ *commutes*, since, by Lemma 2.5, (ii) [applied to the lower horizontal morphism of $(*_3)$]; the discussion immediately preceding the present Lemma 2.6 [applied to the upper horizontal morphism of $(*_3)$]; and the discussion of "Second Cohomology Groups" in the Appendix [applied to the vertical morphisms of $(*_3)$], all of the morphisms of this diagram are compatible with the elements "1" determined by the first Chern class of a line bundle of degree 1. But this implies that *diagram* $(*_1)$ *commutes* (since diagram $(*_2)$ has already been shown to be commutative, and all the arrows in both of these diagrams $(*_1)$, $(*_2)$ are isomorphisms between rank one free $\widehat{\mathbb{Z}}'$-modules). On the other hand, since diagram $(*_1)$ was obtained by applying the functors $H^2_c((\mathcal{I}_i)_{\overline{k}_i}, -)$ (which are manifestly *faithful*, at least when restricted to constant coefficient modules which are finitely generated and free over $\widehat{\mathbb{Z}}'$ — as in the present situation) to the diagram appearing in the statement of Lemma 2.6, we thus conclude that the diagram appearing in the statement of Lemma 2.6 is commutative, as desired. □

The significance of Lemma 2.6 from our point of view is the following: Lemma 2.6 implies that α_K *induces an isomorphism*

$$\alpha_{G_k^{\log}} : G_{k_1}^{\log} \xrightarrow{\sim} G_{k_2}^{\log}$$

which satisfies the hypothesis of Proposition 1.2.4, (iii). Thus, we conclude from Proposition 1.2.4, (iii), that $\alpha_{G_k^{\log}}$ *arises geometrically.* In particular, it follows that *we may apply [Mzk4], Theorem 7.2,* to the commutative diagram

$$\begin{array}{ccc} \Pi^{\mathrm{adm}}_{(X_1)_{K_1}} & \xrightarrow{\sim} & \Pi^{\mathrm{adm}}_{(X_2)_{K_2}} \\ \downarrow & & \downarrow \\ G_{k_1}^{\log} & \xrightarrow{\sim} & G_{k_2}^{\log} \end{array}$$

(where we note that Lemma 2.6 also implies — when translated into the terminology of [Mzk4], §7 — that the *"RT-degree"* associated to this commutative diagram is 1, as is necessary for the application of [Mzk4], Theorem 7.2). In particular, we conclude that the above commutative diagram of fundamental groups *arises geometrically* from a commutative diagram:

$$\begin{array}{ccc} (\mathcal{X}_1^{\log})_{k_1} & \xrightarrow{\sim} & (\mathcal{X}_2^{\log})_{k_2} \\ \downarrow & & \downarrow \\ k_1^{\log} & \xrightarrow{\sim} & k_2^{\log} \end{array}$$

Moreover, the isomorphism exhibited in the upper horizontal arrow of this commutative diagram is easily seen to be *functorial* with respect to *arbitrary finite étale coverings of the* $(X_i)_{K_i}$ (i.e., not just coverings that arise from finite étale coverings of the $(\mathcal{X}_i)_{\mathcal{O}_{K_i}}$). Indeed, this functoriality follows formally from the *uniqueness* assertion in Proposition 1.2.4, (iii), and the fact that *dominant* (i.e., not just finite étale) *morphisms between proper hyperbolic curves in characteristic p may be distinguished* by considering the morphisms that they induce between the respective Jacobians, hence, in particular, by the morphisms that they induce between the *l-power torsion points* (where $l \neq p$) of the respective Jacobians. [Note also that the morphisms between *log structures at the nodes* are uniquely determined by the morphisms between the various *normal bundles* on the branches at the nodes — cf. [Mzk3], §3.7, 3.8, 3.10.] Thus, in summary:

Theorem 2.7 (Group-Theoretic Reconstruction of the Logarithmic Special Fiber of a p-adic Hyperbolic Curve). *Let p be a prime number. For $i = 1, 2$, let K_i be a finite extension of $\mathbb{Q}_p$, and $(X_i)_{K_i}$ a* **hyperbolic curve** *over K_i whose associated pointed stable curve has stable* *reduction over* $\mathcal{O}_{K_i}$. *Denote the resulting "stable model" of* $(X_i)_{K_i}$ *over* $\mathcal{O}_{K_i}$ *by* $(\mathcal{X}_i)_{\mathcal{O}_{K_i}}$. *Assume that we have chosen basepoints of the* $(X_i)_{K_i}$ *(which thus induce basepoints of the* K_i*). Then every isomorphism of profinite groups* $\Pi_{(X_1)_{K_1}} \xrightarrow{\sim} \Pi_{(X_2)_{K_2}}$ *induces commutative diagrams:*

$$\begin{array}{ccc} \Pi_{(X_1)_{K_1}} & \xrightarrow{\sim} & \Pi_{(X_2)_{K_2}} \\ \downarrow & & \downarrow \\ G_{K_1} & \xrightarrow{\sim} & G_{K_2} \end{array} \qquad \begin{array}{ccc} \Pi^{\mathrm{adm}}_{(X_1)_{K_1}} & \xrightarrow{\sim} & \Pi^{\mathrm{adm}}_{(X_2)_{K_2}} \\ \downarrow & & \downarrow \\ G^{\log}_{k_1} & \xrightarrow{\sim} & G^{\log}_{k_2} \end{array}$$

Moreover, the latter commutative diagram (of admissible quotients $\Pi^{\mathrm{adm}}_{(X_i)_{K_i}}$ *of the* $\Pi_{(X_i)_{K_i}}$ *lying over the tame Galois groups* $G^{\log}_{k_i}$ *of the* K_i*) necessarily arises from unique commutative diagrams of log schemes*

$$\begin{array}{ccc} (\widetilde{\mathcal{X}}^{\log}_1)_{k_1} & \xrightarrow{\sim} & (\widetilde{\mathcal{X}}^{\log}_1)_{k_1} \\ \downarrow & & \downarrow \\ (\mathcal{X}^{\log}_1)_{k_1} & \xrightarrow{\sim} & (\mathcal{X}^{\log}_2)_{k_2} \end{array} \qquad \begin{array}{ccc} (k^{\log}_1)^{\sim} & \xrightarrow{\sim} & (k^{\log}_2)^{\sim} \\ \downarrow & & \downarrow \\ k^{\log}_1 & \xrightarrow{\sim} & k^{\log}_2 \end{array}$$

where the commutative diagram on the left lies over the commutative diagram on the right (which is as in Proposition 1.2.4, (iii)). Here, we equip $\mathrm{Spec}(\mathcal{O}_{K_i})$ *(respectively,* $\mathrm{Spec}(k_i)$*;* $(\mathcal{X}_i)_{\mathcal{O}_{K_i}}$*;* $(\mathcal{X}_i)_{k_i}$*) with the log structure determined by the closed point (respectively, determined by restricting the log structure on* $\mathrm{Spec}(\mathcal{O}_{K_i})$*; determined by the monoid of functions invertible on the open subscheme* $(X_i)_{K_i}$*; determined by restricting the log structure on* $(\mathcal{X}_i)_{\mathcal{O}_{K_i}}$*) and denote the resulting log scheme by* $\mathcal{O}^{\log}_{K_i}$ *(respectively,* $k^{\log}_i$*;* $(\mathcal{X}^{\log}_i)_{\mathcal{O}_{K_i}}$*;* $(\mathcal{X}^{\log}_i)_{k_i}$*); the vertical morphisms in the above commutative diagrams of log schemes are the universal coverings induced by the various basepoints chosen.*

Proof. First, note that the additional assumptions that were used in the course of the above discussion — e.g., "$(*^{\mathrm{CSSN}})$"; the assumption that $(\mathcal{X}_i)_{\mathcal{O}_{K_i}}$ arise as some "$(\mathcal{Z}_i)_{\mathcal{O}_{K_i}}$" as in Lemma 2.4 — were applied *only to show that the hypotheses of Proposition 1.2.4, (iii) (and [Mzk4], Theorem 7.2) are satisfied.* Moreover, we observe that although [Mzk4], Theorem 7.2, is only stated in the proper singular case, it *extends immediately to the affine singular, as well as affine nonsingular, cases.* Thus, (cf. Remark 2.5.1; the discussion surrounding "$(*^{\mathrm{CSSN}})$") one concludes that — except when $(X_i)_{K_i}$ is *proper*, with *good reduction* — one may *reconstruct the logarithmic special fiber in a functorial fashion* (i.e., with respect to finite étale coverings of the $(X_i)_{K_i}$), as desired.

In the case that $(X_i)_{K_i}$ is *proper*, but has *good reduction* over $\mathcal{O}_{K_i}$, we may still reconstruct its logarithmic special fiber (despite the fact that [Tama1], Theorem 4.3, is only stated in [Tama1] for *affine* hyperbolic curves!) by arguing as follows: First of all, we observe that in the case of good reduction, the log structure of the special fiber of the curve is obtained by simply *pulling back* the log structure of $k_i^{\log}$. Thus, it suffices to construct the (non-logarithmic, scheme-theoretic) special fiber. Next, we observe that (after possibly enlarging K_i) there exist — cf., e.g., [Mzk4], the first two paragraphs of the proof of Theorem 9.2 — *corresponding finite Galois étale coverings* $(Z_i)_{K_i} \to (X_i)_{K_i}$ (for $i = 1, 2$), where $(Z_i)_{K_i}$ is a hyperbolic curve over K_i with *bad stable reduction* $(\mathcal{Z}_i)_{\mathcal{O}_{K_i}}$ over $\mathcal{O}_{K_i}$. Thus, by applying Theorem 2.7 to $(Z_i)_{K_i}$ allows us to reconstruct the logarithmic special fiber $(\mathcal{Z}_i^{\log})_{k_i}$, together with the *action of the Galois group* $G_i \stackrel{\text{def}}{=} \mathrm{Gal}((Z_i)_{K_i}/(X_i)_{K_i})$. Note that irreducible components of $(\mathcal{Z}_i)_{k_i}$ that *dominate* $(\mathcal{X}_i)_{k_i}$ may be distinguished (group-theoretically!) by the fact that their geometric fundamental groups map *surjectively* onto open subgroups of the geometric fundamental group of $(\mathcal{X}_i)_{k_i}$. Let us choose corresponding (closed, proper) irreducible components

$$C_i \subseteq (\mathcal{Z}_i)_{k_i}$$

that *dominate* (hence surject onto) $(\mathcal{X}_i)_{k_i}$. Denote the *decomposition (respectively, inertia) group* associated to C_i by $D_i \subseteq G_i$ (respectively, $I_i \subseteq D_i \subseteq G_i$). Thus, D_i/I_i acts *faithfully* on C_i, and the *order* $|I_i|$ *of* I_i is a power of p, equal to the *degree of inseparability* of the function field of C_i over the function field of $(\mathcal{X}_i)_{k_i}$. Then we may reconstruct $(\mathcal{X}_i)_{k_i}$ as a *finite flat quotient* of C_i by considering the *subsheaf*

$$(\mathcal{O}_{C_i}^{|I_i|})^{D_i} \subseteq \mathcal{O}_{C_i}$$

(i.e., the D_i-invariants of the subalgebra $\mathcal{O}_{C_i}^{|I_i|} \subseteq \mathcal{O}_{C_i}$, where we use that $|I_i|$ is a *power of* p). By applying the functoriality with respect to finite étale coverings of $(X_i)_{K_i}$ observed in the discussion immediately preceding the statement of Theorem 2.7, we conclude that this construction of $(\mathcal{X}_i)_{k_i}$ is *independent* of the choice of $(Z_i)_{K_i}$, C_i, and itself *functorial* with respect to finite étale coverings of $(X_i)_{K_i}$.

This completes our *reconstruction of the logarithmic special fibers of the* $(X_i)_{K_i}$, in a fashion that is *functorial* with respect to finite étale coverings of the $(X_i)_{K_i}$. Thus, we conclude, in particular, (from this functoriality, applied to covering transformations; the slimness of Lemma 2.2, (i)) that the morphism induced on admissible fundamental groups by the isomorphism constructed between logarithmic special fibers *coincides* with the original given morphism between admissible fundamental groups. This completes the proof of Theorem 2.7. □

Remark 2.7.1. Given data as in Theorem 2.7, one may consider the *outer Galois representation*

$$G_{K_i} \to \mathrm{Out}(\Delta_{X_i})$$

which is known to be *injective* if $r_i > 0$ (cf. Theorem 1.3.6). Thus, at least in the case $r_i > 0$, it is natural to ask:

> *What is the commensurator of* $\mathrm{Im}(G_{K_i})$ *in* $\mathrm{Out}(\Delta_{X_i})$*?*

Although Theorem 2.7 does not give a complete explicit answer to this question, it tells us that, at any rate, *elements of this commensurator* (which define isomorphisms of the sort that are treated in Theorem 2.7) *preserve the logarithmic special fiber*. In particular, (although one does not know whether or not elements of this commensurator induce *"self-isogenies"* of K_i, i.e., are "geometric") one obtains that elements of this commensurator *do* induce *"self-isogenies"* of $k_i^{\log}$. [Here, by a *"self-isogeny of an object"*, we mean an isomorphism between two finite étale coverings of the object.] Moreover, since it follows from Theorem A of [Mzk6] (cf. Theorem 1.3.4, Remark 1.3.6.2) that the *centralizer* of $\mathrm{Im}(G_{K_i})$ in $\mathrm{Out}(\Delta_{X_i})$ consists precisely of those (finitely many) automorphisms that arise *geometrically* (i.e., from automorphisms of $(X_i)_{K_i}$), it follows that a "self-isogeny" of G_{K_i} induced by an element of this commensurator corresponds to (up to finitely many well-understood possibilities) an essentially *unique* element of this commensurator. This motivates the point of view that:

> *The "self-isogenies" of* G_{K_i} *defined by elements of this commensurator — which we shall refer to as* **quasi-conformal self-isogenies of** G_{K_i} *— are natural objects to study in their own right.*

The reason for the choice of the terminology "quasi-conformal" is that those self-isogenies that are "of geometric origin" — i.e., *"conformal"* — are (by the main theorem of [Mzk5]) precisely those which preserve the higher ramification filtration, which is closely related to the *"canonical p-adic metric"* on the local field in question. Thus, quasi-conformal self-isogenies do not preserve the "metric (or conformal) structure" but *do* preserve the "logarithmic special fiber" which one may think of as a sort of p-adic analogue of the "topological type" of the objects in question.

Remark 2.7.2. Note that isomorphisms

$$k_1^{\log} \xrightarrow{\sim} k_2^{\log}$$

(such as those arising from "quasi-conformal isomorphisms" $G_{K_1} \xrightarrow{\sim} G_{K_2}$ as in Theorem 2.7) need not be *"geometric"* from the point of view of characteristic zero (i.e., induced by an isomorphism of fields $K_1 \xrightarrow{\sim} K_2$). For instance, such an isomorphism might take the section of the log structure corresponding to p to some multiple of this section by a root of unity (a situation which could never occur if the isomorphism arose from an isomorphism $K_1 \xrightarrow{\sim} K_2$). Whether or not, however, this sort of phenomenon actually takes place in the case of "quasi-conformal isomorphisms" as in Theorem 2.7 is not clear to the author at the time of writing.

Remark 2.7.3. The theory of the present § prompts the question:

> *Do isomorphisms $\Pi_{(X_1)_{K_1}} \xrightarrow{\sim} \Pi_{(X_2)_{K_2}}$ as in Theorem 2.7* **only** *preserve the logarithmic special fiber or do they preserve* **other information** *as well concerning the liftings $(X_i)_{K_i}$ of the respective logarithmic special fibers?*

Although the author is unable to give a complete answer to this question at the time of writing, it does appear that when the lifting in question is in some sense *"canonical"*, then this canonicality is preserved by isomorphisms as in Theorem 2.7. In a future paper, we hope to discuss this sort of phenomenon — which may be observed, for instance, in the following cases:

(1) *Serre-Tate canonical liftings*;

(2) *"arithmetic hyperbolic curves"*, i.e., hyperbolic curves isogenous to a Shimura curve [that is to say, curves which admit a finite étale covering which is isomorphic to a finite étale covering of some Shimura curve];

(3) *canonical liftings in the sense of "p-adic Teichmüller theory"* (cf. [Mzk1], [Mzk2])

— in more detail. Perhaps this phenomenon should be regarded as a *natural extension* of the phenomenon of preservation of the logarithmic fiber in the sense that *canonical liftings* are, in some sense, liftings that are *"defined over $\mathbb{F}_1$"* — i.e., a hypothetical (but, of course, fictional!) absolute field of constants sitting inside $\mathbb{Z}_p$.

Acknowledgements

I would like to thank *A. Tamagawa* for the time that he so generously shared with me in numerous stimulating discussions, and especially for the following: (i) informing me of the arguments used to prove Lemma 1.1.4 in §1.1; (ii) explaining to me the utility of a theorem of Raynaud in the context of §2 (cf. Lemma 2.4). Also, I would like to thank *F. Oort*, as well as the *referee*, for various useful remarks.

Appendix: Terminology of Graph Theory

The Notion of a Semi-Graph:

We shall refer to as a *semi-graph* Γ the following collection of data:

(1) a set $\mathcal{V}$ — whose elements we refer to as *"vertices"*;

(2) a set $\mathcal{E}$ — whose elements we refer to as *"edges"* — each of whose elements e is a *set of cardinality* 2 satisfying the property "$e \neq e' \in \mathcal{E} \implies e \bigcap e' = \emptyset$";

(3) a collection ζ of maps ζ_e [one for each edge e] — which we refer to as the *"coincidence maps"* — such that $\zeta_e : e \to \mathcal{V} \bigcup \{\mathcal{V}\}$ [where we note that $\mathcal{V} \bigcap \{\mathcal{V}\} = \emptyset$ since $\mathcal{V} \notin \mathcal{V}$] is a map from the set e to the set $\mathcal{V} \bigcup \{\mathcal{V}\}$.

We shall refer to the subset $\zeta_e^{-1}(\mathcal{V}) \subseteq e$ [i.e., the inverse image of the subset $\mathcal{V} \subseteq \mathcal{V} \bigcup \{\mathcal{V}\}$ of elements $\neq \mathcal{V}$] as the *verticial portion* of an edge e; to the restriction of ζ_e to the verticial portion of e as the *verticial restriction* of ζ_e; and to the cardinality of the verticial portion of e as the *verticial cardinality* of e. A *graph* Γ is a semi-graph Γ for which every $e \in \mathcal{E}$ has verticial cardinality precisely 2. We will say that a graph or semi-graph is *finite* if its sets of vertices and edges are finite. We shall refer to an element $b \in e$ as a *branch of the edge* e.

Let $\Gamma = \{\mathcal{V}, \mathcal{E}, \zeta\}$ be a *semi-graph*. If $e \in \mathcal{E}$ is an *edge* of Γ of *verticial cardinality* 2 whose image via ζ_e consists of two (not necessarily distinct) elements v_1, v_2 of $\mathcal{V}$, then we shall say that e *joins* v_1 to v_2. If v is any vertex in the image of ζ_e, then we shall say that e *meets* or *abuts to* v. Thus, an edge of a *graph* always abuts to at least one vertex, while an edge of a semi-graph may abut to no vertices at all.

By thinking of vertices as *points* and edges as *line segments* that join points to points or are "open" at one or both ends, we may think of semi-graphs as defining *topological spaces*. Thus, it makes sense to speak of a semi-graph as being *contractible* (in the sense of algebraic topology). Such a semi-graph will be referred to as a *tree*.

Finally, a *morphism* between semi-graphs

$$\Gamma = \{\mathcal{V}, \mathcal{E}, \zeta\} \to \Gamma' = \{\mathcal{V}', \mathcal{E}', \zeta'\}$$

is a collection of maps $\mathcal{V} \to \mathcal{V}'$; $\mathcal{E} \to \mathcal{E}'$; and for each $e \in \mathcal{E}$ mapping to e', an injection $e \hookrightarrow e'$ — all of which are *compatible* with the *verticial restrictions* of the respective coincidence maps. Thus, here, we allow an edge that abuts to no (respectively, precisely one) vertex to map to an edge that abuts to any number ≥ 0 (respectively, ≥ 1) of vertices.

Semi-Graphs of Profinite Groups:

We shall refer to the following data G:

(i) a finite *semi-graph* Γ;

(ii) for each vertex v of Γ, a *profinite group* G_v;

(iii) for each edge e of Γ, a *profinite group* G_e, together with, for each branch $b \in e$ mapping to a vertex v via ζ_e, a continuous homomorphism $b_* : G_e \to G_v$

as a *semi-graph of profinite groups*. When Γ is a graph, we shall refer to this data G as a *graph of profinite groups.*

Suppose that we are given a *semi-graph of profinite groups* G. Then to G, one may associate (in a natural, functorial fashion) a *profinite group* Π_G — namely, the *profinite analogue* of the well-known construction of the *fundamental group associated to a (semi-)graph of groups* (cf. [Serre1], I, §5.1). One way to describe this profinite analogue is via *Galois categories* (cf., e.g., [SGA1] for an exposition of the theory of Galois categories) as follows: If H is an arbitrary profinite group, then let us write

$$\mathcal{B}(H)$$

for the category of finite sets with continuous H-action. Then H may be recovered as the *"fundamental group"* of the Galois category $\mathcal{B}(H)$. Also, we note that any continuous homomorphism of profinite groups $H \to H'$ determines a natural functor $\mathcal{B}(H') \to \mathcal{B}(H)$ [in the reverse direction]. In particular, the homomorphisms b_* determine functors:

$$b^* : \mathcal{C}_v \stackrel{\text{def}}{=} \mathcal{B}(G_v) \to \mathcal{C}_e \stackrel{\text{def}}{=} \mathcal{B}(G_e)$$

At any rate, to define Π_G, it suffices to define a Galois category $\mathcal{C}_G$ [which we would like to think of as "$\mathcal{B}(\Pi_G)$"]. We define $\mathcal{C}_G$ to be the category of *collections of data:*

$$\{S_v; \alpha_e\}$$

Here, "v" (respectively, "e") ranges over the vertices (respectively, edges of verticial cardinality 2) of Γ; S_v is an object of $\mathcal{C}_v$; and $\alpha_e : b_1^*(S_{v_1}) \cong b_2^*(S_{v_2})$ (where $e = \{b_1, b_2\}$; $v_i \stackrel{\text{def}}{=} \zeta_e(b_i)$, for $i = 1, 2$) is an isomorphism in $\mathcal{C}_e$. Morphisms between such collections of data are defined in the evident way. One then verifies easily that this category $\mathcal{C}_G$ is indeed a *Galois category*, as desired.

Pointed Stable Curves:

Let k be an *algebraically closed field.* Let $g, r \geq 0$ be integers such that $2g - 2 + r > 0$. Let $(\overline{X} \to \mathrm{Spec}(k), D \subseteq \overline{X})$ be an *r-pointed stable curve of genus g* (where $D \subseteq \overline{X}$ is the divisor of marked points) over k, and set:

$$X \stackrel{\mathrm{def}}{=} \overline{X} \backslash D$$

Also, let us write $X' \subseteq X$ for the *complement of the nodes* in X.

Next, we define the *dual graph*

$$\Gamma_X$$

of X. The set of *vertices* of Γ_X is the set of *irreducible components* of X. To avoid confusion, we shall write X_v for the irreducible component corresponding to a vertex v. The set of *edges* of Γ_X is the set of *nodes* of X. To avoid confusion, we shall write ν_e for the node corresponding to an edge e. The node ν_e has two *branches* $\nu_e[1]$ and $\nu_e[2]$, each of which lies on some well-defined irreducible component of X. We take $e \stackrel{\mathrm{def}}{=} \{\nu_e[1], \nu_e[2]\}$ [so e is of verticial cardinality 2], and let ζ_e be the map that sends $\nu_e[i]$ to the irreducible component on which the branch $\nu_e[i]$ lies.

Let us write $X'_v \stackrel{\mathrm{def}}{=} X_v \bigcap X' \subseteq X$, and (for $i = 1, 2$) $X'_{\nu_e[i]}$ for the scheme-theoretic intersection with X' of the completion of the branch $\nu_e[i]$ at the node ν_e. Thus, $X'_{\nu_e[i]}$ is noncanonically isomorphic to $\mathrm{Spec}(k[[t]][t^{-1}])$ (where t is an indeterminate).

In the following discussion, we would like to fix (k-linear) isomorphisms:

$$X'_{\nu_e[1]} \cong X'_{\nu_e[2]}$$

via which we shall identify $X'_{\nu_e[1]}$ with $X'_{\nu_e[2]}$ and denote the resulting object by X'_e. In particular, we have natural morphisms $X'_e \to X'_{v_1}$, $X'_e \to X'_{v_2}$ (where $v_i \stackrel{\mathrm{def}}{=} \zeta_e(\nu_e[i])$, for $i = 1, 2$). One verifies immediately that the induced morphism on *tame algebraic fundamental groups* [i.e., corresponding to coverings which are tamely ramified at the nodes and marked points] is *independent* of the choice of isomorphism. Thus, the dual graph Γ_X, together with the result of applying "$\pi_1^{\mathrm{tame}}(-)$" [more precisely: the tame algebraic fundamental group functor, together with *choices of basepoints* for the various X'_v, X'_e, and *choices of paths* to relate these basepoints via the natural morphisms $X'_e \to X'_{v_i}$ discussed above] to the data $\{X'_v; X'_e; X'_e \to X'_v\}$ determines a *graph of profinite groups* $\mathcal{G}_X$ associated to the stable curve X.

When considering the case of a curve with marked points (i.e., $r > 0$), it is useful to consider the following slightly modified *"data with compact structure"*: Let us denote by Γ_X^{c} the *semi-graph* obtained from Γ_X by appending to Γ_X, for each marked point $x \in \overline{X}$, the following:

> an *edge* $e_x \stackrel{\mathrm{def}}{=} \{x\}$, where $x \in e_x$ is sent by ζ_{e_x} to the vertex v_x corresponding to the irreducible component of $\overline{X}$ that contains x.

We shall refer to the new edges "e_x" that were added to Γ_X to form Γ_X^{c} as the *marked edges* of Γ_X^{c} and to Γ_X^{c} itself as the *dual graph with compact structure* associated to X.

If, moreover, we associate to e_x the scheme X'_x (which is noncanonically isomorphic to $\mathrm{Spec}(k[[t]][t^{-1}])$) obtained by removing x from the completion of $\overline{X}$ at x, and apply "$\pi_1^{\mathrm{tame}}(-)$" to the natural morphism $X'_x \to X'_{v_x}$, then we obtain a natural *semi-graph of profinite groups* $\mathcal{G}_X^{\mathrm{c}}$ with underlying semi-graph Γ_X^{c}. Moreover, one checks easily that when k is *of characteristic* 0, the *profinite group associated* (as described above) to $\mathcal{G}_X$ or $\mathcal{G}_X^{\mathrm{c}}$ is isomorphic to "$\widehat{\Pi}_{g,r}$", i.e., the *profinite completion of the fundamental group of a Riemann surface of genus* g *with* r *points removed.*

In fact, if we take $k = \mathbb{C}$, and we think of the X'_v as *Riemann surfaces* and of the X'_e, X'_x as "copies of the *circle* $\mathbb{S}^1$", then we see that this construction corresponds quite geometrically to *gluing Riemann surfaces with boundary along copies of the circle.*

Finally, let us observe the following (valid for k of *arbitrary characteristic*): The profinite group associated to $\mathcal{G}_X$ or $\mathcal{G}_X^{\mathrm{c}}$ may be identified with the *admissible fundamental group*

$$\Pi_X^{\mathrm{adm}}$$

of $(\overline{X}, D)$ (cf. [Mzk4], §2, 8; [Mzk3], §3). Moreover, if we define semi-graphs of groups $\mathcal{G}_X^{\mathrm{et}}$, $\mathcal{G}_X^{\mathrm{c,et}}$ by taking the profinite group at a vertex v to be the *tame fundamental group* of X_v [i.e., instead of X'_v] and the profinite group at all of the edges to be trivial, then the resulting profinite group may be identified with the *fundamental group*

$$\Pi_X^{\mathrm{et}}$$

associated to the Galois category of *finite étale coverings of* X *which are tamely ramified at the cusps.* In particular, we have a natural surjection $\Pi_X^{\mathrm{adm}} \twoheadrightarrow \Pi_X^{\mathrm{et}}$.

Second Cohomology Groups:

We continue with the notation of the above discussion. Let l be a *prime number* distinct from the characteristic of k. In the following discussion, we shall use the notation "$H^j(-)$" (respectively, "$H^j_c(-)$") to denote the *j-th étale cohomology module* (respectively, *j-th étale cohomology module with compact supports* — cf. [Milne], Chapter III, Proposition 1.29; Remark 1.30) with coefficients in $\mathbb{Z}_l$.

If v is a vertex of Γ_X, write $\overline{X}_v$ for the *normalization* of the corresponding *irreducible component of* $\overline{X}$, so $X'_v = X' \times_{\overline{X}} \overline{X}_v$. Then we have a *natural "push-forward" isomorphism*

$$H^2_c(X'_v) \xrightarrow{\sim} H^2(\overline{X}_v)$$

between *free* $\mathbb{Z}_l$*-modules of rank* 1. Moreover, the natural morphisms $\overline{X}_v \to \overline{X}$ (as v ranges over the vertices of Γ_X) determine natural *restriction morphisms* whose direct sum is easily verified to be an *isomorphism*:

$$H^2(\overline{X}) \xrightarrow{\sim} \bigoplus_v H^2(\overline{X}_v)$$

On the other hand, the composite of the natural "push-forward" morphism $H^2_c(X'_v) \to H^2(\overline{X})$ with a restriction morphism $H^2(\overline{X}) \to H^2(\overline{X}_w)$, where $w \neq v$, is easily verified to be *zero*. [Indeed, this follows by considering the Chern class of the line bundle on $\overline{X}$ associated to a closed point of X'_v: This Chern class is the image of a generator of $H^2_c(X'_v)$ and, moreover, vanishes upon restriction to $\overline{X}_w$ since the restriction to $\overline{X}_w$ of this line bundle is clearly trivial.] In particular, we conclude that the direct sum of the natural "push-forward" morphisms $H^2_c(X'_v) \to H^2(\overline{X})$ yields an *isomorphism*

$$\bigoplus_v H^2_c(X'_v) \xrightarrow{\sim} H^2(\overline{X})$$

which, if we identify $H^2_c(X'_v)$, $H^2(\overline{X}_v)$ via the isomorphism $H^2_c(X'_v) \xrightarrow{\sim} H^2(\overline{X}_v)$, is *inverse* to the isomorphism $H^2(\overline{X}) \xrightarrow{\sim} \bigoplus_v H^2(\overline{X}_v)$.

Put another way, if we identify the cohomology modules $H^2_c(X'_v)$, $H^2(\overline{X}_v)$ via the isomorphism $H^2_c(X'_v) \xrightarrow{\sim} H^2(\overline{X}_v)$, then the natural "push-forward" morphism $H^2_c(X'_v) \to H^2(\overline{X})$ may be *recovered* as the composite

$$H^2(\overline{X}_v) \hookrightarrow \bigoplus_v H^2(\overline{X}_v) \xrightarrow{\sim} H^2(\overline{X})$$

of the natural inclusion of a direct summand with the *inverse* of the isomorphism $H^2(\overline{X}) \xrightarrow{\sim} \bigoplus_v H^2(\overline{X}_v)$ determined by the restriction morphisms.

Finally, let us observe that if $\overline{X}$ is *sturdy* (cf. [Mzk4], Definition 1.1) — i.e., every $\overline{X}_v$ has genus ≥ 2 — then the natural morphisms from profinite group cohomology to étale cohomology give rise to a *commutative diagram*

$$\begin{array}{ccc} H^2(\pi_1^{\text{tame}}(\overline{X}) = \pi_1(\overline{X}), \mathbb{Z}_l) & \longrightarrow & H^2(\overline{X}) \\ \downarrow & & \downarrow \\ H^2(\pi_1^{\text{tame}}(\overline{X}_v) = \pi_1(\overline{X}_v), \mathbb{Z}_l) & \longrightarrow & H^2(\overline{X}_v) \end{array}$$

in which the horizontal morphisms are *isomorphisms* [since arbitrary étale cohomology classes with finite l-power torsion coefficients vanish upon restriction to some finite étale covering of $\overline{X}$ or $\overline{X}_v$].

References

[FC] G. Faltings and C.-L. Chai, "Degenerations of Abelian Varieties," Springer-Verlag, 1990.

[FJ] M. Fried and M. Jarden, "Field Arithmetic," Springer-Verlag, 1986.

[HR] W. Herfort and L. Ribes, *Torsion elements and centralizers in free products of profinite groups*, J. Reine Angew. Math. **358** (1985), 155–161.

[Kato1] K. Kato, *Logarithmic Structures of Fontaine-Illusie*, in "Proceedings of the First JAMI Conference," 191–224, Johns Hopkins Univ. Press, 1990.

[Kato2] K. Kato, *Toric Singularities*, Amer. J. Math. **116** (1994), 1073–1099.

[Knud] F. F. Knudsen, *The Projectivity of the Moduli Space of Stable Curves, II*, Math. Scand. **52** (1983), 161–199.

[Mtmo] M. Matsumoto, *Galois representations on profinite braid groups on curves*, J. Reine Angew. Math. **474** (1996), 169–219.

[Milne] J. S. Milne, "Étale Cohomology," Princeton Mathematical Series, 33. Princeton University Press, 1980.

[Mzk1] S. Mochizuki, *A Theory of Ordinary p-adic Curves*, Publ. of RIMS **32** (1996), 957–1151.

[Mzk2] S. Mochizuki, "Foundations of p-adic Teichmüller Theory," AMS/IP Studies in Advanced Mathematics, 11. American Mathematical Society/International Press, 1999.

[Mzk3] S. Mochizuki, *The Geometry of the Compactification of the Hurwitz Scheme*, Publ. of RIMS **31** (1995), 355–441.

[Mzk4] S. Mochizuki, *The Profinite Grothendieck Conjecture for Closed Hyperbolic Curves over Number Fields*, J. Math. Sci. Univ. Tokyo **3** (1996), 571–627.

[Mzk5] S. Mochizuki, *A Version of the Grothendieck Conjecture for p-adic Local Fields*, The International Journal of Math. **8** (1997), 499–506.

[Mzk6] S. Mochizuki, *The Local Pro-p Anabelian Geometry of Curves*, Invent. Math. **138** (1999), 319–423.

[Mumf] D. Mumford, "Abelian Varieties," Oxford Univ. Press, 1974.

[Naka1] H. Nakamura, *Galois rigidity of pure sphere braid groups and profinite calculus*, J. Math. Sci. Univ. Tokyo **1** (1994), 71–136.

[Naka2] H. Nakamura, *Galois rigidity of profinite fundamental groups* [translation of Sugaku **47** (1995), 1–17], Sugaku Expositions **10** (1997), 195–215.

[NTM] H. Nakamura, A. Tamagawa, and S. Mochizuki, *The Grothendieck Conjecture on the Fundamental Groups of Algebraic Curves*, Sugaku Expositions **14** (2001), 31–53.

[NSW] J. Neukirch, A. Schmidt, K. Wingberg, "Cohomology of number fields," Grundlehren der Mathematischen Wissenschaften, 323. Springer-Verlag, 2000.

[Serre1] J.-P. Serre, "Trees," Springer-Verlag, 1980.

[Serre2] J.-P. Serre, *Local Class Field Theory*, in "Algebraic Number Theory (J. W. S. Cassels and A. Fröhlich, ed.)," 128–161, Academic Press, 1967.

[SGA1] "Revêtement étales et groupe fondamental," Séminaire de Géometrie Algébrique du Bois Marie 1960-1961 (SGA1), dirigé par A. Grothendieck, augmenté de deux exposés de M. Raynaud, Lecture Notes in Mathematics, 224. Springer-Verlag, 1971.

[Tama1] A. Tamagawa, *The Grothendieck Conjecture for Affine Curves*, Compositio Math. **109** (1997), 135–194.

[Tama2] A. Tamagawa, *On the Fundamental Groups of Curves over Algebraically Closed Fields of Characteristic* > 0, International Mathematics Research Notices **16** (1999), 853–873.

[Tate] J. Tate, *p-Divisible Groups*, in "Proceedings of a Conference on Local Fields," 158–183, Driebergen, Springer-Verlag, 1967.

Part II

Galois groups and Galois extensions

Galois Theory and Modular Forms pp 125–142
K. Hashimoto, K. Miyake, and H. Nakamura, Editors
© 2003 Kluwer Academic Publishers

REGULAR GALOIS REALIZATIONS OF $\mathrm{PSL}_2(p^2)$ OVER $\mathbb{Q}(T)$

Takehito SHIINA
Mathematical Institute
Tohoku University,
Sendai Miyagi, 980–8578, Japan
s99m17@math.tohoku.ac.jp

History

The inverse Galois problem which asks whether every finite group occurs as Galois group over $\mathbb{Q}$ is an unsolved problem originally proposed by Hilbert [8] more than 100 years ago. The first approach succeeded in the region of class field theory in the middle of the 20th century: Scholz and Reichardt solved the problem affirmatively for nilpotent groups [15] and Šafarevič extended their result to solvable groups [18]. In contrast, few Galois realizations were known for non-solvable groups until the 1970's except for the symmetric groups S_n and the alternating groups A_n, which had already been realized by Hilbert. The first attack of non-solvable groups was carried out in the 1970's and many important results have been obtained for such groups from then on. For example, using the p-adic Galois representation attached to cusp forms of weight 24, Ribet [17] realized the projective special linear group $\mathrm{PSL}_2(p^2)$ as Galois group over $\mathbb{Q}$ in the case 144169 is non-square modulo p. Recently, this result was widely extended by Reverter-Vila [16] and Dieulefait-Vila [3]. Moreover, studying the Jacobian of a plane curve of genus 2, Mestre [13] showed that the field generated by the p-division points realizes $\mathrm{PSL}_2(p^2)$ *regularly* over $\mathbb{Q}$ for $p \equiv \pm 2 \pmod 5$. Here a group G is said to be realized regularly over $\mathbb{Q}$ if there exists a Galois extension of a rational function field $\mathbb{Q}(T) = \mathbb{Q}(t_1, \ldots, t_r)$, regular over $\mathbb{Q}$, with Galois group G. If G occurs regularly over $\mathbb{Q}$, then Hilbert's irreducibility theorem assures that there exist infinitely many linearly disjoint number fields

[received: May 31, 2002; accepted in revised form: March 1, 2003]

with Galois group G. So it is natural to ask the following more difficult question.

Problem (Regular Inverse Galois Problem). *Does every finite group occur regularly over* $\mathbb{Q}$*?*

With regard to this problem, the most effective approach that has been known is the *rigidity method,* which was established by Belyĭ, Fried, Matzat, Shih and Thompson. This method is based on the theory of Galois coverings of Riemann surfaces. By means of the rigidity method, one can realize a finite group G regularly over $\mathbb{Q}$ if G satisfies certain group theoretical conditions. Many variations of sufficient conditions for a given finite group to apply the rigidity method are known and many positive answers for non-solvable groups including 25 sporadic simple groups have been obtained. Here we make a list of the projective special linear groups $\mathrm{PSL}_2(p^n)$ which have been realized regularly over $\mathbb{Q}$.

(i) $n = 1$	
$p = 2, 3, 5$	(Hilbert, 1892)
$(\frac{l}{p}) = -1,\ {}^{\exists}l \in \{2, 3, 7\}$	(Shih, 1974)
$(\frac{5}{p}) = -1$	(Malle, 1991)
(ii) $n = 2$	
$p = 2, 3$	(Hilbert, 1892)
$(\frac{5}{p}) = -1$	(Feit, 1984 / Mestre, 1987)
$p = 5$	(Przywara, 1988)
(iii) $n = 3$	
$p = 2$	(Matzat, 1987)

The purpose of this paper is to show the following:

Main Theorem. *If* $p \not\equiv \pm 1 \pmod{24}$*, then* $\mathrm{PSL}_2(p^2)$ *occurs regularly over* $\mathbb{Q}$.

This is an extension of our preceding result that $\mathrm{PSL}_2(p^2)$ occurs regularly over $\mathbb{Q}$ for $p \equiv \pm 3 \pmod 8$ [21]. We consider its covering group $\mathrm{P\Sigma L}_2(p^2)$ instead of $\mathrm{PSL}_2(p^2)$, for avoiding some complexity in the process of calculation. The key of the proof was the existence of a certain isolated braid orbit to which we applied Matzat's braid orbit theorem ([10], [12]). By changing some numerical conditions, we will realize $\mathrm{PSL}_2(p^2)$ for the remaining primes $p \equiv \pm 5 \pmod{12}$ in a similar way to [21].

Remark. After obtaining the result above, we were informed that Dettweiler realized the projective symplectic groups $\mathrm{PSp}_{2n}(p^2)$ regularly over $\mathbb{Q}$ for odd primes $p \not\equiv \pm 1 \pmod{24}$ and integers $n \geq 1$. His strategy is to verify the existence of a rigid braid orbit in $\mathrm{PSp}_{2n}(p^2)$ by applying the middle convolution functor to certain tuples of $\mathrm{GL}_2(p^2)$ (see his paper [2] in this volume).

The content of this paper is organized as follows. The first two sections aim to present known results with examples of expository nature: In Section 1, we formulate a system of definitions and notations to state the basic rigidity theorem and in Section 2, we review the braid orbit theorem which is the main tool of our proof of Main Theorem. In Section 3, some group-theoretic properties of $\mathrm{P\Sigma L}_2(p^2)$ are collected, and the use of the braid orbit theorem will be exhibited in some easy cases. In Section 4, we construct a certain nice braid orbit in $\mathrm{P\Sigma L}_2(p^2)$, and settle the proof of Main Theorem.

1. Basic Rigidity Theorem

A finite unramified Galois covering of the punctured projective line

$$\varphi\colon X \longrightarrow \mathbb{P}^1 \setminus \{p_1, \dots, p_r\}$$

gives a basic example of the rigidity method. The meromorphic function field $\tilde{\mathcal{N}} = \mathbb{C}(\hat{X})$ of the compactification $\hat{X}$ of X is a finite Galois extension of the rational function field $\tilde{\mathcal{K}} = \mathbb{C}(\mathbb{P}^1) = \mathbb{C}(t)$ and its Galois group is isomorphic to the covering transformation group of φ. Let $\mathfrak{p}_1, \dots, \mathfrak{p}_r$ be the corresponding prime divisors of $\tilde{\mathcal{N}}/\mathbb{C}$ to the ramification points $p_1, \dots, p_r$ of φ. Then $\tilde{\mathcal{N}}/\tilde{\mathcal{K}}$ is unramified outside $\{\mathfrak{p}_1, \dots, \mathfrak{p}_r\}$ and descends to $\bar{\mathbb{Q}}$. Namely, if $p_1, \dots, p_r$ are algebraic points over $\mathbb{Q}$, then there exists a unique Galois extension $\mathcal{N}$ of $\mathcal{K} = \bar{\mathbb{Q}}(t)$ unramified outside $\{\mathfrak{p}_1, \dots, \mathfrak{p}_r\}$ such that $\mathcal{N} \cdot \mathbb{C} = \mathcal{N} \otimes_{\bar{\mathbb{Q}}} \mathbb{C} = \tilde{\mathcal{N}}$. Moreover, $\mathcal{N}/\mathcal{K}$ descends to the number field k generated by the coefficients of the defining polynomial of $\mathcal{N}/\mathcal{K}$. We are interested in the smallest field of definition k of $\tilde{\mathcal{N}}/\tilde{\mathcal{K}}$.

Example 1.1 (S_n-covering). The branched covering

$$\varphi_0\colon \mathbb{P}^1 \ni x \longmapsto \frac{x^n}{nx - n + 1} = t \in \mathbb{P}^1$$

is unramified outside $\{0, 1, \infty\}$. The corresponding function fields are $\tilde{\mathcal{L}} = \mathbb{C}(x)$ and $\tilde{\mathcal{K}} = \mathbb{C}(t)$ and the corresponding ramified prime divisors $\mathfrak{p}_1, \mathfrak{p}_2, \mathfrak{p}_3$ of $\tilde{\mathcal{K}}/\mathbb{C}$ are decomposed in $\tilde{\mathcal{L}}$ as follows:

$$\mathfrak{p}_1 = \mathfrak{P}_1^n, \qquad \mathfrak{p}_2 = \mathfrak{P}_2^2 \mathfrak{Q}_1 \cdots \mathfrak{Q}_{n-2}, \qquad \mathfrak{p}_3 = \mathfrak{P}_3^{n-1} \mathfrak{R}. \tag{1.1}$$

Here $\mathfrak{P}_1, \mathfrak{P}_2, \mathfrak{P}_3$ and $\mathfrak{R}$ are the prime divisors of $\tilde{\mathcal{L}}/\mathbb{C}$ corresponding to $0, 1, \infty$ and $\frac{n-1}{n}$, respectively. Similarly, $\mathfrak{Q}_1, \ldots, \mathfrak{Q}_{n-2}$ correspond to $n-2$ distinct roots of

$$\{x^n - (nx - n + 1)\}/(x-1)^2 = x^{n-2} + 2x^{n-3} + \cdots + (n-1).$$

Let $\varphi\colon \hat{X} \longrightarrow \mathbb{P}^1$ be the Galois closure of the covering φ_0. Then the meromorphic function field $\tilde{\mathcal{N}} = \mathbb{C}(\hat{X})$ coincides with the splitting field of

$$f(x) = x^n - t(nx - n + 1) \in \mathbb{C}(t)[x]$$

and the Galois group of $\tilde{\mathcal{N}}/\tilde{\mathcal{K}}$ is represented as a permutation group on n distinct roots of $f(x)$. A fundamental fact of the theory of algebraic function fields indicates that the prime divisor decompositions (1.1) express the permutation types of the inertia subgroups corresponding to the ramification points. Namely, generators $\sigma_1, \sigma_2, \sigma_3$ of the inertia subgroups of $\mathfrak{P}_1, \mathfrak{P}_2, \mathfrak{P}_3$ (these groups are cyclic) are respectively an n-cycle, a transposition and an $(n-1)$-cycle. Since these cycles generate the symmetric group S_n, the Galois group of $\tilde{\mathcal{N}}/\tilde{\mathcal{K}}$ is isomorphic to S_n. Moreover, since $f(x)$ has rational coefficients, the smallest field of definition of $\tilde{\mathcal{N}}/\tilde{\mathcal{K}}$ is $\mathbb{Q}$. In other words, the splitting field N of $f(x)$ over $\mathbb{Q}(t)$ is a regular Galois extension of $\mathbb{Q}(t)$ with Galois group S_n.

Let $p_1, \ldots, p_r$ be r distinct algebraic points on $\mathbb{P}^1$ and

$$\Phi\colon \mathcal{U} \longrightarrow \mathbb{P}^1 \setminus \{p_1, \ldots, p_r\}$$

the universal covering. The covering transformation group of Φ is isomorphic to the fundamental group $\Gamma_r := \langle \gamma_1, \ldots, \gamma_r \mid \gamma_1 \cdots \gamma_r = 1 \rangle$ of the punctured projective line. Here γ_i denotes the homotopy class of a sufficiently small closed path counterclockwise around p_i. The property of universal coverings and Galois theory yield a one-to-one correspondence among the following objects.

- The finite unramified Galois coverings of $\mathbb{P}^1 \setminus \{p_1, \ldots, p_r\}$.
- The normal subgroups of Γ_r of finite index.
- The finite Galois extensions of $\bar{\mathbb{Q}}(t)$ unramified outside $\{\mathfrak{p}_1, \ldots, \mathfrak{p}_r\}$.

Therefore the maximal Galois extension $\mathcal{M}_r$ of $\bar{\mathbb{Q}}(t)$ unramified outside $\{\mathfrak{p}_1, \ldots, \mathfrak{p}_r\}$ has Galois group isomorphic to the profinite completion of Γ_r:

$$G(\mathcal{M}_r/\bar{\mathbb{Q}}(t)) = \hat{\Gamma}_r := \varprojlim_N \Gamma_r/N,$$

where the limit is taken over all normal subgroups $N \lhd \Gamma_r$ of finite index. The meromorphic function field $\mathcal{L}/\bar{\mathbb{Q}}(t)$ corresponding to an unramified

Galois covering $\varphi\colon X \longrightarrow \mathbb{P}^1\setminus\{p_1,\dots,p_r\}$ is a subfield of $\mathcal{M}_r$. Then we have the canonical surjective homomorphism

$$\hat{\Gamma}_r = \langle \gamma_1,\dots,\gamma_r \mid \gamma_1\cdots\gamma_r = 1\rangle\hat{} \longrightarrow G(\mathcal{L}/\bar{\mathbb{Q}}(t)) =: G.$$

Let $\sigma_i \in G$ be the image of γ_i and C_i the conjugacy class of σ_i. Then $\sigma_1,\dots,\sigma_r$ generate G and satisfy $\sigma_1\cdots\sigma_r = 1$. The class vector $(C_1,\dots,C_r)$ is called the *ramification structure* of $\mathcal{L}/\bar{\mathbb{Q}}(t)$.

Let G be a finite group with trivial center and $\mathcal{C} = (C_1,\dots,C_r)$ a class vector of G. Consider the set of all generating r-systems of $\mathcal{C}$:

$$\Sigma(\mathcal{C}) := \{(\sigma_1,\dots,\sigma_r) \mid \sigma_i \in C_i,\ \sigma_1\cdots\sigma_r = 1,\ \langle\sigma_1,\dots,\sigma_r\rangle = G\}.$$

The group $G \cong \mathrm{Inn}(G)$ acts on $\Sigma(\mathcal{C})$ by conjugation; denote by $[\boldsymbol{\sigma}] = [\sigma_1,\dots,\sigma_r]$ the $\mathrm{Inn}(G)$-orbit of $\boldsymbol{\sigma} = (\sigma_1,\dots,\sigma_r) \in \Sigma(\mathcal{C})$. The class vector $\mathcal{C}$ is said to be *rigid* if G acts transitively on $\Sigma(\mathcal{C})$, or briefly if $|\Sigma(\mathcal{C})/\mathrm{Inn}(G)| = 1$. In general, a conjugacy class C of a finite group G is said to be *rational* if $C^m = C$ for any integer m prime to $|G|$, or equivalently if all irreducible characters of G take integral values at C. We say the class vector $\mathcal{C}$ *rational* when each C_i is rational. The following is known as a basic theorem in the rigidity method.

Theorem 1.2 (Basic Rigidity Theorem). *Let G be a finite group with trivial center and $\mathcal{C} = (C_1,\dots,C_r)$ a class vector of G. If $\mathcal{C}$ is rational and rigid, then there exists a regular Galois extension $N/\mathbb{Q}(t)$ with Galois group G and ramification structure $\mathcal{C}$.*

See [10] or [22] for proofs.

Remark 1.3. The regular S_n-extension $N/\mathbb{Q}(t)$ in Example 1.1 possesses the ramification structure $\mathcal{C} = (2A, (n-1)A, nA)$, where mA denotes the conjugacy class of m-cycles in S_n. Since each conjugacy class in S_n is rational, the class vector $\mathcal{C}$ is rational. Moreover, one can verify the rigidity of $\mathcal{C}$ by a brief calculation (see for example [19]).

Example 1.4. The projective general semi-linear group $\mathrm{P\Gamma L}_2(p^n)$ is the automorphism group of $\mathrm{PSL}_2(p^n)$. We take a class vector $\mathcal{C} = (2C, 4B, 10A)$ of $\mathrm{P\Gamma L}_2(p^2)$, where we follow the notation in ATLAS [1]. Feit [5] proved that $\mathcal{C}$ is rational and rigid if $p \equiv \pm 2 \pmod 5$. Let $N/\mathbb{Q}(t)$ be a regular Galois extension with Galois group $\mathrm{P\Gamma L}_2(p^2)$ and ramification structure $\mathcal{C}$. The fixed subfield K by the normal subgroup $\mathrm{PSL}_2(p^2)$ of index 4 is an algebraic function field of genus 0. Since the prime divisor corresponding to $2C$ splits completely in $K/\mathbb{Q}(t)$, the subfield K is a rational function field over $\mathbb{Q}$. Therefore the group $\mathrm{PSL}_2(p^2)$ occurs regularly over $\mathbb{Q}$ if $p \equiv \pm 2 \pmod 5$.

2. Braid Orbit Theorem

The projective special semi-linear group $P\Sigma L_2(p^2)$ is one of the normal subgroups of $P\Gamma L_2(p^2)$ of index 2. Przywara [14] showed that there exists a regular Galois extension of $\mathbb{Q}(t)$ with Galois group $G = P\Sigma L_2(25)$ and ramification structure $\mathcal{C} = (2A, 2C, 2D, 12A)$. Here the class vector $\mathcal{C}$ is rational but non-rigid. Actually, one verifies that

$$|\Sigma(\mathcal{C})/\mathrm{Inn}(G)| = 12.$$

Przywara realized $P\Sigma L_2(25)$ by means of the *braid orbit theorem* which had been proposed by Matzat [12] and can be applied to such non-rigid class vectors. Matzat considered the Galois coverings of $\mathbb{P}^1 \setminus \{p_1, \ldots, p_r\}$ with transcendental ramification points $p_1, \ldots, p_r$ and the braid actions on $\Sigma(\mathcal{C})/\mathrm{Inn}(G)$ associated to such Galois coverings. We sketch this theorem restricting to the principal case $r = 4$.

Let $\mathcal{M}$ be the maximal Galois extension of $\bar{\mathbb{Q}}(u)$ unramified outside $\{0, 1, \infty\}$ and $\tilde{\mathcal{M}}$ a maximal regular Galois extension of $\mathcal{M}(t)$ unramified outside $\{0, 1, \infty, u\}$, where u and t are transcendental over $\mathbb{Q}$. Their Galois groups are

$$G(\mathcal{M}/\bar{\mathbb{Q}}(u)) = \widehat{\langle \gamma_1, \gamma_2, \gamma_3 \mid \gamma_1\gamma_2\gamma_3 = 1 \rangle} =: \Gamma$$

and

$$G(\tilde{\mathcal{M}}/\mathcal{M}(t)) = \widehat{\langle \tilde{\gamma}_1, \tilde{\gamma}_2, \tilde{\gamma}_3, \tilde{\gamma}_4 \mid \tilde{\gamma}_1\tilde{\gamma}_2\tilde{\gamma}_3\tilde{\gamma}_4 = 1 \rangle} =: \tilde{\Gamma}.$$

The fields $\mathcal{M}(t)$ and $\tilde{\mathcal{M}}$ are Galois over $\mathbb{Q}(u, t)$ and their Galois groups split as follows:

$$G(\mathcal{M}(t)/\mathbb{Q}(u, t)) \cong \Gamma \rtimes \Gamma_{\mathbb{Q}} \tag{2.1}$$

and

$$G(\tilde{\mathcal{M}}/\mathbb{Q}(u, t)) \cong \tilde{\Gamma} \rtimes G(\mathcal{M}(t)/\mathbb{Q}(u, t)), \tag{2.2}$$

where $\Gamma_{\mathbb{Q}} := G(\bar{\mathbb{Q}}/\mathbb{Q})$ is the absolute Galois group of $\mathbb{Q}$.

Let G be a finite group with trivial center and $\mathcal{C} = (C_1, C_2, C_3, C_4)$ a rational class vector of G. For a given $\boldsymbol{\sigma} = (\sigma_1, \sigma_2, \sigma_3, \sigma_4) \in \Sigma(\mathcal{C})$, consider a surjective homomorphism $\psi_{\boldsymbol{\sigma}} \colon \tilde{\Gamma} \longrightarrow G$ defined by mapping $\tilde{\gamma}_i$ to σ_i. The fixed subfield

$$\mathcal{N}_{\boldsymbol{\sigma}} := \tilde{\mathcal{M}}^{\mathrm{Ker}(\psi_{\boldsymbol{\sigma}})}$$

is a regular Galois extension of $\mathcal{M}(t)$ with Galois group G and ramification structure $\mathcal{C}$. We define an action of $G(\mathcal{M}(t)/\mathbb{Q}(u, t))$ on $\Sigma(\mathcal{C})/\mathrm{Inn}(G)$ by extending the Galois action $G(\mathcal{M}(t)/\mathbb{Q}(u, t)) \curvearrowright \tilde{\Gamma}$ in the semi-direct product (2.2) to G via $\psi_{\boldsymbol{\sigma}}$. Let

$$\Delta_{\boldsymbol{\sigma}} := \{\delta \in G(\mathcal{M}(t)/\mathbb{Q}(u, t)) \mid [\boldsymbol{\sigma}]^{\delta} = [\boldsymbol{\sigma}]\}$$

be the stabilizer of $[\boldsymbol{\sigma}] \in \Sigma(\mathcal{C})/\mathrm{Inn}(G)$ and

$$K_{\boldsymbol{\sigma}}(t) := \mathcal{M}(t)^{\Delta_{\boldsymbol{\sigma}}}$$

the fixed subfield of $\Delta_{\boldsymbol{\sigma}}$. From (2.1), the action of $G(\mathcal{M}(t)/\mathbb{Q}(u,t))$ splits into two sorts of actions; Γ acts on $\Sigma(\mathcal{C})/\mathrm{Inn}(G)$ and $\Gamma_{\mathbb{Q}}$ acts on $(\Sigma(\mathcal{C})/\mathrm{Inn}(G))/\Gamma$. The former is called the braid action because Γ is the normal subgroup of the Hurwitz braid group H_4 of index 2. Denote a Γ-orbit of $[\boldsymbol{\sigma}] \in \Sigma(\mathcal{C})/\mathrm{Inn}(G)$ by $B(\boldsymbol{\sigma})$ and call it a *braid orbit.*

Lemma 2.1. *Under the preceding conditions, we have the following.*
(i) *There exists a regular Galois extension* $N_{\boldsymbol{\sigma}}/K_{\boldsymbol{\sigma}}(t)$ *with Galois group* G *and ramification structure* $\mathcal{C}$.
(ii) $K_{\boldsymbol{\sigma}}$ *is a finite extension of* $k_{\boldsymbol{\sigma}}(u) := K_{\boldsymbol{\sigma}} \cap \bar{\mathbb{Q}}(u)$, *whose degree equals the length of* $B(\boldsymbol{\sigma})$.
(iii) $k_{\boldsymbol{\sigma}}$ *is a finite extension of* $\mathbb{Q}$, *whose degree equals the length of the* $\Gamma_{\mathbb{Q}}$*-orbit of* $B(\boldsymbol{\sigma})$.

Proof. (i) For any $\delta \in G(\tilde{\mathcal{M}}/K_{\boldsymbol{\sigma}}(t))$, there exists $\tau \in G$ such that $\boldsymbol{\sigma}^{\delta} = \boldsymbol{\sigma}^{\tau}$ by the definition of $K_{\boldsymbol{\sigma}}$. This means the Galois group $G = G(\mathcal{N}_{\boldsymbol{\sigma}}/K_{\boldsymbol{\sigma}}(t))$ is invariant under δ and so is $\mathcal{N}_{\boldsymbol{\sigma}}$. Hence $\mathcal{N}_{\boldsymbol{\sigma}}$ is a Galois extension of $K_{\boldsymbol{\sigma}}(t)$ and each element of $G(\mathcal{N}_{\boldsymbol{\sigma}}/K_{\boldsymbol{\sigma}}(t))$ acts as an inner automorphism on G. Let Z be the centralizer of G in $G(\mathcal{N}_{\boldsymbol{\sigma}}/K_{\boldsymbol{\sigma}}(t))$, then

$$G(\mathcal{N}_{\boldsymbol{\sigma}}/K_{\boldsymbol{\sigma}}(t)) = G \cdot Z.$$

Since G has a trivial center, the right-hand side is a direct product. Therefore the fixed subfield $N_{\boldsymbol{\sigma}}$ of Z is a regular Galois extension of $K_{\boldsymbol{\sigma}}(t)$ with $N_{\boldsymbol{\sigma}} \cdot \mathcal{M}(t) = \mathcal{N}_{\boldsymbol{\sigma}}$.
(ii) Let L be the fixed subfield of the stabilizer of $B(\boldsymbol{\sigma})$, then L is contained in $\bar{\mathbb{Q}}(u,t) \cap K_{\boldsymbol{\sigma}}(t) = k_{\boldsymbol{\sigma}}(u,t)$. Furthermore, let $\mathcal{K}_{\boldsymbol{\sigma}}(t)$ be the fixed subfield of $\Gamma_{\boldsymbol{\sigma}} := \Gamma \cap \Delta_{\boldsymbol{\sigma}}$ in $\mathcal{M}(t)$. Here $\Gamma = G(\mathcal{M}(t)/\bar{\mathbb{Q}}(u,t))$ and $G(\mathcal{M}(t)/L)$ act transitively on $B(\boldsymbol{\sigma})$ and the stabilizers of $[\boldsymbol{\sigma}]$ are $\Gamma_{\boldsymbol{\sigma}}$ and $\Delta_{\boldsymbol{\sigma}}$, respectively. Hence

$$\begin{aligned}|B(\boldsymbol{\sigma})| = [K_{\boldsymbol{\sigma}} : L] &= [\mathcal{K}_{\boldsymbol{\sigma}}(t) : \bar{\mathbb{Q}}(u,t)] \\ &= [K_{\boldsymbol{\sigma}}(t) \cdot \bar{\mathbb{Q}}(u,t) : \bar{\mathbb{Q}}(u,t)] \\ &= [K_{\boldsymbol{\sigma}}(t) : K_{\boldsymbol{\sigma}}(t) \cap \bar{\mathbb{Q}}(u,t)] \\ &= [K_{\boldsymbol{\sigma}}(t) : k_{\boldsymbol{\sigma}}(u,t)].\end{aligned}$$

This shows $L = k_{\boldsymbol{\sigma}}(u,t)$ and the degree of $K_{\boldsymbol{\sigma}}(t)/k_{\boldsymbol{\sigma}}(u,t)$ equals $|B(\boldsymbol{\sigma})|$.
(iii) The Galois group $G(\mathcal{M}(t)/\mathbb{Q}(u,t))$ acts transitively on the $\Gamma_{\mathbb{Q}}$-orbit of $B(\boldsymbol{\sigma})$ and the stabilizer of $B(\boldsymbol{\sigma})$ is $G(\mathcal{M}(t)/k_{\boldsymbol{\sigma}}(u,t))$. Hence the degree of $k_{\boldsymbol{\sigma}}(u,t)/\mathbb{Q}(u,t)$ equals the length of the $\Gamma_{\mathbb{Q}}$-orbit of $B(\boldsymbol{\sigma})$. □

The preceding lemma shows that G occurs regularly over $\mathbb{Q}$ if a braid orbit $B(\boldsymbol{\sigma})$ is invariant under the action of $\Gamma_{\mathbb{Q}}$ and the corresponding field of definition $K_{\boldsymbol{\sigma}}$ is a rational function field. Matzat [12] stated a sufficient condition for $K_{\boldsymbol{\sigma}}$ to be a rational function field over $\mathbb{Q}$. Let π_B be the permutation representation of Γ on a braid orbit $B = B(\boldsymbol{\sigma})$ and c_i the number of cycles in $\pi_B(\gamma_i)$. The *genus* of the braid orbit B is defined by

$$g(B) := 1 - |B| + \frac{1}{2}\sum_{i=1}^{3}(|B| - c_i).$$

This coincides with the genus of the algebraic function field $K_{\boldsymbol{\sigma}}/k_{\boldsymbol{\sigma}}$. Furthermore, consider the following *oddness condition.*

(O) There exists i such that $\pi_B(\gamma_i)$ has an odd number of l-cycles for a length l.

We demand this condition for the existence of $k_{\boldsymbol{\sigma}}$-rational points on $K_{\boldsymbol{\sigma}}$. Now the braid orbit theorem is stated as follows.

Theorem 2.2 (Braid Orbit Theorem). *Let G be a finite group with trivial center and $\mathcal{C} = (C_1, C_2, C_3, C_4)$ a rational class vector of G. Suppose $\Sigma(\mathcal{C})/\mathrm{Inn}(G)$ contains a $\Gamma_{\mathbb{Q}}$-invariant braid orbit $B = B(\boldsymbol{\sigma})$ which has genus $g(B) = 0$ and satisfies the oddness condition* (O). *Then the field of definition $K_{\boldsymbol{\sigma}}$ is a rational function field over $\mathbb{Q}$ and hence G occurs regularly over $\mathbb{Q}$.*

See the original paper [12] or a comprehensive book [10]. To apply this theorem to a given finite group G and its class vector $\mathcal{C}$, one needs to calculate the action of Γ on $\Sigma(\mathcal{C})/\mathrm{Inn}(G)$ explicitly. The action of Γ on $\tilde{\Gamma}$ in the semi-direct product $G(\tilde{\mathcal{M}}/\bar{\mathbb{Q}}(u,t)) \cong \tilde{\Gamma} \rtimes \Gamma$ is well-known by the fundamental relations of the Hurwitz braid group. Namely, the action on $\Sigma(\mathcal{C})/\mathrm{Inn}(G)$ reads as follows:

$$\begin{aligned}
[\sigma_1, \sigma_2, \sigma_3, \sigma_4]^{\gamma_1} &= [\sigma_1^{\sigma_2\sigma_3}, \sigma_2, \sigma_3, \sigma_4^{\sigma_2\sigma_3}],\\
[\sigma_1, \sigma_2, \sigma_3, \sigma_4]^{\gamma_2} &= [\sigma_1, \sigma_2^{\sigma_3\sigma_1}, \sigma_3, \sigma_4^{\sigma_1\sigma_3}],\\
[\sigma_1, \sigma_2, \sigma_3, \sigma_4]^{\gamma_3} &= [\sigma_1, \sigma_2, \sigma_3^{\sigma_1\sigma_2}, \sigma_4^{\sigma_1\sigma_2}].
\end{aligned} \tag{2.3}$$

Two braid orbits B and B' have the same length and genus if they are conjugate to each other under the action of $\Gamma_{\mathbb{Q}}$. Hence a braid orbit B is $\Gamma_{\mathbb{Q}}$-invariant if there exist no other braid orbits of the same length and genus in $\Sigma(\mathcal{C})/\mathrm{Inn}(G)$.

3. Groups $\mathrm{P\Sigma L}_2(p^2)$

The projective special linear group $\mathrm{PSL}_2(p^2)$ has few rational conjugacy classes. So we begin with the group $\mathrm{P\Sigma L}_2(p^2)$, which contains $\mathrm{PSL}_2(p^2)$ as normal subgroup of degree 2. In this section, we concretely calculate braid orbits to apply the braid orbit theorem for small p. The p-Frobenius automorphism $\mathbb{F}_{p^2} \ni s \longmapsto \bar{s} := s^p \in \mathbb{F}_{p^2}$ induces an automorphism of $\mathrm{PSL}_2(p^2)$:

$$\varphi\colon \mathrm{PSL}_2(p^2) \ni \begin{pmatrix} s & t \\ u & v \end{pmatrix} \longmapsto \begin{pmatrix} \bar{s} & \bar{t} \\ \bar{u} & \bar{v} \end{pmatrix} \in \mathrm{PSL}_2(p^2).$$

The projective special semi-linear group $\mathrm{P\Sigma L}_2(p^2) =: G$ is defined by the semi-direct product $\mathrm{PSL}_2(p^2) \rtimes \langle \varphi \rangle$. The conjugacy classes $2C, 2D, pA$, pB in G are defined as the classes of

$$\varphi, \quad \begin{pmatrix} 0 & 1 \\ -1 & 0 \end{pmatrix}\varphi, \quad \begin{pmatrix} 1 & 1 \\ 0 & 1 \end{pmatrix}, \quad \begin{pmatrix} 1 & 0 \\ r & 1 \end{pmatrix},$$

respectively, where $r \in \mathbb{F}_{p^2}$ is any non-square element. Since these conjugacy classes are rational, the class vectors $\mathcal{C} := (2C, 2D, pA, pA)$ and $\mathcal{C}' := (2C, 2D, pA, pB)$ are rational. In the following examples, we use the algebraic computation software GAP [6].

Example 3.1 ($p = 3$)**.** The groups $\mathrm{PSL}_2(9)$ and $\mathrm{P\Sigma L}_2(9)$ are isomorphic to A_6 and S_6, respectively. The classes in $\mathrm{P\Sigma L}_2(9)$ are identified with the following classes in S_6:

$$2C = [(1\ 2)], \qquad 2D = [(1\ 2)(3\ 4)(5\ 6)],$$
$$3A = [(1\ 2\ 3)], \qquad 3B = [(1\ 2\ 3)(4\ 5\ 6)].$$

Here $\Sigma(\mathcal{C}')$ divides into two $\mathrm{Inn}(G)$-orbits which are represented by

$$\boldsymbol{\sigma} = ((1\ 2),\ (1\ 3)(2\ 4)(5\ 6),\ (2\ 4\ 5),\ (1\ 2\ 3)(4\ 5\ 6)),$$
$$\boldsymbol{\tau} = ((1\ 2),\ (1\ 3)(2\ 4)(5\ 6),\ (2\ 5\ 3),\ (1\ 4\ 3)(2\ 5\ 6)).$$

Since these orbits are fixed by $\gamma_1 \in \Gamma$ and permuted by $\gamma_2, \gamma_3 \in \Gamma$, the permutation types of $\gamma_1, \gamma_2, \gamma_3$ are $(1)^2$, (2), (2), respectively. Hence the genus of $B' := \{\boldsymbol{\sigma}, \boldsymbol{\tau}\}$ is

$$1 - 2 + \frac{1}{2}(0 + 1 + 1) = 0$$

and there exists a regular Galois extension of $\mathbb{Q}(u, t)$ with Galois group $\mathrm{P\Sigma L}_2(9) \cong S_6$ and ramification structure $(2C, 2D, 3A, 3B)$.

Remark 3.2. The S_6-extension $N_{\boldsymbol{\sigma}}/K_{\boldsymbol{\sigma}}(t)$ with ramification structure $(2C, 2D, 3A, 3B)$ is concretely constructed as follows. Let L be the fixed subfield of a subgroup $S_5 = \{\sigma \in S_6 \mid \sigma(6) = 6\}$ and $\mathfrak{p}_1, \mathfrak{p}_2, \mathfrak{p}_3, \mathfrak{p}_4$ the corresponding prime divisors of $K_{\boldsymbol{\sigma}}(t)$ to the ramification points $0, 1, \infty$ and u. They are decomposed in L as follows:

$$\mathfrak{p}_1 = \mathfrak{P}_1^2\mathfrak{P}_4, \quad \mathfrak{p}_2 = \mathfrak{P}_3^2, \quad \mathfrak{p}_3 = \mathfrak{Q}_1^3\mathfrak{Q}_3, \quad \mathfrak{p}_4 = \mathfrak{Q}_2^3,$$

where the subscripts of $\mathfrak{P}_i$ and $\mathfrak{Q}_j$ coincide with their degrees as divisors. The Riemann-Hurwitz genus formula shows that the genus of L equals 0. Let $x \in L$ be a generator of the divisor $\mathfrak{P}_1\mathfrak{Q}_1^{-1}$, then there exist monic polynomials $f_3, f_4, g_2, g_3 \in K_{\boldsymbol{\sigma}}[X]$ such that

$$\frac{\mathfrak{P}_3}{\mathfrak{Q}_1^3} = (f_3(x)), \quad \frac{\mathfrak{P}_4}{\mathfrak{Q}_1^4} = (f_4(x)), \quad \frac{\mathfrak{Q}_2}{\mathfrak{Q}_1^2} = (g_2(x)), \quad \frac{\mathfrak{Q}_3}{\mathfrak{Q}_1^3} = (g_3(x)),$$

where the subscripts of f_i and g_j coincide with their degrees as polynomials. We may assume that $g_2(X) = X^2 + 2uX + a$, $a \in K_{\boldsymbol{\sigma}}$, by changing x if necessary. These divisors satisfy

$$(t) = \frac{\mathfrak{P}_1^2\mathfrak{P}_4}{\mathfrak{Q}_1^3\mathfrak{Q}_3} = \left(\frac{x^2 f_4(x)}{g_3(x)}\right), \quad (t-1) = \frac{\mathfrak{P}_3^2}{\mathfrak{Q}_1^3\mathfrak{Q}_3} = \left(\frac{f_3(x)^2}{g_3(x)}\right),$$

and

$$(t-u) = \frac{\mathfrak{Q}_2^3}{\mathfrak{Q}_1^3\mathfrak{Q}_3} = \left(\frac{g_2(x)^3}{g_3(x)}\right).$$

Hence there exist certain constants $c_1, c_2, c_3 \in K_{\boldsymbol{\sigma}}$ such that

$$\frac{X^2 f_4(X)}{c_1 g_3(X)} = \frac{f_3(X)^2}{c_2 g_3(X)} + 1 = \frac{g_2(X)^3}{c_3 g_3(X)} + u.$$

Since each polynomial is monic, we have $c_1 = c_2 = c_3 =: c$. From this equation, we obtain a system of equations in 11 undetermined coefficients of f_3, f_4, g_2, g_3 and the unknown constant c. Solving these equations, we have

$$\begin{aligned} f_3(X) &= X^3 + 3uX^2 + 3u(1+v)X + u(1+v)^3, \\ f_4(X) &= X^4 + 6uX^3 + \{12u^2 + 3u(1+v)^2\}X^2 \\ &\quad + 16u^2(1+v)X + 6u^2(1+v)^3, \\ g_2(X) &= X^2 + 2uX + u(1+v)^2, \\ -cg_3(X) &= 4u(1+v)^2X^3 + 9u^2(1+v)^2X^2 \\ &\quad + 6u^2(1+v)^4X + u^2(1+v)^6, \end{aligned}$$

where v denotes a square root of $1-u$. Thus the field of definition $K_{\boldsymbol{\sigma}}$ equals $\mathbb{Q}(\sqrt{1-u})$. The sextic field $L/K_{\boldsymbol{\sigma}}(t)$ is generated by a root of the polynomial $F(t,X) := X^2 f_4(X) - c\,g_3(X)\,t \in \mathbb{Q}(\sqrt{1-u},t)[X]$ and the S_6-extension $N_{\boldsymbol{\sigma}}$ is given as the Galois closure of L. The discriminant of $F(X)$ is computed as

$$\mathrm{Disc}(F) = -2^{13}3^6u^{10}v(1+v)^{23}t(1-t)^3(u-t)^4.$$

Therefore the fixed subfield of A_6 in $N_{\boldsymbol{\sigma}}/\mathbb{Q}(\sqrt{1-u},t)$ is

$$L' = \mathbb{Q}\left(\sqrt{1-u}\,,\sqrt{\frac{-2v(1+v)t}{1-t}}\right),$$

and the specialization $F(s^2/(s^2-2v(1+v)),X)$ is a polynomial with Galois group A_6 over $\mathbb{Q}(v,s)$.

Example 3.3 ($p=5$)**.** For $G = \mathrm{P\Sigma L}_2(25)$, we have

$$|\Sigma(\mathcal{C})/\mathrm{Inn}(G)| = 4 \quad \text{and} \quad |\Sigma(\mathcal{C}')/\mathrm{Inn}(G)| = 2.$$

In either case, the action of Γ is transitive and the oddness condition (O) is satisfied. Actually, the permutation types of $\gamma_1, \gamma_2, \gamma_3$ are (1)(3), (1)(3), (1)(3) on $\Sigma(\mathcal{C})/\mathrm{Inn}(G)$ and $(1)^2$, (2), (2) on $\Sigma(\mathcal{C}')/\mathrm{Inn}(G)$, respectively. In the first case, the genus of $B := \Sigma(\mathcal{C})/\mathrm{Inn}(G)$ is

$$1 - 4 + \frac{1}{2}(2+2+2) = 0$$

and the field of definition $K_{\boldsymbol{\sigma}}/\mathbb{Q}(u)$ is a rational function field whose Galois closure has Galois group A_4 for each $[\boldsymbol{\sigma}] \in B$. In the latter case, the genus of $B' := \Sigma(\mathcal{C}')/\mathrm{Inn}(G)$ is calculated in the same way as the case $p=3$. Consequently, we obtain two types of Galois realization of $\mathrm{P\Sigma L}_2(25)$.

Example 3.4 ($p=7$)**.** For $G = \mathrm{P\Sigma L}_2(49)$, we have

$$|\Sigma(\mathcal{C})/\mathrm{Inn}(G)| = 4 \quad \text{and} \quad |\Sigma(\mathcal{C}')/\mathrm{Inn}(G)| = 8.$$

In either case, the action of Γ is transitive. The permutation types of $\gamma_1, \gamma_2, \gamma_3$ are (1)(3), (1)(3), (1)(3) on $\Sigma(\mathcal{C})/\mathrm{Inn}(G)$ and $(1)^4(2)^2$, $(4)^2$, $(4)^2$ on $\Sigma(\mathcal{C}')/\mathrm{Inn}(G)$, respectively. In the first case, the genus of $B := \Sigma(\mathcal{C})/\mathrm{Inn}(G)$ is calculated in the same way as the case $p=5$. In the latter case, while the genus of $B' := \Sigma(\mathcal{C}')/\mathrm{Inn}(G)$ is

$$1 - 8 + \frac{1}{2}(2+6+6) = 0,$$

the oddness condition (O) is not satisfied. Thus the braid orbit theorem can be applied only to B.

Example 3.5 ($p = 11, 13$). For $\mathcal{C}' = (2C, 2D, pA, pB)$, the cases $p = 11$ and $p = 13$ are similar. In either case, the number of Inn(G)-orbits is

$$|\Sigma(\mathcal{C}')/\mathrm{Inn}(G)| = 18$$

and Γ acts intransitively on $\Sigma(\mathcal{C}')/\mathrm{Inn}(G)$. Namely, $\Sigma(\mathcal{C}')/\mathrm{Inn}(G)$ divides into two braid orbits B'_2 of length 2 and B'_{16} of length 16. The shorter orbit B'_2 is a unique braid orbit of length 2 in $\Sigma(\mathcal{C}')/\mathrm{Inn}(G)$ and Γ acts on it in the same way as the case $p = 3$. On the other hand, for $\mathcal{C} = (2C, 2D, pA, pA)$, we have

$$|\Sigma(\mathcal{C})/\mathrm{Inn}(G)| = \begin{cases} 12 & \text{for } p = 11, \\ 24 & \text{for } p = 13. \end{cases}$$

Here Γ acts transitively on $B := \Sigma(\mathcal{C})/\mathrm{Inn}(G)$. For $p = 11$, the permutation types of all γ_i's are $(1)^2(5)^2$ and hence $g(B) = 1$ and for $p = 13$, the permutation types of all γ_i's are $(1)^3(7)^3$ and hence $g(B) = 4$. Thus the braid orbit theorem can be applied only to B'_2.

Example 3.6 ($p = 17, 19$). For $\mathcal{C} = (2C, 2D, pA, pA)$, the cases $p = 17$ and $p = 19$ are similar. In either case, the number of Inn(G)-orbits is

$$|\Sigma(\mathcal{C})/\mathrm{Inn}(G)| = 40$$

and Γ acts intransitively on $\Sigma(\mathcal{C})/\mathrm{Inn}(G)$. Namely, $\Sigma(\mathcal{C})/\mathrm{Inn}(G)$ divides into two braid orbits B_4 of length 4 and B_{36} of length 36. Since the shorter orbit B_4 is a unique braid orbit of length 4 in $\Sigma(\mathcal{C})/\mathrm{Inn}(G)$ and Γ acts on it in the same way as the cases $p = 5$ and $p = 7$, the braid orbit theorem can be applied to this orbit.

4. Proof of Main Theorem

In the examples of the previous section, we found isolated braid orbits of length 2 for $p = 3, 5, 11, 13$ and those of length 4 for $p = 5, 7, 17, 19$. These facts are generalized as follows:

Proposition 4.1. *Let $\mathcal{C} = (2C, 2D, pA, pA)$ and $\mathcal{C}' = (2C, 2D, pA, pB)$ be the prescribed class vectors of $G = \mathrm{P\Sigma L}_2(p^2)$.*
(i) *If $p \equiv \pm 3 \pmod 8$, then $\Sigma(\mathcal{C}')/\mathrm{Inn}(G)$ contains a unique braid orbit of length 2 and the permutation types of $\gamma_1, \gamma_2, \gamma_3 \in \Gamma$ on it are $(1)^2$, (2), (2), respectively.*

(ii) *If* $p \equiv \pm 5 \pmod{12}$, *then* $\Sigma(\mathcal{C})/\mathrm{Inn}(G)$ *contains a unique braid orbit of length 4 and the permutation types of* $\gamma_1, \gamma_2, \gamma_3 \in \Gamma$ *on it are* (1)(3), (1)(3), (1)(3), *respectively.*

By applying the braid orbit theorem to these orbits, we can realize $\mathrm{PSL}_2(p^2)$ regularly over $\mathbb{Q}$ as follows:

Proof of Main Theorem. By the braid orbit theorem, there exists a regular Galois extension N of $\mathbb{Q}(T) = \mathbb{Q}(u,t)$ with Galois group $\mathrm{P\Sigma L}_2(p^2)$ and ramification structure $\mathcal{C} = (2C, 2D, pA, pA)$ (in the case $p \equiv \pm 5 \pmod{12}$) or $\mathcal{C}' = (2C, 2D, pA, pB)$ (in the case $p \equiv \pm 3 \pmod 8$). The fixed subfield L by the normal subgroup $\mathrm{PSL}_2(p^2)$ is a quadratic extension of $\mathbb{Q}(T)$. Here the two ramification points corresponding to pA or pB are unramified at $L/\mathbb{Q}(T)$ since these classes are contained in $\mathrm{PSL}_2(p^2)$. Therefore, the quadratic extension L is a rational function field over $\mathbb{Q}$, say $L = \mathbb{Q}(T')$. Thus we obtain a regular extension $N/\mathbb{Q}(T')$ with Galois group $\mathrm{PSL}_2(p^2)$. □

Since the first part of Proposition 4.1 has been proved in [21], we will show the remaining part. Hereafter we assume that $p \equiv \pm 5 \pmod{12}$, in other words, $\mathbb{F}_{p^2} = \mathbb{F}_p(\sqrt{3})$. Let P^+ and P^- be the following subgroups of G:

$$P^+ := \left\{ \begin{pmatrix} 1 & a \\ 0 & 1 \end{pmatrix} \middle| \, a \in \mathbb{F}_{p^2} \right\} \cdot \langle \varphi \rangle, \quad P^- := \left\{ \begin{pmatrix} 1 & 0 \\ b & 1 \end{pmatrix} \middle| \, b \in \mathbb{F}_{p^2} \right\} \cdot \langle \varphi \rangle,$$

which are the centralizers of $\left(\begin{smallmatrix} 1 & 1 \\ 0 & 1 \end{smallmatrix}\right)$ and $\left(\begin{smallmatrix} 1 & 0 \\ 1 & 1 \end{smallmatrix}\right)$, respectively. For $\mathcal{C} = (2C, 2D, pA, pA)$, the $\mathrm{Inn}(G)$-orbits in $\Sigma(\mathcal{C})$ are parametrized as follows:

Lemma 4.2. *Each* $\mathrm{Inn}(G)$*-orbit in* $\Sigma(\mathcal{C})$ *possesses the following representative element* $\boldsymbol{\sigma}_{(s,t,u,v)} = (\sigma_1^*, \sigma_2^*, \sigma_3^*, \sigma_4^*)$ *with*

$$\sigma_1^* = \begin{pmatrix} s + u\sqrt{3} & -u\sqrt{3} \\ (2u - sv)\sqrt{3} & s - u\sqrt{3} \end{pmatrix} \varphi, \quad \sigma_2^* = \begin{pmatrix} t + u\sqrt{3} & -t \\ t - s & -t + u\sqrt{3} \end{pmatrix} \varphi,$$

$$\sigma_3^* = \begin{pmatrix} 1 & 1 \\ 0 & 1 \end{pmatrix}, \qquad \sigma_4^* = \begin{pmatrix} 1 & 0 \\ -2 + v\sqrt{3} & 1 \end{pmatrix}.$$

Here $4 - 3v^2$ *and* $4 - 3u^2(4 - 3v^2)$ *are square in* $\mathbb{F}_p$ *with* $u, v \in \mathbb{F}_p$, $(u,v) \neq (0,0)$. *Moreover,* $s, t \in \mathbb{F}_p$ *are the distinct roots of* $X^2 - 3uvX + (3u^2 - 1) \in \mathbb{F}_p[X]$.

Proof. For each $\boldsymbol{\sigma} = (\sigma_1, \sigma_2, \sigma_3, \sigma_4) \in \Sigma(\mathcal{C})$, we must show there exists $\tau \in G$ such that $\boldsymbol{\sigma}^\tau = \boldsymbol{\sigma}_{(s,t,u,v)}$. We may assume that $\sigma_3 = \sigma_3^*$ by taking a suitable conjugate. Then there exists $\tau \in P^+$ such that $\sigma_4^\tau \in P^-$. Rewrite each σ_i^τ as σ_i and put

$$\sigma_1 = \begin{pmatrix} c_1 & c_2 \\ c_3 & \bar{c}_1 \end{pmatrix} \varphi, \; \sigma_2 = \begin{pmatrix} d_1 & d_2 \\ d_3 & -\bar{d}_1 \end{pmatrix} \varphi, \; \sigma_3 = \begin{pmatrix} 1 & 1 \\ 0 & 1 \end{pmatrix}, \; \sigma_4 = \begin{pmatrix} 1 & 0 \\ b & 1 \end{pmatrix},$$

where $c_2, c_3 \in \mathbb{F}_p\sqrt{3} := \{n\sqrt{3} \,|\, n \in \mathbb{F}_p\}$ and $d_2, d_3 \in \mathbb{F}_p$. Solve the equation $\sigma_1\sigma_2\sigma_3\sigma_4 = 1$; we have

$$\begin{aligned} c_1 &= s + u\sqrt{3}, & c_2 &= -u\sqrt{3}, & c_3 &= (2u - sv)\sqrt{3}, \\ d_1 &= t + u\sqrt{3}, & d_2 &= -t, & d_3 &= t - s, \end{aligned}$$

with $3uv = s + t$, $(u, v) \neq (0, 0)$ and $b = -2 + v\sqrt{3}$. Since b is square in $\mathbb{F}_{p^2}$, its norm $4 - 3v^2$ must be square in $\mathbb{F}_p$. Substitute c_i, d_i and b into the equation $\det \sigma_1 = 1$; we have $st = 3u^2 - 1$. Finally, the reducibility of $X^2 - 3uvX + (3u^2 - 1)$ implies that $4 - 3u^2(4 - 3v^2)$ is square in $\mathbb{F}_p$. □

Conversely, the 4-system $\boldsymbol{\sigma}_{(s,t,u,v)}$ as in the lemma belongs to $\Sigma(C)$ for any $(s, t, u, v) \in \mathbb{F}_p^4$ satisfying

$$\begin{aligned} &4 - 3v^2 \in \mathbb{F}_p^{\times 2}, \; 4 - 3u^2(4 - 3v^2) \in \mathbb{F}_p^{\times 2}, \\ &(u, v) \neq (0, 0), \; s + t = 3uv, \; st = 3u^2 - 1. \end{aligned} \tag{4.1}$$

Actually, if $v \neq 0$, then Dickson's theorem ([4], see also [7] for a modern description) shows that $\{\sigma_3^*, \sigma_4^*\}$ generates $\mathrm{PSL}_2(p^2)$ and hence $\boldsymbol{\sigma}_{(s,t,u,v)}$ generates G. On the other hand, if $v = 0$, then $\{\sigma_3^*, \sigma_4^*\}$ generates the subgroup $H := \mathrm{PSL}_2(p)$. By Dickson's classification of the subgroups of $\mathrm{PSL}_2(q)$ in [4], if a maximal subgroup of $\mathrm{PSL}_2(p^2)$ contains $\mathrm{PSL}_2(p)$, it must be isomorphic to $\mathrm{PGL}_2(p)$. Since $\left|\langle H, H^{\sigma_1^*}\rangle : H\right| > 2$, we conclude that $\boldsymbol{\sigma}_{(s,t,u,v)}$ generates G.

Remark 4.3. Under the notation in the previous lemma, $[\boldsymbol{\sigma}_{(s,t,u,v)}] = [\boldsymbol{\sigma}_{(s',t',u',v')}]$ if and only if

$$(s', t', u', v') = (s, t, u, v), (s, t, -u, -v), (-s, -t, -u, v), (-s, -t, u, -v).$$

By elementary number theory, there exist $(p - 2 + \epsilon)/2$ choices for $(s, t, u, v) \in \mathbb{F}_p^4$ satisfying (4.1) and $uv = 0$ up to ± 1, where $\epsilon := (-1)^{(p-1)/2}$. In the case $uv \neq 0$, there exist $(p - 2 + \epsilon)/4$ choices for $v \in \mathbb{F}_p^\times$ with $4 - 3v^2 \in \mathbb{F}_p^{\times 2}$ and $(p - 2 + \epsilon)/4$ choices for $u \in \mathbb{F}_p^\times$ with $4 - 3u^2(4 - 3v^2) \in \mathbb{F}_p^{\times 2}$ up to ± 1. For such $u, v \in \mathbb{F}_p^\times$, there are two choices for distinct roots of $X^2 - 3uvX + (3u^2 - 1)$. Consequently, the cardinality of $\Sigma(\mathcal{C})/\mathrm{Inn}(G)$ is

$$|\Sigma(\mathcal{C})/\mathrm{Inn}(G)| = \frac{p-2+\epsilon}{2} + \left(\frac{p-2+\epsilon}{4}\right)^2 \cdot 2 = \frac{(p+\epsilon)^2 - 4}{8}.$$

Lemma 4.4. *Suppose* $[\boldsymbol{\sigma}] \in \Sigma(\mathcal{C})/\mathrm{Inn}(G)$ *be a fixed point of* γ_3*, then*

$$[\boldsymbol{\sigma}] = \left[\begin{pmatrix} s+u\sqrt{3} & -u\sqrt{3} \\ 2u\sqrt{3} & s-u\sqrt{3} \end{pmatrix}\varphi, \begin{pmatrix} -s+u\sqrt{3} & s \\ -2s & s+u\sqrt{3} \end{pmatrix}\varphi, \begin{pmatrix} 1 & 1 \\ 0 & 1 \end{pmatrix}, \begin{pmatrix} 1 & 0 \\ -2 & 1 \end{pmatrix}\right]$$

with $su \neq 0$ *and* $s^2 + 3u^2 = 1$*. Aside from these trivial fixed points,* γ_3^4 *has no non-trivial fixed points on* $\Sigma(\mathcal{C})/\mathrm{Inn}(G)$*.*

Proof. Take the representative element $\boldsymbol{\sigma} = (\sigma_1, \sigma_2, \sigma_3, \sigma_4)$ of $[\boldsymbol{\sigma}] \in \Sigma(\mathcal{C})/\mathrm{Inn}(G)$ as in Lemma 4.2. If $v = 0$, then it is easily checked that $[\boldsymbol{\sigma}]^{\gamma_3} = [\boldsymbol{\sigma}]$. Suppose $v \neq 0$ and $[\boldsymbol{\sigma}]^{\gamma_3^4} = [\boldsymbol{\sigma}]$, then $\langle \sigma_3, \sigma_4 \rangle = \mathrm{PSL}_2(p^2)$ and $(\sigma_1^{(\sigma_3\sigma_4)^4}, \sigma_2^{(\sigma_3\sigma_4)^4}, \sigma_3, \sigma_4) = \boldsymbol{\sigma}^\tau$ with $\tau \in G$. Since $\mathrm{PSL}_2(p^2)$ has trivial centralizer, τ must be trivial and hence $(\sigma_3\sigma_4)^4$ commutes with σ_1. Here

$$(\sigma_3\sigma_4)^4\sigma_1 = \begin{pmatrix} * & * \\ * & sx - uy\sqrt{3} \end{pmatrix}\varphi, \quad \sigma_1(\sigma_3\sigma_4)^4 = \begin{pmatrix} * & * \\ * & sy - ux\sqrt{3} \end{pmatrix}\varphi$$

with

$$x = 1 - 3v^2 - (2v - 3v^3)\sqrt{3}, \quad y = 1 - 9v^2 + 9v^4 + (2v - 3v^3)\sqrt{3}.$$

So we have $sx - uy\sqrt{3} = \pm(sy - ux\sqrt{3})$, which means $x = \pm y$. If $x = -y$, then

$$0 = 2 - 12v^2 + 9v^4$$

and hence $3v^2(4 - 3v^2) = (2 - 3v^2)^2$, which is a contradiction. Further if $x = y$, then

$$0 = x - y = (2 - 3v^2)(-2 + v\sqrt{3})v\sqrt{3}$$

and hence $v^2 = 2/3$. This contradicts that $4 - 3v^2 = 2$ is square in $\mathbb{F}_p$. Therefore, each fixed point of γ_3^4 is already a fixed point of γ_3. □

Lemma 4.5. *There exist no common fixed points of* γ_3 *and* γ_1^4 *on* $\Sigma(\mathcal{C})/\mathrm{Inn}(G)$*.*

Proof. Let $[\boldsymbol{\sigma}]$ be the fixed point of γ_3 determined in the previous lemma. Suppose $[\boldsymbol{\sigma}]$ be also a fixed point of γ_1^4, then $(\sigma_1^{(\sigma_2\sigma_3)^4}, \sigma_2, \sigma_3, \sigma_4^{(\sigma_2\sigma_3)^4}) = \boldsymbol{\sigma}^\tau$ with $\tau \in G$. Since τ commutes with σ_3 and $(\sigma_2\sigma_3)^4\tau^{-1}$ commutes with σ_4, we have $\tau \in P^+$ and $(\sigma_2\sigma_3)^4\tau^{-1} \in P^-$. This indicates that $(\sigma_2\sigma_3)^4$ is of the form as $\left(\begin{smallmatrix}1 & * \\ * & *\end{smallmatrix}\right)$. By a calculation, we have

$$(\sigma_2\sigma_3)^4 = \begin{pmatrix} s^4-18s^2u^2+9u^4-4su(s^2-3u^2)\sqrt{3} & * \\ * & * \end{pmatrix}.$$

Since $su(s^2-3u^2) \neq 0$, this is a contradiction. □

Proof of (ii) *of Theorem* 4.1. We begin with the following element of $\Sigma(C)/\mathrm{Inn}(G)$:

$$[\boldsymbol{\sigma}_4] = \left[\begin{pmatrix} 1+\sqrt{3} & -\sqrt{3} \\ \sqrt{3} & 1-\sqrt{3} \end{pmatrix}\varphi, \begin{pmatrix} 2+\sqrt{3} & -2 \\ 1 & -2+\sqrt{3} \end{pmatrix}\varphi, \begin{pmatrix} 1 & 1 \\ 0 & 1 \end{pmatrix}, \begin{pmatrix} 1 & 0 \\ -2+\sqrt{3} & 1 \end{pmatrix}\right].$$

Denote by $[\boldsymbol{\sigma}_i]$ $(i = 1, 2, 3)$ the image of $[\boldsymbol{\sigma}_4]$ by γ_i, then

$$[\boldsymbol{\sigma}_1] = \left[\begin{pmatrix} 2+\sqrt{3} & -\sqrt{3} \\ 0 & 2-\sqrt{3} \end{pmatrix}\varphi, \begin{pmatrix} 1+\sqrt{3} & -1 \\ -1 & -1+\sqrt{3} \end{pmatrix}\varphi, \begin{pmatrix} 1 & 1 \\ 0 & 1 \end{pmatrix}, \begin{pmatrix} 1 & 0 \\ -2+\sqrt{3} & 1 \end{pmatrix}\right],$$

$$[\boldsymbol{\sigma}_2] = \left[\frac{1}{2}\begin{pmatrix} 1+\sqrt{3} & -\sqrt{3} \\ 2\sqrt{3} & 1-\sqrt{3} \end{pmatrix}\varphi, \frac{1}{2}\begin{pmatrix} -1+\sqrt{3} & 1 \\ -2 & 1+\sqrt{3} \end{pmatrix}\varphi, \begin{pmatrix} 1 & 1 \\ 0 & 1 \end{pmatrix}, \begin{pmatrix} 1 & 0 \\ -2 & 1 \end{pmatrix}\right],$$

$$[\boldsymbol{\sigma}_3] = \left[\begin{pmatrix} 1 & 0 \\ -\sqrt{3} & 1 \end{pmatrix}\varphi, \begin{pmatrix} -1 & 1 \\ -2 & 1 \end{pmatrix}\varphi, \begin{pmatrix} 1 & 1 \\ 0 & 1 \end{pmatrix}, \begin{pmatrix} 1 & 0 \\ -2+\sqrt{3} & 1 \end{pmatrix}\right].$$

These elements construct a braid orbit $B_4 = \{[\boldsymbol{\sigma}_1], [\boldsymbol{\sigma}_2], [\boldsymbol{\sigma}_3], [\boldsymbol{\sigma}_4]\}$ on which $\gamma_1, \gamma_2, \gamma_3$ act as the following permutations:

$$\begin{aligned} [\boldsymbol{\sigma}_i]^{\gamma_1} &= [\boldsymbol{\sigma}_{\rho_1(i)}],\ \rho_1 = \pi_{B_4}(\gamma_1) = (1\ 2\ 4), \\ [\boldsymbol{\sigma}_i]^{\gamma_2} &= [\boldsymbol{\sigma}_{\rho_2(i)}],\ \rho_2 = \pi_{B_4}(\gamma_2) = (2\ 3\ 4), \\ [\boldsymbol{\sigma}_i]^{\gamma_3} &= [\boldsymbol{\sigma}_{\rho_3(i)}],\ \rho_3 = \pi_{B_4}(\gamma_3) = (1\ 4\ 3). \end{aligned}$$

Let $B \subseteq \Sigma(C)/\mathrm{Inn}(G)$ be a braid orbit of length 4. Lemma 4.4 indicates that γ_3 has no transpositions nor 4-cycles under the permutation representation of Γ on B. Thus $\pi_B(\gamma_3)$ must be the identity or a 3-cycle and hence γ_3 has at least one fixed point $[\boldsymbol{\sigma}] \in B$. Such $[\boldsymbol{\sigma}]$ is determined in Lemma 4.4 and its $\langle\gamma_1\rangle$-orbit must have length 3 by Lemma 4.5. Similar to the proof of Lemma 4.5, $(\sigma_2\sigma_3)^3$ is of the form as $\left(\begin{smallmatrix}1 & * \\ * & *\end{smallmatrix}\right)\varphi$. Since

$$(\sigma_2\sigma_3)^3 = \begin{pmatrix} -s^3+9su^2+3(s^2u-u^3)\sqrt{3} & * \\ * & * \end{pmatrix}\varphi,$$

we have

$$-s^3 + 9su^2 = \pm 1,\ s^2u - u^3 = 0.$$

The unique solution of these equations is $(s, u) = (1/2, 1/2)$ up to ± 1. Thus $[\boldsymbol{\sigma}]$ coincides with $[\boldsymbol{\sigma}_2] \in B_4$. This means $B = B_4$ is a unique braid orbit of length 4. □

Acknowledgments

The author wishes to thank Professors Ki-ichiro Hashimoto, Katsuya Miyake and Hiroaki Nakamura for their invitation to the symposium "Galois Theory and Modular Forms". The author also wishes to thank Professor Toyofumi Takahashi for his continuing encouragement.

References

[1] J. H. Conway et al., "Atlas of finite groups," Clarendon Press, Oxford, 1985.

[2] M. Dettweiler, *Middle convolution and Galois realizations*, to appear in this volume.

[3] L. Dieulefait and N. Vila, *Projective linear groups as Galois groups over* $\mathbb{Q}$ *via modular representations*, J. Symbolic Computation **30** (2000), 799–810.

[4] L. E. Dickson, "Linear groups with an exposition of the Galois field theory," Teubner, Leibzig, 1901.

[5] W. Feit, *Rigidity of* $\mathrm{Aut}(\mathrm{PSL}_2(p^2))$, $p \equiv \pm 2 \pmod 5$, $p \neq 2$, in "Proceedings of the Rutgers group theory year, 1983–1984 (New Brunswick, N. J., 1983–1984)," 351–356, Cambridge Univ. Press, Cambridge, 1984.

[6] The GAP Group, "GAP – Groups, Algorithms, and Programming, Version 4.2," Aachen, St. Andrews, 1999.

[7] D. Gorenstein, "Finite groups," Harper and Row, New York-Evanston-London, 1968.

[8] D. Hilbert, *Ueber die Irreduzibilität ganzer rationaler Funktionen mit ganzzahligen Koeffizienten*, J. Reine Angew. Math. **110** (1892), 104–129.

[9] G. Malle, *Genus zero translates of three point ramified Galois extensions*, Manuscripta Math. **71** (1991) 97–111.

[10] G. Malle and B. H. Matzat, "Inverse Galois theory," Springer-Verlag, Berlin, 1999.

[11] B. H. Matzat, "Konstruktive Galoistheorie," Lecture Notes in Mathematics, 1284. Springer-Verlag, Berlin, 1987.

[12] B. H. Matzat, *Zöpfe und Galoissche Gruppen*, J. Reine Angew. Math. **420** (1991), 99–159.

[13] J.-F. Mestre, *Courbes hyperelliptiques à multiplications réelles*, C. R. Acad. Sci. Paris, **307** (1988), 721–724.

[14] B. Przywara, *Die Operation der Hurwitzschen Zopfgruppe auf den Erzeugendensystemklassen endlicher Gruppen*, Diplomarbeit, Karlsruhe, 1988.

[15] H. Reichardt, Konstruktion von Zahlkörpern mit gegebener Galoisgruppe von Primzahlpotenzordnung, J. Reine Angew. Math. **177** (1937), 1–5.

[16] A. Reverter and N. Vila, *Some projective linear groups over finite fields as Galois groups over* $\mathbb{Q}$, Contemp. Math. **186** (1995), 51–63.

[17] K. A. Ribet, *On l-adic representations attached to modular forms*, Invent. Math. **28** (1975), 245–275.

[18] I. R. Šafarevič, Construction of fields of algebraic numbers with given solvable Galois group, Izv. Akad. Nauk SSSR **18** (1954), 525–578 (Amer. Math. Transl. **4** (1956), 185–237).

[19] J.-P. Serre, "Topics in Galois theory," Jones and Bartlett, Boston, 1992.

[20] K-y. Shih, *On the construction of Galois extensions of function fields and number fields*, Math. Ann. **207** (1974), 99–120.

[21] T. Shiina, *Rigid braid orbits related to* $\mathrm{PSL}_2(p^2)$ *and some simple groups*, to appear in Tohoku Math. J.

[22] H. Völklein, "Groups as Galois groups," Cambridge University Press, Cambridge, 1996.

Galois Theory and Modular Forms pp 143–158
K. Hashimoto, K. Miyake, and H. Nakamura, Editors
© 2003 Kluwer Academic Publishers

MIDDLE CONVOLUTION AND GALOIS REALIZATIONS

Michael DETTWEILER
IWR, Universität Heidelberg,
Im Neuenheimer Feld 368,
69120 Heidelberg, Germany
michael.dettweiler@iwr.uni-heidelberg.de

Abstract The theory of the middle convolution is combined with the theory of curves on Hurwitz spaces. This leads to the following theorem: The projective symplectic groups $\mathrm{PSp}_{2n}(\mathbb{F}_{p^2})$ occur $\mathbb{Q}$-regularly as Galois groups over $\mathbb{Q}(t)$ if p is an odd prime $\not\equiv \pm 1 \bmod 24$.

1. Introduction

The usual approach of realizing a finite group $\mathbb{Q}$-regularly as Galois group over $\mathbb{Q}(t)$ is via rigidity or, more generally, by detecting rational points on the underlying moduli spaces (Hurwitz spaces) using braid groups, see [12], [15] and the references therein. Once the geometry of the Hurwitz spaces becomes more complicated it can be useful to look for rational curves lying on the Hurwitz spaces, instead of studying the whole Hurwitz space, see [3], [4].

In [6], [16], the authors give (different) purely algebraic constructions of Katz' middle convolution functor MC_χ (see [11]). In [6] the resulting analogon of MC_χ is denoted MC_λ, in [16] it is called the braid companion (BC-) functor. As a first application, many families of quasi-simple groups of Lie type were realized regularly as Galois groups over $\mathbb{Q}(t)$, see [5], [6], [17].

The basic idea is the following: Choose a subgroup $H \leq \mathrm{GL}_n(\mathbb{F}_q)$ for which the braid group action on tuples can easily be determined (e.g., a subgroup of the scalars or a dihedral group, viewed as subgroup of $\mathrm{GL}_2(\mathbb{F}_q)$) and the available braid group criteria can be applied. Applying successively the operations of middle convolution and suitable

[received: October 23, 2002; accepted in revised form: March 21, 2003]

multiplication with scalars one can realize groups in higher dimensions regularly as Galois groups over $\mathbb{Q}(t)$.

It is the aim of this work to demonstrate how to combine the approach of [3], [4] with the theory of middle convolution. In fact, T. Shiina [13], [14] recently succeeded in solving the inverse Galois problem for $\mathrm{PSL}_2(p^2)$ for $p \not\equiv \pm 1 \bmod 24$ by using the classical braid orbit theorem due to Matzat, see [12], Chap. III. We show that using our methods one can extend his result to the higher rank Lie-type groups $\mathrm{PSp}_{2n}(p^2)$:

Theorem 1. *The projective symplectic groups* $\mathrm{PSp}_{2n}(\mathbb{F}_{p^2})$ *occur* $\mathbb{Q}$-*regularly as Galois groups over* $\mathbb{Q}(t)$ *if* p *is an odd prime* $\not\equiv \pm 1 \bmod 24$.

For $n = 1$ this reduces to the above mentioned results of Shiina ([13], [14]). For other Galois realizations of symplectic groups, see [7], [12], [17] and the references therein.

The plan of the paper is as follows: In Section 2 the definition of the middle convolution functor MC_λ, as given in [7], is recalled. Also the main properties of MC_λ are restated. In Section 3 we state a result of [8] (Thm. 5), saying that the field of linear moduli is preserved under application of MC_{-1} (compare to [17], Thm. 4.4., where a similar result is proved for the BC-operation). In Sections 4 – 6 the theory of Hurwitz spaces and curves lying on them is revisited. Finally, Thm. 1 is proved in Section 7 (Thm. 11).

2. Convolution

In the following we recall the definition of the middle convolution functor as in [6], [7]:

For $(A_1, \ldots, A_r) \in \mathrm{GL}_n(K)^r$ and $\lambda \in K^\times$, consider the following matrices $B_k \in \mathrm{GL}_{nr}(K)$ for $k = 1, \ldots, r$:

$$B_k := \left(\begin{array}{ccccccc} 1_n & 0 & & \cdots & & & 0 \\ & \ddots & & & & & \\ & & 1_n & & & & \\ \lambda(A_1 - 1_n) & \cdots & \lambda(A_{k-1} - 1_n) & \lambda A_k & (A_{k+1} - 1_n) & \cdots & (A_r - 1_n) \\ & & & & 1_n & & \\ & & & & & \ddots & \\ 0 & & & \cdots & & 0 & 1_n \end{array} \right),$$

where B_k is the identity matrix outside the k-th block row ($1_n \in \mathrm{GL}_n(K)$ denotes the identity matrix). We call $C_\lambda(A_1, \ldots, A_r) = (B_1, \ldots, B_r)$ the *(multiplicative) convolution* of $(A_1, \ldots, A_r)$ with λ. There are the following (left) $\langle B_1, \ldots, B_r \rangle$-invariant subspaces

$$\mathcal{K}_k = \begin{pmatrix} 0 \\ \vdots \\ 0 \\ \ker(A_k - 1_n) \\ 0 \\ \vdots \\ 0 \end{pmatrix} \quad (k\text{-th entry}),\ k = 1, \dots, r,$$

and

$$\mathcal{L} := \bigcap_{i=1}^{r} \ker(B_i - 1_{nr}).$$

Let $\mathcal{K} = \oplus_{k=1}^{r} \mathcal{K}_k$ and m the dimension of the K-vector space $K^{nr}/(\mathcal{K} + \mathcal{L})$. Choose an isomorphism I between $K^{nr}/(\mathcal{K} + \mathcal{L})$ and K^m.

The tuple $MC_\lambda(A_1, \dots, A_r) = (\tilde{B}_1, \dots, \tilde{B}_r) \in \mathrm{GL}_m(K)^r$ which is induced by the action of $(B_1, \dots, B_r)$ on K^m (via I) is called the *middle convolution* of $(A_1, \dots, A_r)$ with λ.

The main properties of the middle convolution are:

Theorem 2. *Let* $(A_1, \dots, A_r) \in \mathrm{GL}_n(K)^r$ *such that* $A_i \neq 1_n$ *for* $i = 1, \dots, r$. *Let further* $\lambda \in K^\times$ *and* $MC_\lambda(A_1, \dots, A_r) = (\tilde{B}_1, \dots, \tilde{B}_r) \in \mathrm{GL}_m(K)^r$.

a) *If* $\lambda \neq 1$, *then*

$$\begin{aligned} m &= \dim(K^{nr}/(\mathcal{K} + \mathcal{L})) \\ &= \sum_{k=1}^{r} \mathrm{rk}(A_k - 1_n) - (n - \mathrm{rk}(\lambda A_1 \cdots A_r - 1_n)). \end{aligned}$$

b) *Assume that* $\langle A_1, \dots, A_r \rangle$ *is an absolutely irreducible subgroup of* $\mathrm{GL}_n(K)$ *and* $r > 1$. *If* $\lambda_1, \lambda_2 \in K$ *with* $\lambda = \lambda_1 \cdot \lambda_2$, *then* $MC_{\lambda_1}(MC_{\lambda_2}(A_1, \dots, A_r)) \in \mathrm{GL}_m(K)^r$ *and*

$$MC_\lambda(A_1, \dots, A_r) \approx MC_{\lambda_1}(MC_{\lambda_2}(A_1, \dots, A_r)),$$

where $\approx$ *means simultaneous conjugation by an element of* $\mathrm{GL}_m(K)$. *Moreover,*

$$MC_1(A_1, \dots, A_r) \approx (A_1, \dots, A_r).$$

c) *If* $\langle A_1, \dots, A_r \rangle$ *is an absolutely irreducible subgroup of* $\mathrm{GL}_n(K)$ *and* $r > 1$, *then* $\langle \tilde{B}_1, \dots, \tilde{B}_r \rangle$ *is an absolutely irreducible subgroup of* $\mathrm{GL}_m(K)$.

d) *Let* $\mathcal{B}_r = \langle Q_1, \dots, Q_{r-1} \rangle$ *be the abstract Artin braid group, where the generators* $Q_1, \dots, Q_{r-1}$ *of* $\mathcal{B}_r$ *act in the following way on tuples*

$(g_1,\ldots,g_r) \in G^r$, *where* G *is any group:*

$$Q_i(g_1,\ldots,g_r) = (g_1,\ldots,g_{i-1},g_ig_{i+1}g_i^{-1},g_i,g_{i+2},\ldots,g_r),\qquad i=1,\ldots,r-1. \quad (1)$$

Then

$$MC_\lambda(Q(A_1,\ldots,A_r)) \approx Q(MC_\lambda(A_1,\ldots,A_r)),\ \forall Q \in \mathcal{B}_r.$$

e) *The convolution* MC_λ *respects hermitian forms. Moreover, if* $(A_1,\ldots,A_r)$ *respects an orthogonal (resp. symplectic) form then* $MC_{-1}(A_1,\ldots,A_r)$ *respects a symplectic (resp. orthogonal) form (the characteristic of* K *is assumed to be* $\neq 2$*).*

Proof. a)-e) follow analogously to [6], Lemma 2.7, Thm. 3.5, Cor. 3.6, Thm. 5.1, Cor. 5.10 (in this order). □

The following lemma is already used implicitly in [6] and will be useful below (the proof follows from Lemma 4.1, loc. cit.):

Lemma 3. *Let* $\lambda \in K^\times$, $(A_1,\ldots,A_r), A_k \in \mathrm{GL}_n(K)$, *such that* $A_k \neq 1_n$, $\langle A_1,\ldots,A_r\rangle$ *is absolutely irreducible and* $r > 1$.
a) *Let* $J_{k,1}\oplus\ldots\oplus J_{k,b_k}$ *be the Jordan canonical form of* A_k, *where* $J_{k,i}$ $(1 \leq i \leq b_k)$ *is the Jordan block with eigenvalue* $a_{k,i}$ *of size* $n_{k,i}$. *Then the Jordan canonical form of* $\tilde{B}_k$ *is determined by the Jordan canonical form of* A_k *in the following way: Every* $J_{k,i}$ *contributes a Jordan block* $\tilde{J}_{k,i}$ *of* $\tilde{B}_k$ *with eigenvalue* $\tilde{a}_{k,i}$ *and size* $m_{k,i}$, *where* $\tilde{a}_{k,i}, m_{k,i}$ *are as follows:*

One has $\tilde{a}_{k,i} = \lambda a_{k,i}$ *and if* $a_{k,i} \neq 1, \lambda^{-1}$ *(resp.* $a_{k,i} = 1$ *or* $a_{k,i} = \lambda^{-1}$*), then* $m_{k,i} = n_{k,i}$ *(resp.* $m_{k,i} = n_{k,i} - 1$, *or* $m_{k,i} = n_{k,i} + 1$*). The other Jordan blocks of* $\tilde{B}_k$ *are blocks of length one having an eigenvalue equal to* 1.
b) *Let* $A_\infty := A_1\cdots A_r$, $\tilde{B}_\infty := \tilde{B}_1\cdots\tilde{B}_r$ *and* $J_{\infty,1}\oplus\ldots\oplus J_{\infty,b_\infty}$ *the Jordan canonical form of* A_∞, *where* $J_{\infty,i}$ $(1 \leq i \leq b_\infty)$ *is the Jordan block with eigenvalue* $a_{\infty,i}$ *of size* $n_{\infty,i}$. *Then the Jordan canonical form of* $\tilde{B}_\infty$ *is determined by the Jordan canonical form of* A_∞ *in the following way: Every* $J_{\infty,i}$ *contributes a Jordan block* $\tilde{J}_{\infty,i}$ *of* $\tilde{B}_\infty$ *with eigenvalue* $\tilde{a}_{\infty,i}$ *and size* $m_{\infty,i}$, *where* $\tilde{a}_{\infty,i}, m_{\infty,i}$ *are as follows:*

One has $\tilde{a}_{\infty,i} = \lambda a_{\infty,i}$ *and if* $a_{\infty,i} \neq 1, \lambda^{-1}$ *(resp.* $a_{\infty,i} = 1$ *or* $a_{\infty,i} = \lambda^{-1}$*), then* $m_{\infty,i} = n_{\infty,i}$ *(resp.* $m_{\infty,i} = n_{\infty,i}+1$, *or* $m_{\infty,i} = n_{\infty,i}-1$*). The other Jordan blocks of* $\tilde{B}_\infty$ *are blocks of length one having an eigenvalue equal to* λ.

3. Convolution and fields of moduli

Let $P_0 := \{p_1, \ldots, p_r\} \subseteq \mathbb{C}$, $|P_0| = r$, $p_0 \in \mathbb{C}$, $p_0 \notin P_0$ and k be a subfield of $\mathbb{C}$ such that $P_0 \subset \bar{k}$.

Let ψ be the Galois cover of $Y := \mathbb{P}^1(\mathbb{C}) \setminus (P_0 \cup \{\infty\})$ associated to the kernel of a homomorphism $\rho : \pi_1(Y, p_0) \to \mathrm{GL}_n(\mathbb{F}_q)$. Let $\gamma_1, \ldots, \gamma_{r+1}$ be generators of $\pi_1(Y, p_0)$ which are chosen in the following way: The element γ_i is the homotopy class of a simple loop in $\mathbb{C}$ which encircles the point p_i in counterclockwise direction and none of the other points p_j, $j \neq i$. (A *simple loop* in a topological space M is by definition a closed path in M which is injective on $]0, 1[$.) We further assume, that $\gamma_{r+1} := (\gamma_1 \cdots \gamma_r)^{-1}$ is the homotopy class of a simple loop in $\mathbb{C}$, encircling the set P_0 in clockwise direction.

If $D_i = \rho(\gamma_i)$, $i = 1, \ldots, r$, then ρ and ψ are uniquely determined by the matrices D_i. Therefore, we call ψ the *Galois cover of* Y *associated to* $(\gamma_1, \ldots, \gamma_r)$ *and* $(D_1, \ldots, D_r) \in \mathrm{GL}_n(\mathbb{F}_q)^r$.

Definition 4. Let ψ be the Galois cover of Y associated to $(\gamma_1, \ldots, \gamma_r)$ and $(D_1, \ldots, D_r) \in \mathrm{GL}_n(\mathbb{F}_q)^r$. Then k is called a *field of moduli* (resp. a *field of linear moduli*) for ψ, if for any $\kappa \in G_k := \mathrm{Gal}(\bar{k}/k)$ there exists an element $\alpha \in \langle D_1, \ldots, D_r \rangle$ (resp. an element $\alpha \in N_{\mathrm{GL}_n(\mathbb{F}_q)}(\langle D_1, \ldots, D_r \rangle)$) such that ψ^κ (κ acting via base change, see [10]) is the Galois cover associated to $(\gamma_1, \ldots, \gamma_r)$ and $(D_1^\alpha, \ldots, D_r^\alpha)$.

Let $\alpha_1, \ldots, \alpha_r$ be generators of $\pi_1(\mathbb{P}^1(\mathbb{C}) \setminus (P_0 \cup \{\infty\}), p_0)$ which go counter-clockwise around $p_1, \ldots, p_r$ as indicated in the following picture:

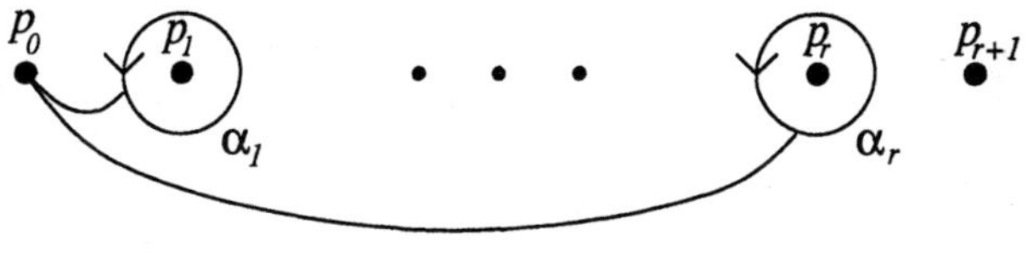

Next, let $p_{r+1} \in \mathbb{C} \setminus (P_0 \cup \{p_0\})$ and $\beta_1, \ldots, \beta_r$ generators of $\pi_1(\mathbb{P}^1(\mathbb{C}) \setminus (P_0 \cup \{\infty\}), p_{r+1})$ which go counter-clockwise around $p_1, \ldots, p_r$ as indicated in the following picture:

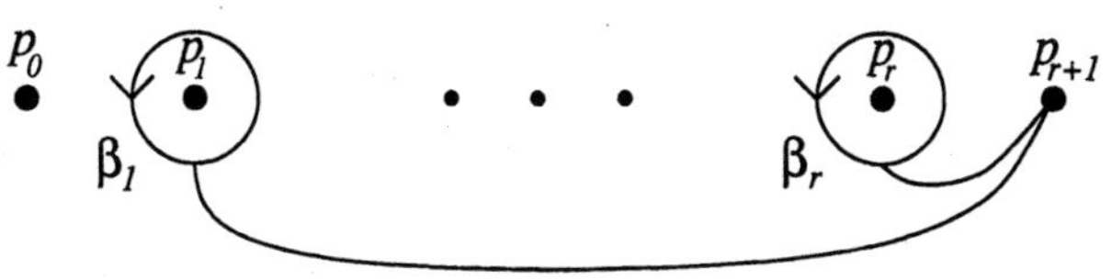

Theorem 5. *Let ϕ be the Galois cover of Y associated to $(\alpha_1,\ldots,\alpha_r)$ and $(A_1,\ldots,A_r) \in \mathrm{GL}_n(\mathbb{F}_q)^r$ (q odd) and let $\tilde{\phi}$ be the Galois cover of Y associated to $(\beta_1,\ldots,\beta_r)$ and $MC_{-1}(A_1,\ldots,A_r) = (\tilde{B}_1,\ldots,\tilde{B}_r) \in \mathrm{GL}_m(\mathbb{F}_q)^r$. Suppose that $\langle A_1,\ldots,A_r\rangle$ is an absolutely irreducible subgroup of $\mathrm{GL}_n(\mathbb{F}_q)$. If k is a field of linear moduli for ϕ, then k is also a field of linear moduli for $\tilde{\phi}$.*

Proof. This is proved in [8] in greater generality. □

Let ψ be the Galois cover of $Y = \mathbb{P}^1(\mathbb{C}) \setminus (P_0 \cup \{\infty\})$ associated to $(\gamma_1,\ldots,\gamma_r)$ and $(D_1,\ldots,D_r) \in \mathrm{GL}_n(\mathbb{F}_q)^r$ and $G := \langle D_1,\ldots,D_r\rangle$. Set $p_{r+1} := \infty$ and let $C_1,\ldots,C_{r+1}$ be the conjugacy classes of the images of the elements $\gamma_1,\ldots,\gamma_{r+1}$ in G. Let

$$[G,P_0^\infty,\mathcal{C}] := [G,P_0\cup\{p_{r+1}=\infty\},(C_{p_1}:=C_1,\ldots,C_{p_{r+1}}:=C_{r+1})]$$

be the ramification type of ψ (see [15], Def.'s 4.31 and 2.12, for the definition of a ramification type of a cover). Then $[G,P_0^\infty,\mathcal{C}]$ is said to be *k-rational*, if for each $p_i \in P_0^\infty$ and for each $\kappa \in G_k$ one has $C_{\kappa(p_i)} = C_{p_i}^m$, where m is an integer such that $\kappa^{-1}(\zeta_n) = \zeta_n^m$.

The following corollary is similar to [17], Cor. 4.6 (b):

Corollary 6. *Let ϕ, $\tilde{\phi}$, be as in Theorem 5 and assume that k is a field of linear moduli for ϕ. Let further $\tilde{C}_1,\ldots,\tilde{C}_r$ be the conjugacy classes of $\tilde{B}_1,\ldots,\tilde{B}_r$ in $\langle\tilde{B}_1,\ldots,\tilde{B}_r\rangle$, $\tilde{C}_{r+1}$ the conjugacy class of $B_{r+1} := (\tilde{B}_1\cdots\tilde{B}_r)^{-1}$ and*

$$S := \{g \in N_{\mathrm{GL}_m(\mathbb{F}_q)}(\langle\tilde{B}_1,\ldots,\tilde{B}_r\rangle) \mid g(\tilde{C}_i) = \tilde{C}_i,\ i=1,\ldots,r+1\}.$$

If $\tilde{\phi}$ has a k-rational ramification type and

$$S = Z(\mathrm{GL}_m(\mathbb{F}_q))\langle\tilde{B}_1,\ldots,\tilde{B}_r\rangle,$$

then k is a field of moduli for $\tilde{\phi}$.

Proof. By the above theorem, k is also a field of linear moduli for $\tilde{\phi}$. By the branch cycle argument (describing the the action of G_k on the conjugacy classes via the cyclotomic character, see [12], Chap. I.4.2, or [15], L. 2.8), the k-rationality of the ramification type and the assumption on S, every $\kappa \in G_k$ acts as an inner automorphism of the cover. □

4. Configuration spaces

The standard reference for configuration spaces and braid groups is the book of [1], where one can find the basic facts on fibrations of configuration spaces and presentations of braid groups given below, see also [9].

The following lemma, which will be useful below, is a special case of the lemma of Schönflies, see [2], Cor. 2.2.6:

Lemma 7. *If* $\Sigma_i : [0,1] \to \mathbb{P}^1$, $i = 1, 2$, *are two simple loops which are piecewise smooth, then there exists an orientation-preserving homeomorphism* $\alpha : \mathbb{P}^1 \to \mathbb{P}^1$, *such that* $\Sigma_2 = \alpha \circ \Sigma_1$.

Let $r \in \mathbb{N}$ with $r \geq 3$. Let further $P_0 := \{p_1, \ldots, p_r\} \subseteq \mathbb{P}^1$ be the base point of

$$\mathcal{U}_r := \{P \subseteq \mathbb{P}^1(\mathbb{C}) \mid |P| = r\}$$

and (P_0, p_0) the base point of

$$U(r+1) := \{(P, p) \in \mathcal{U}_r \times \mathbb{P}^1(\mathbb{C}) \mid p \notin P\}.$$

Set $\mathcal{H}_r := \pi_1(\mathcal{U}_r, P_0)$ (the *Hurwitz braid group*) and $\mathcal{H}(r+1) := \pi_1(\mathcal{U}(r+1), (P_0, p_0))$. It is well known that the group $\mathcal{H}_r$ is generated by elements $H_1, \ldots, H_{r-1}$ subject to the relations

$$[H_i, H_j] = 1, \; |i - j| > 1,$$

$$H_i H_{i+1} H_i = H_{i+1} H_i H_{i+1}, \; i < r - 1,$$

and

$$H_1 \cdots H_{r-2} H_{r-1} H_{r-1} H_{r-2} \cdots H_1 = 1.$$

We call $H_1, \ldots, H_{r-1}$ a *system of standard generators* of $\mathcal{H}_r$.

A system of standard generators can be constructed in the following way: There exists a simple loop $\Sigma : [0,1] \to \mathbb{P}^1(\mathbb{C})$ with initial point p_0 such that there exist $0 < t_1 < \ldots < t_r < 1$ with $\Sigma(t_i) = p_i$, $i = 1, \ldots, r$, and such that Σ is smooth on $]t_i, t_{i+1}[$, $i = 1, \ldots, r-1$. By Lemma 7, there exists an orientation preserving homeomorphism $\alpha : \mathbb{P}^1(\mathbb{C}) \to \mathbb{P}^1(\mathbb{C})$ which maps $\mathbb{R} \cup \{\infty\}$ to $S := \mathrm{im}(\Sigma)$ such that

$$\alpha(i) = p_i, \; i = 0, \ldots, r.$$

We can define H_i to be the homotopy class of the path

$$H_i(t) := \{\alpha(1), \ldots, \alpha(i-1), \hat{p}_i(t), \hat{p}_{i+1}(t), \alpha(i+2), \ldots, \alpha(r)\},$$

where

$$\hat{p}_i(t) := \alpha(\frac{(2i+1) - e^{\pi\sqrt{-1}t}}{2}), \; \hat{p}_{i+1}(t) := \alpha(\frac{(2i+1) + e^{\pi\sqrt{-1}t}}{2})$$

for $t \in [0,1]$ (compare to [4], Section 1.1 and [15], Chap. 10):

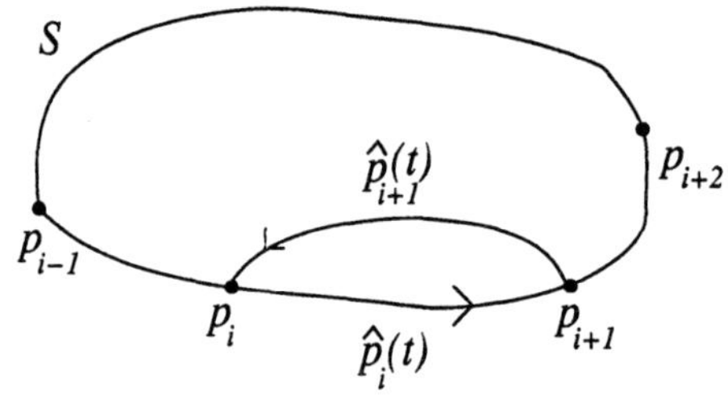

The first projection $\mathrm{pr}_1 : \mathcal{U}(r+1) \to \mathcal{U}_r$ is a locally trivial fibration. The long exact sequence of homotopy groups, associated to a locally trivial fibration, leads then to an exact sequence

$$1 \quad \to \quad \pi_1(\mathcal{F}_r, (P_0, p_0)) \quad \to \quad \mathcal{H}(r+1) \quad \to \quad \mathcal{H}_r \quad \to \quad 1 \;, \tag{2}$$

where $\mathcal{F}_r$ denotes the fibre over P_0. Using the construction of the paths $H_1(t), \ldots, H_{r-1}(t)$ one can easily see that there exists a set of generators $\gamma_1, \ldots, \gamma_r$ of $\pi_1(\mathcal{F}_r, (P_0, p_0))$ such that $\gamma_1 \cdots \gamma_r = 1$ and

$$(\gamma_1^{\hat{H}_i}, \ldots, \gamma_r^{\hat{H}_i}) = (\gamma_1, \ldots, \gamma_{i-1}, \gamma_i\gamma_{i+1}\gamma_i^{-1}, \gamma_i, \gamma_{i+2}, \ldots, \gamma_r), \qquad i = 1, \ldots, r-1, \tag{3}$$

where $\hat{H}_i$ denotes the homotopy class of the path $(H_i(t), p_0)$.

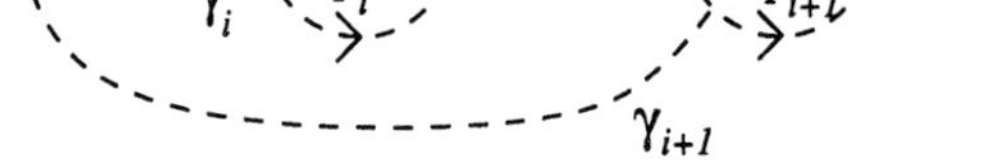

5. Hurwitz spaces

The standard references for the content of this section are [10] and [15].

Let G be a finite group and $\mathcal{C} := (C_1, \ldots, C_r)$ a tuple of conjugacy classes of G. Let

$$\mathcal{N}_r(G) := \{(g_1, \ldots, g_r) \in G^r \mid g_1 \cdots g_r = 1,\ G = \langle g_1, \ldots, g_r \rangle\}/\mathrm{Inn}(G),$$

Inn(G) acting componentwise. Define

$$\mathcal{N}(G,\mathcal{C}) := \{[g_1,\ldots,g_r] \in \mathcal{N}_r(G) \mid \exists \pi \in S_r,\, g_{\pi(i)} \in C_i\}$$

and

$$\tilde{\mathcal{N}}(G,\mathcal{C}) := \{[g_1,\ldots,g_r] \in \mathcal{N}_r(G) \mid g_i \in C_i\}.$$

Denote by $\mathcal{H}_r(G)$ the Hurwitz space which parametrizes, up to inner automorphisms, the Galois covers of the r-punctured projective line with group of deck transformations isomorphic to G.

The map $\Phi : \mathcal{H}_r(G) \to \mathcal{U}_r$, which associates the set of branch points to the equivalence class of a cover, is an unramified covering map. By covering theory, Φ induces an action of the fundamental group $\mathcal{H}_r = \pi_1(\mathcal{U}_r, P_0)$ on $\Phi^{-1}(P_0)$. This action can be made explicit in the following way: Using the same arguments as Fried and Völklein [10], Sections 1.1-1.3, one can construct a bijection between $\Phi^{-1}(P_0)$ and $\mathcal{N}_r(G)$ (by sending the class of the cover, corresponding to the kernel of a homomorphism $\rho : \pi_1(\mathbb{P}^1(\mathbb{C}) \setminus P_0, p_0) \to G$, to $[g_1,\ldots,g_r] := [\rho(\gamma_1),\ldots,\rho(\gamma_r)]$). It follows then (analogously to [10]) from Equation 3 that the action of $\mathcal{H}_r$ on $\Phi^{-1}(P_0) \equiv \mathcal{N}_r(G)$ is given by

$$H_i([g_1,\ldots,g_r]) = [g_1,\ldots,g_{i-1},g_ig_{i+1}g_i^{-1},g_i,g_{i+2},\ldots,g_r], \quad i = 1,\ldots,r-1. \tag{4}$$

We define $\mathcal{H}_r(G,\mathcal{C})$ to be the union of components of $\mathcal{H}_r(G)$, which corresponds to $\mathcal{N}(G,\mathcal{C})$ via covering theory. The spaces $\mathcal{H}_r(G,\mathcal{C})$ are called *Hurwitz spaces.*

6. Curves on Hurwitz spaces

Let $\mathrm{X} \subseteq \mathbb{P}^2$ be a reduced plane curve defined over a subfield k of $\mathbb{C}$. Let $q_1, q_2, q_3 \in \mathbb{P}^2(k)$ be non-collinear points such that the connecting line $\overline{q_2q_3}$ is not contained in X. A line through q_1 which intersects the line $\overline{q_2q_3}$ in a point b is denoted L_b. Such a line is called *exceptional* if it intersects X in a singularity or if it is tangent to X. We always assume that L_{q_2} is an exceptional line, that L_{q_3} is not an exceptional line and that $q_3 \notin \mathrm{X}$. Let $L := L_{b_1} \cup \ldots \cup L_{b_s}$ be the union of all exceptional lines and $\mathcal{B}(\mathrm{X}) := \overline{q_2q_3} \setminus \{b_1,\ldots,b_s\}$.

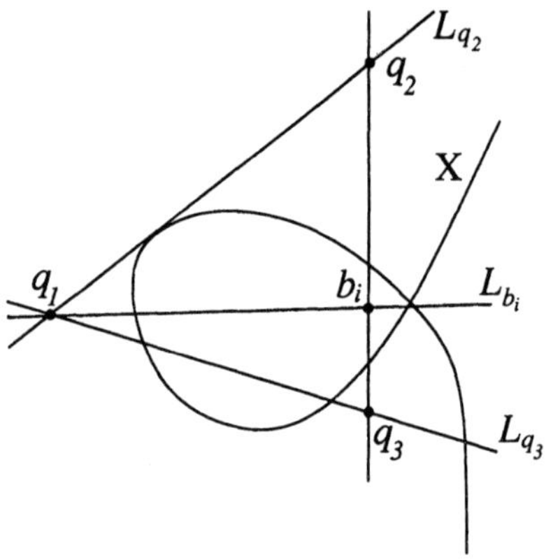

Let further $\Pi_1 : \mathbb{P}^2 \setminus \{q_1\} \to \overline{q_2 q_3}$ (resp. $\Pi_2 : \mathbb{P}^2 \setminus \{q_2\} \to L_{q_3}$) be the central projection with center q_1 (resp. q_2) and $\tau : L_{q_3} \to \mathbb{P}^1$ be a k-isomorphism.

Suppose that generically, a line L_b intersects the curve X in r points (not counting multiplicities). The maps

$$\Psi(r+1) : \mathbb{P}^2 \setminus \mathrm{X} \cup L \to \mathcal{U}(r+1), \; p \mapsto (\tau \circ \Pi_2(L_{\Pi_1(p)} \cap \mathrm{X}), \tau \circ \Pi_2(p))$$

and

$$\Psi : \mathcal{B}(\mathrm{X}) \to \mathcal{U}_r, \; b \mapsto \tau \circ \Pi_2(L_b \cap \mathrm{X})$$

give rise to a commutative diagram

$$\begin{array}{ccc} \mathbb{P}^2 \setminus \mathrm{X} \cup L & \xrightarrow{\Pi_1} & \mathcal{B}(\mathrm{X}) \\ \downarrow & & \downarrow \\ \mathcal{U}(r+1) & \xrightarrow{\mathrm{pr}_1} & \mathcal{U}_r \end{array},$$

whose rows are locally trivial fibrations (see [4], Section 2) and the first vertical arrow maps fibres to fibres. Let $\mathcal{F} := L_{q_3} \setminus \{L_{q_3} \cap (\mathrm{X} \cup L)\}$. The long exact sequences of homotopy groups, associated to locally trivial fibrations, lead then to a commutative diagram

$$\begin{array}{ccccccc} 1 \to & \pi_1(\mathcal{F}, q_3) & \to & \pi_1(\mathbb{P}^2 \setminus \mathrm{X} \cup L, q_3) & \to & \pi_1(\mathcal{B}(\mathrm{X}), q_3) & \to 1 \\ & \downarrow & & \downarrow & & \downarrow & \\ 1 \to & \pi_1(\mathcal{F}_r, (P_0, p_0)) & \to & \mathcal{H}(r+1) & \to & \mathcal{H}_r & \to 1 \end{array}, \tag{5}$$

where (P_0, p_0) (resp. P_0) is the image of q_3 in $\mathcal{U}(r+1)$ (resp. $\mathcal{U}_r$). Since in Diagram 5, the left vertical arrow is surjective, one can choose generators $\tilde{\gamma}_1, \ldots, \tilde{\gamma}_r$ of $\pi_1(\mathcal{F}, q_3)$ such that $\tilde{\gamma}_1, \ldots, \tilde{\gamma}_r$ map to generators $\gamma_1, \ldots, \gamma_r$ of $\pi_1(\mathcal{F}_r, (P_0, p_0))$ which are chosen in a way that Equation 3 holds. Let

$\Psi_* : \pi_1(\mathcal{B}(X), q_3) \to \mathcal{H}_r$ be the right vertical arrow in Diagram 5. Let $\delta_{b_1}, \ldots, \delta_{b_s} \in \pi_1(\mathcal{B}(X), q_3)$ be homotopy classes of simple loops encircling the points $b_1, \ldots, b_s$ (respectively) and let R_{b_i} be the image of δ_{b_i} under Ψ_*. We call the elements $R_{b_1}, \ldots, R_{b_s}$ the *braids of* X with respect to q_1, q_2, q_3.

By a suitable change of coordinates, one can assume that $q_1 = [1, 0, 0]$, $q_2 = [0, 1, 0]$ and one can take

$$\tau : L_{q_3} \to \mathbb{P}^1, \ [x, y, z] \mapsto [x, z]. \tag{6}$$

Under these assumptions it is shown in [3], [4] how one can compute the expression of the braids

$$R_{b_k} = R_{b_k}(H_1, \ldots, H_{r-1}) \in \mathcal{H}_r$$

as words in the standard generators $H_1, \ldots, H_{r-1}$. The following example will be used below:

Example 8. Let $X \subseteq \mathbb{P}^2$ be the plane curve which is the union of the curve defined by the affine equation

$$(y - x^2)x(x - 1) \cdots (x - (r - 4)) = 0$$

and the line at infinity (given by the equation $z = 0$). Let $q_1 = [1, 0, 0]$, $q_2 = [0, 1, 0]$ and $q_3 = [-1, \frac{1}{2}, 1] = (-1, \frac{1}{2})$. Let τ be as in Equation 6, $P_0 = \{p_1, \ldots, p_r\} := \tau(X \cap L_{q_3})$, where $p_1 < p_2 \cdots < p_{r-1}$ (with respect to the real ordering) and $p_r = \infty$. This ordering gives rise to a system of standard generators $H_1, \ldots, H_{r-1}$ of $\mathcal{H}_r$ as described in Section 4 (consider a homeomorphism $\alpha : \mathbb{P}^1 \to \mathbb{P}^1$ which maps $\mathbb{R} \cup \{\infty\}$ to $\mathbb{R} \cup \{\infty\}$, such that $\alpha(i) = p_i$, $i = 1, \ldots, r - 1$ and $\alpha(r) = \infty$, compare also to [4], Section 1). Let $b_i = (-1, y_i)$, $i = 1, \ldots, s - 1$ where we assume that $y_1 < y_2 < \cdots < y_{s-1}$ and let $b_s = [0, 1, 0]$ $(s = r - 2)$. Choose generators $\delta_{b_1}, \ldots, \delta_{b_{s-1}}$ of $\pi_1(\mathcal{B}(X), q_3)$, where δ_{b_i} encircles b_i counter-clockwise as indicated in the following picture:

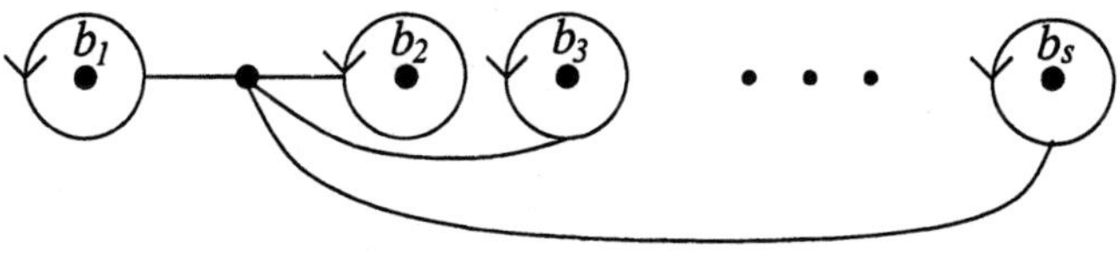

As in [4], Section 2.3, one can see that $R_{b_1} = H_2 H_1 H_2$, $R_{b_2} = H_3^2$ (where $b_1 = (-1, 0)$ and $b_2 = (-1, 1)$), $R_{b_j} = (H_{j+1}^2)^{H_j^{-1} \cdots H_3^{-1}}$, $2 < j < s$, and $R_{b_s} = (R_{b_1} \cdots R_{b_{s-1}})^{-1}$.

Set $\mathcal{S}(\mathrm{X}) := \mathrm{im}(\Psi)$ and let $\mathcal{H}(\mathrm{X}) := \mathrm{im}(\Psi_*)$. It is clear that if Ψ maps $\mathcal{B}(\mathrm{X})$ injectively to $\mathcal{S}(\mathrm{X})$, then the elements $R_{b_1}, \ldots, R_{b_s}$ generate the image of $\pi_1(\mathcal{S}(\mathrm{X}), P_0)$ in $\mathcal{H}_r$ (under the homomorphism which is induced by inclusion of paths). If $\Phi : \mathcal{H}_r(G) \to \mathcal{U}_r$ is as in the last section, then it follows from Equation 3 and Diagram 5 that $R_{b_k} \in \mathcal{H}(\mathrm{X})$ acts (as a word in $H_1, \ldots, H_{r-1}$) according to Equation 4 on $\mathcal{N}_r(G) \equiv \Phi^{-1}(P_0)$. By covering theory, the orbits of $\mathcal{H}(\mathrm{X})$ on $\mathcal{N}_r(G)$ correspond then to the irreducible components of $\Phi^{-1}(\mathcal{S}(\mathrm{X}))$. The genus of such a component is given as the number g below:

An $\mathcal{H}(\mathrm{X})$-orbit $O \subseteq \mathcal{N}_r(G)$ is said to be *of genus* g if

$$g = 1 - |O| + \frac{1}{2}\sum_{i=1}^{s}(|O| - n_i),$$

where n_i is the cardinality of the cycles occurring in the disjoint cycle decomposition of R_{b_i}, acting on O.

Since X is defined over k, $G_k = \mathrm{Gal}(\overline{k}/k)$ permutes the irreducible components of $\Phi^{-1}(\mathcal{S}(\mathrm{X}))$, inducing an action of G_k on the set of orbits of $\mathcal{H}(\mathrm{X})$ on $\mathcal{N}_r(G)$.

Criterion 9. *Let* $\mathrm{X} \subseteq \mathbb{P}^2$ *be a reduced plane curve over* $k \subseteq \mathbb{C}$ *and let* $R_{b_1}, \ldots, R_{b_s} \in \mathcal{H}_r$ *be the braids of* X *with respect to* $q_1, q_2, q_3 \in \mathbb{P}^2(k)$. *Assume that* $r \geq 3$. *Let* G *be a finite group and* $\mathcal{C} = (C_1, \ldots, C_r)$ *a tuple of conjugacy classes of* G *such that the set* $\mathcal{N}(G, \mathcal{C})$ *is non-empty. Then there exists a point* $P_0 = \{p_1, \ldots, p_r\} \in \mathcal{S}(\mathrm{X})$ *and a Galois cover* ϕ *of* $\mathbb{P}^1(\mathbb{C}) \setminus P_0$ *which has* k *as a field of moduli and ramification type* $[G, P_0, \mathcal{C}]$, $C_{p_i} = C_i$, *if the following conditions are satisfied:*
a) *The map* $\Psi : \mathcal{B}(\mathrm{X}) \to \mathcal{S}(\mathrm{X})$ *is injective.*
b) *There exists a* G_k*-stable orbit* O *of* $\mathcal{H}(\mathrm{X})$ *on* $\tilde{\mathcal{N}}(G, \mathcal{C})$ *of genus* 0.
c) *There exists* $i \in \{1, \ldots, s\}$ *with the following property: In the cycle decomposition of* R_{b_i} *on* O *there is one cycle length with odd multiplicity and the number of braids* R_{b_j}, $j = 1, \ldots, s$, *with the same cycle decomposition as* R_{b_i}, *is odd.*

Proof. Under the assumptions, the component of $\Phi^{-1}(\mathcal{S}(\mathrm{X}))$ which belongs to O is a rational curve with a dense set of k-rational points. Every such point corresponds to a cover with ramification type $[G, P, \mathcal{C}]$, $P \in \mathcal{S}(\mathrm{X})$, and field of moduli k, by [10], Thm. 1. □

Remark 10. The G_k-invariance in Criterion 9 of an orbit $O \subset \tilde{\mathcal{N}}(G, \mathcal{C})$ is assured, if there exists a point Q_0 in $\mathcal{S}(\mathrm{X})$ such that $[G, Q_0, \mathcal{C}]$ is a k-rational ramification type and differs from the other orbits of $\mathcal{H}(\mathrm{X})$ on $\tilde{\mathcal{N}}(G, \mathcal{C})$ by its length and/or its genus, compare to [4], Section 3.

7. Middle convolution applied to subgroups of $\mathrm{GL}_2(\mathbb{F}_q)$.

Theorem 11. *The projective symplectic groups* $\mathrm{PSp}_{2n}(\mathbb{F}_{p^2})$ *occur* $\mathbb{Q}$-*regularly as Galois groups over* $\mathbb{Q}(t)$ *if* p *is an odd prime* $\not\equiv \pm 1 \bmod 24$.

Proof. We prove that the claim in two steps:

i) $p \equiv \pm 5 \bmod 12$: Let $H \simeq D_{12}$, where D_{12} denotes the dihedral group of order 24. For $r \geq 5$, consider the class vector

$$\mathcal{C}_r := (2a, 2b, 2a, 2b, 2c, \dots, 2c),$$

where $2a$, $2b$ are the two conjugacy classes of non-central involutions of H and $2c$ denotes the conjugacy class of the central involution, occurring $r-4$-times. Let X and R_{b_i}, $i = 1, \dots, s$ be as in Example 8. Since the last $r - 4$ conjugacy classes consist of central elements, the only elements in $\{R_{b_1}, \dots, R_{b_s}\}$ which act non-trivially on $\tilde{\mathcal{N}}(H, \mathcal{C}_r)$ are $R_{b_1} = H_1 H_2 H_1$, $R_{b_2} = H_3^2$ and R_{b_s} (use Example 8). One can compute that $\tilde{\mathcal{N}}(H, \mathcal{C}_r)$ has cardinality 16, that $\mathcal{H}(\mathrm{X})$ acts transitively on $\tilde{\mathcal{N}}(H, \mathcal{C}_r)$ and that the cycle decompositions of R_{b_1}, R_{b_2}, R_{b_s} are $(2)^8$, $(6)^2(2)^2$, $(6)^1(3)^2(2)^1(1)^2$ (respectively). Using Remark 10, one easily checks that the conditions of Criterion 9 are satisfied for X and $\mathcal{C}_r$. Thus there exists a point $P_0 \in \mathcal{S}(\mathrm{X})$ and a Galois cover ϕ of $\mathbb{P}^1(\mathbb{C}) \setminus P_0$ which has ramification type $[H, P_0, \mathcal{C}_r]$ and $\mathbb{Q}$ as a field of moduli.

Since $12 \mid (p^2 - 1)$, we embed H in the general orthogonal group $\mathrm{GO}_2^+(\mathbb{F}_{p^2})$. Let $\check{r} := r - 1$. Let $\alpha_1, \dots, \alpha_{\check{r}}$ be as in Section 3. Then ϕ is the Galois cover associated to $(\alpha_1, \dots, \alpha_{\check{r}})$ and a tuple $(A_1, \dots, A_{\check{r}}) \in \mathrm{GO}_2^+(\mathbb{F}_{p^2})^{\check{r}}$, where $A_i \in C_i$. Applying the middle convolution MC_{-1} to $(A_1, \dots, A_{\check{r}})$, one obtains a tuple of matrices

$$(\tilde{B}_1, \dots, \tilde{B}_{\check{r}}) \in \mathrm{Sp}_{2n}(\mathbb{F}_{p^2})^{\check{r}},\ n = \check{r} - 3,$$

by Thm. 2, a) and e). Moreover, $\tilde{B}_1, \dots, \tilde{B}_4$ are transvections, $\tilde{B}_5, \dots, \tilde{B}_{\check{r}}$ are double transvections and

$$\tilde{B}_1 \cdots \tilde{B}_{\check{r}} = -1$$

by Lemma 3. By Thm. 2 c), the group $\langle \tilde{B}_1, \dots, \tilde{B}_{\check{r}} \rangle$ is an irreducible subgroup of $\mathrm{GL}_{2n}(\mathbb{F}_{p^2})$. Analogously to [6], Section 8, one concludes that, under the assumptions on p,

$$\langle \tilde{B}_1, \dots, \tilde{B}_{\check{r}} \rangle = \mathrm{Sp}_{2n}(\mathbb{F}_{p^2})$$

(using Thm. 2, b)).

Let $\tilde{C}_1, \dots, \tilde{C}_{\check{r}}$ denote the conjugacy classes of $\tilde{B}_1, \dots, \tilde{B}_{\check{r}}$ and $\tilde{C}_{\check{r}+1}$ the class of $(\tilde{B}_1 \cdots \tilde{B}_{\check{r}})^{-1} = -1$.

We want to show that $\tilde{C}_1 = \tilde{C}_3$: Let $(A'_1, \ldots, A'_4, -1, \ldots, -1) \in (2a, 2b, 2a, 2b, 2c, \ldots, 2c)$ such that $A'_3 = A'_1$ and $A'_4 = \pm {A'_1}^{-1} A'_2 A'_1$. It is easy to check that

$$Q_1^{-1} Q_2 Q_1 (A'_1, \ldots, A'_4, -1, \ldots, -1) = (A'_1, \ldots, A'_4, -1, \ldots, -1),$$

where Q_1, Q_2 are the first two generators of the abstract Artin braid group, acting via Equation 1. Let

$$MC_\lambda(A'_1, \ldots, A'_4, -1, \ldots, -1) = (\tilde{B}'_1, \ldots, \tilde{B}'_{\check{r}}).$$

The normalizer of $\mathrm{Sp}_{2n}(\mathbb{F}_{p^2})$ in $\mathrm{GL}_{2n}(\mathbb{F}_{p^2})$ is the group of symplectic similitudes $\mathrm{GSp}_{2n}(\mathbb{F}_{p^2})$. By Thm. 2 d), there exists an element $B \in \mathrm{GSp}_{2n}(\mathbb{F}_{p^2})$ such that

$$Q_1^{-1} Q_2 Q_1 (\tilde{B}'_1, \ldots, \tilde{B}'_{\check{r}}) = (\tilde{B}'_1, \ldots, \tilde{B}'_{\check{r}})^B.$$

Since $Q_1^{-1} Q_2 Q_1$ leaves the fourth element unchanged, B can be assumed to be contained in $\mathrm{Sp}_{2n}(\mathbb{F}_{p^2})$ (every element $\nu \in \mathrm{GSp}_{2n}(\mathbb{F}_{p^2}) \setminus Z(\mathrm{GSp}_{2n}(\mathbb{F}_{p^2})) \cdot \mathrm{Sp}_{2n}(\mathbb{F}_{p^2})$ permutes the two classes of transvections in $\mathrm{Sp}_{2n}(\mathbb{F}_{p^2})$). It follows now from Thm. 2 d) that $\tilde{C}_1 = \tilde{C}_3$.

Let $\tilde{\phi}$ be the Galois cover associated to $(\beta_1, \ldots, \beta_{\check{r}})$ and $(\tilde{B}_1, \ldots, \tilde{B}_{\check{r}})$. Now, since the classes $\tilde{C}_1, \ldots, \tilde{C}_{\check{r}+1}$ are rational (in the sense that $\tilde{C}_i^l = \tilde{C}_i$ for l coprime to $|\mathrm{Sp}_{2n}(\mathbb{F}_{p^2})|$) and $\tilde{C}_1 = \tilde{C}_3$, $\tilde{\phi}$ has a $\mathbb{Q}$-rational ramification type.

Again, since every element $\nu \in \mathrm{GSp}_{2n}(\mathbb{F}_{p^2}) \setminus Z(\mathrm{GSp}_{2n}(\mathbb{F}_{p^2})) \cdot \mathrm{Sp}_{2n}(\mathbb{F}_{p^2})$ permutes the two classes of transvections in $\mathrm{Sp}_{2n}(\mathbb{F}_{p^2})$, one has

$$Z(\mathrm{GL}_{2n}(\mathbb{F}_{p^2})) \cdot \mathrm{Sp}_{2n}(\mathbb{F}_{p^2}) = \{ g \in N_{\mathrm{GL}_{2n}(\mathbb{F}_{p^2})}(\langle \tilde{B}_1, \ldots, \tilde{B}_{\check{r}} \rangle) \mid g(\tilde{C}_i) = \tilde{C}_i \}.$$

It follows from the last equality and Cor. 6 that the cover $\tilde{\phi}$ associated to $(\beta_1, \ldots, \beta_{\check{r}})$ and $(\tilde{B}_1, \ldots, \tilde{B}_{\check{r}})$ has $\mathbb{Q}$ as a field of moduli. Let $\overline{\tilde{B}}_1, \ldots, \overline{\tilde{B}}_{\check{r}}$ be the images of $\tilde{B}_1, \ldots, \tilde{B}_{\check{r}}$ in $\mathrm{PSp}_{2n}(\mathbb{F}_{p^2})$ and $\overline{\tilde{\phi}}$ be the Galois cover corresponding to the kernel of the homomorphism

$$\pi_1(\mathbb{P}^1(\mathbb{C}) \setminus P_0, p_{r+1}) \to \mathrm{PSp}_{2n}(\mathbb{F}_{p^2}),\ \beta_i \mapsto \overline{\tilde{B}}_i,\ i = 1, \ldots, \check{r}.$$

Of course, $\mathbb{Q}$ is also a field of moduli of $\overline{\tilde{\phi}}$. Since the groups $\mathrm{PSp}_{2n}(\mathbb{F}_{p^2})$ have trivial center, $\overline{\tilde{\phi}}$ is defined over $\mathbb{Q}$, see [15], Prop. 3.6. This proves the claim.

ii) $p \equiv \pm 3 \bmod 8$: The claim follows analogously to i), using the non-central classes $2a, 2b$ and the central class $2c$ of D_8, the class vector $C_r := (2a, 2b, 2a, 2b, 2c, \ldots, 2c)$ and the curve X as above: One checks that

$\mathcal{H}(X)$ acts transitively on $\tilde{\mathcal{N}}(H, C_r)$ and that the cycle decompositions of R_{b_1}, R_{b_2}, R_{b_s} are $(2)^4$, $(4)^2$, $(1)^2(4)^1(2)^1$ (respectively). Proceeding as above, one obtains the result for $p \equiv \pm 3 \bmod 8$. □

Acknowledgments

The author would like to thank Professors K. Hashimoto, K. Miyake and H. Nakamura for the invitation to the conference *Galois Theory and Modular Forms* (held at the Metropolitan University of Tokyo), T. Shiina for sending his papers and Professor B. H. Matzat, S. Reiter and the referee for valuable comments.

References

[1] J. S. Birman, "Braids, Links and Mapping Class Groups," Princeton University Press, Princeton, 1974.

[2] E. Bloch, "A First Course in Geometric Topology and Differential Geometry," Birkhäuser, Boston, 1997.

[3] M. Dettweiler, *Kurven auf Hurwitzräumen und ihre Anwendungen in der Galoistheorie*, Dissertation, Erlangen, 1999.

[4] M. Dettweiler, *Plane curves and curves on Hurwitz spaces*, IWR-Preprint (2001-06).

[5] M. Dettweiler and S. Reiter, *On rigid tuples in linear groups of odd dimension*, J. Algebra **222** (1999), 550–560.

[6] M. Dettweiler and S. Reiter, *An algorithm of Katz and its application to the inverse Galois problem*, J. Symb. Comp. **30** (2000), 761–798.

[7] M. Dettweiler and S. Reiter, *Monodromy of Fuchsian systems*, in preparation.

[8] M. Dettweiler and S. Wewers, *Hurwitz spaces and Shimura varieties*, in preparation.

[9] E. R. Fadell and S. Y. Husseini, "Geometry and Topology of Configuration Spaces," Springer Verlag, Heidelberg, 2001.

[10] M. Fried and H. Völklein, *The inverse Galois problem and rational points on moduli spaces*, Math. Ann. **290** (1991), 771–800.

[11] N. Katz, "Rigid local systems," Princeton University Press, Princeton, 1996.

[12] G. Malle and B. H. Matzat, "Inverse Galois theory," Springer Verlag, Berlin, 1999.

[13] T. Shiina, *Rigid braid orbits related to* $\mathrm{PSL}_2(p^2)$ *and some simple groups*, preprint (2002).

[14] T. Shiina, *Regular Galois realizations of* $\mathrm{PSL}_2(p^2)$ *over* $\mathbb{Q}(T)$, to appear in this volume.

[15] H. Völklein, "Groups as Galois groups," Cambridge Univ. Press, Cambridge, 1996.

[16] H. Völklein, *The braid group and linear rigidity*, Geom. Dedicata **84** (2001), 135–150.

[17] H. Völklein, *A transformation principle for covers of* $\mathbb{P}^1$, J. Reine Angew. Math. **534** (2001), 155–168 .

Galois Theory and Modular Forms pp 159–172
K. Hashimoto, K. Miyake, and H. Nakamura, Editors
© 2003 Kluwer Academic Publishers

ON THE ESSENTIAL DIMENSION OF p-GROUPS

Arne LEDET
Department of Mathematics and Statistic
Texas Tech University
Lubbock, TX 79409-1042
aledet@math.ttu.edu

Abstract We improve the known bounds on the essential dimension of p-groups over (large) fields of characteristic p.

1. Introduction

In [Le], the author provided a construction of generic polynomials with few parameters for p-groups over ground fields of characteristic p, thus giving an (unstated) upper bound on essential dimensions in this case. In the present paper, we improve this bound further, by proving

Theorem 1. *Let K be a field of characteristic $p \neq 0$, and let*

$$1 \to A \to E \to G \to 1$$

be an extension of p-groups, where A is elementary Abelian and maps into the Frattini subgroup $\Phi(E)$ of E. Assume that $|A| \leq |K|$. Then

$$\mathrm{ed}_K\, G \leq \mathrm{ed}_K\, E \leq \mathrm{ed}_K\, G + 1.$$

Here, the *essential dimension* $\mathrm{ed}_K\, G$ of a finite group G over a field K is defined as follows, cf. [BR]: Given a Galois extension M/L with $\mathrm{Gal}(M/L) \simeq G$ and $L \supseteq K$ (which we will refer to as a *G-extension over K*), we consider all 'intermediate' G-extensions $\mathfrak{m}/\mathfrak{l}$:

[received: May 17, 2002; accepted in revised form: October, 16, 2002]

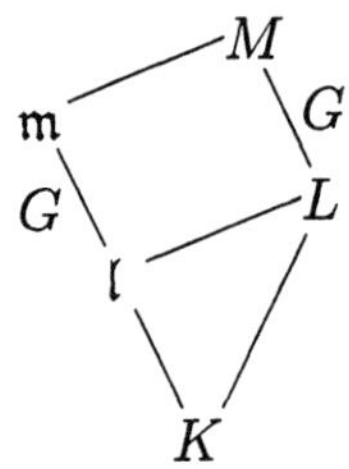

The essential dimension of M/L over K, written $\mathrm{ed}_K(M/L)$, is then defined as the minimal possible transcendence degree of $\mathfrak{l}$ (or $\mathfrak{m}$) over K. Since $\mathfrak{m}$ can be chosen to be finitely generated, this essential dimension is clearly finite, and in fact less than $|G|$: Let $\mathfrak{m} = K(\{\sigma\theta\}_{\sigma\in G})$, where $\{\sigma\theta\}_\sigma$ is a transcendence basis with $\sum_\sigma \sigma\theta = 1$.

Since $\mathrm{ed}_K(M/L)$ is always less than $|G|$, we can now define the essential dimension of the group G over K, written $\mathrm{ed}_K G$, as the maximal value assumed by $\mathrm{ed}_K(M/L)$, when M/L runs through all G-extensions over K.

Thus, the essential dimension gives an upper bound for the number of algebraically independent parameters needed to describe a G-extension M/L over K. Here, the extension M/L is considered 'described' by an intermediate extension $\mathfrak{m}/\mathfrak{l}$ as above, since the Galois theoretical structure of M/L is completely given by the fact that it is obtained from $\mathfrak{m}/\mathfrak{l}$ by extending the scalars from $\mathfrak{l}$ to L.

Example. The only group G that has essential dimension 0 is the trivial group, since essential dimension 0 must mean that every G-extension M/L over K must be described by an intermediate G-extension $\mathfrak{m}/\mathfrak{l}$, where $\mathfrak{l}$ is algebraic over K. And clearly, if $G \neq 1$ we can produce M/L for which this fails.

Over the field $\mathbb{Q}$ of rational numbers, the cyclic groups C_2 and C_3, of order 2 and 3, both have essential dimension 1. This is obvious for C_2: A quadratic extension has the form $L(\sqrt{a})/L$, and we can let $\mathfrak{m} = \mathbb{Q}(\sqrt{a})$ and $\mathfrak{l} = \mathbb{Q}(a)$. For C_3, it requires a little more work: First, we find $x, y \in M$ such that x and y are linearly independent over L, and such that x, y and $-(x+y)$ are conjugate over L. We can then let $\mathfrak{m} = \mathbb{Q}(x/y)$ and $\mathfrak{l} = \mathbb{Q}(x/y)^{C_3}$.

Again over $\mathbb{Q}$, the cyclic groups of order 4 and 5 both have essential dimension 2. We will not prove this here, but only observe that they can easily be proven to have essential dimension *at most* 2. To get equality, it is then necessary to invoke a result implicit in both [BR, Thm. 6.2] and [JLY, Ch. 8]: If $\mathrm{ed}_K G = 1$, then G is isomorphic to a subgroup of $\mathrm{PGL}_2(K)$.

In [BR], Buhler and Reichstein prove a number of results about essential dimensions, including some lower bounds. In general, lower bounds are substantially harder to obtain than upper bounds, and this paper will be concerned only with upper bounds. (It may be of interest to note that in many cases, the *only* lower bound known is the one obtained by proving that the group is *not* a subgroup of $\mathrm{PGL}_2(K)$, i.e., a lower bound of 2.)

Theorem 1 is proved in section 2 below.

In terms of essential dimensions, the results of [Le] amount to proving the inequality

$$\mathrm{ed}_K\, E \leq \mathrm{ed}_K\, G + 1$$

in the case $|A| = p$, and combining it with

Lemma 2. *Let K be a field of characteristic $p \neq 0$, with $|K| \geq p^n$. Then*

$$\mathrm{ed}_K(C_p^n) = 1,$$

where C_p^n is the elementary Abelian group of order p^n.

In characteristic 0, it is proved in [BR] that the essential dimension of C_p^n is at least n, with equality when the ground field contains the nth roots of unity. Intuitively, this makes sense: A C_p^n-extension is the composite of n linearly disjoint C_p-extensions, and we clearly need at least one parameter to describe a C_p-extension. It may therefore come as a small surprise that we need only one parameter in characteristic p (provided that ground field is large enough). At least it initially surprised the author, and led to the realization that p-groups in characteristic p have 'small' essential dimensions.

We will prove Lemma 2 in section 2 below, but refer the reader also to [JLY]. Clearly, Theorem 1 and Lemma 2 provide a very low bound for the essential dimension of a given p-group over 'sufficiently large' fields. In most cases, this bound will be lower than the one obtained in [Le], since we are guaranteed the existence of an A of order at least p^2 if $\Phi(E)$ is non-cyclic, by [Hu, III.7.5 Hilfssatz].

Example. Let $n \in \mathbb{N}$, and consider the group

$$G_n = \left\{ \begin{pmatrix} 1+pa & b \\ pc & 1+pd \end{pmatrix} \,\middle|\, a,b,c,d \in \mathbb{Z}/p^n \right\}$$

of 2×2 matrices over $\mathbb{Z}/p^n$, where p is an odd prime. It is a p-Sylow subgroup of $\mathrm{GL}_2(\mathbb{Z}/p^n)$, and has order p^{4n-3}. Using Theorem 1 and Lemma 2, we find that

$$\mathrm{ed}_K\, G_n \leq n,$$

provided $|K| \geq p^4$. See section 3 below for details.

Example. Let G be an Abelian p-group of exponent p^n, minimally generated by e elements. Then $\operatorname{ed}_K G \leq n$ if $|K| \geq p^e$.

Remark. Both Theorem 1 and Lemma 2 requires the ground field K to be 'big enough'. However, for an arbitrary ground field K and a finite group G we have

$$\operatorname{ed}_K G \leq \operatorname{ed}_{K(t)} G \leq \operatorname{ed}_K G + 1,$$

where t is an indeterminate. Here, the first inequality is clear from the definition of essential dimension: We cannot need more algebraically independent parameters to describe G-extensions over a larger field. The second is also quite easy: If M/L is a G-extension over K, we either have that L/K is algebraic and hence that $\operatorname{ed}_K(M/L) = 0$, or that M/L is a G-extension over $K(t)$. Since $\operatorname{ed}_K G$ is the maximum of all possible $\operatorname{ed}_K(M/L)$'s, the first case does not contribute. And in the second, we have $\operatorname{ed}_K(M/L) \leq \operatorname{ed}_{K(t)}(M/L) + 1$, simply because we can consider t as an extra parameter.

Thus we get a bound for the essential dimension also in the case where K is finite of characteristic p, and G is a p-group that is 'too large' to allow a direct application of Theorem 1 and Lemma 2.

For instance, C_p^n has essential dimension 2 over a ground field K of order less than p^n. (Here, it is clear that the essential dimension is at most 2. To prove that it cannot be 1, it is enough to see that a p-Sylow subgroup of $\mathrm{PGL}_2(K)$ is elementary Abelian of order $|K|$.)

And of course: Unless the group is cyclic, it does not occur as a Galois group over a finite (or absolutely algebraic) field in prime characteristic at all. So in that case, there is no loss in considering the ground field $\mathbb{F}_p(t)$. And a cyclic p-group *is* covered by our results, as well as by the classical theory of Witt vectors, cf. [Wi2].

Remark. It seems a reasonable conjecture that the cyclic group C_{p^n} of order p^n has essential dimension n over any field of characteristic p. Since one of the results of [BR] is that the essential dimension of a subgroup is less than or equal to that of the entire group, this would provide a lower bound for the essential dimensions of p-groups in characteristic p in terms of the exponent of the group, i.e., the maximal order of an element. For instance, the group G_n from the Example above contains an element of order p^n, and would therefore have essential dimension exactly n.

Unfortunately, it is at this point only a conjecture. Also, while it would allow us to determine the essential dimension in many cases, such as G_n or Abelian p-groups, there are p-groups for which the essential dimension is strictly greater than the bound thus given. For instance, the Heisenberg group of order p^3, when p is odd, i.e., the non-Abelian group of order p^3 and exponent p: The exponent provides a trivial lower bound of 1 for the essential dimension, but it is easily seen that the essential dimension cannot be 1.

2. Proof of Theorem 1

Let the assumptions be as in Theorem 1. We first prove that any G-extension M/L over K can be extended to an E-extension. This will establish the inequality

$$\mathrm{ed}_K\, G \leq \mathrm{ed}_K\, E,$$

since any intermediate E-extension gives rise to an intermediate G-extension of the same transcendence degree.*

Next, we prove that any E-extension F/L containing M/L arises in the manner established in the first part, and note that this description involves only one parameter. This, then, gives us

$$\mathrm{ed}_K\, E \leq \mathrm{ed}_K\, G + 1.$$

Part one

Let M/L be a G-extension over K. Also, let $c \in Z^2(G, A)$ be a factor system representing the extension

$$1 \to A \to E \to G \to 1.$$

This extension makes A into a $\mathbb{F}_p[G]$-module in the usual way (by conjugation in E), and we will need an injective $\mathbb{F}_p[G]$-homomorphism $\varphi\colon A \to M^+$:

Let $\varphi'\colon A \hookrightarrow K^+$ be an injective group homomorphism, and let $(\sigma\xi)_{\sigma\in G}$ be a normal basis for M/L. Then we can let

$$\varphi(a) = \sum_{\sigma\in G} \varphi'(\sigma^{-1}a)\sigma\xi, \quad a \in A.$$

We also have an induced homomorphism

$$\varphi^2\colon H^2(G, A) \to H^2(G, M^+).$$

*Note that we cannot simply say 'G is a homomorphic image of E, therefore $\mathrm{ed}_K\, G \leq \mathrm{ed}_K\, E$.' No such result is available (at this time), and in the more general setting of algebraic groups it is known to be untrue, cf. [Re].

By, e.g., [We, Cor. 3–1–4] the cohomology group $H^2(G, M^+)$ is trivial, and so there exists a map $a\colon G \to M$ with

$$\forall \sigma, \tau \in G\colon \varphi(c_{\sigma,\tau}) = a_\sigma + \sigma a_\tau - a_{\sigma\tau}.$$

Let

$$H(X) = \prod_{b \in A}(X - \varphi(b)) \in L[X].$$

This is a *vectorial* polynomial, as defined in [Ab], in that $H(x+y) = H(x) + H(y)$. Now,

$$H(a_\sigma) + \sigma H(a_\tau) - H(a_{\sigma\tau}) = H(\varphi(c_{\sigma,\tau})) = 0,$$

meaning that $\sigma \mapsto H(a_\sigma)$ is a crossed homomorphism. Since, again by [We, Cor. 3–1–4], we also have $H^1(G, M^+) = 0$, there therefore exists an $\omega \in M$ with

$$\forall \sigma \in G\colon \sigma\omega - \omega = H(a_\sigma).$$

Let θ be a root of $H(X) - \omega$. We claim that $M(\theta)/L$ is an E-extension:

The vectorial polynomial $H(X)$ has the form

$$X^{p^n} + d_{n-1}X^{p^{n-1}} + \cdots + d_1 X^p + d_0 X,$$

where $d_0 \neq 0$. It follows that $H(X) - \omega$ is separable. Next, a conjugate of θ (over L) must be a root of $H(X) - \sigma\omega$ for some $\sigma \in G$, i.e., have the form $\theta + a_\sigma + \varphi(b)$ for a $b \in A$. Therefore $M(\theta)/L$ is Galois.

Extend $\sigma \in G$ to $\bar\sigma \in \operatorname{Gal}(M(\theta)/L)$. This extension is then given by

$$\bar\sigma\colon \theta \mapsto \theta + a_\sigma + \varphi(b_\sigma)$$

for some $b_\sigma \in A$.

Clearly, $\operatorname{Gal}(M(\theta)/M)$ can be identified with a subgroup B of A through $\tau \mapsto \varphi^{-1}(\tau\theta - \theta)$. We let $b \in B$ act on θ by $b\colon \theta \mapsto \theta + \varphi(b)$. Then

$$(\sigma b) \circ \bar\sigma = \bar\sigma \circ b,$$

giving the proper action of G on B in

$$1 \to B \to \operatorname{Gal}(M(\theta)/L) \to G \to 1.$$

Also,

$$\bar\sigma\bar\tau = [c_{\sigma,\tau} + (b_\sigma + \sigma b_\tau - b_{\sigma\tau})]\overline{\sigma\tau},$$

allowing us to embed $\operatorname{Gal}(M(\theta)/L)$ into E. Since $\operatorname{Gal}(M(\theta)/L)$ maps onto G, and $A \subseteq \Phi(E)$, this is possible only if $\operatorname{Gal}(M(\theta)/L) = E$.

Ergo: $M(\theta)/L$ is an E-extension, and we may pick $b_\sigma = 0$.

We note that replacing ω by $\omega + t$ for a $t \in L$ will not change the nature of the extension.

Part two

Let again M/L be a G-extension over K, and consider an E-extension F/L containing M/L. We claim that $F = M(\theta_t)$ for some $t \in L$, when θ_t is a root of $H(X) - (\omega + t)$, and ω is obtained as above:

Since F/M is an A-extension, and φ can be considered as a crossed homomorphism $A \to F^+$, we can find $\theta' \in F$ with

$$\forall a \in A\colon a(\theta') = \theta' + \varphi(a).$$

Clearly, then, $F = M(\theta')$. Also, from $\bar{\sigma} \circ a = (\sigma a) \circ \bar{\sigma}$ we get

$$a(\bar{\sigma}\theta') = \bar{\sigma}\theta' + \varphi(a),$$

and hence

$$\bar{\sigma}\theta' - \theta' \in M.$$

Let $b_\sigma = \bar{\sigma}\theta' - \theta'$. Then

$$b_\sigma + \sigma b_\tau = \bar{\sigma}\bar{\tau}\theta' - \theta' = c_{\sigma,\tau}\overline{\sigma\tau}\theta' - \theta' = b_{\sigma\tau} + \varphi(c_{\sigma,\tau}),$$

i.e.,

$$\varphi(c_{\sigma,\tau}) = b_\sigma + \sigma b_\tau - b_{\sigma\tau}.$$

We already have one set of splitting factors for c, namely the a_σ's. Then, $\sigma \mapsto b_\sigma - a_\sigma$ is a crossed homomorphism $G \to M^+$, and therefore

$$b_\sigma = a_\sigma + (\sigma x - x)$$

for some $x \in M$, or

$$\bar{\sigma}(\theta' - x) = (\theta' - x) + a_\sigma.$$

We replace θ' by $\theta'' = \theta' - x$. Then $F = M(\theta'')$ and

$$\bar{\sigma}\theta'' = \theta'' + a_\sigma.$$

This of course means that

$$\sigma H(\theta'') - H(\theta'') = H(\bar{\sigma}\theta'') - H(\theta'') = H(a_\sigma).$$

Let $\omega'' = H(\theta'') \in M$. Then $\sigma\omega'' - \omega'' = H(a_\sigma)$, and hence

$$\omega'' = \omega + t$$

for some $t \in L$. Ergo: $F = M(\theta_t)$.

The details of why $\mathrm{ed}_K\, E \leq \mathrm{ed}_K\, G + 1$ are now clear: We can find φ, ω and the a_σ's inside any intermediate extension $\mathfrak{m}/\mathfrak{l}$ of M/L, and get the intermediate E-extension $\mathfrak{m}(t,\theta_t)/\mathfrak{l}(t)$, where the transcendence degree is obviously at most 1 greater than for $\mathfrak{m}/\mathfrak{l}$.

This completes the proof of Theorem 1. □

Remark. The fact that an embedding problem as in Theorem 1 always has a solution was first proved by Witt in [Wi1, 1936]. This result does not depend on the size of the fields involved, since we can simply restrict ourselves to the case $|A| = p$, applied repeatedly. In fact, Witt's result is that an embedding problem given by a G-extension M/K and a group extension

$$1 \to \Phi(E) \to E \to G \to 1$$

is always solvable, if K has prime characteristic p and the groups E and G are p-groups. This result is an obvious consequence of the argument given in Part One above, and indeed the argument given here is basically a generalization of the proof in [Wi1].

Proof of Lemma 2

Let K be a field in prime characteristic p of order at least p^n, and let M/L be a C_p^n-extension over K. We let $\varphi\colon C_p^n \hookrightarrow K^+$ be an injective group homomorphism, and consider it as a crossed homomorphism $\mathrm{Gal}(M/L) \to M^+$. It is then principal, i.e., there exists $\xi \in M$ with

$$\forall \sigma \in C_p^n\colon \sigma\xi = \xi + \varphi(\sigma),$$

and therefore in particular $M = L(\xi)$. We now let $\mathfrak{m} = K(\xi)$ and $\mathfrak{l} = \mathfrak{m}^{C_p^n}$. Clearly, $\mathfrak{m}$ (or $\mathfrak{l}$) has transcendence degree at most 1 over K, meaning that $\mathrm{ed}_K(C_p^n) \leq 1$. On the other hand, let $M = K(t)$ for an indeterminate t, and define a C_p^n-action on M by $\sigma t = t + \varphi(\sigma)$. Since there are no non-trivial algebraic extensions of K inside M, we must necessarily have $\mathrm{ed}_K(M/M^{C_p^n}) = 1$, and therefore $\mathrm{ed}_K(C_p^n) = 1$. □

Remark. One of the interesting things about the essential dimension for a group G over a field K is that it gives a lower bound for the number of parameters needed in a generic polynomial for G over K, cf. [JLY].

Here, a *generic polynomial* for a finite group G over a field K is a monic polynomial $P(\mathbf{t},X) \in K(\mathbf{t})[X]$, where $\mathbf{t} = (t_1,\dots,t_n)$ and X are indeterminates, such that

(a) $P(\mathbf{t},X)$ has Galois group G over $K(\mathbf{t})$; and

(b) whenever M/L is a G-extension over K, M is the splitting field over L of a specialization $P(\mathbf{a},X)$ of $P(\mathbf{t},X)$ over L.

It is not clear whether the construction above can be made to include a generic polynomial in the general setting. However, if we restrict our attention a little, requiring the successive group extensions – with the exception of the first one – to be *central*, i.e., with $A \subseteq Z(E)$, we *can* produce a generic polynomial as well. For simplicity, we assume K to be 'large':

(1) If we are considering an elementary Abelian group C_p^n, we embed C_p^n into K^+ by an injective homomorphism $\varphi\colon C_p^n \hookrightarrow K^+$, and let $H(X) = \prod_{\sigma \in C_p^n}(X - \varphi(\sigma))$. If M/L is then a C_p^n-extension over K, we have $M = L(\xi)$ for a ξ with $\sigma(\xi) = \xi + \varphi(\sigma)$, and we conclude that $H(X) - t \in K(t)[X]$ is generic, with the single parameter t. (This is essentially the proof of Lemma 2.) Also, the element ξ^{p^n-1} generates a normal basis for M/L: We know that $(\xi^i)_{i=0}^{p^n-1}$ is a basis for M/L, and that

$$(\xi + \varphi(\sigma))^{p^n-1} = \sum_{i=0}^{p^n-1} (-1)^i \varphi(\sigma)^{p^n-1-i} \xi^i,$$

since $\binom{p^n-1}{i} \equiv (-1)^i \pmod{p}$. The matrix transforming $(\xi^i)_i$ into $(\sigma\xi^{p^n-1})_\sigma$ therefore has rows of the form

$$[\varphi(\sigma)^{p^n-1}, -\varphi(\sigma)^{p^n-2}, \ldots, -\varphi(\sigma), 1],$$

meaning that it is basically a Vandermonde matrix (up to a few signs and some swapping of columns), and so has non-zero determinant. This proves that $(\sigma\xi^{p^n-1})_\sigma$ is a basis for M/L. In particular, ξ^{p^n-1} has non-zero trace,[†] and $\xi^{p^n-1}/\operatorname{Tr}_{M/L}(\xi^{p^n-1})$ has trace 1.

(2) If we are considering a group extension as in Theorem 1, with G elementary Abelian, we can proceed as in the proof above. We have a normal basis from step (1), and to get the various elements, i.e., splitting factors etc., we just need to make the results $H^1(G, M^+) = 0$ and $H^2(G, M^+)$ explicit:

Let $\alpha \in M$ have trace 1. If $\psi\colon G \to M^+$ is a crossed homomorphism, we can let $\beta = -\sum_{\sigma \in G} \psi(\sigma)\sigma\alpha$ to get $\psi(\sigma) = \sigma\beta - \beta$, and if $c\colon G \times G \to M^+$ is a factor system, we get $c_{\sigma,\tau} = a_\sigma + \sigma a_\tau - a_{\sigma\tau}$ by letting $a_\sigma = \sum_{\rho \in G} c_{\sigma,\rho}\sigma\rho\alpha$.

Note that these results allow us to construct all necessary quantities in a purely 'mechanical' way, i.e., one that carries over homomorphically under the kind of specialization used with generic polynomials. A new element of trace 1 can then be found for the next step by taking such elements in M/L and in F/M, and multiplying them.

[†]In fact, $p^n - 1$ is the smallest positive exponent k, for which ξ^k has non-zero trace. This is most easily seen by considering the trace of ξ^k as a Newton power sum, since most of the coefficients of $H(X) - t$ are zero. In this way, we show that ξ^k has trace 0 for $0 \leq k < p^n - 1$, and get an expression for the trace of ξ^{p^n-1}.

(3) In the case of a (central) group extension as in Theorem 1, we proceed as in (2). This time, we do not have a normal basis at our disposal, but by assumption we will not need it.

In the general case, the problem lies exactly in the need for G-equivariant embeddings. Depending on the size of K, it may not be necessary to find actual normal bases, but it is required to produce the G-equivariant embeddings in an algorithmic way at each step.

Example. To illustrate the construction of generic polynomials using the procedure described above, we consider the cyclic group C_{p^2} of order p^2 over the field $\mathbb{F}_p$.

Let K be a field in characteristic p, and let M/K be a C_{p^2}-extension. There is then an intermediate C_p-extension L/K. Also, let σ denote a generator for the Galois group $\mathrm{Gal}(M/K) = C_{p^2}$.

The group extension is

$$1 \to C_p \to C_{p^2} \to C_p \to 1,$$

and we choose $\varphi\colon \sigma^p \mapsto 1$ as our injection of C_p into $\mathbb{F}_p$.

There is then, cf. the well-known Artin-Schreier theory, an element $a \in K$, such that $L = K(\theta)$ for a θ with $\sigma\theta = \theta + 1$ and $H(\theta) = \theta^p - \theta = a$.

The trace of θ^{p-1} is -1, and we can pick $-\theta^{p-1}$ as our element of trace 1.

If we let σ' denote the restriction of σ to L, the 'obvious' choice of factor system is given by picking σ^i as the pre-image of σ'^i for $0 \leq i < p$. To get splitting factors, we then only need to find $a_{\sigma'} \in L$, and the only requirement for $a_{\sigma'}$ is that it should have trace 1. Thus, we can let $a_{\sigma'} = -\theta^{p-1}$.

We now seek $\omega \in L$ with $\sigma\omega - \omega = H(a_{\sigma'}) = -H(\theta^{p-1})$. Since $\theta^p = \theta + a$, this means

$$\begin{aligned}
\sigma\omega - \omega &= -[\theta^{p(p-1)} - \theta^{p-1}] = -[(\theta + a)^{p-1} - \theta^{p-1}] \\
&= -\sum_{i=0}^{p-1} \binom{p-1}{i} a^{p-1-i}\theta^i + \theta^{p-1} \\
&= -\sum_{i=0}^{p-2} (-a)^{p-1-i}\theta^i.
\end{aligned}$$

Clearly,

$$\begin{pmatrix} \sigma\theta - \theta \\ \sigma\theta^2 - \theta^2 \\ \vdots \\ \sigma\theta^{p-1} - \theta^{p-1} \end{pmatrix} = \begin{pmatrix} \binom{1}{0} & & & \\ \binom{2}{0} & \binom{2}{1} & & \\ \vdots & & \ddots & \\ \binom{p-1}{0} & \binom{p-1}{1} & \cdots & \binom{p-1}{p-2} \end{pmatrix} \begin{pmatrix} 1 \\ \theta \\ \vdots \\ \theta^{p-2} \end{pmatrix}.$$

Call the matrix $\mathbf{A}$, and note that $\mathbf{A}$ is independent of the specific C_{p^2}-extension M/K. Then, if $\mathbf{A}^{-1}$ has i^{th} row $[a_{i1}, a_{i2}, \ldots, a_{i,p-1}]$, we get

$$\sigma x_{i-1} - x_{i-1} = \sigma(\sum_{j=1}^{p-1} a_{ij}\theta^j) - \sum_{j=1}^{p-1} a_{ij}\theta^j = \theta^{i-1},$$

and thus

$$\omega = -\sum_{i=0}^{p-2} (-a)^{p-1-i} x_i.$$

It follows that $M = L(\xi)$ for a ξ with $\xi^p - \xi = \omega + b$, $b \in K$.

To get a generic polynomial, we consider the two 'parameters' a and b as indeterminates s and t, to get θ and ξ defined over $\mathbb{F}_p(s,t)$. The minimal polynomial for ξ over $\mathbb{F}_p(s,t)$ is then generic, since we can define a homomorphism from $\mathbb{F}_p[s,t]$ to K by $s \mapsto a$, $t \mapsto b$, and extend it to a map $\mathbb{F}_p[s,t,\theta,\xi] \to M$. Thus, the corresponding specialization of the minimal polynomial is well-defined and has splitting field M over K.

In the simplest special case, $p = 2$, we get $\sigma\omega - \omega = a$, and so $\omega = a\theta$. The minimal polynomial for ξ over K is then

$$P(X) = X^4 + (1 + a)X^2 + aX + (a^3 + ab + b^2).$$

It follows that

$$X^4 + (1+s)X^2 + sX + (s^3 + st + t^2) \in \mathbb{F}_2(s,t)[X]$$

is generic for C_4 over $\mathbb{F}_2$.

We note that this is the same polynomial we would get from using Witt vectors: In terms of Witt vectors, $M = K(x,y)$, where $\sigma(x,y) = (x,y) + (1,0) = (x+1, x+y)$. Also, with $(u,v) = \wp(x,y) = (x^2,y^2) - (x,y) = (x^2 - x, y^2 - y + x^3 - x^2)$, the minimal polynomial for y over K is

$$X^4 + (1+u)X^2 + uX + (u^3 + uv + v^2),$$

and in the case of a 'general' Witt vector, i.e., $M = \mathbb{F}_p(x,y)$ with x and y indeterminates, the fixed field is $M^{C_4} = \mathbb{F}_p(u,v)$.

3. The matrix group

We recall that

$$G_n = \left\{ \begin{pmatrix} 1+pa & b \\ pc & 1+pd \end{pmatrix} \;\middle|\; a,b,c,d \in \mathbb{Z}/p^n \right\},$$

and that p is an odd prime. For $n = 1$, this is just a cyclic group of order p, and the essential dimension is known to be 1.

Let therefore $n > 1$. We first note that G_n is then minimally generated by the three matrices

$$A = \begin{pmatrix} 1+p & 0 \\ 0 & 1 \end{pmatrix}, \quad B = \begin{pmatrix} 1 & 1 \\ 0 & 1 \end{pmatrix}, \quad C = \begin{pmatrix} 1 & 0 \\ p & 1 \end{pmatrix}.$$

This becomes clear, once we notice that

$$C^{-1}B^{-1}ACB = \begin{pmatrix} 1 & 0 \\ 0 & 1+p \end{pmatrix}.$$

For $n > 2$, this means that the kernel of the natural map $G_n \twoheadrightarrow G_{n-1}$, i.e.,

$$K_n = \left\{ \begin{pmatrix} 1+p^{n-1}a & p^{n-1}b \\ p^{n-1}c & 1+p^{n-1}d \end{pmatrix} \;\middle|\; a, b, c, d \in \mathbb{Z}/p^n \right\},$$

is contained in $\Phi(G_n)$. Also, a simple computation shows that it is elementary Abelian of order p^4. Therefore,

$$\mathrm{ed}_K\, G_n \leq \mathrm{ed}_K\, G_{n-1} + 1, \quad n > 2,$$

or

$$\mathrm{ed}_K\, G_n \leq \mathrm{ed}_K\, G_2 + (n-2).$$

For $n = 2$, we have that $|\Phi(G_2)| = p^2$. Moreover, $B^p \in \Phi(G_2)$ and

$$BCB^{-1}C^{-1} = \begin{pmatrix} 1+p & -p \\ 0 & 1-p \end{pmatrix} \in G_2' \subseteq \Phi(G_2).$$

Since B^p and $BCB^{-1}C^{-1}$ are not powers of each other, we must have $\Phi(G_2) \simeq C_p^2$, and so

$$\mathrm{ed}_K\, G_2 \leq \mathrm{ed}_K(C_p^3) + 1 = 2,$$

which implies that

$$\mathrm{ed}_K\, G_n \leq n.$$

Remark. A simple calculation shows that

$$Z(G_n) = \left\{ \begin{pmatrix} 1+pa & p^{n-1}b \\ p^{n-1}c & 1+pa \end{pmatrix} \;\middle|\; a, b, c \in \mathbb{Z}/p^n \right\},$$

and hence that the subgroup

$$Z_n = \left\{ \begin{pmatrix} 1+p^{n-1}a & p^{n-1}b \\ p^{n-1}c & 1+p^{n-1}a \end{pmatrix} \;\middle|\; a, b, c \in \mathbb{Z}/p^n \right\}$$

is central of exponent p, and contained in K_n. Since K_n/Z_n is normal in G_n/Z_n of order p, it must be central as well. Thus, by replacing the single group extension

$$1 \to K_n \to G_n \to G_{n-1} \to 1, \quad n > 2,$$

with two extensions

$$1 \to K_n/Z_n \to G_n/Z_n \to G_{n-1} \to 1$$

and

$$1 \to Z_n \to G_n \to G_n/Z_n \to 1,$$

we get G_n 'built up' through central extensions. Consequently, by the Remark in section 2, it is possible to construct generic polynomials by means of these extensions. And assuming that $|K| \geq p^3$, we get for the number of parameters that

$$\#(\text{parameters for } G_2) = 2$$

and

$$\#(\text{parameters for } G_n) \leq \#(\text{parameters for } G_{n-1}) + 2,$$

i.e., a generic polynomial for G_n can be constructed with $2n-2$ parameters.

Remark. The other Example, of an Abelian p-group, can be treated by means of Witt vectors in a way very similar to the proof of Lemma 2:

If A is an Abelian p-group of exponent p^n, generated by e elements, and K is a field of characteristic p and order at least p^e, we can find an injective group homomorphism $\varphi\colon A \hookrightarrow W_n(K)^+$, where W_n denotes the ring of n-dimensional Witt vectors, cf. [Wi2]. In fact, if $a_1, \dots, a_e \in K$ are linearly independent over $\mathbb{F}_p$, the Witt vectors $(a_1, 0, \dots, 0)$, ..., $(a_e, 0, \dots, 0)$ will generate a subgroup of $W_n(K)^+$ isomorphic to $C_{p^n}^e$.

Now, let M/L be an A-extension over K. Then φ can be considered as a crossed homomorphism from A into $W_n(M)^+$, where A acts on $W_n(M)$ coordinate-wise. The cohomology group $H^1(A, W_n(M)^+)$ is trivial, as can be easily shown in general in the same way as for $n = 1$ (done in section 2 above), and so there exists a $\boldsymbol{\theta} \in W_n(M)$ with $\sigma\boldsymbol{\theta} = \boldsymbol{\theta} + \varphi(\sigma)$ for all σ. This means that $M = L(\boldsymbol{\theta})$ (with $\boldsymbol{\theta}$ considered as a set on n elements), and letting $\mathfrak{m} = K(\boldsymbol{\theta})$ we see that $\operatorname{ed}_K A \leq n$.

Moreover, by one of the results in [Le] we have: If $\mathbf{t} = (t_0, \dots, t_{n-1})$ is a set of n indeterminates, and we let A act on $K(\mathbf{t})$ by $\sigma\mathbf{t} = \mathbf{t} + \varphi(\sigma)$, the fixed field $K(\mathbf{t})^A$ is a rational (purely transcendental) extension of K. Consequently, if K is *infinite*, we get a generic polynomial with

n parameters: If $K(\mathbf{t})^A = K(\mathbf{s})$, with $\mathbf{s} = (s_0, \ldots, s_{n-1})$, any monic polynomial $P(\mathbf{s}, X) \in K(\mathbf{s})[X]$ with splitting field $K(\mathbf{t})$ will be generic, cf. [JLY, Prop. 1.1.5].

References

[Ab] S. S. Abhyankar, *Galois embeddings for linear groups*, Trans. Amer. Math. Soc. **352** (2000), 3881–3912.

[BR] J. Buhler and Z. Reichstein, *On the essential dimension of a finite group*, Compositio Mathematica **106** (1997), 159–179.

[Hu] B. Huppert, "Endliche Gruppen I," Grundlehren der mathematischen Wissenschaften, 134. Springer-Verlag, 1967.

[JLY] C. U. Jensen, A. Ledet and N. Yui, "Generic Polynomials," MSRI Publications, 45. Cambridge University Press, 2002.

[Le] A. Ledet, *On p-group in characteristic p*, preprint, 2001.

[Re] Z. Reichstein, *On the notion of essential dimension for algebraic groups*, Transform. Groups **5** (2000), 265–304.

[We] E. Weiss, "Cohomology of Groups," Pure and applied mathematics, 34. Academic Press, New York, 1969.

[Wi1] E. Witt, *Konstruktion von galoisschen Körpern der Charakteristik p zu vorgegebener Gruppe der Ordnung* p^f, J. Reine Angew. Math. **174** (1936), 237–245.

[Wi2] E. Witt, *Zyklische Körper und Algebren der Charakteristik p vom Grad* p^n, J. Reine Angew. Math. **176** (1937), 126–140.

Galois Theory and Modular Forms pp 173–194
K. Hashimoto, K. Miyake, and H. Nakamura, Editors
© 2003 Kluwer Academic Publishers

EXPLICIT CONSTRUCTIONS OF GENERIC POLYNOMIALS FOR SOME ELEMENTARY GROUPS

Yūichi RIKUNA
Department of Mathematical Sciences,
Waseda University,
3-4-1 Ohkubo, Shinjuku-ku, Tokyo
169-8555, Japan
rikuna@gm.math.waseda.ac.jp

Introduction

For a finite group G and a field k, we call a G-Galois extension over k by G/k-extension. Whether a G/k-extension exists or not is the first version of inverse Galois problem. Especially the case when $k = \mathbf{Q}$, the rational number field, plays an important role in the study of the absolute Galois Group of $\mathbf{Q}$. By many mathematicians, the existence of $G/\mathbf{Q}$-extensions has been shown for a lot of finite groups G by now (cf. Malle-Matzat [14], Serre [19], etc.)

Recently the next stage of inverse Galois problem, so-called *constructive* inverse Galois problem, has begun to be studied. If there exists a G/k-extension, the problem asks for constructing all G-extensions over the base field k concretely. It, however, is extremely difficult for us to construct a general theory for this problem, because this contains the construction problem of non-commutative class field theory. Therefore constructive inverse Galois problem has been studied only for relatively small or simple-structured groups until now. But we should remark here that even study for the cyclic groups brings considerable progresses for algebraic number theory.

Here we study explicit constructions of generic polynomials for some simple-structured groups. A *k-generic G-polynomial* F is defined as a G-polynomial over a rational function field $k(t_1, \dots, t_n)$ such that every G-extension L/K with $K \supset k$ and $\#K = \infty$ is obtained from F by a specialization of parameters $t_1, \dots, t_n$. To construct such polynomials, we

[received: Jul. 03, 2002; accepted in revised form: Oct. 14, 2002]

compute an explicit positive answer of linear Noether(-Lüroth)'s problem. This method is indicated by Kemper [8; 9] and Kemper-Mattig [11]. G. Kemper also gave explicit expressions of generic polynomials for the generalized quaternion groups in his thesis [8]. By now, this method has been applied for some other important groups, for example, the cyclic groups, the modular 2-groups, the quasi(semi)-dihedral groups, ..., etc. (cf. Hashimoto-Rikuna [6], Jensen-Ledet-Yui [7], and Rikuna [16; 18]).

In this paper we construct, among others, generic polynomials for the dihedral groups with even degree as a new result by using the above method. These calculation is based on some results about constructing cyclic polynomials from linear fractional transformations developed by Miyake [15], Hashimoto-Miyake [5], and Rikuna [17]. We also study two ways of descending for generic polynomials, namely, "arithmetic descent" and "geometric descent". The former way makes the base field k smaller. The latter way gives a generic polynomials for a subgroup of G, which is a concrete example of the descent-generic property, which is defined in Kemper [10]. Finally we give generic polynomials for almost all finite groups of order up to sixteen by using data of corresponding Noether's extensions.

Before beginning the following sections, the author wishes to express his deep appreciation for the referee of this paper who gave him very careful advices and very warm encouragements.

1. Generic polynomials derived from linear Noether's problem

Let k be a field. Put $k(\boldsymbol{x}) := k(x_1, \ldots, x_n)$, the rational function field with n variables $x_1, \ldots, x_n$, and let $\mathbf{Aut}_k k(\boldsymbol{x})$ be its k-automorphism group. For $\mathbf{GL}_n(k)$, the n-dimensional general linear group over k, define an injective homomorphism $\mathbf{GL}_n(k) \longrightarrow \mathbf{Aut}_k k(\boldsymbol{x})$ by $(a_{ij}) \longmapsto (x_j \mapsto \sum_{i=1}^{n} a_{ij}x_i, 1 \leq j \leq n)$. We denote its image by $\mathbf{Aut}_k^{\text{linear}} k(\boldsymbol{x})$ and we call an element in $\mathbf{Aut}_k^{\text{linear}} k(\boldsymbol{x})$ a *linear k-automorphism* of the field $k(\boldsymbol{x})$.

Let G be a finite subgroup of $\mathbf{Aut}_k^{\text{linear}} k(\boldsymbol{x})$. We denote by $k(\boldsymbol{x})^G$ the fixed field of $k(\boldsymbol{x})$ under G. The extension $k(\boldsymbol{x})/k(\boldsymbol{x})^G$ is G-Galois. Hence there exists a finite G-stable subset R_G of $k(\boldsymbol{x})$ such that $k(\boldsymbol{x}) = k(\boldsymbol{x})^G(R_G)$. (We call a subset $R \subset k(\boldsymbol{x})$ *G-stable* if R satisfies $\{\alpha(x) \mid x \in R, \alpha \in G\} = R$.) This implies that

$$f_{R_G}(X) := \prod_{x \in R_G} (X - x) = \sum_j c_j X^j \in k(\boldsymbol{x})^G[X]$$

is a defining polynomial of the extension $k(\boldsymbol{x})/k(\boldsymbol{x})^G$. Our main purpose is to construct a k-generic G-polynomial from $f_{R_G}(X)$. Therefore the following question is essential for us:

Is the fixed field $k(\boldsymbol{x})^G$ rational, i.e., purely transcendental, over k ?

This problem is called *linear Noether(-Lüroth)'s problem* for $G/k(\boldsymbol{x})$. We now assume that this problem has a positive answer, that is, there is a k-isomorphism $\phi_G : k(\boldsymbol{t}) \longrightarrow k(\boldsymbol{x})^G$, where $k(\boldsymbol{t}) := k(t_1, \ldots, t_n)$ is the rational function field with n variables $t_1, \ldots, t_n$. Then ϕ_G^{-1} is extended naturally to an isomorphism between polynomial rings $k(\boldsymbol{x})^G[X] \longrightarrow k(\boldsymbol{t})[X]$, and

$$F_G(\boldsymbol{t}; X) := \phi_G^{-1}(f_{R_G}(X)) = \sum_j \phi_G^{-1}(c_j)X^j \in k(\boldsymbol{t})[X] \tag{1.1}$$

has the Galois group G over $k(\boldsymbol{t})$.

Example 1. Let $N \geq 2$ be an integer and ζ_N is a primitive N-th root of unity. We define an one-dimensional linear $\mathbf{Q}(\zeta_N)$-automorphism $\alpha \in \mathbf{Aut}^{\text{linear}}_{\mathbf{Q}(\zeta_N)}\mathbf{Q}(\zeta_N)(x_1)$ by $\alpha(x_1) := \zeta_N x_1$, and we put $G := \langle\alpha\rangle$. This group G is isomorphic to C_N, the cyclic group of order N. In this case, we can choose $t_1 \longmapsto x_1^N$ as a $\mathbf{Q}(\zeta_N)$-isomorphism $\phi_G : \mathbf{Q}(\zeta_N)(t_1) \longrightarrow \mathbf{Q}(\zeta_N)(x_1)^G$ and $\{\zeta_N^j x_1 \mid 1 \leq j \leq N\}$ as R_G. Then we have $f_{R_G}(X) = X^N - x_1^N$ and $F_G(t_1; X) = X^N - t_1 \in \mathbf{Q}(t_1)[X]$. Kummer theory shows that this is a $\mathbf{Q}(\zeta_N)$-generic C_N-polynomial.

In general, we can show the following result by using normal basis theorem:

Theorem 1.1 (theorem 7 in Kemper-Mattig [11]). *The polynomial $F_G(\boldsymbol{t}; X) \in k(\boldsymbol{t})[X]$ defined by* (1.1) *is a k-generic G-polynomial.*

To construct the k-generic G-polynomial $F_G(\boldsymbol{t}; X)$, we have to

1 construct explicitly an affirmative answer of the corresponding linear Noether's problem,

2 find a suitable R_G and compute the corresponding defining polynomial $f_{R_G}(X)$,

3 calculate the coefficients $\phi_G^{-1}(c_j) \in k(\boldsymbol{t})$.

Remark 2. The separability of the extension $k(\boldsymbol{x})/k(\boldsymbol{x})^G$ implies that there exists an element $x \in k(\boldsymbol{x})$ such that R_G can be chosen as $\mathbf{Orb}_G(x)$, the single G-orbit of x. The polynomial $f_{R_G}(X)$ is irreducible over $k(\boldsymbol{x})^G$ under this selection, hence $F_G(\boldsymbol{t}; X)$ is also irreducible over $k(\boldsymbol{t})$. (See the next example.) The definition of generic polynomials does not require their irreducibility. It is, however, important to construct irreducible generic polynomials in the viewpoint of applicability to algebraic number theory.

Example 3. Put $G := \langle \alpha_1, \alpha_2 \rangle < \mathbf{Aut}_{\mathbf{Q}}^{\text{linear}} \mathbf{Q}(x_1, x_2)$, where

$$\alpha_1 : \begin{cases} x_1 \longmapsto -x_1 \\ x_2 \longmapsto -x_2 \end{cases} \quad \text{and} \quad \alpha_2 : \begin{cases} x_1 \longmapsto x_2 \\ x_2 \longmapsto x_1. \end{cases}$$

Then G is isomorphic to $C_2 \times C_2$ and we can choose a $\mathbf{Q}$-isomorphism $\phi_G : \mathbf{Q}(t_1, t_2) \longrightarrow \mathbf{Q}(x_1, x_2)^G$ by

$$t_1 \longmapsto \frac{x_1^2 + x_2^2}{x_1 x_2} \quad \text{and} \quad t_2 \longmapsto \frac{(x_1 x_2)^2}{x_1^2 + x_2^2}.$$

We take $R_1 := \mathbf{Orb}_G(x_1) = \{\pm x_1, \pm x_2\}$ and $R_2 := \mathbf{Orb}_G(x_1 + x_2) \cup \mathbf{Orb}_G(x_1 - x_2) = \{\pm(x_1 + x_2), \pm(x_1 - x_2)\}$, and hence both $f_{R_1}(X)$ and $f_{R_2}(X)$ are defining polynomials of the extension $\mathbf{Q}(x_1, x_2)/\mathbf{Q}(x_1, x_2)^G$. Therefore we obtain that $\phi_G^{-1}(f_{R_1}(X))$ and $\phi_G^{-1}(f_{R_2}(X))$ are $\mathbf{Q}$-generic $C_2 \times C_2$-polynomials. The former is irreducible over $\mathbf{Q}(t_1, t_2)$, but the latter is not. Indeed, we have

$$\begin{cases} \phi_G^{-1}(f_{R_1}(X)) = X^4 - t_1^2 t_2 X^2 + t_1^2 t_2^2, \\ \phi_G^{-1}(f_{R_2}(X)) = (X^2 - (t_1 + 2) t_1 t_2)(X^2 - (t_1 - 2) t_1 t_2). \end{cases}$$

(Naturally, $(X^2 - t_1)(X^2 - t_2) \in \mathbf{Q}(t_1, t_2)[X]$ is a $\mathbf{Q}$-generic $C_2 \times C_2$-polynomial.)

Remark 4. For the non-linear cases, their defining polynomials have not always generic. Whether these polynomials are generic or not is very difficult problem. Only a few examples is known (e.g., Hashimoto [3; 4], Ledet [12; 13], Jensen-Ledet-Yui [7], etc.)

2. Arithmetic descent

Keep the notation above. In this section we also suppose that linear Noether's problem for $G/k(\boldsymbol{x})$ has a positive answer, that is, there exists a k-isomorphism $\phi_G : k(\boldsymbol{t}) \longrightarrow k(\boldsymbol{x})^G$. And we choose a finite subset R_G of $k(\boldsymbol{x})$ so that $k(\boldsymbol{x}) = k(\boldsymbol{x})^G(R_G)$. Then $f_{R_G}(X) \in k(\boldsymbol{x})^G[X]$ is a defining polynomial of the G-extension $k(\boldsymbol{x})/k(\boldsymbol{x})^G$, and $F_G(\boldsymbol{t}; X) \in k(\boldsymbol{t})[X]$ is a k-generic G-polynomial.

Let $\rho : k(\boldsymbol{y}) \longrightarrow k(\boldsymbol{x})$ to be a linear k-isomorphism between rational function fields, i.e., ρ is a k-isomorphism which is defined by $(y_j \mapsto$

$\sum_{i=1}^{n} a_{ij}x_i, 1 \le j \le n)$ with some $(a_{ij}) \in \mathbf{GL}_n(k)$. And put

$$G' := \{\rho^{-1} \circ \alpha \circ \rho \mid \alpha \in G\} < \mathbf{Aut}_k^{\mathrm{linear}} k(\boldsymbol{y}).$$

This group is isomorphic to G. We fix a subfield k' of k which satisfies that $G' < \mathbf{Aut}_{k'}^{\mathrm{linear}} k'(\boldsymbol{y})$. Then the extension $k'(\boldsymbol{y})/k'(\boldsymbol{y})^{G'}$ is G'-Galois.

Here we suppose that

$$k'(\boldsymbol{y})^{G'} = (\rho^{-1} \circ \phi_G)(k'(\boldsymbol{t})). \qquad \text{(Assumption 1)}$$

This implies that $\phi_{G'} := \rho^{-1} \circ \phi_G|_{k'(\boldsymbol{t})}$ is a k'-isomorphism between $k'(\boldsymbol{t})$ and $k'(\boldsymbol{y})^{G'}$, namely, linear Noether's problem for $G'/k'(\boldsymbol{y})$ has a positive answer $\phi_{G'}$.

To obtain a k'-generic G'-polynomial from $F_G(\boldsymbol{t}; X)$, we accept the following assumption to make use of $f_{R_G}(X)$:

$$\rho^{-1}(R_G) \subset k'(\boldsymbol{y}). \qquad \text{(Assumption 2)}$$

Then $R_{G'} := \rho^{-1}(R_G)$ is G'-stable and generates $k'(\boldsymbol{y})$ over $k'(\boldsymbol{y})^{G'}$. Hence we have

$$\begin{aligned} f_{R_{G'}}(X) &= \prod_{y \in \rho^{-1}(R_G)} (X - y) = \prod_{x \in R_G} (X - \rho^{-1}(x)) \\ &= \sum_j \rho^{-1}(c_j) X^j \in k'(\boldsymbol{y})^{G'}[X] \end{aligned}$$

is a defining polynomial of the G'-extension $k'(\boldsymbol{y})/k'(\boldsymbol{y})^{G'}$. Moreover, we obtain that

$$\begin{aligned} F_{G'}(\boldsymbol{t}; X) &= \phi_{G'}^{-1}(f_{R_{G'}}(X)) = \sum_j (\phi_{G'}^{-1} \circ \rho^{-1})(c_j) X^j \\ &= \sum_j \phi_G^{-1}(c_j) X^j = F_G(\boldsymbol{t}; X). \end{aligned}$$

From the above observation, we have:

Proposition 5. *Under (Assumption 1 and 2), a k-generic G-polynomial $F_G(\boldsymbol{t}; X)$ has the coefficients in $k'(\boldsymbol{t})$ and it is k'-generic for G.*

3. Geometric descent

This section is independent of the previous section but continued from section 1. Kemper [10] shows that a k-generic G-polynomial $F_G(\boldsymbol{t}; X) \in k(\boldsymbol{t})[X]$ is also *descent-generic* over k, that is, for every subgroup H of

G, every H-extension M/K with $K \supset k$ and $\#K = \infty$ is obtained from $F_G(\boldsymbol{t}; X)$ by a specialization of parameters $\boldsymbol{t}$. This implies that $F_G(\boldsymbol{t}; X)$ defined by our construction is also k-generic for every subgroup of G. Here we study how to obtain a k-generic H-polynomial from $F_G(\boldsymbol{t}; X)$ for a subgroup H of G. We assume that $k(\boldsymbol{x})^H$ is k-rational. Then there exists an k-isomorphism $\phi_H : k(\boldsymbol{u}) \longrightarrow k(\boldsymbol{x})^H$, where $k(\boldsymbol{u}) := k(u_1, \ldots, u_n)$ is the rational function field with n variables $u_1, \ldots, u_n$. Define $\pi : k(\boldsymbol{x})^G \longrightarrow k(\boldsymbol{x})^H$ to be the inclusion, and put $\Phi := \phi_H^{-1} \circ \pi \circ \phi_G$. We note that the k-rational map Φ describes a relation between the generators $\{\phi_G(t_1), \ldots, \phi_G(t_n)\}$ and $\{\phi_H(u_1), \ldots, \phi_H(u_n)\}$. There exists a H-stable subset $R_H \subset R_G$ such that $k(\boldsymbol{x}) = k(\boldsymbol{x})^H(R_H)$. Then

$$g_{R_H}(X) := \prod_{x \in R_H} (X - x) \in k(\boldsymbol{x})^H[X]$$

is a defining polynomial of the H-extension $k(\boldsymbol{x})/k(\boldsymbol{x})^H$, and

$$F_H(\boldsymbol{u}; X) := \phi_H^{-1}(g_{R_H}(X)) \in k(\boldsymbol{u})[X]$$

is a k-generic H-polynomial. Hence we obtain a relation between two polynomials $F_G(\boldsymbol{t}; X)$ and $F_H(\boldsymbol{u}; X)$.

Theorem 3.1. *For a k-generic H-polynomial $F_H(\boldsymbol{u}; X) \in k(\boldsymbol{u})[X]$ obtained by the above construction, we have*

$$F_G(\Phi(t_1), \ldots, \Phi(t_n); X) = F_H(\boldsymbol{u}; X)\phi_H^{-1}\left(\prod_{x \in R_G \setminus R_H} (X - x)\right),$$

that is, $\Phi(F_G(\boldsymbol{t}; X))$ contains $F_H(\boldsymbol{u}; X)$ as a factor. Especially if $R_H = R_G$ then we have $F_H(\boldsymbol{u}; X) = F_G(\Phi(t_1), \ldots, \Phi(t_n); X)$.

Proof. From $k(\boldsymbol{x})^G \subset k(\boldsymbol{x})^H$, we have $\pi(c_j) = c_j$ for every $1 \leq j \leq n$ and

$$\begin{aligned}\Phi(F_G(\boldsymbol{t}; X)) &= \Phi(\phi_G^{-1}(f_{R_G}(X))) = (\phi_H^{-1} \circ \pi \circ \phi_G)(\phi_G^{-1}(f_{R_G}(X))) \\ &= \phi_H^{-1}(\pi(f_{R_G}(X))) = \phi_H^{-1}(f_{R_G}(X)).\end{aligned}$$

We obtain that $F_H(\boldsymbol{u}; X)$ divides $\phi_H^{-1}(f_{R_G}(X))$, because $g_{R_H}(X)$ divides $\pi(f_{R_G}(X)) = f_{R_G}(X)$ in $k(\boldsymbol{x})^H[X]$. □

Example 6. Take $G = \langle \alpha_1, \alpha_2 \rangle < \mathbf{Aut}^{\mathrm{linear}}_{\mathbf{Q}}\mathbf{Q}(x_1, x_2)$ as example 3. The group $G \cong C_2 \times C_2$ has three non-trivial subgroups isomorphic to

C_2, namely, $H_1 := \langle \alpha_1 \rangle$, $H_2 := \langle \alpha_2 \rangle$, and $H_{12} := \langle \alpha_1\alpha_2 \rangle$. We can choose

$$\begin{cases} \phi_G(t_1) = \dfrac{x_1^2+x_2^2}{x_1x_2}, \\ \phi_G(t_2) = \dfrac{(x_1x_2)^2}{x_1^2+x_2^2}, \end{cases} \qquad \begin{cases} \phi_{H_1}(u_1) = \dfrac{x_1}{x_2}, \\ \phi_{H_1}(u_2) = \phi_G(t_2), \end{cases}$$

$$\begin{cases} \phi_{H_2}(v_1) = \phi_G(t_1), \\ \phi_{H_2}(v_2) = \dfrac{x_1x_2}{x_1+x_2}, \end{cases} \qquad \begin{cases} \phi_{H_{12}}(w_1) = \phi_G(t_1), \\ \phi_{H_{12}}(w_2) = \dfrac{x_1x_2}{x_1-x_2}, \end{cases}$$

as the corresponding answers of linear Noether's problems for them. They have the following relations:

$$\begin{cases} \phi_G(t_1) = \phi_{H_1}(u_1) + \phi_{H_1}(u_1)^{-1}, \\ \phi_G(t_2) = \dfrac{(\phi_{H_2}(v_1)+2)\phi_{H_2}(v_2)^2}{\phi_{H_2}(v_1)} = \dfrac{(\phi_{H_{12}}(w_1)-2)\phi_{H_{12}}(w_2)^2}{\phi_{H_{12}}(w_1)}. \end{cases}$$

Put $R_G := \mathbf{Orb}_G(x_1) = \{\pm x_1, \pm x_2\}$, $R_{H_1} := \mathbf{Orb}_{H_1}(x_1) = \{\pm x_1\}$, $R_{H_2} := \mathbf{Orb}_{H_2}(x_1) = \{x_1, x_2\}$, and $R_{H_{12}} := \mathbf{Orb}_{H_{12}}(x_1) = \{x_1, -x_2\}$, and hence these are generating sets of the corresponding extensions $\mathbf{Q}(x_1,x_2)/\mathbf{Q}(x_1,x_2)^G, \ldots, \mathbf{Q}(x_1,x_2)/\mathbf{Q}(x_1,x_2)^{H_{12}}$. Then we obtain that

$$\begin{cases} F_G(t_1,t_2;X) = X^4 - t_1^2t_2X^2 + t_1^2t_2^2, \\ F_{H_1}(u_1,u_2;X) = X^2 - (u_1^2+1)u_2, \\ F_{H_2}(v_1,v_2;X) = X^2 - (v_1+2)v_2X + (v_1+2)v_2^2, \\ F_{H_{12}}(w_1,w_2;X) = X^2 - (w_1-2)w_2X - (w_1-2)w_2^2, \end{cases}$$

are the corresponding $\mathbf{Q}$-generic polynomials. And they have the following relations:

$$\begin{cases} F_G(u_1+u_1^{-1},u_2;X) = F_{H_1}(u_1,u_2;X) \times (X^2 - (u_1^{-2}+1)u_2), \\ F_G(v_1,(v_1+2)v_1^{-1}v_2^2;X) = F_{H_2}(v_1,v_2;X) \\ \qquad\qquad \times (X^2 + (v_1+2)v_2X + (v_1+2)v_2^2), \\ F_G(w_1,(w_1-2)w_1^{-1}w_2^2;X) = F_{H_{12}}(w_1,w_2;X) \\ \qquad\qquad \times (X^2 + (w_1-2)w_2X - (w_1-2)w_2^2). \end{cases}$$

4. Construction of generic even-dihedral polynomials

In this section we apply the above construction to the even-dihedral groups. Let $N \geq 4$ be an even integer, k be a field whose characteristic does not divide N, and assume that k contains ζ, a primitive N-th root of unity in a fixed algebraic closure of k. We define two linear k-automorphisms of $k(\boldsymbol{x}) = k(x_1, x_2)$ by

$$\begin{cases} \alpha(x_1) = \zeta x_1, \\ \alpha(x_2) = \zeta^{-1} x_2, \end{cases} \quad \text{and} \quad \begin{cases} \beta(x_1) = x_2, \\ \beta(x_2) = x_1. \end{cases}$$

They satisfy $\alpha^N = \beta^2 = 1$ and $\beta^{-1}\alpha\beta = \alpha^{-1}$, hence $G := \langle \alpha, \beta \rangle < \mathbf{Aut}_k^{\mathrm{linear}} k(x_1, x_2)$ is isomorphic to $\mathcal{D}_{2N}$, the dihedral group of order $2N$. Our aim is to find a k'-generic $\mathcal{D}_{2N}$-polynomial, where k' is a subfield of k contains $\omega := \zeta + \zeta^{-1}$.

We first show that linear Noether's problem for $G/k(x_1, x_2)$ has an affirmative answer. Let $S \lhd G$ be the scalar subgroup, i.e.,

$$S := \{\sigma \in G \mid \sigma(x_1/x_2) = x_1/x_2\}.$$

Then G/S acts naturally on $k(x_1/x_2)$, and $k(x_1/x_2)/k(x_1/x_2)^{G/S}$ is a G/S-extension. Because the transcendental degree of $k(x_1/x_2)^{G/S}$ over k is one, there exists an G/S-invariant $\theta_1 \in k(x_1/x_2)^{G/S}$ such that $k(x_1/x_2)^{G/S} = k(\theta_1)$ by Lüroth's theorem.

Lemma 7 (section 1.1 in Kemper [9]). *If $\theta_2 \in k(x_1, x_2)^G$ is a quotient f/g of homogeneous G-invariants $f, g \in k[x_1, x_2]^G$ with $|\deg f - \deg g| = \#S$, then we have $k(x_1, x_2)^G = k(x_1/x_2)^{G/S}(\theta_2) = k(\theta_1, \theta_2)$.*

By using this lemma, we define a k-isomorphism ϕ_G by

$$\phi_G(t_1) = \theta_1 \quad \text{and} \quad \phi_G(t_2) = \theta_2.$$

We give two pairs of (θ_1, θ_2) explicitly in the following two subsections. The former choice, which has very simple form, is not applicable for arithmetic descent. But the latter choice matches our arithmetic descent method for k'.

4.1 Trivial construction

From $\alpha(x_1/x_2) = \zeta^2(x_1/x_2)$ and $\beta(x_1/x_2) = x_2/x_1$, one can easily show that $S = \{1, \alpha^{N/2}\}$ and that we can choose θ_1 and θ_2 by

$$\begin{cases} \theta_1 := \left\{ \left(\dfrac{x_1}{x_2}\right)^{N/2} + \left(\dfrac{x_2}{x_1}\right)^{N/2} \right\}^{-1} = \dfrac{(x_1 x_2)^{N/2}}{x_1^N + x_2^N} \\ \theta_2 := \dfrac{x_1^N + x_2^N}{(x_1 x_2)^{(N-2)/2}}. \end{cases}$$

And we choose $R_G := \mathbf{Orb}_G(x_1) = \{\zeta^j x_1, \zeta^{-j} x_2 \mid 1 \le j \le N\}$. Then we have

$$\begin{aligned} f_{R_G}(X) &= \prod_{j=1}^{N} (X - \zeta^j x_1)(X - \zeta^{-j} x_2) \\ &= X^{2N} - (x_1^N + x_2^N) X^N + (x_1 x_2)^N \in k(x_1, x_2)^G[X]. \end{aligned}$$

By using two relations $x_1 x_2 = \theta_1 \theta_2$ and $x_1^N + x_2^N = \theta_1^{(N-2)/2} \theta_2^{N/2}$, we obtain

$$\begin{aligned} F_G(t_1, t_2; X) &= \phi_G^{-1}(f_{R_G}(X)) \\ &= X^{2N} - t_1^{(N-2)/2} t_2^{N/2} X^N + (t_1 t_2)^N \in k(t_1, t_2)[X] \end{aligned}$$

as a k-generic $\mathcal{D}_{2N}$-polynomial. But we cannot apply our arithmetic descent method, which is stated in section 2, to this polynomial.

4.2 Non-trivial construction

For our purpose, we replace the previous choice for θ_1, θ_2, and R_G by another.

Lemma 8. *We can choose two generators θ_1 and θ_2 of $k(x_1, x_2)^G$ over k as follows:*

$$\theta_1 := \frac{(\omega^2 - 4)(-x_1 x_2)^{N/2}}{(x_1^{N/2} - (-x_2)^{N/2})^2} \quad \text{and} \quad \theta_2 := \frac{(x_1^{N/2} - (-x_2)^{N/2})^2}{(\omega^2 - 4)(-x_1 x_2)^{(N-2)/2}}. \tag{4.1}$$

Proof. Clearly θ_2 satisfies the condition in lemma 7. Therefore it is sufficient to show that θ_1 generates $k(x_1/x_2)$ over $k(x_1/x_2)^{G/S}$. We have $k(\theta_1) \subset k(x_1/x_2)^{G/S}$ and $[k(x_1/x_2) : k(x_1/x_2)^{G/S}] = N$, hence $[k(x_1/x_2) : k(\theta_1)] \ge N$. On the other hand, x_1/x_2 is a root of a equation $X^N - (-1)^{N/2}(2\theta_1 + \omega^2 - 4)\theta_1^{-1} X^{N/2} + 1 = 0$ of degree N, hence $[k(x_1/x_2) : k(\theta_1)] \le N$. This shows the assertion. □

We choose $R_G := \mathbf{Orb}_G(x_1 + x_2) = \{\zeta^j x_1 + \zeta^{-j} x_2 \mid 1 \le j \le N\}$. Then we have

$$f_{R_G}(X) = \prod_{j=1}^{N} (X - (\zeta^j x_1 + \zeta^{-j} x_2)),$$

and the next lemma calculates the explicit expansion of this.

Lemma 9 (lemma 2 in Hashimoto-Rikuna [6]). *We put* $B(N,j) := \binom{N-j-1}{j-1} + \binom{N-j}{j}$, *then we have*

$$\prod_{j=1}^{N}(X - (\zeta^j x_1 + \zeta^{-j} x_2))$$
$$= X^N + \sum_{j=1}^{(N-2)/2} B(N,j)(-x_1x_2)^j X^{N-2j} - (x_1^{N/2} - (-x_2)^{N/2})^2.$$

The definition (4.1) implies

$$-x_1x_2 = \theta_1\theta_2 \qquad \text{and} \qquad (x_1^{N/2} - (-x_2)^{N/2})^2 = (\omega^2-4)\theta_1^{(N-2)/2}\theta_2^{N/2}.$$

Hence we obtain another k-generic $\mathcal{D}_{2N}$-polynomial:

Theorem 4.1. *A polynomial*

$$\begin{aligned} &F_G(t_1,t_2;X) \\ &= \phi_G^{-1}(f_{R_G}(X)) \\ &= X^N + \sum_{j=1}^{(N-2)/2} B(N,j)(t_1t_2)^j X^{N-2j} - (\omega^2-4)t_1^{(N-2)/2}t_2^{N/2} \end{aligned} \tag{4.2}$$

in $k(t_1,t_2)[X]$ *is a* k*-generic* $\mathcal{D}_{2N}$*-polynomial.*

4.3 Arithmetic descent

We define a linear k-isomorphism $\rho : k(y_1,y_2) \longrightarrow k(x_1,x_2)$ by

$$\rho(y_1) = x_1 + x_2 \qquad \text{and} \qquad \rho(y_2) = \zeta x_1 + \zeta^{-1}x_2, \tag{4.3}$$

then we have

$$\begin{cases} (\rho^{-1}\circ\alpha\circ\rho)(y_1) = y_2, \\ (\rho^{-1}\circ\alpha\circ\rho)(y_2) = -y_1 + \omega y_2, \end{cases} \quad \text{and} \quad \begin{cases} (\rho^{-1}\circ\beta\circ\rho)(y_1) = y_1, \\ (\rho^{-1}\circ\beta\circ\rho)(y_2) = \omega y_1 - y_2. \end{cases}$$

Hence we have $G' = \langle \rho^{-1}\circ\alpha\circ\rho, \rho^{-1}\circ\beta\circ\rho\rangle < \mathbf{Aut}_{k'}^{\text{linear}} k'(y_1,y_2)$.

Lemma 10. *The linear* k*-isomorphism* ρ *defined by* (4.3) *satisfies (Assumption 1), that is,* ρ *satisfies* $k'(y_1,y_2)^{G'} = \rho^{-1}(\phi_G(k'(t_1,t_2)))$.

Proof. We have

$$\begin{cases} (\rho^{-1}\circ\phi_G)(t_1) = \rho^{-1}(\theta_1) = \dfrac{(\omega^2-4)^2(y_1^2-\omega y_1y_2+y_2^2)^{N/2}}{((y_1-\zeta y_2)^{N/2}-(y_1-\zeta^{-1}y_2)^{N/2})^2}, \\ (\rho^{-1}\circ\phi_G)(t_2) = \rho^{-1}(\theta_2) = \dfrac{((y_1-\zeta y_2)^{N/2}-(y_1-\zeta^{-1}y_2)^{N/2})^2}{(\omega^2-4)^2(y_1^2-\omega y_1y_2+y_2^2)^{(N-2)/2}}, \end{cases}$$

and we obtain $((y_1-\zeta y_2)^{N/2}+(y_1-\zeta^{-1}y_2)^{N/2})^2\in k'[y_1,y_2]$ by using

$$\begin{aligned}(y_1-\zeta y_2)^N&+(y_1-\zeta^{-1}y_2)^N\\&=\sum_{j=0}^{N}(-1)^j\binom{N}{j}(\zeta^j+\zeta^{-j})y_1^{N-j}y_2^j\in k'[y_1,y_2].\end{aligned}$$

We put

$$\begin{cases}\eta_{\mathrm{nu}}:=((y_1-\zeta y_2)^{N/2}-(y_1-\zeta^{-1}y_2)^{N/2})^2,\\ \eta_{\mathrm{de}}:=(\omega^2-4)^2(y_1^2-\omega y_1y_2+y_2^2)^{(N-2)/2},\end{cases}$$

so that $\rho^{-1}(\theta_2)=\eta_{\mathrm{nu}}/\eta_{\mathrm{de}}$. Then $\eta_{\mathrm{nu}},\eta_{\mathrm{de}}\in k'(y_1,y_2)^{G'}$. And they are coprime in $k'[y_1,y_2]$. Hence $\rho^{-1}(\theta_2)$ satisfies the condition in lemma 7. We can show that $k'(y_1/y_2)^{G'}=k'(\rho^{-1}(\theta_1))$ by the same way as the proof of lemma 8. Therefore we obtain that $k'(y_1,y_2)^{G'}=k'(\rho^{-1}(\theta_1),\rho^{-1}(\theta_2))$. □

By the definition of R_G, we have $\rho^{-1}(R_G)=\rho^{-1}(\mathbf{Orb}_G(x_1+x_2))=\mathbf{Orb}_{G'}(y_1)\subset k'(y_1,y_2)$. Hence (Assumption 2) is also satisfied.

Therefore we obtain our main result from proposition 5.

Theorem 4.2. *The k-generic $\mathcal{D}_{2N}$-polynomial $F_G(t_1,t_2;X)$ obtained by (4.2) is also a k'-generic $\mathcal{D}_{2N}$-polynomial.*

4.4 Geometric descent

Put $H':=\langle\rho^{-1}\circ\alpha\circ\rho\rangle<G'$. Then H' is cyclic of order N. Here we consider a relation between the k'-generic $\mathcal{D}_{2N}$-polynomial $F'_G(t_1,t_2;X)$ ($=F_G(t_1,t_2;X)$) and the k'-generic $\mathcal{C}_N$-polynomial $F_{H'}(u_1,u_2;X)$ which is obtained by Hashimoto-Rikuna [6].

From Hashimoto-Rikuna [6, section 3], we can define a k'-isomorphism $\phi_{H'}:k'(u_1,u_2)\longrightarrow k'(y_1,y_2)^{H'}$ by

$$\begin{cases}\phi_{H'}(u_1)=\dfrac{\zeta^{-1}(y_1-\zeta y_2)^{N/2}-\zeta(y_1-\zeta^{-1}y_2)^{N/2}}{(y_1-\zeta y_2)^{N/2}-(y_1-\zeta^{-1}y_2)^{N/2}},\\[2ex] \phi_{H'}(u_2)=\dfrac{((y_1-\zeta y_2)^{N/2}-(y_1-\zeta^{-1}y_2)^{N/2})^2}{(\omega^2-4)^2(y_1^2-\omega y_1y_2+y_2^2)^{(N-2)/2}},\end{cases}$$

as an answer of linear Noether's problem for $H'/k'(y_1,y_2)$. In this case we obtain that

$$\phi_{G'}(t_1)=\phi_{H'}(u_1)^2-\omega\phi_{H'}(u_1)+1\qquad\text{and}\qquad\phi_{G'}(t_2)=\phi_{H'}(u_2),$$

that is, $\Phi(t_1)=u_1^2-\omega u_1+1$ and $\Phi(t_2)=u_2$. Since $\mathbf{Orb}_{\langle\alpha\rangle}(x_1+x_2)=R_G$, we can choose $R_{H'}:=R_{G'}$, and hence $f_{R_{H'}}(X)=f_{R_{G'}}(X)$. Therefore theorem 3.1 implies the following result:

Proposition 11. *For the k'-generic $\mathcal{D}_{2N}$-polynomial $F_{G'}(t_1,t_2;X)$ obtained above,*

$$F_{G'}(u_1^2-\omega u_1+1,u_2;X)\in k'(u_1,u_2)[X]$$

is a k'-generic $\mathcal{C}_N$-polynomial.

5. Generic polynomials for small groups

In this section we display generic polynomials for groups of order up to sixteen obtained by our construction with characteristic zero. We use computer manipulations with GAP (GAP [1]) and Mathematica (Wolfram [22]). We denote a primitive n-th root of unity by ζ_n and put $\omega_n := \zeta_n + \zeta_n^{-1}$.

Order 2

There is one group (isomorphism class) of order two.

2/1	$\mathcal{C}_2$ **over Q**
action	$\begin{cases} x_1 \longmapsto x_2 \\ x_2 \longmapsto x_1 \end{cases}$
generating set R_G	$\{x_1,x_2\}$
answer of linear Noether's problem	$\phi_G^{-1}(\boldsymbol{t})=\{\theta_1,\theta_2,\dots\}$ $=\left\{\frac{(x_1+x_2)^2}{x_1x_2},\frac{x_1x_2}{x_1+x_2}\right\}$
generic polynomial	$X^2-t_1X+t_1$

The first row means the isomorphism class of G indexed by the ordering in Thomas-Wood [21] and the base field k. The second and third row express a generator of $G<\mathbf{Aut}_k^{\mathrm{linear}}k(\boldsymbol{x})$ and a generating set R_G of the extension $k(\boldsymbol{x})/k(\boldsymbol{x})^G$ respectively. The fourth and fifth row show an explicit answer of linear Noether's problem for $G/k(\boldsymbol{x})$ and the corresponding k-generic G-polynomial respectively.

Remark 12. Assume that the group G has trivial scalar subgroup, i.e., $k(x_1/x_2)/k(x_1/x_2)^G$ has the same Galois group as $k(x_1,x_2)/k(x_1,x_2)^G$. Then $F_G(t_1,1;X)$ is also a k-generic G-polynomial if $k(x_1/x_2)^G=k(\theta_1)$ (see Kemper-Mattig [11, theorem 7]). With the above construction, we obtain that $X^2-t_1t_2X+t_1t_2^2\in\mathbf{Q}(t_1,t_2)[X]$ is a $\mathbf{Q}$-generic $\mathcal{C}_2$-polynomial. Hence $X^2-t_1X+t_1=(X^2-t_1t_2X+t_1t_2^2)|_{t_2=1}$ is also $\mathbf{Q}$-generic for $\mathcal{C}_2$ from this observation.

Order 3

There is one group of order three.

3/1	C_3 **over Q**
act.	$\begin{cases} x_1 \longmapsto x_2 \\ x_2 \longmapsto -x_1 - x_2 \end{cases}$
R_G	$\{x_1, -x_1 - x_2, x_2\}$
ans.	$\left\{\frac{x_1^3-3x_1x_2^2-x_2^3}{x_1x_2(x_1+x_2)}, \frac{x_1x_2(x_1+x_2)}{x_1^2+x_1x_2+x_2^2}\right\}$
poly.	$X^3 - (t_1^2 + 3t_1 + 9)X + t_1^2 + 3t_1 + 9$
note	See remark 12. Shanks' "the simplest cubic" (cf. Shanks [20]), $X^3 - t_1X^2 - (t_1 + 3)X - 1$ is well-known as a **Q**-generic C_3-polynomial.

Order 4

There are two groups of order four.

4/1	C_4 **over Q**
poly.	$X^4 + 4(t_1^2 + 1)t_2X^2 + 4(t_1^2 + 1)t_2^2$
note	This is obtained in section 4.

4/2	$C_2 \times C_2$ **over Q**
act.	$\begin{cases} x_1 \longmapsto x_2 \\ x_2 \longmapsto x_1 \end{cases}$ $\begin{cases} x_1 \longmapsto -x_2 \\ x_2 \longmapsto -x_1 \end{cases}$
R_G	$\{\pm x_1, \pm x_2\}$
ans.	$\left\{\frac{x_1^2+x_2^2}{x_1x_2}, \frac{(x_1x_2)^2}{x_1^2+x_2^2}\right\}$
poly.	$X^4 - t_1^2t_2X^2 + t_1^2t_2^2$

Order 5

There is one group of order five.

5/1	C_5 **over** $\mathbf{Q}(\omega_5)$
poly.	$X^5-5t_1X^4+10(\omega_5t_1-1)X^3+10\omega_5(t_1+1)X^2-5(t_1-\omega_5)X-1$
note	This is given by Hashimoto-Miyake [5] and Rikuna [17].

Order 6

There are two groups of order six.

6/1	C_6 **over Q**
poly.	$X^6+6(t_1^2-t_1+1)t_2X^4+9(t_1^2-t_1+1)^2t_2^2X^2+3(t_1^2-t_1+1)^2t_2^3$
note	This is obtained in section 4.

6/2	$\mathfrak{S}_3$, **the symmetric group of degree 3,** $\cong \mathcal{D}_6$ **over Q**
act.	$\begin{cases} x_1 \longmapsto x_2 \\ x_2 \longmapsto -x_1 - x_2 \end{cases}$ $\begin{cases} x_1 \longmapsto x_2 \\ x_2 \longmapsto x_1 \end{cases}$
R_G	$\{x_1, x_1 + x_2, x_2\}$
ans.	$\left\{ \frac{(x_1^2+x_1x_2+x_2^2)^3}{x_1^2x_2^2(x_1+x_2)^2}, \frac{x_1x_2(x_1+x_2)}{x_1^2+x_1x_2+x_2^2} \right\}$
poly.	$X^3 - t_1X + t_1$
note	This is obtained in section 4. (And see remark 12.) By a geometric descent $t_1 \longmapsto t_1^2+3t_1+9$, we have a **Q**-generic $\mathcal{C}_3$-polynomial displayed above.

Order 7

There is one group of order seven.

7/1	$\mathcal{C}_7$ **over** $\mathbf{Q}(\omega_7)$
poly.	$X^7 - 7t_1X^6 + 21(\omega_7t_1 - 1)X^5 - 35((\omega_7^2 - 1)t_1 - \omega_7)X^4$ $-35(\omega_7^2-1)(t_1+1)X^3+21(\omega_7t_1-\omega_7^2+1)X^2-7(t_1-\omega_7)X-1$
note	This is given by Hashimoto-Miyake [5] and Rikuna [17].

Order 8

There are five groups of order eight.

It is well-known that there is no **Q**-generic $\mathcal{C}_8$-polynomial (cf. Jensen-Ledet-Yui [7], etc.)

8/1	$\mathcal{C}_8$ **over** $\mathbf{Q}(\omega_8) = \mathbf{Q}(\sqrt{2})$
poly.	$X^8 + \sum_{j=1}^{3}\{\binom{7-j}{j-1} + \binom{8-j}{j}\}(t_1^2 - \sqrt{2}t_1 + 1)t_2)^jX^{8-2j}$ $+2(t_1^2 - \sqrt{2}t_1 + 1)^3t_2^4$
note	This is obtained in section 4.

8/1	$\mathcal{C}_8$ **over** $\mathbf{Q}(\sqrt{-1})$
act.	$\begin{cases} x_1 \longmapsto x_2 \\ x_2 \longmapsto \sqrt{-1} - x_1 \end{cases}$
R_G	$\{\pm x_1, \pm\sqrt{-1}x_1, \pm x_2, \pm\sqrt{-1}x_2\}$
ans.	$\left\{ \frac{x_1^2-\sqrt{-1}x_2^2}{x_1x_2}, \frac{(x_1x_2)^4}{x_1^4+x_2^4} \right\}$
poly.	$X^8 - (t_1^2 + 4\sqrt{-1})t_1^2t_2X^4 + (t_1^2 + 4\sqrt{-1})t_1^2t_2^2$
note	This polynomial comes from a $\mathbf{Q}(\sqrt{-1})$-generic $\mathcal{M}od_{16}$-polynomial, where $\mathcal{M}od_{16}$ is the modular 2-group of order 16, in Rikuna [16, section 3] by a geometric descent.

8/1	$\mathcal{C}_8$ **over** $\mathbf{Q}(\sqrt{-2})$
poly.	$F_{\mathcal{QD}_{16}}\left((t_1^2+\sqrt{-2}t_1+1)^{-1}+2, t_2; X\right)$, where $F_{\mathcal{QD}_{16}}(t_1,t_2;X) := X^8 - \frac{24(t_1-2)^2t_2^2}{t_1}X^4 - \frac{32(t_1-1)^3t_2^4}{t_1}X^2 - \frac{8(t_1-1)^5t_2^4}{t_1^2}$.
note	We have that $F_{\mathcal{QD}_{16}}(t_1,t_2;X)$ is a $\mathbf{Q}(\sqrt{-2})$-generic $\mathcal{QD}_{16}$-polynomial, where $\mathcal{QD}_{16}$ is the quasi-dihedral group of order 16, from Rikuna [18],

8/2	$\mathcal{C}_2 \times \mathcal{C}_4$ **over** $\mathbf{Q}(\sqrt{-1})$
act.	$\begin{cases} x_1 \longmapsto \sqrt{-1}x_2 \\ x_2 \longmapsto \sqrt{-1}x_1 \end{cases}$ $\begin{cases} x_1 \longmapsto x_2 \\ x_2 \longmapsto x_1 \end{cases}$
R_G	$\{\pm x_1, \pm\sqrt{-1}x_1, \pm x_2, \pm\sqrt{-2}x_2\}$
ans.	$\left\{\frac{x_1^2+x_2^2}{x_1x_2}, \frac{x_1^8+x_2^8}{x_1^4+x_2^4}\right\}$
poly.	$X^8 - \frac{(t_1^2-2)^2t_2}{t_1^4-4t_1^2+2}X^4 + \frac{(t_1^2-2)^2t_2^2}{(t_1^4-4t_1^2+2)^2}$
note	We can also get a $\mathcal{Q}$-generic $\mathcal{C}_2\times\mathcal{C}_4$-polynomial. See the next remark.

Remark 13. Assume that $F_{G_1}(\boldsymbol{t};X)$ and $F_{G_2}(\boldsymbol{u};Y)$ are irreducible generic polynomials for G_1 and G_2 over an infinite field k obtained by linear Noether's problem. And let r_1 and r_2 be roots of there polynomials respectively. Then there exists a constant $c \in k$ such that the minimal polynomial of r_1+cr_2 over $k(\boldsymbol{t},\boldsymbol{u})$ is a k-generic $G_1\times G_2$-polynomial. But the polynomial obtained by this construction has terribly complicated expression in general.

8/3	$\mathcal{C}_2 \times \mathcal{C}_2 \times \mathcal{C}_2$ **over** $\mathbf{Q}$
act.	$\{x_1 \longmapsto -x_1$ $\{x_2 \longmapsto -x_2$ $\{x_3 \longmapsto -x_3$
R_G	$\{\pm x_1, \pm x_2, \pm x_3\}$
ans.	$\left\{\frac{x_1^2}{x_3^2}, \frac{x_2^2}{x_3^2}, \frac{x_1^4+x_2^4+x_3^4}{x_1^2+x_2^2+x_3^2}\right\}$
poly.	$X^8 - \frac{4(t_1+t_2+1)^2t_3}{t_1^2+t_2^2+1}X^6 + \frac{2(t_1+t_2+1)^2(3t_1^2+3t_2^2+2t_1t_2+2t_1+2t_2+3)t_3^2}{(t_1^2+t_2^2+1)^2}X^4 - \frac{4(t_1+t_2+1)^2(t_1^4+t_2^4-2t_1^2t_2^2+8t_1^2t_2+8t_1t_2^2-2t_1^2-2t_2^2+8t_1t_2+1)t_3^3}{(t_1^2+t_2^2+1)^3}X^2 + \frac{(t_1+t_2+1)^4(t_1^2+t_2^2-t_1t_2-t_1-2t_2+1)^2t_3^4}{(t_1^2+t_2^2+1)^4}$

8/4	$\mathcal{D}_8$ **over** $\mathbf{Q}$
poly.	$X^4 + 4t_1t_2X^2 + 4t_1t_2^2$
note	This is obtained in section 4. By a geometric descent $t_1 \longmapsto t_1^2+1$, we have a $\mathbf{Q}$-generic $\mathcal{C}_4$-polynomial displayed above.

8/5	$\mathcal{Q}_8$, **the generalized quaternion group of order eight, over** $\mathbf{Q}(\sqrt{-1})$
act.	$\begin{cases} x_1 \longmapsto \sqrt{-1}x_1 \\ x_2 \longmapsto -\sqrt{-1}x_2 \end{cases}$ $\begin{cases} x_1 \longmapsto -x_2 \\ x_2 \longmapsto x_1 \end{cases}$
R_G	$\{\pm x_1, \pm\sqrt{-1}x_1, \pm x_2, \pm\sqrt{-1}x_2\}$
ans.	$\left\{\frac{(x_1^2+x_2^2)^2}{(x_1x_2)^2}, \frac{(x_1x_2)^2}{x_1^2+x_2^2}\right\}$
poly.	$X^8 - \frac{t_1t_2^2}{t_1^2-4}X^4 + \frac{t_2^4}{(t_1^2-4)^2}$

Remark 14. This is equivalent to a $\mathbf{Q}(\sqrt{-1})$-generic $\mathcal{Q}_8$-polynomial obtained by Kemper [8]. Some $\mathbf{Q}$-generic $\mathcal{Q}_8$-polynomials are also given by Kemper [8; 9], Gröbner [2], etc.

Order 9

There are two groups of order nine.

9/1	$\mathcal{C}_9$ **over** $\mathbf{Q}(\omega_9)$
poly.	$X^9 - 9t_1X^8 + 36(\omega_9t_1 - 1)X^7 - 84((\omega_9^2 - 1)t_1 - \omega_9)X^6$ $+126(\omega_9 - 1)(t_1 - \omega_9 - 1)X^5 + 126(\omega_9 + 1)(t_1 + 1)X^4$ $-84(\omega_9 - 1)((\omega_9 + 1)t_1 - 1)X^3 + 36(\omega_9t_1 - \omega_9^2 + 1)X^2$ $-9(t_1 - \omega_9)X - 1$
note	This is given by Hashimoto-Miyake [5] and Rikuna [17].

9/2	$\mathcal{C}_3 \times \mathcal{C}_3$ **over** $\mathbf{Q}(\zeta_3)$
act.	$\{x_1 \longmapsto \zeta_3x_1$ $\{x_2 \longmapsto \zeta_3^{-1}x_2$
R_G	$\{\zeta_3^ix_1 + \zeta_3^jx_2 \mid 1 \leq i, j \leq 3\}$
ans.	$\left\{\frac{x_1^3}{x_2^3}, \frac{(x_1x_2)^3}{x_1^3+x_2^3}\right\}$
poly.	$X^9 - \frac{3(t_1+1)^2t_2}{t_1}X^6 + \frac{3(t_1+1)^2(t_1^2-7t_1+1)t_2^2}{t_1^2}X^3 - \frac{3(t_1+1)^2t_2}{t_1}$
note	We can also get a $\mathbf{Q}$-generic $\mathcal{C}_3 \times \mathcal{C}_3$-polynomial. See remark 13.

Order 10

There are two groups of order ten.

10/1	$\mathcal{C}_{10}$ **over** $\mathbf{Q}(\omega_{10})$
poly.	$X^{10} + \sum_{j=1}^4\{\binom{9-j}{j-1} + \binom{10-j}{j}\}((t_1^2 - \omega_{10}t_1 + 1)t_2)^jX^{10-2j}$ $-(\omega_{10}^2 - 4)(t_1 - \omega_{10}t_1 + 1)^4t_2^5$
note	This is obtained in section 4.

10/2	$\mathcal{D}_{10}$ **over** $\mathbf{Q}(\omega_5)$
poly.	$X^5 - (3\omega_5 - 1)X^4 - (5\omega_5 - 3)X^3 - (3\omega_5 - 2)X^2 + t_1$
note	This polynomial is given by Hashimoto-Miyake [5].

Order 11

There is one group of order eleven.

11/1	$\mathcal{C}_{11}$ **over** $\mathbf{Q}(\omega_{11})$
poly.	$X^{11} - 11t_1X^{10} + 55(\omega_{11}t_1 - 1)X^9 - 165((\omega_{11}^2 - 1)t_1 - \omega_{11})X^8$ $+330(\omega_{11}(\omega_{11}^2 - 2)t_1 - \omega_{11}^2 + 1)X^7$ $-462((\omega_{11}^4 - 3\omega_{11}^2 + 1)t_1 - \omega_{11}^3 + 2\omega_{11})X^6$ $-462(\omega_{11}^4 - 3\omega_{11}^2 + 1)(t_1 + 1)X^5$ $+330(\omega_{11}(\omega_{11}^2 - 2)t_1 - \omega_{11}^4$ $+3\omega_{11}^2 - 1)X^4 - 165((\omega_{11}^2 - 1)t_1 - \omega_{11}^3$ $+2\omega_{11})X^3 + 55(\omega_{11}t_1 - \omega_{11}^2 + 1)X^2 - 11(t_1 - \omega_{11})X - 1$
note	This is given by Hashimoto-Miyake [5] and Rikuna [17].

Order 12

There are five groups of order twelve.

12/1	$\mathcal{C}_{12}$ **over** $\mathbf{Q}(\omega_{12}) = \mathbf{Q}(\sqrt{3})$
poly.	$X^{12} + \sum_{j=1}^{5}\{\binom{11-j}{j-1} + \binom{12-j}{j}\}((t_1^2 - \sqrt{3}t_1 + 1)t_2)^j X^{12-2j}$ $+(t_1 - \sqrt{3}t_1 + 1)^5 t_2^6$
note	This is obtained in section 4.

12/2	$\mathcal{C}_6 \times \mathcal{C}_2$ **over** $\mathbf{Q}(\zeta_6)$
act.	$\{x_1 \longmapsto \zeta_6 x_1 \qquad \{x_2 \longmapsto -x_2$
R_G	$\{\zeta_6^i x_1 \pm x_2 \mid 1 \leq i \leq 6\}$
ans.	$\left\{\frac{x_1^6}{x_2^6}, \frac{x_1^6}{x_2^4}\right\}$
poly.	$X^{12} - \frac{6t_2}{t_1}X^{10} + \frac{15t_2^2}{t_1^2}X^8 - \frac{2(t_1+10)t_2^3}{t_1^3}X^6 - \frac{15(2t_1-1)t_2^4}{t_1^4}X^4$ $-\frac{6(5t_1+1)t_2^5}{t_1^5}X^2 + \frac{(t_1-1)^2t_2^6}{t_1^6}$
note	We can also get a $\mathbf{Q}$-generic $\mathcal{C}_6 \times \mathcal{C}_2$-polynomial. See remark 13.

12/3	$\mathcal{D}_{12}$ **over Q**
poly.	$X^6 + 6t_1t_2X^4 + 9t_1^2t_2^2X^2 + 3t_1^2t_2^3$
note	This is obtained in section 4.

12/4	$\mathfrak{A}_4$**, the alternating group of degree four, over Q**
poly.	$X^4 - \frac{6T_1}{T_2}X^2 - 8X + \frac{9T_1^2 - 12(t_1^3 - t_2^3 + 27)T_2}{T_2^2}$, where $T_1 := t_1^3 - t_2^3 - 9t_2^2 - 27t_2 - 54$ and $T_2 := t_1^3 - 3t_1t_2^2 + 2t_2^3 - 9t_1t_2 + 9t_2^2 - 27t_1 + 27t_2 + 27$.
note	This polynomial is introduced in Jensen-Ledet-Yui [7].

12/5	$\mathcal{Q}_{12}$, **the quaternion type group of order twelve, over** $\mathbf{Q}(\zeta_6)$
act.	$\begin{cases} x_1 \longmapsto \zeta_6 x_1 \\ x_2 \longmapsto \zeta_6^{-1} x_2 \end{cases}$ $\begin{cases} x_1 \longmapsto -x_2 \\ x_2 \longmapsto x_1 \end{cases}$
R_G	$\{\zeta_6^i x_1, \zeta_6^i x_2 \mid 1 \leq i \leq 6\}$
ans.	$\left\{ \frac{x_1^6 - x_2^6}{(x_1 x_2)^3}, \frac{x_1 x_2 (x_1^6 - x_2^6)}{x_1^6 + x_2^6} \right\}$
poly.	$X^{12} - \frac{(t_1^2+4)^2 t_2^3}{t_1^3} X^6 + \frac{(t_1^2+4)^3 t_2^6}{t_1^6}$
note	G. Kemper also gave a $\mathbf{Q}(\omega_6)$-generic $\mathcal{Q}_{12}$-polynomial by using dimension four version of our method in Kemper [8; 9].

Order 13

There is one group of order thirteen.

13/1	$\mathcal{C}_{13}$ **over** $\mathbf{Q}(\omega_{13})$
poly.	$X^{13} - 13t_1 X^{12} + 78(\omega_{13} t_1 - 1) X^{11} - 286((\omega_{13}^2 - 1)t_1 - \omega_{13}) X^{10}$ $+715(\omega_{13}(\omega_{13}^2 - 2)t_1 - \omega_{13}^2 + 1)X^9$ $-1287((\omega_{13}^4 - 3\omega_{13}^2 + 1)t_1 - \omega_{13}^3 + 2\omega_{13})X^8$ $+1716(\omega_{13}^3 - 3\omega_{13} - 1)((\omega_{13}^2 - 1)t_1 - \omega_{13})X^7$ $+1716\omega_{13}(\omega_{13}^4 - 4\omega_{13}^2 + 3)(t_1 + 1)X^6$ $-1287((\omega_{13}^4 - 3\omega_{13}^2 + 1)t_1 - \omega_{13}^5 + 4\omega_{13}^3 - 3\omega_{13})X^5$ $+715(\omega_{13}(\omega_{13}^2 - 2)t_1 - \omega_{13}^4 + 3\omega_{13}^2 - 1)X^4$ $-286((\omega_{13}^2 - 1)t_1 - \omega_{13}^3 + 2\omega_{13})X^3 + 78(\omega_{13} t_1 - \omega_{13}^2 + 1)X^2$ $-13(t_1 - \omega_{13})X - 1$
note	This is given by Hashimoto-Miyake [5] and Rikuna [17].

Order 14

There are two groups of order fourteen.

14/1	$\mathcal{C}_{14}$ **over** $\mathbf{Q}(\omega_{14})$
poly.	$X^{14} + \sum_{j=1}^{6} \{ \binom{13-j}{j-1} + \binom{14-j}{j} \} ((t_1^2 - \omega_{14} t_1 + 1) t_2)^j X^{14-2j}$ $-(\omega_{14}^2 - 4)(t_1 - \omega_{14} t_1 + 1)^6 t_2^7$
note	This is obtained in section 4.

14/2	$\mathcal{D}_{14}$ **over** $\mathbf{Q}(\omega_7)$
poly.	$X^7 + (3\omega_7^2 - 5\omega_7 - 2)X^6 + (27\omega_7^2 - 19\omega_7 - 14)X^5$ $+(80\omega_7^2 - 65\omega_7 - 45)X^4 + (117\omega_7^2 - 94\omega_7 - 65)X^3$ $+(65\omega_7^2 - 52\omega_7 - 36)X^2 + t_1$
note	This polynomial is given by Hashimoto-Miyake [5].

Order 15

There is one group of order fifteen.

15/1	C_{15} **over** $\mathbf{Q}(\omega_{15})$
poly.	$X^{15}-15t_1X^{14}+105(\omega_{15}t_1-1)X^{13}-455((\omega_{15}^2-1)t_1-\omega_{15})X^{12}$ $+1365(\omega_{15}(\omega_{15}^2-2)t_1-\omega_{15}^2+1)X^{11}$ $-3003\omega_{15}((\omega_{15}^2+\omega_{15}-4)t_1-\omega_{15}^2+2)X^{10}$ $+5005((\omega_{15}^3-2\omega_{15}-1)t_1-\omega_{15}^3-\omega_{15}^2+4\omega_{15})X^9$ $-6435(\omega_{15}^2-\omega_{15}-1)(t_1-\omega_{15}-1)X^8$ $-6435(\omega_{15}^2-\omega_{15}-1)(t_1+1)X^7$ $+5005(\omega_{15}^2-\omega_{15}-1)((\omega_{15}+1)t_1-1)X^6$ $-3003((\omega_{15}^3+\omega_{15}^2-4\omega_{15})t_1-\omega_{15}^3+2\omega_{15}+1)X^5$ $+1365\omega_{15}((\omega_{15}^2-2)t_1-\omega_{15}^2-\omega_{15}+4)X^4$ $-455((\omega_{15}^2-1)t_1-\omega_{15}^3+2\omega_{15})X^3+105(\omega_{15}t_1-\omega_{15}^2+1)X^2$ $-15(t_1-\omega_{15})X-1$
note	This is given by Hashimoto-Miyake [5] and Rikuna [17].

Order 16

There are fourteen groups of order sixteen. Type **16/9** and **16/13** have no faithful linear representation of dimension two. So the author has not obtained these generic polynomials yet.

16/1	C_{16} **over** $\mathbf{Q}(\omega_{16})$
poly.	$X^{16}+\sum_{j=1}^{7}\{\binom{15-j}{j-1}+\binom{16-j}{j}\}((t_1^2-\omega_{16}t_1+1)t_2)^jX^{16-2j}$ $-(\omega_{16}^2-4)(t_1-\omega_{16}t_1+1)^9t_2^8$
note	This is obtained in section 4.

16/1	C_{16} **over** $\mathbf{Q}(\zeta_8)$
poly.	$F_{\mathcal{M}od_{32}}(\zeta_8t_1^2-2,t_2;X)$
note	From Rikuna [18], we have that $F_{\mathcal{M}od_{32}}(t_1,t_2;X):=X^{16}-(t_1^2-4)t_1^2t_2X^8-(t_1^2-4)(t_1t_2)^2$ is a $\mathbf{Q}(\zeta_8)$-generic $\mathcal{M}od_{32}$-polynomial.

16/1	C_{16} **over** $\mathbf{Q}(\nu_{16}:=\zeta_{16}-\zeta_{16}^{-1})$
poly.	$F_{\mathcal{QD}_{32}}\left(\frac{\nu_{16}^2+4}{2(t_1^2+\nu_{16}t_1-1)}+2,t_2;X\right)$
note	From Rikuna [18], we have that $F_{\mathcal{QD}_{32}}(t_1,t_2;X)$ $:=X^{16}-\frac{48(t_1-1)^2t_2^2}{t_1}X^{12}+\frac{592(t_1-1)^4t_2^4}{t_1^2}X^8-\frac{128(t_1-1)^5t_2^8}{t_1^2}X^6$ $-\frac{1408(t_1-1)^6t_2^6}{t_1^3}X^4-\frac{512(t_1-1)^7t_2^8}{t_1^3}X^2-\frac{32(t_1-1)^9t_2^8}{t_1^4}$ is a $\mathbf{Q}(\nu_{16})$-generic $\mathcal{QD}_{32}$-polynomial.

16/2	$\mathcal{C}_8 \times \mathcal{C}_2$ **over** $\mathbf{Q}(\zeta_8)$
act.	$\{x_1 \longmapsto \zeta_8 x_1 \qquad \{x_2 \longmapsto -x_2$
R_G	$\{\zeta_8^i x_1 \pm x_2 \mid 1 \le i \le 8\}$
ans.	$\left\{\frac{x_1^8}{x_2^8}, \frac{x_1^8}{x_2^6}\right\}$
poly.	$X^{16} - \frac{8t_2}{t_1}X^{14} + \frac{28t_2^2}{t_1^2}X^{12} - \frac{56t_2^3}{t_1^3}X^{10} - \frac{2(t_1-35)t_2^4}{t_1^4}X^8 - \frac{56(t_1+1)t_2^5}{t_1^5}X^6 - \frac{28(5t_1-1)t_2^6}{t_1^6}X^4 - \frac{8(7t_1+1)t_2^7}{t_1^7}X^2 + \frac{(t_1-1)^2t_2^8}{t_1^8}$

16/3	$\mathcal{C}_4 \times \mathcal{C}_4$ **over** $\mathbf{Q}(\sqrt{-1})$
act.	$\{x_1 \longmapsto \sqrt{-1}x_1 \qquad \{x_2 \longmapsto -\sqrt{-1}x_2$
R_G	$\{\sqrt{-1}^i x_1 + \sqrt{-1}^j x_2 \mid 1 \le i, j \le 4\}$
ans.	$\left\{\frac{x_1^4}{x_2^4}, \frac{(x_1x_2)^4}{x_1^4+x_2^4}\right\}$
poly.	$X^{16} - \frac{4(t_1+1)^2t_2}{t_1}X^{12} + \frac{2(t_1+1)^2(3t_1^2-62t_1+3)t_2^2}{t_1^2}X^8 - \frac{4(t_1+1)^4(t_1^2+30t_1+1)t_2^3}{t_1^3}X^4 + \frac{(t_1^2-1)^4t_2^4}{t_1^4}$
note	We can also obtain a $\mathbf{Q}$-generic $\mathcal{C}_4 \times \mathcal{C}_4$-polynomial. See remark 13.

16/4	$\mathcal{C}_4 \times \mathcal{C}_2 \times \mathcal{C}_2$ **over Q**
note	See remark 13.

16/5	$\mathcal{C}_2 \times \mathcal{C}_2 \times \mathcal{C}_2 \times \mathcal{C}_2$ **over Q**
note	See remark 13.

16/6	$\mathcal{D}_8 \times \mathcal{C}_2$ **over Q**
note	See remark 13.

16/7	$\mathcal{Q}_8 \times \mathcal{C}_2$ **over Q**
note	See remark 13.

16/8	${}_8\mathcal{T}_{11}$, **the eleventh transitive subgroup of** $\mathfrak{S}_8$, **over** $\mathbf{Q}(\sqrt{-1})$
act.	$\begin{cases} x_1 \longmapsto \sqrt{-1}x_1 \\ x_2 \longmapsto \sqrt{-1}x_2 \end{cases} \quad \begin{cases} x_1 \longmapsto x_2 \\ x_2 \longmapsto x_1 \end{cases} \quad \{x_2 \longmapsto -x_2$
R_G	$\{\pm x_1, \pm\sqrt{-1}x_1, \pm x_2, \pm\sqrt{-1}x_2\}$
ans.	$\left\{\frac{x_1^4+x_2^4}{(x_1x_2)^2}, \frac{x_1^8+x_2^8}{x_1^4+x_2^4}\right\}$
poly.	$X^8 - \frac{t_1^2t_2}{t_1^2-2}X^4 + \frac{t_1^2t_2^2}{(t_1^2-2)^2}$

16/10	$\mathcal{QD}_{16}$ **over** $\mathbf{Q}(\sqrt{-2})$
note	See $\mathcal{C}_8$ over $\mathbf{Q}(\sqrt{-2})$.

16/11	$\mathcal{M}od_{16}$ **over** $\mathbf{Q}(\sqrt{-1})$
poly.	$X^8 + (t_1^2 - 4)t_2X^4 - (t_1^2 - 4)t_2^2$
note	This polynomial is given by Rikuna [16].

16/12	$\mathcal{D}_{16}$ **over** $\mathbf{Q}(\omega_8) = \mathbf{Q}(\sqrt{2})$
poly.	$X^8 + 8(t_1t_2)X^6 + 20(t_1t_2)^2X^4 + 16(t_1t_2)^3X^2 + 2t_1^3t_2^4$
note	This is obtained in section 4.

16/14	$\mathcal{Q}_{16}$ **over** $\mathbf{Q}(\zeta_8)$
poly.	$X^{16} - \frac{t_1^5t_2^4}{(t_1^2-4)^2}X^8 + \frac{(t_1t_2)^8}{(t_1^2-4)^4}$
note	By the same way as $\mathcal{Q}_8$ and $\mathcal{Q}_{16}$, we have this polynomial. See also remark 14.

Acknowledgments

The author would like to thank the organizers of the two conferences in Saga and Tokyo who gave such a good opportunity of speaking and a discussion for him. And he also expresses gratitude to Armand Brumer, Ki-ichiro Hashimoto, Gregor Kemper, Arne Ledet, B. Heinrich Matzat, Katsuya Miyake, and many participants in the conferences who obtained a lot of kind advices. The author is a Research Fellow of the Japan Society for the Promotion of Science, and this study was supported by Grant-in-Aid for JSPS Fellows.

References

[1] The GAP Group, *GAP— Groups, Algorithms, and Programming*, Version 4.3, 2002, `http://www.gap-system.org`.

[2] W. Gröbner, *Minimalbasis der Quaternionengruppe*, Monatshefte f. Math. und Physik **41** (1934), 78–84.

[3] K. Hashimoto, *Inverse Galois problems related to dihedral groups* (Japanese), Algebraic number theory and related topics (Japanese) (Kyoto, 2000), Sūrikaisekikenkyūsho Kōkyūroku No. 1154 (2000), 125–136.

[4] K. Hashimoto, *On Brumer's family of RM-curves of genus two*, Tohoku Math. J. **52** (2000), no. 4, 475–488.

[5] K. Hashimoto and K. Miyake, *Inverse Galois problem for dihedral groups*, in "Number theory and its applications (S. Kanemitsu and K. Gyory, ed.)," 165–181, Developments in Mathematics, 2. Kluwer Acad. Publ., Doredrecht, 1999.

[6] K. Hashimoto and Y. Rikuna, *On generic families of cyclic polynomials with even degree*, Manuscripta Math. **107** (2002), 283–288.

[7] C. Jensen, A. Ledet, and N. Yui, "Generic polynomials, constructive aspects of the inverse Galois problem," Mathematical Sciences Research Institute Publications, Cambridge, 2003.

[8] G. Kemper, *Das Noethersche Problem und generische Polynome*, dissertation, Universität Heidelberg, 1994.

[9] G. Kemper, *A consructive approach to Noether's problem*, Manuscripta Math. **90** (1996), 343–363.

[10] G. Kemper, *Generic polynomials are descent-generic*, Manuscripta Math. **105** (2001), 139–141.

[11] G. Kemper and E. Mattig, *Generic polynomials with few parameters*, J. Symbolic Comp. **30** (2000), 843–857.

[12] A. Ledet, *Generic and explicit realisation of small p-groups*, J. Symbolic Comp. **30** (2000), 859–865.

[13] A. Ledet, *Generic extensions and generic polynomials*, J. Symbolic Comp. **30** (2000), 867–872.

[14] G. Malle and B. H. Matzat, "Inverse Galois theory," Springer Monographs in Mathematics, Springer-Verlag, Berlin, 1999.

[15] K. Miyake, *Linear Fractional Transformations and Cyclic Polynomials*, Algebraic number theory (Hapcheon/Saga, 1996). Adv. Stud. Contemp. Math. (Pusan) **1**, (1999), 137–142.

[16] Y. Rikuna, *On generic polynomials for the modular 2-groups*, Proc. Japan Acad. Ser. A Mathe. Sci. **78** (2002), no. 3, 33–35.

[17] Y. Rikuna, *On simple families of cyclic polynomials*, Proc. Amer. Math. Soc. **130** (2002), no 8, 2215–2218.

[18] Y. Rikuna, *On generic quasi-dihedral polynomials*, preprint.

[19] J.-P. Serre, "Topics in Galois Theory," Research Notes in Mathematics, 1. Jones and Bartlett Publ., 1992.

[20] D. Shanks, *The simplest cubic fields*, Math. Comp. **28** (1974), 1137–1152.

[21] A. D. Thomas and G. V. Wood, "Group tables," Shiva Publishing Limited, 1980.

[22] S. Wolfram, "The Mathematica Book, 4th ed.," Wolfram Media/Cambridge University Press, 1999.

Galois Theory and Modular Forms pp 195–220
K. Hashimoto, K. Miyake, and H. Nakamura, Editors
© 2003 Kluwer Academic Publishers

ON DIHEDRAL EXTENSIONS AND FROBENIUS EXTENSIONS

Masafumi IMAOKA
and Yasuhiro KISHI*
Department of Mathematics
Tokyo Metropolitan University,
Minami-ohsawa 1-1 Hachioji-shi, Tokyo,
192-0397 Japan
imaokam@comp.metro-u.ac.jp
ykishi@comp.metro-u.ac.jp

Introduction

The two quadratic fields $\mathbb{Q}(\sqrt{d})$ and $\mathbb{Q}(\sqrt{-3d})$ are known to be closely related. A. Scholz [Sc] gave the "Spiegelung" relation between the 3-ranks of their ideal class groups. In Y. Kishi [Ki1], we gave a precise Spiegelung relation by a constructive approach: every cubic polynomial which generates a cyclic cubic extension E of $\mathbb{Q}(\sqrt{d})$ unramified outside 3 over $\mathbb{Q}(\sqrt{d})$ with $\mathrm{Gal}(E/\mathbb{Q}) \cong S_3$ is constructed by making use of an element of the associated field $\mathbb{Q}(\sqrt{-3d})$; here S_3 is the symmetric group on three symbols. In 1972, on the other hand, G. Gras [Gr1] showed that all S_3-extensions of $\mathbb{Q}$ containing a fixed quadratic field $\mathbb{Q}(\sqrt{d})$ correspond to some elements of $\mathbb{Q}(\sqrt{-3d})$ (see Example 3.4 (1)). In this paper, we extend it for an arbitrary odd prime p over an algebraic number field k with $k \cap \mathbb{Q}(\zeta) = \mathbb{Q}$, where ζ is a primitive p-th root of unity (see Section 2).

For an odd prime p and for a quadratic extension L_1/k of algebraic number fields with $L_1 \cap \mathbb{Q}(\zeta) = \mathbb{Q}$, we define an extension L_2 as follows: L_2 is the proper subextension of $L_1(\zeta)/k(\omega)$ other than $k(\zeta)$ and $L_1(\omega)$, $\omega = \zeta + \zeta^{-1}$. Let us call L_2 *the associated field with* L_1. The main purpose of this paper is to show the following relation between L_1 and L_2: all D_p-extensions (resp. certain types of F_p-extensions) of k containing L_1 (resp. L_2) correspond to some elements of L_2 (resp. L_1), where D_p and

*This author was supported by JSPS Research Fellowships for Young Scientists.
[received: Jun. 25, 2002; accepted in revised form: Oct. 15, 2002]

F_p denote the dihedral group of order $2p$ and the Frobenius group of order $p(p-1)$, respectively.

Before we prove our main theorem, we need a classical result which may be regarded as the cyclic version of the main theorem.

Theorem (Theorem 1.2). *Let K be a field with* $\mathrm{char}(K) \neq p$. *Let $\mathcal{M}(K)$ be a set of elements of $K(\zeta)$ with certain properties (to be specified later).*
(1) For every $\gamma \in \mathcal{M}(K)$, the field $K(\zeta)(\sqrt[p]{\gamma^{t(K)}})$ is a cyclic extension of K of degree pm, where $t(K)$ is a certain element of the group ring $\mathbb{Z}[\mathrm{Gal}(K(\zeta)/K)]$. (Hence $K(\zeta)(\sqrt[p]{\gamma^{t(K)}})$ contains only one cyclic extension of K of degree p. We denote it by $F(K,\gamma)$.)
(2) Let E be a cyclic extension of K of degree p. Then there exists an element $\gamma \in \mathcal{M}(K)$ such that $E = F(K,\gamma)$.

Main Theorem (Theorem 2.1). *Let π be a permutation of the set $\{1,2\}$. Let $\mathcal{M}(L_{\pi(1)})$ be a set of elements of $L_{\pi(2)}$ with certain properties (to be specified later), and let $F(L_{\pi(1)},\delta)$ with $\delta \in \mathcal{M}(L_{\pi(1)})$ be the field defined as in the above theorem. Moreover, let $G_1 = D_p$ and $G_2 = F_p$.*
(1) For every $\delta \in \mathcal{M}(L_{\pi(1)})$, $F(L_{\pi(1)},\delta)$ is a $G_{\pi(1)}$-extension of k containing $L_{\pi(1)}$.
(2) Let E be a Galois extension of k and contains $L_{\pi(1)}$. Suppose that the Galois group of the extension E/k is isomorphic to $G_{\pi(1)}$ whose generators have some special relations if π is not trivial. Then there exists an element $\delta \in \mathcal{M}(L_{\pi(1)})$ such that $E = F(L_{\pi(1)},\delta)$.

The case where π is trivial and $k = \mathbb{Q}$ in this theorem was shown by Gras [Gr1, Section 3].

Moreover, we give two results (Theorem 1.5 and Proposition 2.5) concerning the generator of E. They are very effective for constructing polynomials of degree p which generate C_p, D_p or F_p-extensions.

We treat the cyclic case in Section 1 and the dihedral and the Frobenius cases in Section 2. In Section 3, we give some examples for $p=3$ and 5. Finally, we give in an appendix, using a cohomological approach, a relation between the p-rank of the congruent ideal class group modulo a power of p of a quadratic field L_1 and the p-rank of the absolute ideal class group of the associated cyclic field L_2 with L_1 of degree $p-1$. Its relation is given by Gras' result [Gr2, Theorem 7.11] also (see Remark 4.2).

We list here those symbols which will be used throughout this article.

Let $\mathbb{Q}$ denote the field of rational numbers and $\mathbb{Z}$ denote the ring of rational integers.

For an extension L/K, denote the norm map and the trace map of L/K by $N_{L/K}$ and by $\mathrm{Tr}_{L/K}$, respectively. For a Galois extension L/K, we denote the Galois group of L/K by $\mathrm{Gal}(L/K)$. For $\sigma, \tau \in \mathrm{Gal}(L/K)$, the action of the product $\sigma\tau$ on L is as usual, namely,

$$\alpha^{\sigma\tau} = (\alpha^\sigma)^\tau \qquad (\alpha \in L).$$

For an integer n, let C_n denote the cyclic group of order n. For a prime p, let D_p and F_p denote the dihedral group of order $2p$ and the Frobenius group of order $p(p-1)$, respectively:

$$D_p = \langle \sigma, \iota \mid \sigma^p = \iota^2 = 1,\ \iota^{-1}\sigma\iota = \sigma^{-1} \rangle,$$
$$F_p = \langle \sigma, \iota \mid \sigma^p = \iota^{p-1} = 1,\ \iota^{-1}\sigma\iota = \sigma^a \rangle,$$

where a is a primitive root modulo p.

1. Preliminaries (Cyclic case)

The aim of this section is to construct the polynomials with cyclic groups of prime degree as Galois groups using classical method. This is a review of [Co, Chapter 5].

Let p be a fixed prime and ζ be a primitive p-th root of unity. Let K be a field with $\mathrm{char}(K) \neq p$. Fix a generator τ of $\mathrm{Gal}(K(\zeta)/K)$, and take an element $l(\tau) \in \mathbb{Z}$ so that we have $\zeta^\tau = \zeta^{l(\tau)}$. Put $m := [K(\zeta) : K]$ and define an element $t(K)$ of a group ring $\mathbb{Z}[\mathrm{Gal}(K(\zeta)/K)]$ by

$$t(K) := \sum_{i=0}^{m-1} \tau^i l(\tau)^{m-1-i}.$$

We define a natural surjection $\phi : \mathbb{Z} \longrightarrow \mathbb{Z}/p\mathbb{Z}$. We use the same notation ϕ for the induced map $\mathbb{Z}[\mathrm{Gal}(K(\zeta)/K)] \longrightarrow (\mathbb{Z}/p\mathbb{Z})[\mathrm{Gal}(K(\zeta)/K)]$. Moreover, let $T(K)$ denote the following subset of $\mathbb{Z}[\mathrm{Gal}(K(\zeta)/K)]$:

$$T(K) := \{t'(K) \in \mathbb{Z}[\mathrm{Gal}(K(\zeta)/K)] \mid \phi(t'(K)) = n\phi(t(K)) \text{ for some } n \in (\mathbb{Z}/p\mathbb{Z})^\times\}.$$

The element $t(K)$ depends on τ and $l(\tau)$, but the set $T(K)$ does not. Moreover we easily see

$$K(\zeta)(\sqrt[p]{\gamma^{t(K)}}) = K(\zeta)(\sqrt[p]{\gamma^{t'(K)}})$$

for every $\gamma \in K(\zeta)$ and for every $t'(K) \in T(K)$.

We first introduce a result of H. Miki.

Proposition 1.1 ([Mik, Corollary of Proposition 2]). *Let the notation be as above. Put $F := K(\zeta)(\sqrt[p]{y})$ with $y \in K(\zeta)$ and $y \notin K(\zeta)^p$. Then the following three statements are equivalent:*

(a) F/K *is cyclic;*

(b) $y^{\tau - l(\tau)} \in K(\zeta)^p$;

(c) $F = K(\zeta)(\sqrt[p]{y^{t(K)}})$.

We define subsets $\mathcal{M}(K)$ and $\mathcal{N}(K)$ of $K(\zeta)^\times$ as follows:

$$\mathcal{M}(K) := \{\gamma \in K(\zeta)^\times \mid \gamma^{t(K)} \notin K(\zeta)^p\},$$
$$\mathcal{N}(K) := K(\zeta)^\times \setminus \mathcal{M}(K).$$

By this proposition, we obtain the following:

Theorem 1.2. *Let the notation be as above. Then the following statements hold.*
(1) *For every $\gamma \in \mathcal{M}(K)$, the field $K(\zeta)(\sqrt[p]{\gamma^{t(K)}})$ is a cyclic extension of K of degree pm. (Hence $K(\zeta)(\sqrt[p]{\gamma^{t(K)}})$ contains only one cyclic extension of K of degree p. We denote it by $F(K,\gamma)$.)*
(2) *If E is a cyclic extension of K of degree p, then there exists an element $\gamma \in \mathcal{M}(K)$ such that $E = F(K,\gamma)$.*

Proof. (1) Note that the order of τ is equal to m and $l(\tau)^m - 1$ is divisible by p. Then in $\mathbb{F}_p[\mathrm{Gal}(K(\zeta)/K)]$, we see that

$$\phi(t(K)(\tau - l(\tau))) = \phi(\tau^m - l(\tau)^m) = 0.$$

Hence we have $\gamma^{t(K)(\tau - l(\tau))} \in K(\zeta)^p$. If $\gamma \in \mathcal{M}(K)$, therefore, $K(\zeta)(\sqrt[p]{\gamma^{t(K)}})/K$ is a cyclic extension by Proposition 1.1. It is clear that $K(\zeta)(\sqrt[p]{\gamma^{t(K)}})/K$ is of degree pm.
(2) Since $E(\zeta)$ is a Kummer extension of $K(\zeta)$, there exists an element $\gamma \in K(\zeta) \setminus K(\zeta)^p$ such that $E(\zeta) = K(\zeta)(\sqrt[p]{\gamma^{t(K)}})$. Then we see that $\gamma \in \mathcal{M}(K)$ and $E = F(K,\gamma)$. □

We now state a criterion for two fields $F(K,\gamma_1)$ and $F(K,\gamma_2)$ with $\gamma_1, \gamma_2 \in \mathcal{M}(K)$ to coincide with each other.

Proposition 1.3. *For elements $\gamma_1, \gamma_2 \in \mathcal{M}(K)$, the following statements are equivalent:*

(a) $F(K,\gamma_1) = F(K,\gamma_2)$;

(b) $\gamma_1^n/\gamma_2 \in \mathcal{N}(K)$ *for some* $n \in \mathbb{Z} \setminus p\mathbb{Z}$.

Proof. By the definition of $F(K,\gamma_i)$ $(i = 1,2)$, we have $F(K,\gamma_1) = F(K,\gamma_2)$ if and only if $K(\zeta)(\alpha_1) = K(\zeta)(\alpha_2)$ with $\alpha_i^p = \gamma_i^{t(K)}$ $(i = 1,2)$.

By Kummer theory, it is equivalent to $(\gamma_1^{t(K)})^n/\gamma_2^{t(K)} = (\gamma_1^n/\gamma_2)^{t(K)} \in K(\zeta)^p$ for some $n \in \mathbb{Z} \setminus p\mathbb{Z}$. Therefore the statement follows from the definition of $\mathcal{N}(K)$. □

Corollary 1.4. *For every $\rho \in \mathrm{Gal}(K(\zeta)/K)$ and $\gamma \in \mathcal{M}(K)$, we have $F(K,\gamma) = F(K,\gamma^\rho)$.*

Proof. For $\gamma \in \mathcal{M}(K)$, we have $\gamma^{t(K)(\tau - l(\tau))} \in K(\zeta)^p$ as we have seen in the proof of Theorem 1.2 (1). Then

$$\begin{aligned}(\gamma^{l(\tau)^i}/\gamma^{\tau^i})^{t(K)} &= \gamma^{t(K)(l(\tau)^i - \tau^i)} \\ &= \gamma^{-t(K)(\tau - l(\tau))(\tau^{i-1} + \tau^{i-2}l(\tau) + \cdots + \tau l(\tau)^{i-2} + l(\tau)^{i-1})} \\ &\in K(\zeta)^p\end{aligned}$$

for every i, $0 \le i \le m-1$. Therefore we have $\gamma^{l(\tau)^i}/\gamma^{\tau^i} \in \mathcal{N}(K)$. By Proposition 1.3, we see that $F(K,\gamma) = F(K,\gamma^\rho)$ for every $\rho \in \mathrm{Gal}(K(\zeta)/K)$. □

Next, we give a generator of a cyclic extension of K of degree p by making use of elements of $K(\zeta)$.

Theorem 1.5. *Suppose $\gamma \in \mathcal{M}(K)$ and put $E = F(K,\gamma)$. Then we have*

$$E = K\left(\mathrm{Tr}_{E(\zeta)/E}(\sqrt[p]{\gamma^{t'(K)}})\right)$$

for every $t'(K) \in T(K)$.

Proof. It follows from Theorem 1.2 that $E(\zeta)/K$ is cyclic of degree pm. We extend τ to an automorphism of $E(\zeta)/E$. We put $\alpha = \sqrt[p]{\gamma^{t'(K)}}$ with $t'(K) \in T(K)$. Then, as we saw in the proof of Corollary 1.4, we have $\alpha^{\tau^i - l(\tau)^i} \in K(\zeta)^\times$ for every i, $0 \le i \le m-1$. On the other hand, 1, α, $\alpha^{l(\tau)}$, $\alpha^{l(\tau)^2}$, $\cdots$, $\alpha^{l(\tau)^{m-1}}$ are linearly independent over $K(\zeta)$. Therefore so are 1, α, α^τ, α^{τ^2}, $\cdots$, $\alpha^{\tau^{m-1}}$. Hence the trace of α,

$$\mathrm{Tr}_{E(\zeta)/E}(\alpha) = \alpha + \alpha^\tau + \alpha^{\tau^2} + \cdots + \alpha^{\tau^{m-1}},$$

does not belong to $K(\zeta)$. Since E/K is a cyclic extension of prime degree p, we see $E = K(\mathrm{Tr}_{E(\zeta)/E}(\alpha))$. The proof is completed. □

By this theorem, we can construct a polynomial of which E is the minimal splitting field over K. (See Proposition 1.10 and Examples 3.1, 3.2 and 3.3.) Indeed, we give the polynomial in the case $[K(\zeta) : K] = 2$ as follows.

Definition 1.6. For an odd integer s, we define polynomials $g_s(X;a)$ inductively:

$$g_1(X;a) := X,$$

$$g_s(X;a) := X^s - \sum_{j=1}^{(s-1)/2} a^j \binom{s}{j} g_{s-2j}(X;a), \quad (s \geq 3),$$

where $\binom{s}{j}$ is the binomial coefficient:

$$\binom{s}{j} = \frac{s(s-1)\cdots(s-j+1)}{j!}.$$

Remark 1.7. By the definition of $g_s(X;a)$, the coefficient of X^n is equal to 0 when n is even.

Lemma 1.8. *Let s be an odd integer and let $c(s,\ n)$ denote the coefficient of X^n of $g_s(X;a)$. Then*

$$c(s,\ s-2g) = (-a)^g \frac{s}{s-2g}\binom{s-g-1}{g} \qquad (0 \leq g \leq \frac{s-1}{2}). \qquad (1.1)$$

Proof. It is clear that

$$c(s,\ n) = \begin{cases} 1, & \text{if } n = s, \\ 0, & \text{if } n > s. \end{cases}$$

By the definition of $g_s(X;a)$, we immediately have

$$\begin{aligned} c(s,\ n) &= -\sum_{j=1}^{(s-1)/2} a^j \binom{s}{j} c(s-2j,\ n) \\ &= -\sum_{j=1}^{(s-n)/2} a^j \binom{s}{j} c(s-2j,\ n) \qquad (1.2) \end{aligned}$$

for $n < s$.

Both sides of (1.1) are equal to 1 for $s = 1$. Assume that the relation (1.1) holds for all s, $s \leq l-2$, and for all g, $0 \leq g \leq (s-1)/2$. We will show that the relation (1.1) holds for $s = l$ and for all g, $0 \leq g \leq (l-1)/2$, by mathematical induction on g.

It is clear that (1.1) holds in case $g = 0$. Assume that (1.1) holds for all g, $0 \le g \le k$. Then by (1.2) and the assumptions,

$$\begin{aligned}
c(l,\ l-2(k+1)) &= c(l,\ l-2k-2) \\
&= -\sum_{j=1}^{(l-(l-2k-2))/2} a^j \binom{l}{j} c(l-2j,\ l-2k-2) \\
&= -\sum_{j=1}^{k+1} a^j \binom{l}{j} c(l-2j,\ l-2j-2(k+1-j)) \\
&= -\sum_{j=1}^{k+1} a^j \binom{l}{j} (-a)^{k+1-j} \\
&\quad \times \frac{l-2j}{l-2j-2(k+1-j)} \binom{l-2j-(k+1-j)-1}{k+1-j} \\
&= \frac{(-1)^k a^{k+1}}{l-2k-2} \sum_{j=1}^{k+1} (-1)^j (l-2j) \binom{l}{j} \binom{l-k-2-j}{k+1-j}.
\end{aligned}$$

Now we need the following equation.

Lemma 1.9. *For all t, $0 \le t \le k$, we have*

$$\begin{aligned}
&\sum_{j=k-t+1}^{k+1} (-1)^j (l-2j) \binom{l}{j} \binom{l-k-2-j}{k+1-j} \\
&\quad = (-1)^{k-t+1} \frac{(t+1)(k-t+1)}{k+1} \binom{l}{k-t+1} \binom{l-2k-2+t}{t+1}.
\end{aligned}$$

Proof. A straightforward calculation, by mathematical induction on t, gives this equation. □

Take $t = k$ in Lemma 1.9. Then we have

$$\begin{aligned}
\sum_{j=1}^{k+1} (-1)^j (l-2j) \binom{l}{j} \binom{l-k-2-j}{k+1-j} &= -\frac{(k+1)\cdot 1}{k+1} \binom{l}{1} \binom{l-k-2}{k+1} \\
&= -l \binom{l-k-2}{k+1}.
\end{aligned}$$

Therefore

$$\begin{aligned}
c(l,\ l-2(k+1)) &= \frac{(-1)^{k+1} a^{k+1}}{l-2k-2} l \binom{l-k-2}{k+1} \\
&= (-a)^{k+1} \frac{l}{l-2(k+1)} \binom{l-(k+1)-1}{k+1}.
\end{aligned}$$

The proof of Lemma 1.8 is completed. □

Proposition 1.10. (cf. [Co, Chapter 5, Exercise 6].) *Let the notation be as in Theorem 1.5. Suppose that K contains $\omega = \zeta + \zeta^{-1}$ and does not contain ζ. Let $\alpha \in E(\zeta)$ be a root of $X^p - \gamma^{(p+1)/2}(\gamma^\tau)^{(p-1)/2} = 0$, $\gamma \in \mathcal{M}(K)$. Then $\mathrm{Tr}_{E(\zeta)/E}(\alpha)$ generates E over K: $E = K(\mathrm{Tr}_{E(\zeta)/E}(\alpha))$. Moreover, the polynomial*

$$\begin{aligned} &f_p(X;\gamma)\\ &= \sum_{k=0}^{(p-1)/2} (-N(\gamma))^k \frac{p}{p-2k}\binom{p-k-1}{k} X^{p-2k} - N(\gamma)^{(p-1)/2}\mathrm{Tr}(\gamma) \end{aligned}$$

is the minimal polynomial of $\mathrm{Tr}_{E(\zeta)/E}(\alpha)$ over K, where $N = N_{K(\zeta)/K}$ and $\mathrm{Tr} = \mathrm{Tr}_{K(\zeta)/K}$.

Proof. By the assumptions, we have $m = [K(\zeta) : K] = 2$ and $l(\tau) = -1$; and hence $t(K) = \tau - 1$. Since

$$\phi(\frac{p+1}{2} + \frac{p-1}{2}\tau) = \frac{p-1}{2}\phi(\tau - 1),$$

we have, from Theorem 1.5,

$$E = K(\mathrm{Tr}_{E(\zeta)/E}(\alpha)). \tag{1.3}$$

We extend τ to an automorphism of $E(\zeta)/E$. Let us show that

$$\alpha \cdot \alpha^\tau = N(\gamma). \tag{1.4}$$

Since

$$(\alpha \cdot \alpha^\tau)^p = \gamma^{(p+1)/2}(\gamma^\tau)^{(p-1)/2} \cdot (\gamma^\tau)^{(p+1)/2}\gamma^{(p-1)/2} = (\gamma \cdot \gamma^\tau)^p = N(\gamma)^p,$$

we have $\alpha \cdot \alpha^\tau = \zeta^i N(\gamma)$ for some i, $0 \leq i \leq p-1$. Note that $\alpha \cdot \alpha^\tau \in E$. Since neither E nor K contains ζ, it must be $i = 0$, that is, $\alpha \cdot \alpha^\tau = N(\gamma)$. By (1.4), therefore, we have a relation

$$(\alpha + \alpha^\tau)^s = \alpha^s + (\alpha^\tau)^s + \sum_{j=1}^{(s-1)/2} N(\gamma)^j \binom{s}{j} (\alpha^{s-2j} + (\alpha^\tau)^{s-2j}) \tag{1.5}$$

for an odd integer s. On the other hand, we have

$$\begin{aligned} \alpha + \alpha^\tau &= \mathrm{Tr}_{E(\zeta)/E}(\alpha),\\ \alpha^p + (\alpha^\tau)^p &= \gamma^{(p+1)/2}(\gamma^\tau)^{(p-1)/2} + \gamma^{(p-1)/2}(\gamma^\tau)^{(p+1)/2}\\ &= \gamma^{(p-1)/2}(\gamma^\tau)^{(p-1)/2}(\gamma + \gamma^\tau)\\ &= N(\gamma)^{(p-1)/2}\mathrm{Tr}(\gamma). \end{aligned}$$

Put $s = p$ in the relation (1.5); then we see that

$$\mathrm{Tr}_{E(\zeta)/E}(\alpha)^p - \sum_{j=1}^{(p-1)/2} N(\gamma)^j \binom{p}{j} (\alpha^{p-2j} + (\alpha^\tau)^{p-2j}) - N(\gamma)^{(p-1)/2}\mathrm{Tr}(\gamma) = 0. \quad (1.6)$$

Next we will show that

$$g_s(\mathrm{Tr}_{E(\zeta)/E}(\alpha); N(\gamma)) = \alpha^s + (\alpha^\tau)^s \quad (1.7)$$

for all odd integers s. It is clear that the equation (1.7) holds in case $s = 1$. By the definition of $g_s(X; a)$ and the relation (1.5), we see that

$$\begin{aligned} &g_s(\mathrm{Tr}_{E(\zeta)/E}(\alpha); N(\gamma)) \\ &= \mathrm{Tr}_{E(\zeta)/E}(\alpha)^s - \sum_{j=1}^{(s-1)/2} N(\gamma)^j \binom{s}{j} g_{s-2j}(\mathrm{Tr}_{E(\zeta)/E}(\alpha); N(\gamma)) \\ &= \alpha^s + (\alpha^\tau)^s \\ &\quad + \sum_{j=1}^{(s-1)/2} N(\gamma)^j \binom{s}{j} (\alpha^{s-2j} + (\alpha^\tau)^{s-2j} - g_{s-2j}(\mathrm{Tr}_{E(\zeta)/E}(\alpha); N(\gamma)). \end{aligned}$$

Hence if the equation (1.7) holds in all cases of odd integers less than s, then we have $g_s(\mathrm{Tr}_{E(\zeta)/E}(\alpha); N(\gamma)) = \alpha^s + (\alpha^\tau)^s$.

We define a polynomial $f_p(X; \gamma)$ as follows:

$$f_p(X; \gamma) := g_p(X; N(\gamma)) - N(\gamma)^{(p-1)/2}\mathrm{Tr}(\gamma).$$

Then $f_p(X; \gamma)$ has a root $\mathrm{Tr}_{E(\zeta)/E}(\alpha)$. Indeed, by (1.6) and (1.7), we have

$$\begin{aligned} &f_p(\mathrm{Tr}_{E(\zeta)/E}(\alpha); \gamma) \\ &= \mathrm{Tr}_{E(\zeta)/E}(\alpha)^p \\ &\quad - \sum_{j=1}^{(p-1)/2} N(\gamma)^j \binom{p}{j} g_{p-2j}(\mathrm{Tr}_{E(\zeta)/E}(\alpha); N(\gamma)) - N(\gamma)^{(p-1)/2}\mathrm{Tr}(\gamma) \\ &= \mathrm{Tr}_{E(\zeta)/E}(\alpha)^p \\ &\quad - \sum_{j=1}^{(p-1)/2} N(\gamma)^j \binom{p}{j} (\alpha^{p-2j} + (\alpha^\tau)^{p-2j}) - N(\gamma)^{(p-1)/2}\mathrm{Tr}(\gamma) \\ &= 0. \end{aligned}$$

Hence by the condition (1.3), $f_p(X;\gamma)$ is the minimal polynomial of $\mathrm{Tr}_{E(\zeta)/E}(\alpha)$ over K. Therefore the statement of the proposition follows from Lemma 1.8. □

2. The cases D_p and F_p

In this section, we construct D_p-polynomials and F_p-polynomials by utilizing the results in Section 1.

Let p be an odd prime. Let ζ be a primitive p-th root of unity and put $\omega := \zeta + \zeta^{-1}$. Let k be an algebraic number field and L_1 be a quadratic extensions of k which are linearly disjoint with $\mathbb{Q}(\zeta)$ over $\mathbb{Q}$. Then there exists a unique proper subextension of the bicyclic biquadratic extension $L_1(\zeta)/k(\omega)$ other than $k(\zeta)$ and $L_1(\omega)$. We denote it by L_2 (see Figure 1). L_2 is called *the associated field with* L_1. Moreover, we define m_1 and m_2 by

$$m_1 := 2\ (= [L_1 : k] = [L_2(\zeta) : L_2]),$$
$$m_2 := p - 1\ (= [L_2 : k] = [L_1(\zeta) : L_1]),$$

respectively.

Figure 1.

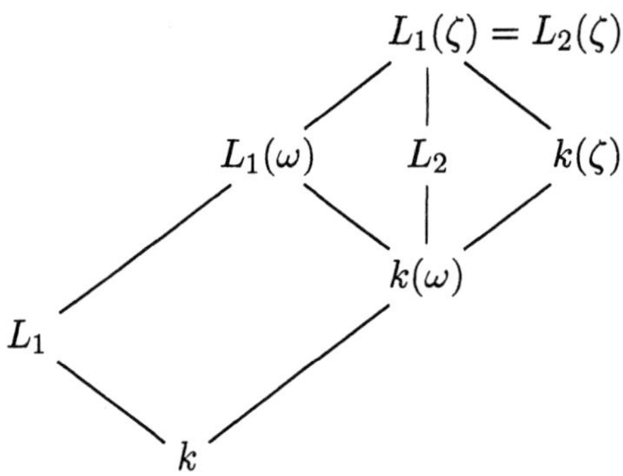

For an algebraic number field L, let $t(L)$ and $\mathcal{M}(L)$ be the same notation as in Section 1. For an element $\gamma \in \mathcal{M}(L)$, let the definition of the field $F(L,\gamma)$ be as in Theorem 1.2 (1). Our main theorem states

Theorem 2.1. *Let π be either one of permutations of two elements $\{1, 2\}$. Fix a generator ρ of $\mathrm{Gal}(L_{\pi(1)}(\zeta)/L_{\pi(2)})$, and take an element $l(\rho) \in \mathbb{Z}$ so that we have $\zeta^\rho = \zeta^{l(\rho)}$.*

(1) *For every* $\delta \in \mathcal{M}(L_{\pi(1)}) \cap L_{\pi(2)}$, $F(L_{\pi(1)}, \delta)$ *is a Galois extension of* k *of degree* $pm_{\pi(1)}$ *containing* $L_{\pi(1)}$. *Furthermore, let* σ *and* ι *be generators of* $\mathrm{Gal}(F(L_{\pi(1)}, \delta)/k)$ *which satisfy the following two relations:*

(i) $\iota|_{L_{\pi(1)}} = \rho|_{L_{\pi(1)}}$;

(ii) $\sigma^p = \iota^{m_{\pi(1)}} = 1$.

Then we have

(iii) $\iota^{-1}\sigma\iota = \sigma^{l(\rho)}$.

(2) *Let* E *be a Galois extension of* k *of degree* $pm_{\pi(1)}$ *and contains* $L_{\pi(1)}$. *Let* σ *be a* k*-automorphism of* E *of order* p. *Suppose that the Galois group of the extension* E/k *is*

$$\mathrm{Gal}(E/k) = \langle \sigma, \iota \mid \sigma^p = \iota^{m_{\pi(1)}} = 1, \ \iota^{-1}\sigma\iota = \sigma^{l(\rho)} \rangle$$

with $\iota|_{L_{\pi(1)}} = \rho|_{L_{\pi(1)}}$. *Then there exists an element* $\delta \in \mathcal{M}(L_{\pi(1)}) \cap L_{\pi(2)}$ *such that* $E = F(L_{\pi(1)}, \delta)$.

First we prove a general proposition.

Proposition 2.2. *Let* p *be a prime and* K *be a field with* $\mathrm{char}(K) \neq p$. *Put* $m := [K(\zeta) : K]$. *Let* L *be the minimal splitting field of* $X^p - \alpha$ $(\alpha \in K \setminus K^p)$ *over* K. *And let* $\langle \rho \rangle$ *and* $\langle \sigma \rangle$ *denote the subgroups of* $\mathrm{Gal}(L/K)$ *corresponding to* $K(\sqrt[p]{\alpha})$ *and* $K(\zeta)$, *respectively. Then we have*

$$\mathrm{Gal}(L/K) = \langle \sigma, \rho \mid \sigma^p = \rho^m = 1, \ \rho^{-1}\sigma\rho = \sigma^b \rangle,$$

where b *is an element of* $(\mathbb{Z}/p\mathbb{Z})^\times$ *with* $\zeta^\rho = \zeta^b$.

Proof. Let us write $G = \mathrm{Gal}(L/K)$ for simplicity. Since

$$(|\langle \sigma \rangle|, \ |\langle \rho \rangle|) = (p, \ m) = 1$$

and since $\langle \sigma \rangle$ is normal in G, we have

$$G = \langle \sigma, \rho \rangle.$$

Since $\langle \sigma \rangle$ is a normal subgroup of G, we have $\rho^{-1}\sigma\rho = \sigma^b$ for some b, $1 \leq b \leq p - 1$.

Now take c and d be integers satisfying

$$(\sqrt[p]{\alpha})^\sigma = \zeta^c \sqrt[p]{\alpha} \qquad (1 \leq c \leq p-1),$$
$$\zeta^\rho = \zeta^d \qquad (1 \leq d \leq p-1).$$

Since we have $\zeta^\sigma = \zeta$ and $(\sqrt[p]{\alpha})^\rho = \sqrt[p]{\alpha}$, we see that

$$(\sqrt[p]{\alpha})^{\rho^{-1}\sigma\rho} = (\sqrt[p]{\alpha})^{\sigma\rho} = (\zeta^c\sqrt[p]{\alpha})^\rho = \zeta^{cd}\sqrt[p]{\alpha}.$$

On the other hand, we have

$$(\sqrt[p]{\alpha})^{\rho^{-1}\sigma\rho} = (\sqrt[p]{\alpha})^{\sigma^b} = \zeta^{cb}\sqrt[p]{\alpha}.$$

Hence we must have $\zeta^{cd} = \zeta^{cb}$. Since $(c,\ p) = 1$, we have $b \equiv d \pmod p$, and hence $\zeta^\rho = \zeta^b$. The proof is completed. □

By this proposition, we immediately have the following corollary.

Corollary 2.3. *In the situation of Proposition* 2.2, *we have*

$$\mathrm{Gal}(L/K) \cong \begin{cases} C_p, & \textit{if } m = 1, \\ D_p, & \textit{if } m = 2, \\ F_p, & \textit{if } m = p-1. \end{cases}$$

Proof of Theorem 2.1. (1) Suppose $\delta \in \mathcal{M}(L_{\pi(1)}) \cap L_{\pi(2)}$. Let τ is the generator of $\mathrm{Gal}(L_{\pi(1)}(\zeta)/L_{\pi(1)})$. Since $L_{\pi(1)} \cap L_{\pi(2)} = k$, the restriction map $\tau|_{L_{\pi(2)}}$ is an automorphism of $L_{\pi(2)}/k$. Then we have $\delta^{t(L_{\pi(1)})} \in L_{\pi(2)}$. Put

$$E' := L_{\pi(1)}(\zeta)(\sqrt[p]{\delta^{t(L_{\pi(1)})}}) = L_{\pi(2)}(\zeta)(\sqrt[p]{\delta^{t(L_{\pi(1)})}});$$

then E' is normal over $L_{\pi(2)}$ of degree $pm_{\pi(1)}$. Let σ' and ι' be generators of $\mathrm{Gal}(E'/L_{\pi(1)}(\zeta))$ and $\mathrm{Gal}(E'/L_{\pi(2)}(\sqrt[p]{\delta^{t(L_{\pi(1)})}}))$, respectively. Then by Proposition 2.2, the Galois group $\mathrm{Gal}(E'/L_{\pi(2)})$ is

$$\mathrm{Gal}(E'/L_{\pi(2)}) = \langle \sigma', \iota' \mid \sigma'^p = \iota'^{m_{\pi(1)}} = 1,\ \iota'^{-1}\sigma'\iota' = \sigma'^b \rangle,$$

where b is an element of $(\mathbb{Z}/p\mathbb{Z})^\times$ with $\zeta^{\iota'} = \zeta^b$. From the natural restriction map, we see that $\mathrm{Gal}(E'/L_{\pi(2)}(\sqrt[p]{\delta^{t(L_{\pi(1)})}}))$ is isomorphic to $\mathrm{Gal}(L_{\pi(1)}(\zeta)/L_{\pi(2)})$, and $\iota'|_{L_{\pi(1)}(\zeta)}$ is a generator of $\mathrm{Gal}(L_{\pi(1)}(\zeta)/L_{\pi(2)})$. Then we have $\rho = (\iota'|_{L_{\pi(1)}(\zeta)})^c$ for some integer c with $(c, m_{\pi(1)}) = 1$. Therefore we have $\zeta^\rho = \zeta^{\iota'^c} = \zeta^{b^c}$, and hence $b^c \equiv l(\rho) \pmod p$. Moreover, σ' and ι'^c generate $\mathrm{Gal}(E'/L_{\pi(2)})$ and satisfy

$$(\iota'^c)^{-1}\sigma'\iota'^c = \sigma'^{b^c} = \sigma'^{l(\rho)}.$$

Now let us show that E' is normal over k. Let ν be a k-isomorphism from E' to the algebraic closure of E'. Then $(E')^\nu$ is the minimal splitting field of $X^p - (\delta^{t(L_{\pi(1)})})^\nu$ over $(L_{\pi(1)}(\zeta))^\nu$. Since $L_{\pi(1)}(\zeta)/k$ is abelian, we have $(L_{\pi(1)}(\zeta))^{\nu} = L_{\pi(1)}(\zeta)$ and $(\delta^{t(L_{\pi(1)})})^{\nu} = (\delta^{\nu})^{t(L_{\pi(1)})}$. Therefore we have

$$(E')^{\nu} = L_{\pi(1)}(\zeta)(\sqrt[p]{\delta^{\nu t(L_{\pi(1)})}}) = F(L_{\pi(1)}, \delta^{\nu})(\zeta).$$

Then by Corollary 1.4, we have $(E')^{\nu} = E'$. Hence E' is normal over k. Since $(L_{\pi(1)})^{\nu} = L_{\pi(1)}$ and $F(L_{\pi(1)}, \delta)$ is the unique cyclic subextension of order p of $E'/L_{\pi(1)}$, we see $(F(L_{\pi(1)}, \delta))^{\nu} = F(L_{\pi(1)}, \delta)$. Therefore $F(L_{\pi(1)}, \delta)$ is normal over k. From the natural restriction map, we see that $\mathrm{Gal}(E'/L_{\pi(2)})$ is isomorphic to $\mathrm{Gal}(F(L_{\pi(1)}, \delta)/k)$. Then $F(L_{\pi(1)}, \delta)$ is normal over k of degree $pm_{\pi(1)}$. Put $\sigma := \sigma'|_{F(L_{\pi(1)},\delta)}$ and $\iota := (\iota'|_{F(L_{\pi(1)},\delta)})^{c}$; then we see that σ and ι generate $\mathrm{Gal}(F(L_{\pi(1)}, \delta)/k)$ and satisfy the relations

$$\begin{aligned} \iota|_{L_{\pi(1)}} &= (\iota'|_{L_{\pi(1)}})^{c} = \rho|_{L_{\pi(1)}}, \\ \sigma^{p} &= \iota^{m_{\pi(1)}} = 1, \\ \iota^{-1}\sigma\iota &= \sigma^{l(\rho)}. \end{aligned}$$

The last relation does not depend on the choice of σ and ρ. Indeed, if we change σ to σ^{i} satisfying the condition (ii), then we have

$$\iota^{-1}\sigma^{i}\iota = (\iota^{-1}\sigma\iota)^{i} = (\sigma^{l(\rho)})^{i} = (\sigma^{i})^{l(\rho)},$$

and if we change ρ to ρ^{j}, then we have

$$\iota^{j}|_{L_{\pi(1)}} = \rho^{j}|_{L_{\pi(1)}}$$

and

$$\iota^{-j}\sigma\iota^{j} = \sigma^{l(\rho)^{j}} = \sigma^{l(\rho^{j})}.$$

It is also clear that ι is determined uniquely by the choice of ρ.

(2) Let E be a Galois extension of k of degree $pm_{\pi(1)}$ which contains $L_{\pi(1)}$. Since $E/L_{\pi(1)}$ is cyclic of order p, there exists $\gamma \in \mathcal{M}(L_{\pi(1)})$ such that $E = F(L_{\pi(1)}, \gamma)$ and $E(\zeta) = L_{\pi(1)}(\zeta)(\alpha)$ with $\alpha^{p} = \gamma^{t(L_{\pi(1)})}$, by Theorem 1.2. From the natural restriction map, we see that $\mathrm{Gal}(E(\zeta)/L_{\pi(2)})$ is isomorphic to $\mathrm{Gal}(E/k)$; namely, the elements σ' and ι' which satisfy $\sigma'|_{E} = \sigma$ and $\iota'|_{E} = \iota$ generate $\mathrm{Gal}(E(\zeta)/L_{\pi(2)})$. Then we may assume that $\alpha^{\sigma'} = \zeta\alpha$. Denote the fixed field of $\langle\iota'\rangle$ by E_1. Put $\beta := N_{E(\zeta)/E_1}(\alpha) \in E_1$ and $\delta := N_{L_{\pi(1)}(\zeta)/L_{\pi(2)}}(\gamma) \in L_{\pi(2)}$; then we have

$$\begin{aligned} \beta^{p} &= N_{E(\zeta)/E_1}(\alpha^{p}) = N_{E(\zeta)/E_1}(\gamma^{t(L_{\pi(1)})}) \\ &= N_{E(\zeta)/E_1}(\gamma)^{t(L_{\pi(1)})} = N_{L_{\pi(1)}(\zeta)/L_{\pi(2)}}(\gamma)^{t(L_{\pi(1)})} = \delta^{t(L_{\pi(1)})}. \end{aligned}$$

Since

$$\beta^{\sigma'} = \alpha^{(\sum_{s=0}^{m_{\pi(1)}-1} \iota'^{s})\sigma'} = \alpha^{\sum_{s=0}^{m_{\pi(1)}-1} \sigma'^{l(\rho)^{-s}} \iota'^{s}}$$
$$= \prod_{s=0}^{m_{\pi(1)}-1} (\zeta^{l(\rho)^{-s}}\alpha)^{\iota'^{s}} = \beta \prod_{s=0}^{m_{\pi(1)}-1} \zeta^{l(\rho)^{s} l(\rho)^{-s}} = \beta\zeta^{m_{\pi(1)}}$$

and $m_{\pi(1)} \not\equiv 0 \pmod{p}$, we have $\beta^{\sigma'} \neq \beta$. Then we get $\beta \notin L_{\pi(1)}(\zeta)$ and $E_1 = L_{\pi(2)}(\beta)$. Therefore we have $E(\zeta) = L_{\pi(1)}(\zeta)(\beta)$, and hence $E = F(L_{\pi(1)}, \delta)$. This completes the proof of Theorem 2.1. □

Remarks 2.4. (1) When π is the identity in Theorem 2.1, we see that every D_p-extension of k containing L_1 is given as $F(L_1, \delta)$ for some $\delta \in \mathcal{M}(L_1) \cap L_2$. This was shown in the case $k = \mathbb{Q}$ by G. Gras [Gr1, Section 3].
(2) When π is not trivial in Theorem 2.1, we see that a certain F_p-extension of k containing L_2 is given as $F(L_2, \delta)$ for some $\delta \in \mathcal{M}(L_2) \cap L_1$; in this case, generators σ and ι of the Galois group $\mathrm{Gal}(F(L_2, \delta))/k)$ satisfying the conditions (i) and (ii) of Theorem 2.1 must also satisfy the condition (iii) of Theorem 2.1. In general, however, there are $\varphi(p-1)$ primitive roots modulo p, where φ is the Euler function. Hence a natural question arises: Are there F_p-extensions E of k containing L_2 whose Galois group have generators σ and ι satisfying the conditions (i), (ii) of Theorem 2.1, and

(iii′) $\iota^{-1}\sigma\iota = \sigma^{b}$,

where b is any one of primitive roots modulo p except $l(\rho)$?

Assume that p satisfies $p \equiv 1 \pmod{4}$. Then there exists a quadratic extension of k other than L_1 which is contained in $L_1(\zeta)$ but is not contained in $k(\zeta)$. We denote it by L_1'. Let τ be a generator of the Galois group $\mathrm{Gal}(L_2(\zeta)/L_2)$. Then we have $\zeta^{\tau} = \zeta^{-1}$ and L_1' is the fixed field of $\langle \tau\rho \rangle$. For $\delta' \in \mathcal{M}(L_2) \cap L_1'$, therefore, $E = F(L_2, \delta')$ contains L_2 and the Galois group $\mathrm{Gal}(E/k)$ has generators σ and ι satisfying the conditions

(i) $\iota|_{L_2} = (\tau\rho)|_{L_2} = \rho|_{L_2}$,

(ii) $\sigma^{p} = \iota^{p-1} = 1$,

and

(iii″) $\iota^{-1}\sigma\iota = \sigma^{-l(\rho)}$

because

$$\zeta^{l(\tau\rho)} = \zeta^{\tau\rho} = (\zeta^{-1})^{\rho} = \zeta^{-l(\rho)}.$$

We note that $-l(\rho)$ is also a primitive root modulo p.

The authors have not yet verified whether or not there exists an F_p-extension other than these. However, it seems that there exist such F_p-extensions because the p-rank of the ideal class group of L_1 relates only to the p-rank of the χ^*-component (χ^* is a suitable character of $\mathrm{Gal}(L_1(\zeta)/k)$) of the ideal class group of L_2, as we will study in Section 4.

Using Theorem 1.5 we have, for $\delta \in \mathcal{M}(L_{\pi(1)}) \cap L_{\pi(2)}$,

$$E := F(L_{\pi(1)}, \delta) = L_{\pi(1)}(\Psi),$$

where

$$\Psi = \mathrm{Tr}_{E(\zeta)/E}(\sqrt[p]{\delta^{t'(L_{\pi(1)})}}).$$

One of our aims is to construct a polynomial of degree p with coefficients in k which generates E. As we see the following proposition, it can be constructed by making use of Ψ.

Proposition 2.5. *Let the notation be as in Theorem 2.1. For $\delta \in \mathcal{M}(L_{\pi(1)}) \cap L_{\pi(2)}$, we put $E = F(L_{\pi(1)}, \delta)$. For $t'(L_{\pi(1)}) \in T(L_{\pi(1)})$, let $\beta \in E(\zeta)$ be a root of $X^p - \delta^{t'(L_{\pi(1)})} = 0$ and put $\Psi := \mathrm{Tr}_{E(\zeta)/E}(\beta)$. Then we have*

$$[k(\Psi) : k] = [L_{\pi(1)}(\Psi) : L_{\pi(1)}] = p. \tag{2.1}$$

Moreover, the minimal polynomial of Ψ over k is irreducible over $L_{\pi(1)}$ and its minimal splitting field over k coincides with E.

Proof. For the first assertion it is sufficient to show $[k(\Psi) : k] = [L_{\pi(1)}(\Psi) : L_{\pi(1)}]$. Let τ be a generator of $\mathrm{Gal}(E(\zeta)/E)$ of degree $m_{\pi(2)}$. Then we have $\Psi = \sum_{i=0}^{m_{\pi(2)}-1} \beta^{\tau^i}$. Since $\delta^{t'(L_{\pi(1)})} \in L_{\pi(2)}$, $L_{\pi(2)}(\beta)/L_{\pi(2)}$ is of degree p and $E(\zeta)/L_{\pi(2)}(\beta)$ is cyclic of degree $m_{\pi(1)}$. We extend ρ to an automorphism of $E(\zeta)/L_{\pi(2)}(\beta)$. Since $E(\zeta) = E \cdot L_{\pi(2)}$ and $E \cap L_{\pi(2)} = k$, the map

$$\vartheta : \mathrm{Gal}(E(\zeta)/k) \longrightarrow \mathrm{Gal}(E/k) \times \mathrm{Gal}(L_{\pi(2)}/k)$$

defined by $\vartheta(\sigma) := (\sigma|_E,\ \sigma|_{L_{\pi(2)}})$ is an isomorphism. Then $\vartheta(\tau) = (1,\ \tau|_{L_{\pi(2)}})$ and $\vartheta(\rho) = (\rho|_E,\ 1)$. Therefore τ and ρ commute with each other. Then we get

$$\Psi^\rho = \left(\sum_{i=0}^{m_{\pi(2)}-1} \beta^{\tau^i}\right)^\rho = \sum_{i=0}^{m_{\pi(2)}-1} \beta^{\tau^i \rho} = \sum_{i=0}^{m_{\pi(2)}-1} \beta^{\rho\tau^i} = \sum_{i=0}^{m_{\pi(2)}-1} \beta^{\tau^i} = \Psi,$$

and hence Ψ is contained in the fixed field $E^{\langle\rho\rangle}$ of $\langle\rho\rangle$ of E. Since $k = L_{\pi(1)}^{\langle\rho\rangle}$, we have

$$[L_{\pi(1)}(\Psi) : L_{\pi(1)}] \le [k(\Psi) : k] \le [E^{\langle\rho\rangle} : k] \le [L_{\pi(1)}(\Psi) : L_{\pi(1)}]$$

Therefore we have $[k(\Psi):k]=[L_{\pi(1)}(\Psi):L_{\pi(1)}]$.

The second assertion is obvious from (2.1) together with the fact that the order of every proper normal subgroup of D_p or F_p is divisible by p. □

Note that we have the following diagram (Figure 2) by this proposition. When π is not trivial in Proposition 2.5, by using Proposition 1.10, we have the following corollary as a construction of F_p-polynomials.

Figure 2.

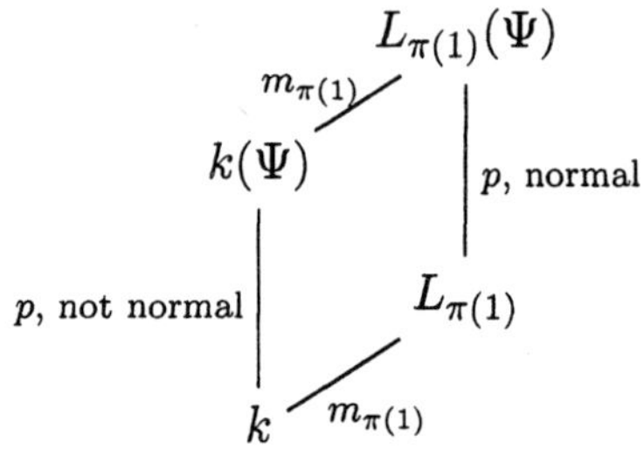

Corollary 2.6. *Let the notation be as in Theorem 2.1. Let $f_p(X;\delta)$ be a polynomial defined in Proposition 1.10, where $N=N_{L_1/k}$ and $\mathrm{Tr}=\mathrm{Tr}_{L_1/k}$. Then for $\delta\in\mathcal{M}(L_2)\cap L_1$, the minimal splitting field of $f_p(X;\delta)$ over k is an F_p-extension of k containing L_2. Conversely, every Galois extension E of k containing L_2 with the Galois group*

$$\mathrm{Gal}(E/k)=\langle\sigma,\iota\mid\sigma^p=\iota^{p-1}=1,\ \iota^{-1}\sigma\iota=\sigma^{l(\rho)}\rangle$$

with $\iota|_{L_2}=\rho|_{L_2}$ is given as the minimal splitting field of $f_p(X;\delta)$ over k for some $\delta\in\mathcal{M}(L_2)\cap L_1$.

In [Ki2] and [Ki3], by utilizing this corollary, Y. Kishi gave infinitely many cyclic fields of degree $p-1$ containing $\mathbb{Q}(\omega)$ whose class numbers are divisible by p for each odd prime p.

Theorem 2.7. *Let k be the rational number field $\mathbb{Q}$ and L_1 be a real quadratic field. Then the associated field L_2 with L_1 is an imaginary cyclic field of degree $p-1$. Assume that there exist a unit ε of L_1 with $\varepsilon\notin L_1^p$ which satisfies the conditions either*

$$\mathrm{Tr}_{L_1/\mathbb{Q}}(\varepsilon)\equiv 0 \pmod{p^2}$$

or

$$\begin{cases} N_{L_1/\mathbb{Q}}(\varepsilon)=1,\\ \mathrm{Tr}_{L_1/\mathbb{Q}}(\varepsilon)\equiv\pm 2 \pmod{p^3}. \end{cases}$$

Then the minimal splitting field of $f_p(X;\delta)$ over $\mathbb{Q}$ is an unramified cyclic extension of L_2 of degree p. Hence the class number of L_2 is divisible by p.

3. Examples

Let the notation be as in Section 1 and suppose that $\zeta \notin K$. First we construct C_p-polynomials over K for $p = 3$ or 5 explicitly. For $\gamma \in \mathcal{M}(K)$, we put $E := F(K,\gamma)$. For $t'(K) \in T(K)$, let $h_i(X;t'(K),\gamma)$ $(1 \le i \le 3)$ in the following examples denote the minimal polynomial of $\Psi := \mathrm{Tr}_{E(\zeta)/E}(\sqrt[p]{\gamma^{t'(K)}})$ over K.

Example 3.1. (cf. [Gr1, p. 182], [Ch, Proposition 2], [Co, Chapter 5, Examples (1), p. 253].) The case $p = 3$. We have $[K(\zeta) : K] = 2$. Since 2 is a primitive root modulo 3, we may take the generator τ of $\mathrm{Gal}(K(\zeta)/K)$ with $\zeta^\tau = \zeta^2$. Then we have $t(K) = 2 + \tau$, and hence

$$h_1(X;t(K),\gamma) = X^3 - 3N(\gamma)X - N(\gamma)\mathrm{Tr}(\gamma)$$

for $\gamma \in \mathcal{M}(K)$ where $N = N_{K(\zeta)/K}$ and $\mathrm{Tr} = \mathrm{Tr}_{K(\zeta)/K}$. This polynomial is immediately obtained from Proposition 1.10.

Example 3.2. (cf. [Co, Chapter 5, Examples (3), p. 254].) The case $p = 5$ and $[K(\zeta) : K] = 2$. Since 4 is of order 2 modulo 5, we may take the generator τ of $\mathrm{Gal}(K(\zeta)/K)$ with $\zeta^\tau = \zeta^4$. Therefore $t(K) = 4 + \tau$. Put $t'(K) = 3 + 2\tau$; then we have $t'(K) \in T(K)$ because $t'(K) \equiv 2t(K) \pmod 5$. Hence by Proposition 1.10, we have

$$h_2(X;t'(K),\gamma) = X^5 - 5N(\gamma)X^3 + 5N(\gamma)^2X - N(\gamma)^2\mathrm{Tr}(\gamma)$$

for $\gamma \in \mathcal{M}(K)$ where $N = N_{K(\zeta)/K}$ and $\mathrm{Tr} = \mathrm{Tr}_{K(\zeta)/K}$.

Example 3.3. (cf. [Ha, p. 433], [Gr1, p. 182], [Co, Chapter 5, Examples (2), p. 253].) The case $p = 5$ and $[K(\zeta) : K] = 4$. Since 2 is a primitive root modulo 5, we may take the generator τ of $\mathrm{Gal}(K(\zeta)/K)$ with $\zeta^\tau = \zeta^2$. Therefore $t(K) = 8 + 4\tau + 2\tau^2 + \tau^3$. Put $t'(K) = 3 + 4\tau + 2\tau^2 + \tau^3$; then we have $t'(K) \in T(K)$ because $t(K) \equiv t'(K) \pmod 5$. Hence we have

$$\begin{aligned}
&h_3(X;t'(K),\gamma)\\
&= X^5 - 10N_{K(\zeta)/K}(\gamma)X^3 - 5N_{K(\zeta)/K}(\gamma)N_{K(\omega)/K}(\mathrm{Tr}_{K(\zeta)/K(\omega)}(\gamma))X^2\\
&\quad + 5N_{K(\zeta)/K}(\gamma)\{N_{K(\zeta)/K}(\gamma) - N_{K(\omega)/K}(\mathrm{Tr}_{K(\zeta)/K(\omega)}(\gamma^{1+\tau}))\}X\\
&\quad - N_{K(\zeta)/K}(\gamma)N_{K(\omega)/K}(\mathrm{Tr}_{K(\zeta)/K(\omega)}(\gamma^{2+\tau}))
\end{aligned}$$

for $\gamma \in \mathcal{M}(K)$.

Next we give examples for the dihedral and Frobenius cases. From now on, let the notation be as in Section 2.

Example 3.4. Let k be an algebraic number field. Put $L_1 := k(\sqrt{d})$, $d \in k$. Then we have

$$L_2 = \begin{cases} k(\sqrt{-3d}), & \text{if } p = 3, \\ k\left(\sqrt{-10d + 2d\sqrt{5}}\right), & \text{if } p = 5. \end{cases}$$

(1) Take $K = L_{\pi(1)}$ in Example 3.1. Then by Proposition 2.5, we see that every $S_3(= D_3 = F_3)$-extension of k containing $L_{\pi(1)}$ is given by the roots of the polynomial $h_1(X; t(L_{\pi(1)}), \delta)$ for some $\delta \in \mathcal{M}(L_{\pi(1)}) \cap L_{\pi(2)}$, where $N = N_{L_{\pi(2)}/k}$ and $\mathrm{Tr} = \mathrm{Tr}_{L_{\pi(2)}/k}$.
(2) Take $K = L_2$ in Example 3.2. Since $p = 5 \equiv 1 \pmod 4$, as we have seen in Remark 2.4, there exist two quadratic extensions L_1 and $L_1' = k(\sqrt{5d})$ of k associated with L_2. On the other hand, there exist two primitive roots modulo 5. By Proposition 2.5, therefore, we see that every F_5-extension of k containing L_2 is given by the roots of the polynomial $h_2(X; t'(L_2), \delta)$ for some $\delta \in \mathcal{M}(L_2) \cap L_1$ or $\delta \in \mathcal{M}(L_2) \cap L_1'$, where $N = N_{L_1/k}$, $\mathrm{Tr} = \mathrm{Tr}_{L_1/k}$ or $N = N_{L_1'/k}$, $\mathrm{Tr} = \mathrm{Tr}_{L_1'/k}$.
(3) Take $K = L_1$ in Example 3.3. Then by Proposition 2.5, we see that every D_5-extension of k containing L_1 is given by the roots of the polynomial $h_3(X; t'(L_1), \delta)$ for some $\delta \in \mathcal{M}(L_1) \cap L_2$, where $N_{K(\zeta)/K} = N_{L_2/k}$, $\mathrm{Tr}_{K(\zeta)/K} = \mathrm{Tr}_{L_2/k}$ and $N_{K(\omega)/K}\mathrm{Tr}_{K(\zeta)/K(\omega)} = N_{k(\omega)/k}\mathrm{Tr}_{L_2/k(\omega)}$.

Example 3.5. H. Hasse gave a polynomial which generates an unramified cyclic quintic extension E of $L_1 = \mathbb{Q}(\sqrt{-47})$ and a D_5-extension of $\mathbb{Q}$ (see [Ha, p. 434]). This polynomial is given as our $h_3(-X; t'(L_1), \delta_1)$ with $\zeta = e^{2\pi\sqrt{-1}/5}$ and

$$\delta_1 = \frac{1}{2}\left((2 - \sqrt{5})\sqrt{-47} + \frac{25 - 11\sqrt{5}}{2}(\zeta - \zeta^{-1})\right).$$

It has integral coefficients, but δ_1 is not an element of $L_2 = \mathbb{Q}(\sqrt{-47}(\zeta - \zeta^{-1}))$. Take

$$\begin{aligned} \delta_2 &:= N_{L_1(\zeta)/L_2}(\delta_1) \\ &= \frac{1}{2}\left(424 - 189\sqrt{5} - \frac{105 - 47\sqrt{5}}{2}\sqrt{-47}(\zeta - \zeta^{-1})\right) \in L_2. \end{aligned}$$

Then we have $\delta_2 = -(\delta_1)^2$, and hence

$$h_3(X; t'(L_1), \delta_2) = X^5 - 10X^3 - 5855X^2 - 1368890X - 88365871$$

generates the same E over $\mathbb{Q}$ by Proposition 1.3.

On the basis of Example 3.4 (1), one of the authors, M. Imaoka [Im], will give a family of S_3-polynomials with negative discriminants whose roots are fundamental units of the cubic fields. We will state his results.

We note the fact that for every $\delta \in \mathbb{Q}(\sqrt{-3d})$, there exists a rational number $m \in \mathbb{Q}$ such that the minimal splitting field of $h_1(X; \mathbb{Q}(\sqrt{d}), \delta)$ coincides with the minimal splitting field of $h_1(X; \mathbb{Q}(\sqrt{d}), \delta')$ with

$$\delta' = \frac{1}{2} + \left(\frac{1}{9}(m-5) + \frac{1}{2}\right)\sqrt{-3d}.$$

Put $X = (Y-1)/3$ in $h_1(X; \mathbb{Q}(\sqrt{d}), \delta')$ and multiply it by 3^3. Then we have

$$f_{d,m}(Y) := Y^3 - 3Y^2 - (A(m) - 3)Y - 1,$$

where

$$A(m) := \frac{d(2m-1)^2 + 27}{4}.$$

Then the discriminant of $f_{d,m}(Y)$ is $d(2m-1)^2 A(m)^2$.

Assume that d is a square free negative integer with $d \equiv 1 \pmod 4$ and m is a positive integer. Then $A(m)$ is an rational integer. Moreover, we assume that $A(m)$ is cube-free. Then $f_{d,m}(Y)$ is irreducible over $\mathbb{Q}$ and has only one real root. We denote it by θ and put $K_{d,m} := \mathbb{Q}(\theta)$.

Theorem 3.6. *Let the notation and the assumptions be as above. Then θ is the fundamental unit of $K_{d,m}$ except for $(d,m) = (-3,3)$, $(-11,1)$, $(-23,1)$, $(-31,1)$. For exceptional cases, we have*

$$\theta = \begin{cases} \xi_{d,m}^2, & \text{if } (d,m) = (-3,3),\ (-11,1), \\ \xi_{d,m}^3, & \text{if } (d,m) = (-23,1),\ (-31,1), \end{cases}$$

and

$$g_{d,m}(X) = \begin{cases} X^3 + 3X^2 + 3X - 1, & \text{if } (d,m) = (-3,3), \\ X^3 - X^2 - X - 1, & \text{if } (d,m) = (-11,1), \\ X^3 - X - 1, & \text{if } (d,m) = (-23,1), \\ X^3 + X - 1, & \text{if } (d,m) = (-31,1), \end{cases}$$

where $\xi_{d,m}$ is the fundamental unit of $K_{d,m}$ and $g_{d,m}(X)$ is the minimal polynomial of $\xi_{d,m}$ over $\mathbb{Q}$.

By using of this theorem, we have

Theorem 3.7. *Under the same situation as that in Theorem 3.6, we assume in addition that $A(m)$ is square-free. Then there are exactly* *nine pairs (d, m) such that the class number of $K_{d,m}$ is one: their pairs are $(d, m) = (-3, 1)$, $(-7, 1)$, $(-15, 1)$, $(-19, 1)$, $(-23, 1)$, $(-31, 1)$, $(-35, 1)$, $(-39, 1)$, $(-47, 1)$.*

4. Appendix

Let p be a fixed odd prime and ζ be a primitive p-th root of unity.

For an algebraic number field K and an integral divisor $\mathfrak{f}$ of K, we denote the congruent ideal class group modulo $\mathfrak{f}$ of K by $\mathrm{Cl}_K(\mathfrak{f})$. For simplicity, we denote Cl_K if $\mathfrak{f} = 1$; that is, Cl_K is the absolute ideal class group of K. Moreover S_K denote the set of prime divisors lying over p in K, and G_{K,S_K} (resp. G^{ab}_{K,S_K}) denote the Galois group of the maximal extension (resp. maximal abelian extension) of K unramified outside S_K over K.

For a character χ of a finite abelian group G with $(\#G, p) = 1$ and an $\mathbb{F}_p[G]$-module M, we denote the χ-component of M by M^χ_G; that is, $M^\chi_G = e_\chi M$ where

$$e_\chi = \frac{1}{\#G} \sum_{\sigma \in G} \chi(\sigma)\sigma^{-1}.$$

Let L_1 be a quadratic field with $L_1 \cap \mathbb{Q}(\zeta) = \mathbb{Q}$ and L_2 be the associated field with L_1. (For the meaning of "associated field," see Section 2.) Let $\tilde{\Delta}$ be the Galois group of $L_1(\zeta)/\mathbb{Q}$, and χ be the character of $\tilde{\Delta}$ corresponding to L_1. Let $\Delta(\subset \tilde{\Delta})$ be the Galois group of $L_1(\zeta)/L_1$. Let ω be the Teichmüller character of Δ and put $\chi^* := \omega\chi^{-1}$. Then χ^* corresponds to L_2.

We define positive integers ρ and ρ' by

$$\rho := \dim_{\mathbb{F}_p}(G^{\mathrm{ab}}_{L_1,S_{L_1}}/pG^{\mathrm{ab}}_{L_1,S_{L_1}}),$$
$$\rho' := \dim_{\mathbb{F}_p}(\mathrm{Cl}_{L_2}/p\mathrm{Cl}_{L_2})^{\chi^*}_{\tilde{\Delta}}.$$

We note that by Class Field Theory we have

$$\rho = \dim_{\mathbb{F}_p}(\mathrm{Cl}_{L_1}(p^n)/p\mathrm{Cl}_{L_1}(p^n))$$

for sufficiently large n.

The aim of this appendix is to prove the following theorem:

Theorem 4.1 (M. Kurihara). *Let the notation and the assumptions be as above. Then we have*

$$\rho = \begin{cases} \rho' + \delta + 1, & \text{if } p \geq 5, \\ \rho' + \delta + \varepsilon + \eta + 1, & \text{if } p = 3, \end{cases}$$

where

$$\delta := \begin{cases} 0, & \text{if } L_1 \text{ is real,} \\ 1, & \text{if } L_1 \text{ is imaginary;} \end{cases} \qquad \varepsilon := \begin{cases} 0, & \text{if } [\mathfrak{P}] \in p\mathrm{Cl}_{L_2}, \\ -1, & \text{if } [\mathfrak{P}] \notin p\mathrm{Cl}_{L_2}; \end{cases}$$

$$\eta := \begin{cases} 0, & \text{if } \#S_{L_2} = 1, \\ 1, & \text{if } \#S_{L_2} = 2, \end{cases}$$

and $\mathfrak{P} \in S_{L_2}$.

Remark 4.2. A result of G. Gras [Gr2, Theorem 7.11] immediately leads to this theorem. However, we will give an another proof using cohomological approach. The first version of this paper had an erroneous choice of characters. The authors wish to express their gratitude to Professor G. Gras for pointing out the error which led to improvement of this appendix.

Proof of Theorem 4.1. First we introduce the following J. Tate's theorem on which Theorem 4.1 is based.

Theorem 4.3 ([Ta1], [Ta2, Theorem 2.2]). *Let K be an algebraic number field and M be a finite G_{K,S_K}-module. Assume that the order of M is a power of p. Then the following equation holds:*

$$\frac{\#H^0(G_{K,S_K}, M) \cdot \#H^2(G_{K,S_K}, M)}{\#H^1(G_{K,S_K}, M)} = \left(\frac{\#M^{\mathrm{Gal}(\mathbb{C}/\mathbb{R})}}{\#M}\right)^{r_1} \cdot \left(\frac{1}{\#M}\right)^{r_2}, \tag{4.1}$$

where r_1 and r_2 denote the number of real and imaginary embeddings of K into $\mathbb{C}$, respectively.

Let us apply Theorem 4.3 to the case $K = L_1$ and $M = \mathbb{Z}/p\mathbb{Z}$. Then the Galois group $G_{L_1,S_{L_1}}$ acts trivially on $\mathbb{Z}/p\mathbb{Z}$, and the following relations hold:

$$\begin{aligned} H^0(G_{L_1,S_{L_1}}, \mathbb{Z}/p\mathbb{Z}) &= \mathbb{Z}/p\mathbb{Z}, \\ H^1(G_{L_1,S_{L_1}}, \mathbb{Z}/p\mathbb{Z}) &= \mathrm{Hom}_{conti}(G_{L_1,S_{L_1}}, \mathbb{Z}/p\mathbb{Z}) \end{aligned}$$

(cf. [Se, Chapter I, 2.3]). Then we have

$$\#H^0(G_{L_1,S_{L_1}}, \mathbb{Z}/p\mathbb{Z}) = p, \tag{4.2}$$

$$\#H^1(G_{L_1,S_{L_1}}, \mathbb{Z}/p\mathbb{Z}) = \#(\mathbb{Z}/p\mathbb{Z})^{\oplus\rho} = p^\rho. \tag{4.3}$$

From now on, we calculate $\#H^2(G_{L_1,S_{L_1}}, \mathbb{Z}/p\mathbb{Z})$. Since the order of Δ is relatively prime to p and $\mathbb{Z}/p\mathbb{Z}$ is isomorphic to the group μ_p of the

p-th roots of unity as a $G_{L_1(\zeta),S_{L_1(\zeta)}}$-module, we have

$$\begin{aligned} H^2(G_{L_1,S_{L_1}}, \mathbb{Z}/p\mathbb{Z}) &= H^2(G_{L_1(\zeta),S_{L_1(\zeta)}}, \mathbb{Z}/p\mathbb{Z})^{\Delta} \\ &= (H^2(G_{L_1(\zeta),S_{L_1(\zeta)}}, \mu_p) \otimes \mu_p^{\otimes(-1)})^{\Delta}, \end{aligned} \tag{4.4}$$

where $\mu_p^{\otimes -1}$ satisfies $\mu_p \otimes \mu_p^{\otimes -1} = \mathbb{Z}/p\mathbb{Z}$.

Denote the ring of integers of $L_1(\zeta)$ by $\mathcal{O}_{L_1(\zeta)}$. Then we have

$$H^q(G_{L_1(\zeta),S_{L_1(\zeta)}}, \mu_p) = H^q_{\text{et}}(\text{Spec } \mathcal{O}_{L_1(\zeta)}[\frac{1}{p}], \mu_p) \tag{4.5}$$

for each $q \geq 0$, where H^q_{et} is an étale cohomology. For simplicity, we denote

$$H^q_{\text{et}}(\text{Spec } \mathcal{O}_{L_1(\zeta)}[\frac{1}{p}], *) =: H^q(*).$$

For an étale cohomology, the following lemma is well-known.

Lemma 4.4. *Let $\mathbb{G}_m$ be the multiplicative group. Then we have an exact sequence*:

$$0 \longrightarrow H^1(\mathbb{G}_m)/pH^1(\mathbb{G}_m) \longrightarrow H^2(\mu_p) \longrightarrow H^2(\mathbb{G}_m)[p] \longrightarrow 0. \tag{4.6}$$

Proof. We have an exact sequence:

$$0 \longrightarrow \mu_p \longrightarrow \mathbb{G}_m \xrightarrow{[p]} \mathbb{G}_m \longrightarrow 0,$$

where $[p]$ denotes the endomorphism $x \mapsto x^p$. From this, we have the long exact sequence

$$H^1(\mathbb{G}_m) \xrightarrow{[p]} H^1(\mathbb{G}_m) \longrightarrow H^2(\mu_p) \longrightarrow H^2(\mathbb{G}_m) \xrightarrow{[p]} H^2(\mathbb{G}_m),$$

and hence we get the exact sequence (4.6). □

We go back to the proof of Theorem 4.1. We have

$$H^1(\mathbb{G}_m) \simeq \text{Pic}(\mathcal{O}_{L_1(\zeta)}[\frac{1}{p}]) \simeq \text{Cl}_{L_1(\zeta)}/\langle [\mathfrak{P}] \mid \mathfrak{P} \in S_{L_1(\zeta)} \rangle =: \text{Cl}_{L_1(\zeta),S_{L_1(\zeta)}}$$

(cf. [Mil2, Chapter II, Proposition 2.1]). Moreover we have

$$H^2(\mathbb{G}_m) \simeq \text{Br}(\mathcal{O}_{L_1(\zeta)}[\frac{1}{p}])$$

(cf. [Mil1, Chapter IV]). By Class Field Theory, we have

$$\begin{aligned} H^2(\mathbb{G}_m)[p] &\simeq \ker(\text{Br}(L_1(\zeta)) \longrightarrow \bigoplus_{v \in S_{L_1(\zeta)}} \text{Br}(L_1(\zeta)_v))[p] \\ &= \ker(\bigoplus_{v \in S_{L_1(\zeta)}} \mathbb{Z}/p\mathbb{Z} \xrightarrow{\Sigma} \mathbb{Z}/p\mathbb{Z}), \end{aligned}$$

where Σ is the map $(a_v) \mapsto \sum_{v \in S_{L_1(\zeta)}} a_v$. Then by Lemma 4.4, we have an exact sequence

$$0 \longrightarrow \mathrm{Cl}_{L_1(\zeta),S_{L_1(\zeta)}}/p\mathrm{Cl}_{L_1(\zeta),S_{L_1(\zeta)}} \longrightarrow H^2(\mu_p) \longrightarrow \ker(\bigoplus_{v \in S_{L_1(\zeta)}} \mathbb{Z}/p\mathbb{Z} \xrightarrow{\Sigma} \mathbb{Z}/p\mathbb{Z}) \longrightarrow 0.$$

Then the sequence

$$0 \longrightarrow ((\mathrm{Cl}_{L_1(\zeta),S_{L_1(\zeta)}}/p\mathrm{Cl}_{L_1(\zeta),S_{L_1(\zeta)}}) \otimes \mu_p^{\otimes -1})^{\Delta} \longrightarrow (H^2(\mu_p) \otimes \mu_p^{\otimes -1})^{\Delta} \longrightarrow (\ker(\bigoplus_{v \in S_{L_1(\zeta)}} \mathbb{Z}/p\mathbb{Z} \xrightarrow{\Sigma} \mathbb{Z}/p\mathbb{Z}) \otimes \mu_p^{\otimes -1})^{\Delta} \longrightarrow 0 \quad (4.7)$$

is also exact.

Now in general for an $\mathbb{F}_p[\tilde{\Delta}]$-module M, we have

$$(M \otimes \mu_p^{\otimes -1})^{\Delta} = \bigoplus_{i=0}^{p-2} (M_{\Delta}^{\omega^i} \otimes \mu_p^{\otimes -1})^{\Delta} = M_{\Delta}^{\omega} \otimes \mu_p^{\otimes -1} = (M_{\tilde{\Delta}}^{\omega} \oplus M_{\tilde{\Delta}}^{\chi^*}) \otimes \mu_p^{\otimes -1}.$$

Therefore we have

$$((\mathrm{Cl}_{L_1(\zeta),S_{L_1(\zeta)}}/p\mathrm{Cl}_{L_1(\zeta),S_{L_1(\zeta)}}) \otimes \mu_p^{\otimes -1})^{\Delta} = T \otimes \mu_p^{\otimes -1},$$
$$(\ker(\bigoplus_{v \in S_{L_1(\zeta)}} \mathbb{Z}/p\mathbb{Z} \xrightarrow{\Sigma} \mathbb{Z}/p\mathbb{Z}) \otimes \mu_p^{\otimes -1})^{\Delta} = U \otimes \mu_p^{\otimes -1},$$

where

$$T = (\mathrm{Cl}_{L_1(\zeta),S_{L_1(\zeta)}}/p\mathrm{Cl}_{L_1(\zeta),S_{L_1(\zeta)}})_{\tilde{\Delta}}^{\omega} \oplus (\mathrm{Cl}_{L_1(\zeta),S_{L_1(\zeta)}}/p\mathrm{Cl}_{L_1(\zeta),S_{L_1(\zeta)}})_{\tilde{\Delta}}^{\chi^*},$$
$$U = (\ker(\bigoplus_{v \in S_{L_1(\zeta)}} \mathbb{Z}/p\mathbb{Z} \xrightarrow{\Sigma} \mathbb{Z}/p\mathbb{Z}))_{\tilde{\Delta}}^{\omega} \oplus (\ker(\bigoplus_{v \in S_{L_1(\zeta)}} \mathbb{Z}/p\mathbb{Z} \xrightarrow{\Sigma} \mathbb{Z}/p\mathbb{Z}))_{\tilde{\Delta}}^{\chi^*}.$$

From this, together with (4.4), (4.5), and the exact sequence (4.7), we have

$$\begin{aligned} \# H^2(G_{L_1,S_{L_1}}, \mathbb{Z}/p\mathbb{Z}) &= \#(H^2(\mu_p) \otimes \mu_p^{\otimes -1})^{\Delta} \\ &= \#(T \otimes \mu_p^{\otimes -1}) \cdot \#(U \otimes \mu_p^{\otimes -1}) \\ &= \#T \cdot \#U. \end{aligned} \quad (4.8)$$

Let us calculate $\#T$. Since $(\mathrm{Cl}_{\mathbb{Q}(\zeta),S_{\mathbb{Q}(\zeta)}}/p\mathrm{Cl}_{\mathbb{Q}(\zeta),S_{\mathbb{Q}(\zeta)}})_{\tilde{\Delta}}^{\chi^*} = 0$, we have

$$T \simeq (\mathrm{Cl}_{L_2,S_{L_2}}/p\mathrm{Cl}_{L_2,S_{L_2}})_{\tilde{\Delta}}^{\chi^*}.$$

Let $\mathfrak{P}$ be a prime divisor lying over p in L_2: $\mathfrak{P} \in S_{L_2}$.

The case $[\mathfrak{P}] \in p\mathrm{Cl}_{L_2}$. It follows from $\mathrm{Cl}_{L_2,S_{L_2}} = \mathrm{Cl}_{L_2}/\langle[\mathfrak{P}]\rangle$ that $\mathrm{Cl}_{L_2,S_{L_2}}/p\mathrm{Cl}_{L_2,S_{L_2}} = \mathrm{Cl}_{L_2}/p\mathrm{Cl}_{L_2}$. Hence we have

$$\#T = \#(\mathrm{Cl}_{L_2}/p\mathrm{Cl}_{L_2})_{\tilde{\Delta}}^{\chi^*} = p^{\rho'}.$$

The case $[\mathfrak{P}] \notin p\mathrm{Cl}_{L_2}$. Since

$$0 \longrightarrow (\langle[\mathfrak{P}]\rangle/\langle[\mathfrak{P}]\rangle \cap p\mathrm{Cl}_{L_2})_{\tilde{\Delta}}^{\chi^*} \longrightarrow (\mathrm{Cl}_{L_2}/p\mathrm{Cl}_{L_2})_{\tilde{\Delta}}^{\chi^*} \longrightarrow (\mathrm{Cl}_{L_2,S_{L_2}}/p\mathrm{Cl}_{L_2,S_{L_2}})_{\tilde{\Delta}}^{\chi^*} \longrightarrow 0$$

is exact and

$$\#(\langle[\mathfrak{P}]\rangle/\langle[\mathfrak{P}]\rangle \cap p\mathrm{Cl}_{L_2})_{\tilde{\Delta}}^{\chi^*} = \begin{cases} 1, & \text{if } p \geq 5, \\ p, & \text{if } p = 3, \end{cases}$$

we have

$$\#T = \begin{cases} \#(\mathrm{Cl}_{L_2}/p\mathrm{Cl}_{L_2})_{\tilde{\Delta}}^{\chi^*} = p^{\rho'}, & \text{if } p \geq 5, \\ p^{-1} \cdot \#(\mathrm{Cl}_{L_2}/p\mathrm{Cl}_{L_2})_{\tilde{\Delta}}^{\chi^*} = p^{\rho'-1}, & \text{if } p = 3. \end{cases}$$

Next, we calculate $\#U$. Since p is totally ramified in $\mathbb{Q}(\zeta)/\mathbb{Q}$, we have $\#S_{L_2} = 1$ or 2 and

$$U \simeq (\ker(\bigoplus_{v \in L_2} \mathbb{Z}/p\mathbb{Z} \xrightarrow{\Sigma'} \mathbb{Z}/p\mathbb{Z}))_{\tilde{\Delta}}^{\chi^*},$$

where Σ' is the map $(a_v) \mapsto \sum_{v \in S_{L_2}} a_v$.

The case $\#S_{L_2} = 1$. Since $\ker(\mathbb{Z}/p\mathbb{Z} \xrightarrow{\Sigma'} \mathbb{Z}/p\mathbb{Z}) = 0$, we have $\#U = \#(0)_{\tilde{\Delta}}^{\chi^*} = 1$.

The case $\#S_{L_2} = 2$. Let τ be the generator of Δ. In this case, we have $\tau(\mathfrak{P}) = \mathfrak{P}'$ with $\mathfrak{P}, \mathfrak{P}' \in S_{L_2}$, $\mathfrak{P} \neq \mathfrak{P}'$. Therefore τ acts on $(\ker(\mathbb{Z}/p\mathbb{Z} \oplus \mathbb{Z}/p\mathbb{Z} \xrightarrow{\Sigma'} \mathbb{Z}/p\mathbb{Z}))_{\tilde{\Delta}}^{\chi^*}$ by multiplication by -1. Then we have

$$(\ker(\bigoplus_{v \in L_2} \mathbb{Z}/p\mathbb{Z} \xrightarrow{\Sigma'} \mathbb{Z}/p\mathbb{Z}))_{\tilde{\Delta}}^{\chi^*} = (\mathbb{Z}/p\mathbb{Z})_{\tilde{\Delta}}^{\chi^*} = \begin{cases} 0, & \text{if } p \geq 5, \\ \mathbb{Z}/p\mathbb{Z}, & \text{if } p = 3, \end{cases}$$

and hence

$$\#U = \begin{cases} 1, & \text{if } p \geq 5, \\ p, & \text{if } p = 3. \end{cases}$$

For the right side hand of (4.1), we can get

$$\frac{\#(\mathbb{Z}/p\mathbb{Z})^{\mathrm{Gal}(\mathbb{C}/\mathbb{R})}}{\#(\mathbb{Z}/p\mathbb{Z})} = \frac{\#(\mathbb{Z}/p\mathbb{Z})}{\#(\mathbb{Z}/p\mathbb{Z})} = 1,$$
$$\frac{1}{\#(\mathbb{Z}/p\mathbb{Z})} = \frac{1}{p}.$$

From this, together with (4.1), (4.2), (4.3), and (4.8), we have

$$\frac{p \cdot \#T \cdot \#U}{p^{\rho}} = \begin{cases} 1, & \text{if } L_1 \text{ is real,} \\ p^{-1}, & \text{if } L_1 \text{ is imaginary.} \end{cases}$$

Hence by

$$\#T = \begin{cases} p^{\rho'}, & \text{if } p \geq 5 \text{ or } [\mathfrak{P}] \in p\mathrm{Cl}_{L_2}, \\ p^{\rho'-1}, & \text{if } p = 3 \text{ and } [\mathfrak{P}] \notin p\mathrm{Cl}_{L_2}; \end{cases}$$
$$\#U = \begin{cases} 1, & \text{if } p \geq 5 \text{ or } \#S_{L_2} = 1, \\ p, & \text{if } p = 3 \text{ and } \#S_{L_2} = 2, \end{cases}$$

as we saw above, we obtain the desired conclusion. □

Acknowledgments

The authors would like to thank Professor Ryotaro Okazaki whose suggestions led to the improvement of this paper. They would also like to thank Professor Masato Kurihara for his suggestions and comments on Section 4.

References

[Ch] R. J. Chapman, *Automorphism polynomials in cyclic cubic extensions*, J. Number Theory **61** (1996), 283–291.

[Co] H. Cohen, "Advanced topics in computational number theory," Graduate Texts in Mathematics, 193. Springer-Verlag, New York, 2000.

[Gr1] G. Gras, *Extensions abéliennes non ramifiées de degré premier d'un corps quadratique*, Bull. Soc. Math. France **100** (1972), 177–193.

[Gr2] G. Gras, *Théorèmes de réflexion*, J. Théor. Nombres Bordeaux **10** (1998), 399–499.

[Ha] H. Hasse, *Über den Klassenkörper zum quadratischen Zahlkörper mit der Diskriminante* −47, Acta Arith. **9** (1964), 419–434.

[Im] M. Imaoka, *Fundamental units in a family of cubic fields with negative discriminants*, preprint.

[Ki1] Y. Kishi, *A constructive approach to Spiegelung relations between 3-ranks of absolute ideal class groups and congruent ones modulo* $(3)^2$ *in quadratic fields*, J. Number Theory **83** (2000), 1–49.

[Ki2] Y. Kishi, *Imaginary cyclic fields of degree* $p-1$ *whose relative class numbers are divisible by* p, Proc. Japan Acad. **77A** (2001), 55–58.

[Ki3] Y. Kishi, *On imaginary cyclic fields of degree* $p-1$ *which have* p*-ranks of the ideal class groups greater than one*, preprint.

[Mik] H. Miki, *On* $\mathbb{Z}_p$*-extensions of complete* p*-adic power series fields and function fields*, J. Fac. Sci. Univ. Tokyo Sect. IA Math. **21** (1974), 377–393.

[Mil1] J. S. Milne, "Étale cohomology," Princeton Mathematical Series, 33. Princeton University Press, Princeton, N.J., 1980.

[Mil2] J. S. Milne, "Arithmetic duality theorems," Perspectives in Mathematics, 1. Academic Press, Inc., Boston, MA, 1986.

[Sc] A. Scholz, *Über die Beziehung der Klassenzahl quadratischer Körper zueinander*, J. Reine Angew. Math. **166** (1932), 201–203.

[Se] J.-P. Serre, "Galois cohomology," Translated from the French by Patrick Ion and revised by the author. Corrected reprint of the 1997 English edition. Springer Monographs in Mathematics, Springer-Verlag, Berlin, 2002.

[Ta1] J. Tate, *Duality theorems in Galois cohomology over number fields*, in "Proc. Internat. Congr. Mathematicians (Stockholm, 1962)," 288–295, Inst. Mittag-Leffler, Djursholm, 1963.

[Ta2] J. Tate, *On the conjectures of Birch and Swinnerton-Dyer and a geometric analog*, in "Séminaire Bourbaki, Vol. 9," Exp. No. 306, 415–440, Soc. Math. France, Paris, 1995.

Galois Theory and Modular Forms pp 221–232
K. Hashimoto, K. Miyake, and H. Nakamura, Editors
© 2003 Kluwer Academic Publishers

ON THE NON-EXISTENCE OF CERTAIN GALOIS EXTENSIONS

Hyunsuk MOON*
Graduate School of Mathematics
Kyushu University 33
Fukuoka 812-8581, Japan
moon@math.kyushu-u.ac.jp

Introduction

In this article, we give a survey of my results on the non-existence and finiteness of certain Galois extensions of the rational number field $\mathbb{Q}$ with prescribed ramification. The detail has been (will be) published in [8], [9], [10], [11], [12].

First, we begin with Serre's conjecture, which is our motivation, and explain how this conjecture implies the non-existence and finiteness of certain Galois extensions of $\mathbb{Q}$. Second, we recall what is known on the non-existence together with the main ingredients in the proofs, and sketch the proof of some of my results presented at the Conference. Third, we state a generalized finiteness problem and summarize my results on this problem.

1. Serre's conjecture

Let $G_{\mathbb{Q}}$ be the absolute Galois group $\mathrm{Gal}(\overline{\mathbb{Q}}/\mathbb{Q})$ and $\overline{\mathbb{F}}_p$ an algebraic closure of the finite field $\mathbb{F}_p$ of p elements. Serre ([14]) conjectured that every odd and irreducible continuous Galois representation $\rho : G_{\mathbb{Q}} \to \mathrm{GL}_2(\overline{\mathbb{F}}_p)$ would be isomorphic to the representaion ρ_f associated to a mod p cuspidal eigenform f. More strongly, Serre also conjectured that such an eigenform exists of a certain "minimal" type $(N(\rho), k(\rho), \varepsilon(\rho))$.

Now we recall the definition of $N(\rho)$ and $k(\rho)$. Consider a representation $\rho : G_{\mathbb{Q}} \to \mathrm{GL}(V) \simeq \mathrm{GL}_2(\overline{\mathbb{F}}_p)$, where V is a two-dimensional vector space over $\overline{\mathbb{F}}_p$. Let $K/\mathbb{Q}$ be a finite Galois extension such that ρ factors

*The author was supported by the JSPS Postdoctoral Fellowship for Foreign Researchers.
[received: May 26, 2002; accepted: Oct. 14, 2002]

through $\mathrm{Gal}(K/\mathbb{Q})$. Then *the Artin conductor* $N(\rho)$ *outside* p is defined as follows:

$$N(\rho) := \prod_{l \neq p} l^{n(l,\rho)},$$

where l runs through the prime numbers different from p, and for each l,

$$n(l,\rho) := \sum_{i=0}^{\infty} \frac{1}{(G_0 : G_i)} \dim_{\overline{\mathbb{F}}_p}(V/V^{G_i}),$$

where V^{G_i} is the fixed part of V by the ith ramification subgroup G_i of the decomposition group of a prime of K lying above l. Hence the level $N(\rho)$ reflects the ramification away from p. For example, $n(l,\rho) = 0$ means that ρ is unramified at l and $n(l,\rho) = 1$ means ρ is tamely ramified at l.

The weight $k(\rho)$ depends only on the restriction of ρ to some inertia subgroup $I_p \subset G_{\mathbb{Q}}$ at p. Hence the weight reflects the ramification at p. Let χ be the mod p cyclotomic character. If ρ is wildly ramified at p, then its restriction to I_p has the following form:

$$\rho|_{I_p} \sim \begin{pmatrix} \chi^{\beta} & * \\ 0 & \chi^{\alpha} \end{pmatrix} \qquad \text{with} \quad * \neq 0, \tag{1.1}$$

where $\sim$ denotes the equivalence relation of representations of I_p. Take the integers α and β (uniquely) so that $0 \leq \alpha \leq p-2$ and $1 \leq \beta \leq p-1$. We set $a = \min(\alpha,\beta)$, $b = \max(\alpha,\beta)$, and define

$$k(\rho) := \begin{cases} 1 + pa + b + p - 1 & \text{if } \beta - \alpha = 1 \text{ and } \rho \otimes \chi^{-\alpha} \text{ is not finite,} \\ 1 + pa + b & \text{otherwise.} \end{cases}$$

For other cases, see [4], [14] (if $p = 2$, we adopt the definition in [4] rather than the original one of Serre [14]). According to the definition, we have $k(\rho) \leq p^2 - 1$. After twisting ρ by a power of the cyclotomic character, we can reduce the weight so that $k(\rho) \leq p + 1$; indeed if we put $\tilde{k}(\rho) := k(\rho \otimes \chi^{-\alpha})$ with α as above, then we have $\tilde{k}(\rho) \leq p + 1$.

This is a very strong conjecture, and there are many consequences of it. In particular, we are interested in the following ones:

• For $p = 2, 3, 5, 7$, there are no Galois representations $\rho : G_{\mathbb{Q}} \to \mathrm{GL}_2(\overline{\mathbb{F}}_p)$ which are odd, irreducible and unramified outside p, since cusp forms for $\mathrm{SL}_2(\mathbb{Z})$ do not exist for $k < 12$.

• For a given integer N, there exist only finitely many Galois representations $\rho : G_{\mathbb{Q}} \to \mathrm{GL}_2(\overline{\mathbb{F}}_p)$ up to isomorphism which are odd, irreducible and $N(\rho)|N$, since normalized eigenforms for $\Gamma_1(N)$ are finite in number if their weights are bounded.

In the following sections 2 and 3, we will treat the relating subjects to the above assertions respectively.

2. The non-existence

We recall what is known about the above-mentioned non-existence implied by Serre's conjecture. Tate ([16]) showed the non-existence in the case of $p = 2$, and Serre also remarked that this can be extended to the case of $p = 3$ by the same method. Recently Brueggeman ([2]) showed the non-existence in the case of $p = 5$ assuming the Generalized Riemann Hypothesis (GRH). All these results are proved basically by the same principle. That is by comparing two esimates for discriminants of opposite directions. We let $K/\mathbb{Q}$ be the corresponding field to $\mathrm{Ker}(\rho)$. Using class field theory, we estimate from above the discriminant of the field K. On the other hand, we have a lower bound of the discriminant by Odlyzko. Then the non-existence follows from the contradiction of these two inequalities. This method is vital. For example, using this method, Fontaine ([5]) showed the non-existence of abelian schemes over $\mathbb{Z}$. Along this line, we show the non-existence of some representations having non-trivial Artin conductor $N(\rho)$ outside p ([10]):

Theorem 1. (1) *There exist no non-trivial Galois representations $\rho : G_{\mathbb{Q}} \to \mathrm{GL}_2(\overline{\mathbb{F}}_2)$ which are semisimple and have $N(\rho)$ dividing 3.*
(2) *Assume the GRH. Let $\rho : G_{\mathbb{Q}} \to \mathrm{GL}_2(\overline{\mathbb{F}}_3)$ be a non-trivial semisimple representation with $N(\rho)$ dividing 2. Then $\rho \simeq 1 \oplus \chi_3$ or $\chi_3 \oplus \chi_3$, where χ_3 is the mod 3 cyclotomic character.*

By using a discriminant estimate, it is impossible (even under the GRH) to prove the conjecture of Serre for $p = 7$ with $N(\rho) = 1$ (cf. [2]). However, for *even* mod 7 representations, we can show:

Theorem 2. *Assume the GRH. Let $\rho : G_{\mathbb{Q}} \to \mathrm{GL}_2(\overline{\mathbb{F}}_7)$ be a semisimple even representation which is unramified outside 7. Then $\rho \simeq \chi_7^a \oplus \chi_7^b$ with $a + b =$ even, where χ_7 is the mod 7 cyclotomic character.*

Here, we give a sketch of the proof of Theorem 1(2). Our proof is divided into two parts. If $\mathrm{Im}(\rho)$ is solvable, we use a structure theorem on solvable subgroups of $\mathrm{GL}_2(\overline{\mathbb{F}}_p)$ and class field theory. Then we find that our representation is abelian. If $\mathrm{Im}(\rho)$ is non-solvable, we compare two different estimations of the discriminant of the kernel field of ρ to deduce contradiction.

Let $\rho : G_{\mathbb{Q}} \to \mathrm{GL}_2(\overline{\mathbb{F}}_3)$ with $N(\rho) \mid 2$. We first note that $N(\rho) = 2$ means that we allow tame ramification at 2. Let $K/\mathbb{Q}$ be the fixed field by $\mathrm{Ker}(\rho)$ and $G = \mathrm{Im}(\rho)$ its Galois group.

• *Solvable case:*

Suppose G is solvable. When G is a reducible subgroup of $\mathrm{GL}_2(\overline{\mathbb{F}}_3)$, simisimplicity implies that the non-trivial representations are precisely those listed in Theorem 1(2).

When G is an irreducible subgroup, we use the follwing Lemma ([15]):

Lemma 3. *Let G be an irreducible solvable subgroup of $\mathrm{GL}_2(\overline{\mathbb{F}}_p)$ and H its maximal abelian normal subgroup. Let A be a normal subgroup of G such that A/H is a maximal abelian normal subgroup of G/H. Then G is either of the following types:*

$$1 \longrightarrow H \longrightarrow G \longrightarrow \mathbb{Z}/2 \longrightarrow 1, \quad H \subset (\overline{\mathbb{F}}_p^\times)^2 \tag{G1}$$

or

$$\begin{aligned} &1 \longrightarrow A \longrightarrow G \longrightarrow \overline{G} \longrightarrow 1, \quad \overline{G} \subset S_3\ (\simeq \mathrm{SL}_2(\mathbb{F}_2)), \\ &1 \longrightarrow H \longrightarrow A \longrightarrow \overline{A} \longrightarrow 1, \quad H \subset \overline{\mathbb{F}}_p^\times, \quad \overline{A} \subset (\mathbb{Z}/2)^2. \end{aligned} \tag{G2}$$

Furthermore, if G is maximal, then all the inclusions are the equalities. If $p = 2$, only type (G1) is possible. Also, $\overline{G}$ acts on $\overline{A}$ by conjugation, and $\overline{G}$ injects into $\mathrm{Aut}(\overline{A})$, since $\overline{A}$ is its own centralizer in G/H.

Apply this Lemma to our G. If G is of type (G1), then K is an abelian extension of degree prime to 3 over the quadratic field $F = \mathbb{Q}(\sqrt{-3})$. This field F has class number 1 and $\mathcal{O}_{F,3}^\times = \mathbb{F}_3^\times \times$ (pro-3 group), $k_{F,2}^\times = \mathbb{F}_4^\times$, $\mathcal{O}_F^\times \simeq \mathbb{Z}/6$. Here $\mathcal{O}_F$ is the ring of integers of F, $\mathcal{O}_{F,p} := \mathcal{O}_F \otimes_{\mathbb{Z}} \mathbb{Z}_p$ and $k_{F,p}^\times$ is the maximal prime-to-p quotient of $\mathcal{O}_{F,p}^\times$. Hence, by class field theory, there is no abelian extension K/F of degree prime to 3. Thus $K = F = \mathbb{Q}(\sqrt{-3})$.

Next, we suppose that G is of type (G2). The quotient group $\overline{G}$ of G is contained in S_3. We consider each of the cases $\overline{G} = S_3$, C_3, C_2 and trivial. For example, if $\overline{G} = C_3$, then we have just one C_3-extension of $\mathbb{Q}$, that is $\mathbb{Q}(\zeta_9 + \zeta_9^{-1})$. This field does not have a $(\mathbb{Z}/2)^2$-extension unramified outside 3. Indeed, such an extension would produce a 2-extension of $\mathbb{Q}(\zeta_9)$ unramified outside 3, which does not exist. Since $\overline{G} \hookrightarrow \mathrm{Aut}(\overline{A})$, it is impossible that $\overline{G} = C_3$. Hence G is abelian.

• *Non-solvable case:*

First, we introduce two estimates for discriminants of opposite directions:

Lemma 4 (Tate, [16]). *Let $K/\mathbb{Q}$ be a finite Galois extension of degree n with Galois group G. Suppose that there is an embedding $\rho : G \hookrightarrow \mathrm{GL}_2(\overline{\mathbb{F}}_p)$. Then the p-part $d_{K/\mathbb{Q}}^{(p)}$ of the discriminant $d_{K/\mathbb{Q}}$ of $K/\mathbb{Q}$ satisfies the following estimates:*

(i) *If $K/\mathbb{Q}$ is at most tamely ramified, then*

$$|d_{K/\mathbb{Q}}^{(p)}| < p^n.$$

(ii) *If $K/\mathbb{Q}$ has wild ramification of degree p^m, $m \geq 1$, then*

$$|d_{K/\mathbb{Q}}^{(p)}| \leq p^{n(2+1/p-1/(p-1)p^{m-1})}. \tag{2.1}$$

On the other hand, we have

Lemma 5 (Odlyzko bound, [13]). *Let K be an algebraic number field of degree n, with r_1 real conjugate fields and $2r_2$ complex conjugate fields. Then*

$$\frac{1}{n}\log|d_K| \geq \begin{cases} \gamma + \log 4\pi + 1 - 8.317302n^{-\frac{2}{3}} & \textit{if } r_1 = n, \\ \gamma + \log 4\pi - 6.860404n^{-\frac{2}{3}} & \textit{if } r_1 = 0. \end{cases} \tag{2.2}$$

Assume the GRH. Then

$$\frac{1}{n}\log|d_K| \geq \gamma + \log 8\pi + \frac{r_1}{n}\frac{\pi}{2} - \frac{2\pi^2(\lambda + \frac{r_1}{n}\beta)}{(\log n)^2} - \frac{16\pi^2(1+\frac{1}{n})}{(\log n)^3(1+\frac{\pi^2}{(\log n)^2})^2}, \tag{2.3}$$

where $\lambda = 1.0517997903$, $\beta = 0.9689461463$. Note that these bounds are monotone increasing in n if either $r_1 = n$ or 0.

We assume $G = \mathrm{Gal}(K/\mathbb{Q})$ is non-solvable. Then $|G| \geq 60$. From (2.2) for $n = 60$, we know that $|d_K|^{1/n} > 14.3050$. If K is tamely ramified at 3, by Lemma 4 we have $|d_K|^{1/n} < 3 \cdot 2 = 6$. Contradiction. If K has wild ramification, we consider the composite map $\bar{\rho} : G_{\mathbb{Q}} \to \mathrm{GL}_2(\overline{\mathbb{F}}_p) \to \mathrm{PGL}_2(\overline{\mathbb{F}}_p)$. We let $\widetilde{G}$ be the image of G and $L/\mathbb{Q}$ be the field corresponding to $\mathrm{Ker}(\bar{\rho})$. Then $\widetilde{G}$ is also non-solvable. Suppose $L/\mathbb{Q}$ has wild ramification of index p^m. Then by [3], we have

Lemma 6. *Let $\widetilde{G}$ be a subgroup of $\mathrm{PGL}_2(\overline{\mathbb{F}}_p)$. If its p-Sylow subgroups have order p^m, $m \geq 1$, and are not normal in $\widetilde{G}$, then $\widetilde{G}$ is isomorphic to either $\mathrm{PGL}_2(\mathbb{F}_{p^m})$ or $\mathrm{PSL}_2(\mathbb{F}_{p^m})$.*

Apply this Lemma to our $\widetilde{G} \subset \mathrm{PGL}_2(\overline{\mathbb{F}}_3)$. Note that we have $m \geq 2$ because $\widetilde{G}$ is solvable if $m = 1$.

From (2.1), we have

$$|d_L|^{1/n} < \begin{cases} 21.6169 & \text{if } m = 2, \\ 25.9605 & \text{if } m \geq 3. \end{cases} \tag{2.4}$$

We assume the GRH. Then from (2.3), we have

$$|d_L|^{1/n} > \begin{cases} 19.5441 & \text{if } n \geq 360 = |\mathrm{PSL}_2(\mathbb{F}_9)|, \\ 22.5793 & \text{if } n \geq 1440 = 4 \times |\mathrm{PSL}_2(\mathbb{F}_9)|, \\ 29.7470 & \text{if } n \geq 9828 = |\mathrm{PSL}_2(\mathbb{F}_{27})|. \end{cases}$$

We have contradiction if $m \geq 3$. For the case $m = 2$, if ρ is odd, we can show easily that the order of G is greater than or equal to $4 \times |\mathrm{PSL}_2(\mathbb{F}_9)| = 1440$ (cf. [2], §3). Thus we have contradiction for the case $m = 2$. If ρ is even, since $p \neq 2$, K is a totally real or CM field, and L is totally real. From (2.2) we have

$$|d_L|^{1/n} > 51.6187 \qquad \text{if } n \geq 360.$$

This contradicts (2.4).

3. The Finiteness Problem

3.1 Problem

It will be natural to generalize the second assertion of Section 1:

(FP). *Let K be an algebraic number field of finite degree over $\mathbb{Q}$ and G_K be its absolute Galois group* $\mathrm{Gal}(\overline{K}/K)$. *Fix an integer $d \geq 1$ and a nonzero integral ideal N of K. Then do there exist only finitely many isomorphism classes of continuous semisimple representations $\rho : G_K \longrightarrow \mathrm{GL}_d(\overline{\mathbb{F}}_p)$ whose Artin conductor $N(\rho)$ outside p divides N?*

(Here, $N(\rho)$ is a suitable generalization of that of §1 ([8]), but we use the same notation.) This problem, together with the proof of some special cases (Th. 7 below), was presented in [8]. Khare ([6]) also considers the same problem independently (He states this as a conjecture).

The case of $d = 1$ of (FP) follows from global class field theory, i.e., given an ideal N of K, there exist only finitely many abelian extensions of K with conductor dividing N. It also follows from this that there are only finitely many isomorphism classes of continuous representations $\rho : G_K \to \mathrm{GL}_d(\overline{\mathbb{F}}_p)$ with $N(\rho)|N$ which surject onto $\mathrm{GL}_d(\mathbb{F}_q)$ for some finite field $\mathbb{F}_q$ with q elements. The case of $d = 2$, $K = \mathbb{Q}$, odd, as we said, follows from Serre's conjecture. The case of $d \geq 3$, $K = \mathbb{Q}$, "odd" in some sense is related to a conjecture of Ash and Sinnott ([1]), which is an analogue of Serre's conjecture formulated by using cohomology classes instead of modular forms. This (FP) is reduced to the case $K = \mathbb{Q}$ by the induction of representations ([8], §4).

Now we give some remarks to explain why the conditions in the Problem are necessary:

Remarks. (1) Without any restriction on ramification outside p, we cannot expect the finiteness. For example, the set of isomorphism classes of $\rho : G_{\mathbb{Q}} \longrightarrow \mathrm{GL}_1(\overline{\mathbb{F}}_p)$ unramified outside $l(\neq p)$ is infinite, since we have, for each $n \geq 1$, the representation $\rho_n : G_{\mathbb{Q}} \longrightarrow \mathbb{Z}/l^n\mathbb{Z} \hookrightarrow \overline{\mathbb{F}}_p^{\times}$ corresponding to the n-th layer of the cyclotomic $\mathbb{Z}_l$-extension of $\mathbb{Q}$.
(2) If we replace $\overline{\mathbb{F}}_p$ by a finite field $\mathbb{F}_{p^m}$, the finiteness follows from the Hermite-Minkowski Theorem saying that there exist only finitely many algebraic number fields which are of a given degree and unramified outside a given set of primes.
(3) The assumption of semisimplicity is necessary. In fact, there may be infinitely many (mutually unisomorphic) non-semisimple representations of a finite group G.

We can consider such non-existence and finiteness also in the context of inverse Galois problem. Then our Finiteness Problem is equivalent to the following:

(FP′). *Given K, d, p, N, do there exist only finitely many Galois extensions L/K such that its Galois group* $\mathrm{Gal}(L/K)$ *embeds semisimply in* $\mathrm{GL}_d(\overline{\mathbb{F}}_p)$ *and the prime-to-p part of the different $\mathfrak{D}'_{L/K}$ of L/K divides N?*

For comparison, we mention the following two things:

Theorem (Hermite-Minkowski). *Given any constant $N > 0$, there exist only finitely many algebraic number fields $L/\mathbb{Q}$ such that the discriminant $|d_L| \leq N$.*

Class field tower. *For "large" N, there exist infinitely many finite Galois extensions $L/\mathbb{Q}$ such that $|d_L|^{1/[L:\mathbb{Q}]} \leq N$.*

Note that in such an example of infinite family of $L/\mathbb{Q}$ coming from an infinite class field tower, their Galois groups cannot all be embedded into $\mathrm{GL}_d(\overline{\mathbb{F}}_p)$ with fixed p and d.

3.2 Results

In this section, we survey my results on (FP) with a brief sketch of the proofs. First, we discuss the Finiteness Problem in the case where $K = \mathbb{Q}$, $N = 1$ and, p and d are small. Let $X_{p,d}$ be the set of isomorphism classes of continuous semisimple representations $\rho : G_{\mathbb{Q}} \rightarrow \mathrm{GL}_d(\overline{\mathbb{F}}_p)$ unramified outside p, and $X_{p,d}^{\mathbb{R}}$ the subset of $X_{p,d}$ consisting of those ρ's of which the field corresponding to the kernel is totally real (in other words, ρ is unramified also at ∞). Then we have ([8]):

Theorem 7. (1) *The set $X_{2,4}^{\mathbb{R}}$ is finite. Under GRH, the sets $X_{2,4}$ and $X_{2,8}^{\mathbb{R}}$ are finite.*

(2) *Under GRH, the set* $X^{\mathbb{R}}_{3,4}$ *is finite.*
(3) *The sets* $X_{2,2}$, $X_{3,2}$, $X^{\mathbb{R}}_{5,2}$ *are finite. Under GRH, the sets* $X_{5,2}$, $X^{\mathbb{R}}_{7,2}$, $X^{\mathbb{R}}_{11,2}$, $X^{\mathbb{R}}_{13,2}$ *are finite.*

The proof of this Theorem is based on the same principle as in §2 (except that we use an asymptotic lower bound of discriminants (the limit of Lemma 5 as $n \to \infty$)). The key point of the proof is an extension of a part of Tate's estimation (§2, Lemma 4) in terms of the invariant "p-length" of its Galois group:

Lemma 8. *Let* $K/\mathbb{Q}$ *be a finite Galois extension of degree* n *with Galois group* G. *Suppose that there is an embedding* $\rho : G \hookrightarrow \mathrm{GL}_d(\overline{\mathbb{F}}_p)$. *Then the* p*-part of the discriminant of* $K/\mathbb{Q}$ *divides* p^{cn} *where*

$$c < \ulcorner \log_2 d \urcorner + 1 + \frac{\ulcorner \log_2 d \urcorner}{p-1}$$

Here, $\ulcorner \alpha \urcorner$ *denotes the integer satisfying* $\alpha \leq \ulcorner \alpha \urcorner < \alpha + 1$.

Further, if we compare this bound with Lemma 5, we can bound the order of image of representations; two estimates contradicts if $n \geq 100$. Hence there exists no totally real Galois extension $L/\mathbb{Q}$ unramified outside 2 such that $\mathrm{Gal}(L/\mathbb{Q}) \simeq \mathrm{GL}_4(\mathbb{F}_{2^m})$, $\mathrm{SL}_4(\mathbb{F}_{2^m})$ or $\mathrm{PSL}_4(\mathbb{F}_{2^m})$ for any m, because these groups have order > 100.

Next result is on the Finiteness Problem for general K, d, p, N under the condition that their images are solvable ([11]):

Theorem 9. *Given an integer* $d \geq 1$ *and a nonzero integral ideal* N *of* K, *there exist only finitely many isomorphism classes of continuous semisimple representations* $\rho : G_K \longrightarrow \mathrm{GL}_d(\overline{\mathbb{F}}_p)$ *with solvable image and with* $N(\rho)$ *dividing* N.

The finiteness statement holds true also for classical Artin representations, i.e., if we replace $\overline{\mathbb{F}}_p$ by the complex number field $\mathbb{C}$ and $N(\rho)$ by the usual Artin conductor. Theorem 9 is proved by combining the Hermite-Minkowski theorem on discriminants and class field theory (finiteness of abelian extensions of bounded conductor) with a theorem of Mal'cev and Kolchin ([15]) on the structure of finite solvable subgroups of $\mathrm{GL}_d(\overline{\mathbb{F}}_p)$. Also, in [11], the Finiteness Problem is reduced to a special case in which the image of ρ is a finite simple group of Lie type in characteristic p. This is based on a theorem of Larsen and Pink ([7]) on the structure of finite subgroups of $\mathrm{GL}_d(\overline{\mathbb{F}}_p)$. Furthermore, these results hold also for function fields K in one variable over a finite field, under the reasonable condition that there are no constant field extensions.

Now we consider quantitative and effective versions of (FP). For mod p representations, to give a quantitative result is to give an explicite upper bound for the number of representations and to give an effective result is to give an upper bound of the image of representations. This problem for the general case seems very difficult. So, we consider the set of n-dimensional monomial mod p representations of $G_{\mathbb{Q}}$ with bounded conductor. We say that a representation $\rho : G_{\mathbb{Q}} \to \mathrm{GL}_n(\overline{\mathbb{F}}_p)$ is *monomial* if it is of the form $\rho = \mathrm{Ind}_K^{\mathbb{Q}}\chi$, i.e. if it is induced from a character $\chi : G_K \to \overline{\mathbb{F}}_p^{\times}$ of the absolute Galois group G_K of an algebraic number field K of degree n over $\mathbb{Q}$. From the construction together with the Hermite-Minkowski theorem and the finiteness of ray class groups, it follows easily that this set is finite. In the following theorem, (i) is a quantitative result and (ii) is an effective result:

Theorem 10. *Fix positive integers n and N. Consider n-dimensional monomial mod p Galois representations $\rho : G_{\mathbb{Q}} \longrightarrow \mathrm{GL}_n(\overline{\mathbb{F}}_p)$ with $N(\rho)|N$.*
(i) *The number of isomorphism classes of such ρ's is bounded by*

$$\frac{2^{n^2+n+1} \cdot (11.1)}{\pi^n}\left(2 + \frac{1}{2}n^n p^{n-1} N\right)^n p^{2n-1} N^n.$$

(ii) *The order of the image of such a ρ is bounded by*

$$\frac{2^{n(n+1)}(11.1)^n}{\pi^{n^2}} n! n^{n^2} p^{n(2n-1)} N^{n^2}.$$

A sharper estimate is given in [9]. The outline of the proof is: First, we bound the discriminant of K and the conductor of χ when the conductor of $\rho = \mathrm{Ind}_K^{\mathbb{Q}}\chi$ is given. We give an upper bound of the number of algebraic number fields K of degree n and discriminant (outside p) dividing D in terms of n, p and D. For a given K, we give an upper bound for the number of characters χ of G_K with a given Artin conductor N. Combining these results together, we obtain the above Theorem (i). Finally, we deduce the estimate of the order of the image of $\mathrm{Ind}_K^{\mathbb{Q}}\chi$ from that of the image of χ by group theoretic arguments.

Recently, we refined Tate's bound in terms of the reduced Serre weight $\tilde{k}(\rho)$ of §1 ([12]).

Lemma 11. *Suppose $\rho : G_{\mathbb{Q}_p} \to \mathrm{GL}_2(\overline{\mathbb{F}}_p)$ is wildly ramified, with α, β as in (1,1). Let $k = \tilde{k}(\rho)$ be the reduced Serre weight of ρ. Put $d := (\alpha, \beta, p-1) = (\alpha, k-1, p-1)$. Let p^m be the wild ramification* *index of the kernel field $K/\mathbb{Q}_p$. Then we have*

$$v_p(\mathcal{D}_{K/\mathbb{Q}_p}) = \begin{cases} 1 + \dfrac{k-1}{p-1} - \dfrac{k-1+d}{(p-1)p^m} & \text{if } 2 \le k \le p, \\ 2 + \dfrac{1}{(p-1)p} - \dfrac{2}{(p-1)p^m} & \text{if } k = p+1. \end{cases}$$

Then, using this Lemma, we can prove that there exist no continuous irreducible representations $\rho : G_{\mathbb{Q}} \to \mathrm{GL}_2(\overline{\mathbb{F}}_p)$ unramified outside p for $p \le 19$ and with small Serre weight k. For example, there exist no such representations in the cases $p = 17$ $(k = 2, 3, 5, 7)$, and $p = 19$ $(k = 2, 3, 5, 7)$.

References

[1] A. Ash and W. Sinnott, *An analogue of Serre's conjecture for Galois representations and Hecke eigenclasses in the mod-p cohomology of* $\mathrm{GL}(n, \mathbb{Z})$, Duke Math. J. **105** (2000), 1–24.

[2] S. Brueggeman, *The nonexistence of certain Galois extensions unramified outside* 5, J. Number Theory **75** (1999), 47–52.

[3] L. E. Dickson, "Linear Groups," Dover, New York, 1958.

[4] B. Edixhoven, *Serre's Conjectures*, in "Modular forms and Fermat's Last Theorem," 209–242, Springer-Verlag, 1997.

[5] J.-M. Fontaine, *Il n'y a pas de variété abélienne sur* $\mathbb{Z}$, Invent. Math. **81** (1985), 515–538.

[6] C. Khare, *Conjectures on finiteness of mod p Galois representations*, J. Ramanujan Math. Soc. **15** (2000), 23–42.

[7] M. J. Larsen and R. Pink, *Finite subgroups of algebraic groups*, preprint (1998).

[8] H. Moon, *Finiteness results on certain mod p Galois representations*, J. Number Theory **84** (2000), 156–165.

[9] H. Moon, *The number of monomial mod p Galois representations with bounded conductor*, Tohoku Math. J. **55** (2003), 89–98.

[10] H. Moon, *The non-existence of certain mod p Galois representations*, to appear in Bulletin of Korean Math. Soc.

[11] H. Moon and Y. Taguchi, *Mod p Galois representations of solvable image*, Proc. Amer. Math. Soc. **129** (2001), 2529–2534.

[12] H. Moon and Y. Taguchi, *Refinement of Tate's discriminant bound and nonexistence theorems for mod p Galois representations*, to appear.

[13] G. Poitou, *Sur les petits discriminants*, in "Séminaire Delange-Pisot-Poitou, 18e année: (1976/77), Théorie des nombres, Fasc. 1," Exp. No. 6, Secrétariat Math., Paris, 1977.

[14] J.-P. Serre, *Sur les représentations modulaires de degré* 2 *de* $\mathrm{Gal}(\overline{\mathbb{Q}}/\mathbb{Q})$, Duke Math. J. **54** (1987), 179–230.

[15] D. A. Suprunenko, "Matrix Groups," American Mathematical Society, Providence, R.I., 1976.

[16] J. Tate, *The non-existence of certain Galois extensions of* $\mathbb{Q}$ *unramified outside* 2, Contemp. Math. **174** (1994), 153–156.

Galois Theory and Modular Forms pp 233–268
K. Hashimoto, K. Miyake, and H. Nakamura, Editors
© 2003 Kluwer Academic Publishers

FROBENIUS MODULES AND GALOIS GROUPS

B. Heinrich MATZAT
Interdisziplinäres Zentrum
für Wissenschaftliches Rechnen (IWR)
der Universität Heidelberg,
Im Neuenheimer Feld 368
69120 Heidelberg, Germany
matzat@iwr.uni-heidelberg.de

Contents:

Introduction

In these notes some basic facts on Frobenius modules are collected. Frobenius modules are finite-dimensional vector spaces over fields with a Frobenius endomorphism ϕ, provided with an injective ϕ-semilinear Frobenius operator Φ.

The first part is devoted to Frobenius modules M over fields F of positive characteristic. It is shown that for any such Frobenius module M

[received: September 11, 2002; accepted in revised form: February 25, 2003]

there exists an extension field E containing a full system of Φ-invariant solutions. The solution field E is a finite Galois extension of F, and any finite Galois extension of F can be obtained in this way. Further in case F is a field with an iterative derivation, M becomes in a natural way a differential module, whose Picard-Vessiot field coincides with the Frobenius solution field (in the case of an algebraically closed field of constants). This correspondence in mind we obtain an upper bound for the Galois group given by the smallest connected linear algebraic group in which a representing matrix of Φ is contained. Since lower bounds can be obtained by specialization in analogy to the classical Dedekind criterion, this technique promises to be a useful tool for the construction of finite Galois extension with given (connected) Galois group in positive characteristic.

In the second part, we study Frobenius modules over p-adic Frobenius fields and p-adic Frobenius rings, respectively. The latter are called here integral p-adic Frobenius modules (in contrast to "bounded p-adic Frobenius modules" as in [13] or [8]). It is shown that integral Frobenius modules over p-adic number fields F always have a solution field but that these in general are not algebraic over F in contrast to the previous case. If the base field F is a differential field, again any integral Frobenius module M over F is a differential module, and a solution field of the Frobenius module contains the corresponding PV-extension. On the other hand after an unramified extension of the field of constants a solution field of the differential module contains a full system of solutions of the underlying Frobenius module, proving the existence of Frobenius solution fields. The automorphism group over F of such a Frobenius solution field is a p-adic analytic group, in which the subgroup of elements commuting with the Frobenius operator, the so-called Frobenius automorphism group, is dense (at least in the case of connected groups). Finally, it is shown that the residue module of an integral p-adic Frobenius module (obtained by reducing modulo the valuation ideal of F) becomes a Frobenius module in positive characteristic as studied in the first part. Further, the Galois group of the reduced module is a subgroup of the quotient of the Frobenius Galois group by its principal congruence subgroup (at least in the connected case). This essentially proves Conjecture 8.5 in [9] in the case of Frobenius modules.

Partial results are contained in Taguchi [16] in the characteristic p case and in Tsuzuki [17] in the p-adic case (see also the recent Bourbaki article of Colmez [4]).

I. Finite Frobenius Modules

1. The Solution Field of a Finite Frobenius Module

1.1. Let F be a field containing the finite field $\mathbb{F}_q$ of $q = p^d$ elements. Such a field together with the Frobenius endomorphism

$$\phi_q : F \to F, \ a \mapsto a^q \tag{1}$$

is called here a *finite Frobenius field* for short (because the field fixed by ϕ_q is finite). Correspondingly a finite-dimensional F-vector space M with an injective ϕ_q-semilinear map

$$\Phi_q : M \to M, \ x \mapsto \Phi_q(x) \tag{2}$$

is called a *finite Frobenius module* (over F). The set of fixed points of the Frobenius endomorphism Φ_q on M

$$\mathrm{Sol}^{\Phi}(M) := \{x \in M \mid \Phi_q(x) = x\} \tag{3}$$

is called the *solution space of* (M,Φ_q). Further (M,Φ_q) is said to be *trivial* if $\mathrm{Sol}^{\Phi}(M)$ contains an F-basis of M. Obviously for a Frobenius field (E,ϕ_q^E) with $E \geq F$ and Frobenius endomorphism ϕ_q^E extending ϕ_q the tensor product $M_E := E \otimes_F M$ in a natural way becomes a Frobenius module over E. By abuse of notation

$$\mathrm{Sol}_E^{\Phi}(M) := \mathrm{Sol}^{\Phi}(E \otimes_F M) \tag{4}$$

is called the *solution space of* M *over* E and M is *trivial over* E if $\mathrm{Sol}_E^{\Phi}(M)$ contains an E-basis of M_E.

1.2. The following theorem is proved for iterative differential fields F in [9], Prop. 5.6 and [8], Thm. 2.18, respectively, but it holds more generally for any finite Frobenius field (compare [16], Lemma 3.2 for a categorical treatment).

Theorem 1.1. *Let* (M, Φ_q) *be a Frobenius module over a finite Frobenius field* (F, ϕ_q)*.*

(a) *For every Frobenius extension field* (E, ϕ_q^E) *the solution space* $\mathrm{Sol}_E^{\Phi}(M)$ *is an* $\mathbb{F}_q$*-vector space of dimension*

$$\dim_{\mathbb{F}_q}(\mathrm{Sol}_E^{\Phi}(M)) \leq \dim_E(E \otimes_F M) = \dim_F(M). \tag{5}$$

(b) *There exists a finite Frobenius field extension E/F such that*

$$\dim_{\mathbb{F}_q}(\mathrm{Sol}_E^{\Phi}(M)) = \dim_F(M). \tag{6}$$

(c) *The minimal Frobenius field extension E/F admitting a full solution space is unique (inside a given algebraic closure of F) and Galois over F.*

Proof. Obviously $\mathrm{Sol}_E^{\Phi}(M)$ is an $\mathbb{F}_q$-vector space. We assume that $\dim_{\mathbb{F}_q}(\mathrm{Sol}_E^{\Phi}(M)) > m := \dim_F(M)$. Then there a exists a minimal number n of elements $x_i \in \mathrm{Sol}_E^{\Phi}(M)$ which are linearly independent over $\mathbb{F}_q$ but linearly dependent over E. This implies a nontrivial relation $\sum_{i=1}^{n} a_i x_i = 0$ over E with $a_n = 1$. Because of $\sum_{i=1}^{n} \phi_q(a_i) x_i = 0$ with $\phi_q(a_n) = 1$ the difference of the two equations gives a nontrivial relation of $x_1, x_2, \ldots, x_{n-1}$ over E in contrast to the choice of n.

In order to prove (b), we choose a basis $B = \{b_1, b_2, \ldots, b_m\}$ of M. Then by $\Phi_q(b_j) = \sum_{i=1}^{m} d_{ij} b_i$ with $d_{ij} \in F$ we obtain a representing matrix $D := D_B(\Phi_q) \in F^{m\times m}$ of Φ_q with respect to B. Now let $B\boldsymbol{y} = \sum_{i=1}^{m} b_i y_i$ with $\boldsymbol{y} = (y_1, \ldots, y_m)^{tr} \in E^m$ be a solution of M over E. Then

$$B\boldsymbol{y} = \Phi_q(B\boldsymbol{y}) = \Phi_q(B)\phi_q(\boldsymbol{y}) = B \cdot D \cdot \phi_q(\boldsymbol{y}) \tag{7}$$

holds with $\phi_q(\boldsymbol{y}) = (y_1^q, \ldots, y_m^q)^{tr}$. This equation defines a system of algebraic equations $\boldsymbol{y} - D\boldsymbol{y}^q = 0$ over F for y_i, where the Jacobian matrix equals the identity matrix. Hence by the Theorem of Bézout this system has q^n different solutions (with multiplicity 1) in an algebraic closure $\overline{F}$ of F. The field E/F generated by these solutions has finite degree over F and admits a full system of solutions.

Obviously E is minimal inside $\overline{F}$ and does not depend on the chosen basis B. Since all the zeroes of $\boldsymbol{y} - D\boldsymbol{y}^q = 0$ are simple, E/F is separable and normal, hence Galois. □

Corollary 1.2. *Every Frobenius module over $\overline{\mathbb{F}}_p$ and more generally over any separably closed field containing $\mathbb{F}_p$ is trivial.*

1.3. The unique Galois extension E in Theorem 1.1(c) is called the *solution field of* (M, Φ_q) and a fundamental system $Y \in \mathrm{GL}_m(E)$ of solutions $\boldsymbol{y} \in E^m$ is called a *fundamental solution matrix.* It is characterized by the property $\phi_q^E(Y) = D^{-1}Y$. From this observation we obtain the following characterization of solution fields.

Corollary 1.3. *The field E is the solution field of the finite Frobenius module (M, Φ_q) over F if and only if the following holds:*

(a) *There exists a matrix $Y = (y_{ij})_{i,j=1}^m \in \mathrm{GL}_m(E)$ with $\phi_q^E(Y) = D^{-1}Y$.*

(b) *E/F is generated by the coefficients y_{ij} of Y, i.e., $E = F(Y)$.*

By a change of basis $B \mapsto B \cdot C$ (and $Y \mapsto C^{-1}Y$ respectively) with $C \in \mathrm{GL}_m(F)$, the matrix $D = D_B(\Phi_q) \in \mathrm{GL}_m(F)$ transforms into $C^{-1}D\phi_q(C)$. Such matrices are called *Frobenius equivalent* in the following. Frobenius equivalent matrices define the same solution field.

2. Cyclic Frobenius Modules

2.1. A Frobenius module (M, Φ_q) of dimension m over a Frobenius field (F, ϕ_q) is called *cyclic* if it possesses a Φ_q-basis, i.e., an element $b \in M$ such that $\{b, \Phi_q(b), \ldots, \Phi_q^{m-1}(b)\}$ forms a basis of M.

Theorem 2.1. *Let (F, ϕ_q) be a finite Frobenius field and (M, Φ_q) a Frobenius module over F with $\dim_F(M) = m$. Assume F contains more than $\binom{q^m}{2}$ elements, then M is cyclic.*

Proof. Let E/F be a solution field of M and $x = By = \sum_{i=1}^m b_i y_i \in \mathrm{Sol}_E^\Phi(M)$, i.e., $y = D\phi_q(y) = Dy^q$. Since F possesses more than $\binom{q^m}{2}$ elements, there exists an F-linear combination $z = \sum_{i=1}^m c_i y_i$ in normal position with $c_m = 1$ (compare [6], Prop. 3.7.22). Then the Buchberger algorithm applied with the variables $y_1, \ldots, y_{m-1}, z$ in lexicographical ordering leads to a polynomial equation $f(z) = 0$ where $f(X) \in F[X]$ is monic separable and additive of degree q^m, i.e., $f(X) = \sum_{i=0}^m a_i X^{q^i}$ with $a_m = 1$.

Now let $z_1, \ldots, z_m$ be a basis of the $\mathbb{F}_q$-vector space of zeroes of f in $\overline{F}$. Then the Moore determinant

$$\Delta_q(z_1, \ldots, z_m) = \det \begin{pmatrix} z_1 & z_2 & \cdots & z_m \\ z_1^q & z_2^q & \cdots & z_m^q \\ \vdots & \vdots & \ddots & \vdots \\ z_1^{q^{m-1}} & z_2^{q^{m-1}} & \cdots & z_m^{q^{m-1}} \end{pmatrix} = \det(Z) \quad (8)$$

is different from zero ([5], Lemma 1.3.3). All z_j are F-linear combinations of the y_i. Because of $\phi_q(Y) = D^{-1}Y$ for the given solution matrix Y the same holds for the powers $z_j^{q^l}$. Hence there exists a matrix $C \in \mathrm{GL}_m(F)$ such that $Z = CY$. From this equation we obtain

$$D = Y\phi_q(Y)^{-1} = C^{-1}Z\phi_q(Z)^{-1}\phi_q(C) = C^{-1}\widetilde{D}\phi_q(C) \tag{9}$$

with

$$\widetilde{D}^{-1} = \begin{pmatrix} 0 & 1 & 0 & \cdots & 0 \\ \vdots & \ddots & \ddots & \ddots & \vdots \\ \vdots & & \ddots & \ddots & 0 \\ 0 & \cdots & \cdots & 0 & 1 \\ -a_0 & \cdots & \cdots & \cdots & -a_{m-1} \end{pmatrix}, \tag{10}$$

where the entries a_i are the coefficients of the additive polynomial $f(X)$. Thus D is Frobenius equivalent to $\widetilde{D}$. The image of the basis $\widetilde{B} := BC^{-1} = \{\widetilde{b}_1, \ldots, \widetilde{b}_m\}$ is

$$\begin{aligned} \Phi_q(\widetilde{B}) &= BC^{-1}CD\phi_q(C)^{-1} = \widetilde{B}\widetilde{D} \\ &= (\widetilde{b}_2 - \frac{a_1}{a_0}\widetilde{b}_1, \ldots, \widetilde{b}_m - \frac{a_{m-1}}{a_o}\widetilde{b}_1, -\frac{1}{a_0}\widetilde{b}_m). \end{aligned} \tag{11}$$

Hence $b := \widetilde{b}_1$ defines a Φ_q-basis of M. □

Another proof for Theorem 2.1 with F infinite can be found in [16], Lemma 3.3. The polynomial $f(X) \in F[X]$ in the proof only depends on the Φ_q-basis $\{b\}$ of M and is called the *characteristic polynomial* f_b *of* M *with respect to* b. Obviously, the splitting field of f_b over F coincides with the solution field of (M, Φ_q).

2.2. Given a cyclic Frobenius module with Φ_q-basis $\{b\}$ and with corresponding fundamental solution matrix Z as in the proof of Theorem 2.1 (with respect to the basis $\widetilde{B}$ with $\widetilde{b}_1 = b$, $\widetilde{b}_2 = \Phi_q(\widetilde{b}_1) + \frac{a_1}{a_0}\widetilde{b}_1, \ldots, \widetilde{b}_m = \Phi_q(\widetilde{b}_{m-1}) + \frac{a_{m-1}}{a_0}\widetilde{b}_1$), we obtain the characteristic polynomial f_b in the following way (compare [5], Lemma 1.3.6).

Corollary 2.2. *Let* (M, Φ_q) *be a cyclic Frobenius module with* Φ_q*-basis* $\{b\}$. *Then the characteristic polynomial of* M *with respect to* b *equals*

$$f_b(X) = \frac{\Delta_q(z_1, \ldots, z_m, X)}{\Delta_q(z_1, \ldots, z_m)}, \quad \textit{where } Z = \begin{pmatrix} z_1 & \cdots & z_m \\ z_1^q & & z_m^q \\ \vdots & & \vdots \\ z_1^{q^{m-1}} & \cdots & z_m^{q^{m-1}} \end{pmatrix} \tag{12}$$

denotes the corresponding fundamental solution matrix.

3. The Associated ID-Module

3.1. In this section let $F \geqslant \mathbb{F}_q, q = p^d$, be a field with an *iterative derivation* $\partial_F^* = (\partial_F^{(k)})_{k\in\mathbb{N}}$. This is a field with a family of maps $\partial_F^{(k)}$: $F \to F$ with $\partial_F^{(0)} = \mathrm{id}_F$ and the following properties for $a, b \in F$ and $i, j, k \in \mathbb{N}$:

$$\partial_F^{(k)}(a+b) = \partial_F^{(k)}(a) + \partial_F^{(k)}(b), \tag{13}$$

$$\partial_F^{(k)}(a\cdot b) = \sum_{i+j=k} \partial_F^{(i)}(a)\,\partial_F^{(j)}(b), \tag{14}$$

$$\partial_F^{(i)}\partial_F^{(j)} = \binom{i+j}{i}\partial_F^{(i+j)}. \tag{15}$$

Such a field sometimes is called an ID-*field* for short ([8], §1.1). It defines a series of subfields $F_l := \bigcap_{j<l} \mathrm{Ker}(\partial_F^{(p^j)})$ whose intersection K is the field of constants of F, i.e.,

$$K = \bigcap_{l\in\mathbb{N}} F_l = \{a \in F \mid \partial_F^{(k)}(a) = 0 \quad \text{for} \quad k > 0\}. \tag{16}$$

On F we have a natural Frobenius endomorphism $\phi_p : F \to F, a \mapsto a^p$. In case $F_l = \phi_p^l(F)$ we call $(F, \partial_F^*, \phi_p)$ an ID-field with *Frobenius structure* or an IDF-*field* for short. Obviously, ID-fields which are function fields of one variable over a perfect field are always IDF-fields (compare [9], §2.1 or [8], §1.5).

3.2. A vector space M over an ID-field (F, ∂_F^*) with a family of maps $\partial_M^* = \left(\partial_M^{(k)}\right)_{k\in\mathbb{N}}$ is called an *iterative differential module* (or ID-*module* for short) if ∂_M^* fulfills (13), (15) and for $a \in F, x \in M$

$$\partial_M^{(k)}(a\cdot x) = \sum_{i+j=k} \partial_F^{(i)}(a)\partial_M^{(j)}(x) \tag{17}$$

for the scalar multiplication (instead of (14)). Now we want to show that any finite Frobenius module over an IDF-field in a natural way becomes an ID-module.

Proposition 3.1. *Let (M, Φ_q), $q = p^d$, be a finite Frobenius module over an* IDF*-field $(F, \partial_F^*, \phi_p)$.*

(a) *There exists a unique iterative derivation* $\partial_M^* = (\partial_M^{(k)})_{k\in\mathbb{N}}$ *on* M *such that*

$$\Phi_q^l(M) = \bigcap_{j<dl} \mathrm{Ker}(\partial_M^{(p^j)}). \tag{18}$$

(b) *If* $\widetilde{\varphi}_l : M \to M$ *denotes the* F*-linear extension of the* $F_{d(l+1)}$*-linear embedding* $\varphi_l : \Phi_q^{l+1}(M) \to \Phi_q^l(M)$ *we have*

$$\partial_M^{(k)} = \widetilde{\varphi}_0 \circ \cdots \circ \widetilde{\varphi}_{l-1} \circ \partial_F^{(k)} \circ \widetilde{\varphi}_{l-1}^{-1} \circ \cdots \circ \widetilde{\varphi}_0^{-1} \quad \text{for } k < p^{dl}, \tag{19}$$

where $\partial_F^{(k)}$ *is defined on the coefficients with respect to an* F_{dl}*-basis of* $\Phi_q^l(M)$.

Proof. First let ∂_M^* be an iterative derivation with the property in (a). Then with respect to a basis $B = \{b_1, \dots, b_m\}$ of $\Phi_q^l(M)$ for any $x \in M$ we have

$$\partial_M^{(k)}(x) = \partial_M^{(k)}(\sum_{i=1}^m b_i y_i) = \sum_{i=1}^m b_i \partial_M^{(k)}(y_i) \quad \text{for} \quad k < p^{dl}. \tag{20}$$

This shows the uniqueness and (b). On the other hand the formula in (b) defines an iterative derivation ∂_M^* on M with the property $\Phi_q^l(M) = \bigcap_{j<dl} \mathrm{Ker}(\partial_M^{(p^j)})$ (compare [9], Prop. 5.1 or [8], Thm. 2.8). □

The ID-module $(M, \partial_M^*, \Phi_q)$ in Proposition 3.1 is called the ID-*module associated to* (M, Φ_q), and the triple $(M, \partial_M^*, \Phi_q)$ is an ID-*module with Frobenius structure* or an IDF-module for short.

3.3. For an ID-module (M, ∂_M^*) over an ID-field (F, ∂_F^*) the solution space is defined by

$$\mathrm{Sol}(M) := \bigcap_{1\le k\in\mathbb{N}} \mathrm{Ker}(\partial_M^{(k)}). \tag{21}$$

If E/F is a field extension with an iterative derivation ∂_E^* extending ∂_F^*, the tensor product $M_E := E \otimes_F M$ becomes an ID-module over E by

$$\partial_{M_E}^{(k)}(a \otimes x) = \sum_{i+j=k} \partial_E^{(i)}(a) \otimes \partial_M^{(j)}(x). \tag{22}$$

As in Section 1.1 the solution space of M_E

$$\mathrm{Sol}_E(M) := \mathrm{Sol}(E \otimes M) \tag{23}$$

is called the solution space of M over E. This is a K-vector space of dimension $\dim_K(\mathrm{Sol}_E(M)) \leq \dim_F(M)$. It is well known that there exists a smallest ID-field extension E/F such that $\dim_K(\mathrm{Sol}_E(M)) = \dim_F(M)$ called an *iterative Picard-Vessiot field* (IPV-field) *for* M (compare [9], Lemma 3.4 or [8], Thm. 3.4).

In case (M, ∂_M^*) is an ID-module associated to a finite Frobenius module (M, Φ_q), the corresponding solution spaces are related in the following way (compare [8], Thm. 3.12 in the special case $d = 1$).

Theorem 3.2. *Let $(F, \partial_F^*, \phi_q)$ be an* IDF*-field with algebraically closed field of constants K and $(M, \partial_M^*, \Phi_q)$ an* IDF*-module over F with $q = p^d$. Then the following holds:*

(a) *The solution field E of the Frobenius module (M, Φ_q) is an* IPV-*field E of (M, ∂_M^*) and*

$$\mathrm{Sol}_E(M) = K \otimes_{\mathbb{F}_q} \mathrm{Sol}_E^{\Phi}(M). \tag{24}$$

(b) *The* IPV-*field E of the* ID-*module (M, ∂_M^*) is a solution field of (M, Φ_q) and the solution space of the Frobenius module (M, Φ_q) can be obtained as the $\mathbb{F}_q$-vector space of Φ_q-invariants in* $\mathrm{Sol}_E(M)$, *i.e.,*

$$\mathrm{Sol}_E^{\Phi}(M) = \mathrm{Sol}_E(M)^{\Phi_q}. \tag{25}$$

Proof. Let $B = \{b_1, \ldots, b_m\}$ be a basis of the Frobenius module (M, Φ_q), let E be the solution field of (M, Φ_q) and $B\boldsymbol{y} \in \mathrm{Sol}_E^{\Phi}(M)$. Then by definition $D^{-1}\boldsymbol{y} = \phi_q(\boldsymbol{y})$ belongs to $\phi_q(E)^m = E_d^m$, where $D = D_B(\Phi_q)$ is the matrix representing Φ_q with respect to B. By induction we obtain

$$\phi_q^{l-1}(D^{-1}) \cdots \phi_q(D^{-1}) D^{-1}\boldsymbol{y} = \phi_q^l(\boldsymbol{y}) \in \phi_q^l(E)^m = E_{dl}^m. \tag{26}$$

Hence $B\boldsymbol{y}$ belongs to $\mathrm{Sol}_E(M)$ by the criterion [8], Thm. 2.11(2). This implies that $K \otimes_{\mathbb{F}_q} \mathrm{Sol}_E^{\Phi}(M) \leq \mathrm{Sol}_E(M)$. Equality holds since by the proof of Theorem 1.1(a) a basis of $\mathrm{Sol}_E^{\Phi}(M)$ remains linearly independent over E.

Now let E be an IPV-field of the ID-module (M, ∂_M^*) with basis B and $Y \in \mathrm{GL}_m(E)$ a fundamental solution matrix. Then $\widetilde{B} = BY$ is a K-basis of $\mathrm{Sol}_E(M)$. Because of

$$\Phi_q(\widetilde{B}E) = \Phi_q(BY)\phi_q(E) = BD\phi_q(Y)E_d = \widetilde{B}Y^{-1}D\phi_q(Y)E_d \tag{27}$$

with $Y^{-1}D \in \mathrm{GL}_m(E_d)$ and $\phi_q(Y) \in \mathrm{GL}_m(E_d)$ we obtain by induction for all $l \in \mathbb{N}$

$$\Phi_q^l(M) = \Phi_q^l(\widetilde{B}E) = \bigoplus_{i=1}^{m} E_{dl}\widetilde{b_i}. \tag{28}$$

This implies

$$\bigcap_{l\in\mathbb{N}}\Phi_q^l(M)=\bigoplus_{i=1}^{m}K\widetilde{b_i}=\mathrm{Sol}_E(M). \tag{29}$$

Hence Φ_q acts on $\mathrm{Sol}_E(M)$ and $\mathrm{Sol}_E(M)$ is a Frobenius module over K which is trivial by Corollary 1.2. □

4. The Galois Group of a Finite Frobenius Module

4.1. Let (M,Φ_q) be a finite Frobenius module of dimension m over a Frobenius field (F,ϕ_q) and let E be the solution field of (M,Φ_q) over F. By Theorem 1.1(c), E/F is a finite Galois extension whose Galois group is called the *Galois group* $\mathrm{Gal}^\Phi(M)$ *of* (M,Φ). In the case F is an ID-field with algebraically closed field of constants, this group $\mathrm{Gal}^\Phi(M)$ further coincides with the ID-Galois group $\mathrm{Gal}_{\mathrm{ID}}(M)$ of the associated ID-module (M,∂_M*) (compare [9], §4.1 or [8], §3, respectively).

First we note that $\mathrm{Gal}^\Phi(M)$ has a faithful linear representation in $\mathrm{GL}_m(\mathbb{F}_q)$. This observation is based on

Proposition 4.1. *Let (M,Φ_q) be a finite Frobenius module over a finite Frobenius field (F,ϕ_q) with solution field E and let $Y,\widetilde{Y}\in\mathrm{GL}_m(E)$ be two fundamental solution matrices of (M,Φ_q). Then there exists a matrix $C\in\mathrm{GL}_m(\mathbb{F}_q)$ such that $\widetilde{Y}=YC$.*

Proof. Obviously there exists a $C\in\mathrm{GL}_m(E)$ with $\widetilde{Y}=YC$. Then from $Y=D\phi_q(Y)$ and $\widetilde{Y}=D\phi_q(\widetilde{Y})$ we obtain $C=\phi_q(C)$ implying $C\in\mathrm{GL}_m(\mathbb{F}_q)$. □

Corollary 4.2. *The Galois group $\mathrm{Gal}^\Phi(M)$ of a finite Frobenius module (M,Φ_q) of dimension m is isomorphic to a subgroup of $\mathrm{GL}_m(\mathbb{F}_q)$.*

Proof. Let E be a solution field of (M,Φ_q) and $Y\in\mathrm{GL}_m(E)$ be a fundamental solution matrix. Then for any $\gamma\in\mathrm{Gal}^\Phi(M)=\mathrm{Gal}(E/F)$ the matrix $\gamma(Y)$ again is a fundamental solution matrix of (M,Φ_q). Hence by Proposition 4.1 there exists a matrix $C_\gamma\in\mathrm{GL}_m(\mathbb{F}_q)$ with $\gamma(Y)=YC_\gamma$. Then

$$\Gamma:\mathrm{Gal}^\Phi(M)\to\mathrm{GL}_m(\mathbb{F}_q),\gamma\mapsto C_\gamma \tag{30}$$

defines a linear representation which is faithful because of $E=F(Y)$. □

4.2. From the representing matrix $D = D_B(\Phi_q)$ of the Frobenius operator of a finite Frobenius module (M, Φ_q) we obtain bounds for the Galois group $\mathrm{Gal}^{\Phi}(M)$ in a similar way as in the case of ID-modules.

Theorem 4.3. *Let (M, Φ_q) be a finite Frobenius module over a finite Frobenius field (F, Φ_q). Let $\mathcal{G}$ be a reduced connected linear group defined over $\mathbb{F}_q$. Assume that there exists a basis B such that $D_B(\Phi_q) \in \mathcal{G}(F)$. Then*

$$\mathrm{Gal}^{\Phi}(M) \leq \mathcal{G}(\mathbb{F}_q). \tag{31}$$

Proof. The proof relies on the surjectivity of the Lang isogeny

$$\pi : \mathcal{G}(F^{sep}) \to \mathcal{G}(F^{sep}), (x_{ij})_{i,j=1}^{m} \mapsto \phi_q(x_{ij})(x_{ij})^{-1} \tag{32}$$

([15], Thm. 4.4.17). This proves the existence of a matrix $Y \in \mathcal{G}(F^{sep})$ with $\phi_q(Y)Y^{-1} = D^{-1}$. By Corollary 1.3, $E := F(Y)$ is a solution field of (M, Φ_q). Now for any $\gamma \in \mathrm{Gal}^{\Phi}(M) = \mathrm{Gal}(E/F)$ the representing matrix $C_\gamma = Y^{-1}\gamma(Y)$ in Corollary 4.2 belongs to $\mathcal{G}(E)$ and to $\mathrm{GL}_m(\mathbb{F}_q)$, so C_γ is an element of $\mathcal{G}(\mathbb{F}_q)$. □

In the case $q = p$, i.e., $d = 1$, Theorem 4.3 already is contained in [9], Prop. 5.4(3) or [8] Cor. 5.3, respectively.

4.3. Lower bounds for the Galois group $\mathrm{Gal}^{\Phi}(M)$ can be obtained by an analogue of the Dedekind criterion for linear groups. This is based on

Proposition 4.4. *Let (M, Φ_q) be a finite Frobenius module over a finite field $F \geq \mathbb{F}_q$ with $[F : \mathbb{F}_q] =: f$ and $D = D_B(\Phi_q)$ a representing matrix of Φ_q with respect to a basis B of M. Then*

$$\mathrm{Gal}^{\Phi}(M) \cong < \widehat{D} >, \quad \textit{where } \widehat{D} := D\phi_q(D)\cdots\phi_q^{f-1}(D). \tag{33}$$

Proof. Let $Y \in \mathrm{GL}_m(\overline{\mathbb{F}}_p)$ be a fundamental solution matrix and $E := F(Y)$ the solution field of (M, Φ_q). Then $F = E^{<\phi>}$ for $\phi = \phi_q^f$. From $Y = D\phi_q(Y)$ we derive $Y = \widehat{D}\phi(Y)$. Since ϕ generates $\mathrm{Gal}(E/F)$, there exists a matrix $C_\phi \in \mathrm{GL}_m(\mathbb{F}_q)$ with $\phi(Y) = YC_\phi$. By the above $C_\phi = Y^{-1}\widehat{D}^{-1}Y$ is conjugate to $\widehat{D}^{-1}$ inside $\mathrm{GL}_m(E)$. □

The proof of Proposition 4.4 shows more precisely that the linear group $\Gamma(\mathrm{Gal}^{\Phi}(M)) \leq \mathrm{GL}_m(\overline{\mathbb{F}}_p)$ given by (31) contains a subgroup H which inside $\mathrm{GL}_m(E)$ is conjugate to $\langle \widehat{D} \rangle$. This is the key of the following "Linear Dedekind Criterion".

Theorem 4.5. *Let (M, Φ_q) be a finite Frobenius module over a global field (F, ϕ_q) with $F \geq \mathbb{F}_q$. Let $\mathcal{O}_Q \leq F$ be a valuation ring with valuation ideal $\mathcal{P}_Q$ and with residue field $F_Q := \mathcal{O}_Q/\mathcal{P}_Q \leq \overline{\mathbb{F}}_p$. Assume $D = D_B(\Phi_q)$ belongs to $\mathrm{GL}_m(\mathcal{O}_Q)$ with residue matrix $D_Q \in \mathrm{GL}_m(F_Q)$. Then $\mathrm{Gal}^\Phi(M) \leq \mathrm{GL}_m(\mathbb{F}_q)$ contains a matrix C_Q in $\mathrm{GL}_m(F_Q)$ conjugate to*

$$\widehat{D}_Q := D_Q\phi_q(D_Q)\cdots\phi_q^{f_Q-1}(D_Q) \quad \textit{with } f_Q := [F_Q : \mathbb{F}_q]. \tag{34}$$

Proof. Let E/F be a solution field of (M, Φ_q) and $Y \in \mathrm{GL}_m(E)$ a fundamental solution matrix. Then from $\phi_q(Y) = D^{-1}Y$ with $D \in \mathrm{GL}_m(\mathcal{O}_Q)$ we obtain $y_{ij} \in \mathcal{O}_{\tilde{Q}}$ for all coefficients y_{ij} of Y and all valuation ideals $\mathcal{P}_{\tilde{Q}}$ in E extending $\mathcal{P}_Q$ and in addition $\det(Y) \in \mathcal{O}_{\tilde{Q}}^\times$. Hence the reduction $Y_{\tilde{Q}}$ of Y is well defined and belongs to the linear group $\mathrm{GL}_m(E_{\tilde{Q}})$ over the residue field $E_{\tilde{Q}} := \mathcal{O}_{\tilde{Q}}/\mathcal{P}_{\tilde{Q}}$. Over $E_{\tilde{Q}}$ the reduced equation $Y_{\tilde{Q}} = D_Q\phi_q(Y_{\tilde{Q}})$ holds. Hence $E_{\tilde{Q}}$ contains the solution field E_Q of the reduced Frobenius module (M_Q, Φ_Q) whose representing matrix of Φ_Q with respect to the reduced basis coincides with D_Q. From $E_{\tilde{Q}} = F_Q(Y_{\tilde{Q}})$ finally follows $E_{\tilde{Q}} = E_Q$.

Since F_Q is a finite field, the Galois extension E_Q/F_Q is cyclic and its Galois group is generated by the power $\phi_Q = \phi_q^{f_Q}$ of the Frobenius automorphism. By Proposition 4.4 the matrix C_Q defined by $\phi_Q(Y_{\tilde{Q}}) = Y_{\tilde{Q}}C_Q$ fulfills $C_Q = Y_{\tilde{Q}}^{-1}\widehat{D}_Q^{-1}Y_{\tilde{Q}}$. Now let $\phi \in \mathrm{Gal}(E/F)$ be a pre-image of ϕ_Q which is trivial on the decomposition field $T_{\tilde{Q}}$ of $\mathcal{P}_{\tilde{Q}}/\mathcal{P}_Q$. Then there exists a matrix $C_\phi \in \mathrm{GL}_m(\mathbb{F}_q)$ with $\phi(Y) = YC_\phi$. Reducing modulo $\mathcal{P}_{\tilde{Q}}$ we obtain $\phi_Q(Y_{\tilde{Q}}) = Y_{\tilde{Q}}C_\phi$ from which $C_\phi = C_Q$ follows. This proves the theorem. □

From the uniqueness of C_ϕ in the above proof we obtain the uniqueness of the lift ϕ of ϕ_Q inside $\mathrm{Gal}(E/T_{\tilde{Q}})$. This implies that under the condition $D \in \mathrm{GL}_m(\mathcal{O}_Q)$ all extensions of valuation ideals $\mathcal{P}_{\tilde{Q}} \supseteq \mathcal{P}_Q$ are unramified in E/F.

4.4. The following simple example shows how Theorems 4.3 and 4.5 can be combined to determine the Galois group of a finite Frobenius module. Let $M = Fb_1 \oplus Fb_2$ be a 2-dimensional Frobenius module over $F = \mathbb{F}_p(t)$ with $D = D_B(\Phi_p) = \begin{pmatrix} t & 1 \\ 0 & t^{-1} \end{pmatrix}$. Obviously D is an element of the connected Borel subgroup $\mathcal{B}_2(F) \leq \mathrm{SL}_2(F)$. Hence by Theorem 4.2 $\mathrm{Gal}^\Phi(M) \leq \mathcal{B}_2(\mathbb{F}_p)$. The latter is a group of order $p(p-1)$. Reducing modulo $\mathcal{P}_Q = (t-1)$ we obtain $D_Q = \begin{pmatrix} 1 & 1 \\ 0 & 1 \end{pmatrix} \in \mathrm{SL}_2(\mathbb{F}_p)$ of order p, reducing modulo $(t-w)$ with a generating element w of $\mathbb{F}_p^\times$ we obtain the element $D_Q = \begin{pmatrix} w & 1 \\ 0 & w^{-1} \end{pmatrix} \in \mathrm{SL}_2(\mathbb{F}_p)$ of order divisible by $(p-1)$. This implies $\mathrm{Gal}^\Phi(M) \cong \mathcal{B}_2(\mathbb{F}_p)$.

A more substantial example is elaborated in Malle [7], giving equations for $\mathrm{G}_2(\mathbb{F}_q)$-extensions over $\mathbb{F}_q(t)$.

5. Effective Finite Frobenius Modules

5.1. Next we are interested in the question for which finite Frobenius modules the upper bound for the Galois group given in Theorem 4.3 is sharp. First we show that the assumption that the Galois group is connected is necessary.

Proposition 5.1. *Let (M,Φ_q) be a finite Frobenius module over a finite Frobenius field (F,ϕ_q) with $F \geq \overline{\mathbb{F}}_p$. Assume there exists a linear algebraic group $\mathcal{G}$ defined over $\mathbb{F}_q$ with $D(\Phi_q) \in \mathcal{G}(F)$ and $\mathrm{Gal}^\Phi(M) = \mathcal{G}(\mathbb{F}_q)$. Then $\mathcal{G}$ is connected.*

Proof. Let E/F be a solution field of (M,Φ_q) and $Y \in \mathrm{GL}_m(E)$ a fundamental solution matrix, i.e., $Y = D\phi_q^E(Y)$. By a theorem of Borel-Serre [1] and Platonov (see [18], Lemma 10.10), the connected component $\mathcal{G}^0$ of $\mathcal{G}$ has a finite supplement $\mathcal{H}$ in $\mathcal{G}$. Hence there exist matrices $D_0 \in \mathcal{G}^0(F)$ and $C_0 \in \mathcal{H}(F)$ with $D = D_0C_0$. Since $\mathcal{H}$ is finite we obtain $\mathcal{H}(F) = \mathcal{H}(\overline{\mathbb{F}}_p)$. Now as in the proof of Theorem 4.3 the surjectivity of the Lang isogeny proves the existence of a matrix $C \in \mathrm{GL}_m(\overline{\mathbb{F}}_p)$ with $C_0^{-1} = \phi_q(C)C^{-1}$. We obtain

$$C^{-1}Y = C^{-1}D\phi_q^E(Y) = C^{-1}D_0C_0\phi_q^E(Y) = C^{-1}D_0C\phi_q^E(C^{-1}Y). \quad (35)$$

Because of $C^{-1}D_0C \in \mathcal{G}^0(F)^C \cong \mathcal{G}^0(F)$, Theorem 4.3 shows that $\mathrm{Gal}^\Phi(M) \leq \mathcal{G}^0(\mathbb{F}_q)$. □

5.2. Now let (M,Φ_q) be a finite Frobenius module over an arbitrary finite Frobenius field (F,ϕ_q). Then (M,Φ_q) is called an *effective Frobenius module* if there exists a reduced connected linear algebraic group $\mathcal{G}$ defined over $\mathbb{F}_q$ such that the representing matrix $D = D_B(\Phi_q)$ belongs to $\mathcal{G}(F)$ (for a suitable basis B of M) and in addition $\mathrm{Gal}^\Phi(M) = \mathcal{G}(\mathbb{F}_q)$. Correspondingly a finite Galois extension E/F is called an *effective Galois extension* if E is the solution field of an effective Frobenius module over F.

The following theorem shows the existence of effective Galois extensions in some interesting cases:

Theorem 5.2 (Nori). *Let $\mathcal{G}$ be a semisimple simply connected linear algebraic group defined over $\mathbb{F}_q$. Then there exists an effective Galois extension $E/\mathbb{F}_q(t)$ with $\mathrm{Gal}(E/\mathbb{F}_q(t)) \cong \mathcal{G}(\mathbb{F}_q)$ which is unramified at all finite places (i.e., outside (t^{-1})).*

The proof follows from [10] or [2], Ch. 14.3, respectively.

5.3. It is an interesting question if for a given field F all Galois extensions E/F with a given Galois group are effective. A positive answer could help to construct generic additive polynomials for these groups (in the sense of Saltman [14]) as characteristic polynomials of Frobenius modules. But here up to now we can only present a conjecture.

Conjecture 5.3. *Let (F, ϕ_q) be a finite Frobenius field of cohomological dimension $\mathrm{cd}(F) \leq 2$ and E/F a finite Galois extension. Assume $\mathrm{Gal}(E/F)$ is isomorphic to the group of $\mathbb{F}_q$-rational points of a semisimple simply connected linear algebraic group. Then E/F is an effective Galois extension.*

The conjecture would give an analogue for finite Frobenius modules of the following known result for ID-modules ([8]], Ch. 5.2): Every ID-module over an ID-field F with $\mathrm{cd}(F) \leq 1$ and with connected ID-Galois group $\mathcal{G}$ is effective. The proof of the last result is based on the Theorem of Springer and Steinberg stating $H^1(F, \mathcal{G}_F) = 0$.

II. p-adic Frobenius Modules

6. Integral p-adic Frobenius Modules

6.1. In the second part of these notes we study Frobenius modules over p-adic Frobenius fields. Let us first fix our notation. $\mathbb{Z}_q$ denotes the Witt ring $\mathbb{W}(\mathbb{F}_q)$ of the finite field $\mathbb{F}_q$ of $q = p^d$ elements and by $\mathbb{Q}_q$ its quotient field. This is the unique unramified extension of degree d of the field of p-adic numbers $\mathbb{Q}_p = \mathrm{Quot}(\mathbb{W}(\mathbb{F}_p))$. The union $\mathbb{T}_p := \bigcup_{1 \leq d \in \mathbb{N}} \mathbb{Q}_{p^d}$ of all these fields is the maximal unramified algebraic extension of $\mathbb{Q}_p$. Its completion $\widehat{\mathbb{T}}_p$ coincides with the quotient field of the Witt ring $\mathbb{W}(\overline{\mathbb{F}}_p)$.

For each d the field $\mathbb{T}_p$ possesses a unique Frobenius automorphism $\phi_q = \phi_{p^d}$ whose reduction $\widetilde{\phi}_q$ to the residue field $\overline{\mathbb{F}}_p$ gives the q-power Frobenius automorphism. It is continuous with respect to the p-adic valuation and generates $\mathrm{Gal}(\mathbb{T}_p/\mathbb{Q}_q)$. It follows that the fixed field of the unique continuous extension $\widehat{\phi}_q$ of ϕ_q to $\widehat{\mathbb{T}}_p$ still remains $\mathbb{Q}_q$.

The algebraic closure of $\mathbb{Q}_p$ is denoted by $\overline{\mathbb{Q}}_p = \mathbb{Q}_p^{\text{alg}}$ and its completion, the p-adic universe, as usual by $\mathbb{C}_p$. The unique Frobenius automorphism ϕ_p on $\mathbb{T}_p$ can be extended (in a non-unique way) to an automorphism $\overline{\phi}_p$ of $\overline{\mathbb{Q}}_p/\mathbb{Q}_p$ and further it can be continuously extended to $\mathbb{C}_p$.

In the following a complete non-archimedian field F containing $\mathbb{Q}_q$ with a continuous Frobenius endomorphism $\phi_q^F : F \to F$ reducing to the q-power Frobenius endomorphism is called a *p-adic Frobenius field.* Examples of such Frobenius fields are $(\mathbb{Q}_{q^f}, \phi_q), (\widehat{\mathbb{T}}_p, \widehat{\phi}_q)$ and $(\mathbb{C}_p, \overline{\phi}_q)$ where the first two of them additionally are discrete valued. Notice that a subfield L of a p-adic Frobenius field (F, ϕ_q^F) with the restricted Frobenius endomorphism $\phi_q^F|_L$ may fail to be a Frobenius field (in the case $\phi_q^F(L) \not\leq L$).

6.2. Now let (F, ϕ_q^F) be any p-adic Frobenius field with $F \geq \mathbb{Q}_q$. Then a finite-dimensional F-vector space with an injective ϕ_q^F-semilinear map $\Phi_q^F : M \to M$ is called a *p-adic Frobenius module.* The solution spaces $\text{Sol}^\Phi(M)$ and $\text{Sol}_E^\Phi(M) := \text{Sol}^\Phi(E \otimes_F M)$ for a p-adic Frobenius extension field E/F are defined exactly as in the finite case in Section 1.1, in particular M is called trivial over E if $\text{Sol}_E^\Phi(M)$ contains a basis of the E-vector space $M_E := E \otimes_F M$.

Proposition 6.1. *Let (M, Φ_q^F) be a p-adic Frobenius module over a p-adic Frobenius field (F, ϕ_q^F) with $F \geq \mathbb{Q}_p$. Then the following holds:*

(a) *For all p-adic Frobenius extension fields E/F the solution space $\text{Sol}_E^\Phi(M)$ of M over E is a vector space over the field of ϕ_q^E-invariants $E^\phi := E^{<\phi_q^E>}$ with*

$$\dim_{E^\phi}(\text{Sol}_E^\Phi(M)) \leq \dim_F(M) =: m. \tag{36}$$

(b) *Equality holds in* (a) *if and only if there exists a matrix $Y \in \text{GL}_m(E)$ such that $Y = D_B(\Phi_q^F)\phi_q^E(Y)$ with the representing matrix $D_B(\Phi_q^F)$ of Φ_q^F with respect to a basis B of M.*

(c) *Any two matrices $Y, \widetilde{Y} \in \text{GL}_m(E)$ with the property in* (b) *only differ by a matrix $C \in \text{GL}_m(E^\phi)$, i.e., $\widetilde{Y} = Y \cdot C$.*

The proofs are exactly the same as in the finite case (compare Thm. 1.1 and Prop. 4.1) with $\mathbb{F}_q$ substituted by E^ϕ.

A Frobenius field $(E, \phi_q^E) \geq (F, \phi_q^F)$ containing a *fundamental solution matrix* Y as in Proposition 6.1(b) is called a *solution field of* (M, Φ_q^F), if E/F is generated by the entries y_{ij} of Y. The p-adic Frobenius field $(\widehat{E}, \widehat{\phi}_q^E)$ obtained by the completion of E and continuous extension of ϕ_q^E is called a *complete solution field.* In contrast to the finite case the existence of (complete) solution fields is not obvious.

6.3. A p-adic Frobenius module (M, Φ_q^F) is called an *integral Frobenius module* (or an *étale Frobenius module* according to [17], Prop. 3.2.1) if it contains a Φ_q^F-stable $\mathcal{O}_F$-lattice Λ of full rank, i.e., with $\mathrm{rank}(\Lambda) = \dim_F(M)$, and

$$\mathcal{O}_F \Phi_q^F(\Lambda) = \Lambda. \tag{37}$$

The corresponding lattice Λ is then called a *Frobenius lattice.*

Theorem 6.2. *Let (M, Φ_q^F) be an integral p-adic Frobenius module over a p-adic Frobenius field (F, ϕ_q^F) with $[F : \mathbb{Q}_q] < \infty$. Then (M, Φ_q^F) has a solution field E inside $\mathbb{C}_p$ and unramified over F.*

Proof. Let F_T denote the intersection $F \cap \mathbb{T}_p$ and let $\widehat{E}$ be the completion of $F \otimes_{F_T} \mathbb{T}_p$ with the induced Frobenius endomorphism $\phi_q^{\widehat{E}}$. Then F/F_T is totally ramified of finite degree and $\widehat{E}/F$ is unramified. As before, let $\mathcal{O}_F$ be the valuation ring of F, $r\mathcal{O}_F = \mathcal{P}_F$ its valuation ideal and $\widetilde{F} = \mathcal{O}_F/\mathcal{P}_F$ its residue field which is finite.

By assumption M contains a Frobenius lattice Λ with basis B and with $D := D_B(\Phi_q^F) \in \mathrm{GL}_m(\mathcal{O}_F)$. Hence the residue matrix $\widetilde{D}$ of D belongs to $\mathrm{GL}_m(\widetilde{F})$. The surjectivity of the Lang isogeny $\pi : \mathrm{GL}_m(\overline{\mathbb{F}}_p) \to \mathrm{GL}_m(\overline{\mathbb{F}}_p)$ already used in Section 4.2 gives a matrix $\widetilde{D}_0 \in \mathrm{GL}_m(\overline{\mathbb{F}}_p)$ with $\widetilde{D}^{-1} = \phi_q(\widetilde{D}_0)\widetilde{D}_0^{-1}$. In fact, the entries of $\widetilde{D}_0$ belong to some finite extension $\widetilde{F}_0/\widetilde{F}$. Thus there exists a Frobenius field $(F_0, \phi_q^{F_0})$ unramified and of finite degree over (F, ϕ_q^F) and a matrix $D_0 \in \mathrm{GL}_m(\mathcal{O}_{F_0})$ such that

$$D^{-1} = \phi_q^{F_0}(D_0)(I + rG_0)D_0^{-1} \quad \text{with} \quad G_0 \in \mathcal{O}_{F_0}^{m \times m}. \tag{38}$$

Now we want to refine the resulting congruence $D^{-1} \equiv \phi_q^{F_0}(D_0)D_0^{-1} \pmod{r}$ modulo higher powers of r. The next such approximation step with $D_1 = I + rH_1$ and $\phi_q^F(r) = er$ would lead to the congruence

$$\begin{aligned} I + rG_0 &\equiv \phi_q(D_1)(I + r^2G_1)D_1^{-1} \\ &\equiv (I + \phi_q(rH_1))(I - rH_1) \\ &\equiv I + er\phi_q(H_1) - rH_1 \pmod{r^2}. \end{aligned} \tag{39}$$

Since the reduced equation $\widetilde{G}_0 = \widetilde{e}\widetilde{\phi}_q(\widetilde{H}_1) - \widetilde{H}_1$ with the q-power Frobenius endomorphism $\widetilde{\phi}_q$ over $\widetilde{F}_0$ has a solution matrix $\widetilde{H}_1$ over a finite extension $\widetilde{F}_1/\widetilde{F}_0$ there exists a Frobenius field $(F_1, \phi_q^{F_1})$ unramified and of finite degree over $(F_0, \phi_q^{F_0})$ and a matrix $D_1 = I + rH_1 \in \mathrm{GL}_m(F_1)$ such that

$$D^{-1} = \phi_q^{F_0}(D_0)\phi_q^{F_1}(D_1)(I + r^2G_1)D_1^{-1}D_0^{-1} \quad \text{with} \quad G_1 \in \mathcal{O}_{F_1}^{m\times m}. \tag{40}$$

Thus by induction we obtain a tower of unramified field extensions $F \leq F_0 \leq F_1 \leq \cdots \leq F_l$ inside $\widehat{E}$ and matrices $D_l \in \mathrm{GL}_m(F_l)$ such that

$$D^{-1} \equiv \phi_q^{F_0}(D_0)\cdots\phi_q^{F_l}(D_l)D_l^{-1}\cdots D_0^{-1} \pmod{r^{l+1}}. \tag{41}$$

Because of $D_l = I + r^l H_l \in \mathrm{GL}_m(\widehat{E})$, the product $D_0 \cdots D_l$ converges in $\mathrm{GL}_m(\widehat{E})$. Hence there exists a matrix $Y \in \mathrm{GL}_m(\widehat{E})$ with $D^{-1} = \phi_q^{\widehat{E}}(Y)Y^{-1}$. By Proposition 6.1(b), Y is a fundamental solution matrix of (M, Φ_q^F). □

Since any p-adic Frobenius module over $(\overline{\mathbb{Q}}_p, \overline{\phi}_q)$ is already defined over a finite extension $F/\mathbb{Q}_p$, from Theorem 6.2 immediately follows

Corollary 6.3. *Every integral p-adic Frobenius module over the Frobenius field $(\overline{\mathbb{Q}}_p, \overline{\phi}_p)$ becomes trivial over $\mathbb{C}_p$.*

6.4. Now we want to show by examples that integral p-adic Frobenius modules over p-adic number fields, i.e., finite extensions of $\mathbb{Q}_p$, may have transcendental solutions. For this purpose set $F := \mathbb{Q}_p$ and $M := Fb$ with Frobenius endomorphism given by $\Phi_p^F(b) = \mu b$ and $\mu \in \mathbb{Z}_p^\times$. Then by Theorem 6.2, the solution field E/F is generated by an element $y \in \widehat{\mathbb{T}}_p$ which satisfies the functional equation $\phi_p^E(y) = \mu^{-1}y$. Since ϕ_p^E is an automorphism of E/F, the field extension E/F is algebraic if and only if μ is a root of unity.

7. The Associated D-Module

7.1. Let (K, ϕ_q) be a discrete valued p-adic Frobenius field with $K \geq \mathbb{Q}_q$. Then on the rational function field $K(t)$, the Gauss norm $\| \cdot \|$ with $\| t \| = 1$ defines a valuation extending $|\cdot|$ on K. In the following, this will again be denoted by $|\cdot|$. The completion $F = \widehat{K(t)}$ of $K(t)$ with respect to the Gauss valuation becomes a Frobenius field by defining $\phi_q^F(t) = t^q$ and by extending ϕ_q^F continuously. As before the valuation ring and the valuation ideal of F are denoted by $\mathcal{O}_F$ and $\mathcal{P}_F$, respectively. Then the residue field $\widetilde{F} := \mathcal{O}_F/\mathcal{P}_F$ is isomorphic to the field of rational functions over the residue field $\widetilde{K} = \mathcal{O}_K/\mathcal{P}_K$, i.e., $\widetilde{F} = \widetilde{K}(t)$.

The p-adic Frobenius field $F = \widehat{K(t)}$ with the derivation $\widehat{\partial}_t$ obtained from $\partial_t = \frac{d}{dt}$ by continuous extension in addition is a differential field (D-field) with field of constants K. In the following an arbitrary p-adic Frobenius field (F, ϕ_q^F) with a derivation ∂_F is called a *p-adic differential field with Frobenius structure* or *a p-adic* DF*-field* for short if the D-structure and the Frobenius structure are related by the formula

$$\partial_F \circ \phi_q^F = z_F \cdot \phi_q^F \circ \partial_F \quad \text{with } z_F = \frac{\partial_F(\phi_q^F(t))}{\phi_q^F(\partial_F(t))} \in \mathcal{P}_F \tag{42}$$

for a non-constant $t \in F$. Obviously z_F is independent of the choice of t. For the powers of $\phi_q := \phi_q^F$ we derive

$$\partial_F \circ \phi_q^l = z_F \phi_q(z_F) \cdots \phi_q^{l-1}(z_F) \phi_q^l \circ \partial_F \quad \text{for } 1 \leq l \in \mathbb{N} \tag{43}$$

implying

$$\partial_F(\phi_q^l(\mathcal{O}_F)) \subseteq r^l \mathcal{O}_F \quad \text{for } r \in \mathcal{P}_K \text{ with } |r| = |z_F|. \tag{44}$$

Then the quotient field $F_l := \phi_q^l(F)$ of $\mathcal{O}_l := \phi_q^l(\mathcal{O}_F)$ is something like a "congruence constant field". In particular, $(F, \partial_F, \phi_q^F) = (\widehat{K(t)}, \widehat{\partial}_t, \phi_q^F)$ is an example of a "rational" p-adic DF-field with $F_l = \widehat{K(t^{q^l})}$.

7.2. Next we want to show that on an integral Frobenius module M over a p-adic DF-field in a natural way a derivation ∂_M can be defined which is compatible with the Frobenius action in the above sense.

Proposition 7.1. *Let $(F, \partial_F, \phi_q^F)$ be a discrete valued p-adic* DF*-field with $F \geq \mathbb{Q}_q$ and let (M, Φ_q^F) be an integral p-adic Frobenius module over F with Frobenius lattice Λ.*

(a) *For $1 \leq l \in \mathbb{N}$ the Frobenius image $\Lambda_l := \Phi_q^l(\Lambda)$ of Λ is an $\mathcal{O}_l$-lattice inside Λ containing an $\mathcal{O}_F$-basis of Λ.*

(b) *The $\mathcal{O}_{l+1}$-linear embeddings $\varphi_l : \Lambda_{l+1} \hookrightarrow \Lambda_l$ define a projective system $(\Lambda_l, \varphi_l)_{l \in \mathbb{N}}$.*

(c) *Any φ_l in* (b) *can be extended uniquely to an $\mathcal{O}_F$-isomorphism $\widetilde{\varphi}_l : \Lambda = \mathcal{O}_F \otimes_{\mathcal{O}_{l+1}} \Lambda_{l+1} \to \mathcal{O}_F \otimes_{\mathcal{O}_l} \Lambda_l = \Lambda$.*

Proof. We only have to verify (a): Λ_l is an $\mathcal{O}_l$-lattice because of $\Phi_q^l(ax) = \phi_q^l(a)\Phi_q^l(x)$ for $a \in F$ and $x \in M$ and it contains a basis of Λ since $\mathcal{O}_F(\Phi_q^F(\Lambda)) = \Lambda$ by (37). □

7.3. Now we are able to define a differential structure on M in a way very similar to the finite case (compare [17], Thm. 3.3.2, for fields of power series).

Theorem 7.2. *Let (M, Φ_q^F) be an integral p-adic Frobenius module with Frobenius lattice Λ over a discrete valued p-adic* DF*-field $(F, \partial_F, \phi_q^F)$ with $F \geq \mathbb{Q}_q$.*

(a) *Then there exists a unique derivation ∂_M on M with congruence solution lattices Λ_l, i.e.,*

$$\partial_M(\Lambda_l) \equiv 0 \pmod{r^l\Lambda} \quad \textit{for} \quad 1 \leq l \in \mathbb{N}. \tag{45}$$

(b) *This derivation is characterized by the congruences*

$$\partial_M \equiv \widetilde{\varphi}_0 \circ \cdots \circ \widetilde{\varphi}_{l-1} \circ \partial_F \circ \widetilde{\varphi}_{l-1}^{-1} \circ \cdots \circ \widetilde{\varphi}_0^{-1} \pmod{r^l\Lambda} \quad \textit{for} \quad 1 \leq l \in \mathbb{N}. \tag{46}$$

Proof. First we assume the existence of a derivation ∂_M on M and Λ with congruence solution lattices $\Lambda_l = \Phi_q^l(\Lambda)$ as in (a). Then for any basis $B_l = \{b_{l1}, \ldots, b_{lm}\}$ of Λ_l we get $\partial_M(b_{lj}) \equiv 0 \pmod{r^l\Lambda}$. Hence for $x = \sum_{j=1}^{m} a_j b_{lj} \in M$ with $a_j \in F$ we obtain

$$\partial_M(x) = \sum_{j=1}^{m} (\partial_F(a_j) b_{lj} + a_j \partial_M(b_{lj})) \equiv \sum_{j=1}^{m} \partial_F(a_j) b_{lj} \pmod{r^l\Lambda}. \tag{47}$$

This shows the uniqueness of ∂_M and the formula in (b).

Now we want to use the congruences in (b) for the definition of ∂_M. For this purpose we let $D = D_B(\Phi_q) \in \mathrm{GL}_m(\mathcal{O}_F)$ denote the representing matrix of $\Phi_q = \Phi_q^F$ with respect to a basis $B = \{b_1, \ldots, b_m\}$ of Λ. By induction we see that $B_l := BD_0 \cdots D_{l-1}$ with $D_k = \phi_q^k(D) \in \mathrm{GL}_m(\mathcal{O}_k)$ is an $\mathcal{O}_l$-basis of Λ_l. In view of the congruences in (b) we now define

$$\delta_l(x) := BD_0 \cdots D_{l-1} \partial_F(D_{l-1}^{-1} \cdots D_0^{-1} \boldsymbol{y}) \in \Lambda \tag{48}$$

for $x = \sum_{j=1}^{m} b_j y_j = B\boldsymbol{y} \in \Lambda$. Because of $\partial_F(D_l^{-1}) \equiv 0 \pmod{r^l}$ the coefficients of $\delta_l(x)$ converge in F, hence

$$\partial_\Lambda(x) := \lim_{l \to \infty} (\delta_l(x)) \in \Lambda \tag{49}$$

is well defined. It is easy to verify that the map $\partial_\Lambda : \Lambda \to \Lambda$ is additive with $\partial_\Lambda(ax) = \partial_F(a)x + a\partial_\Lambda(x)$ for $a \in \mathcal{O}_F$. Hence ∂_Λ is a derivation on Λ with congruence solution lattices Λ_l by construction. Finally, we obtain the desired derivation ∂_M on $M = F \otimes_{\mathcal{O}_F} \Lambda$ by setting $\partial_M := \partial_F \otimes \mathrm{id}_\Lambda + \mathrm{id}_F \otimes \partial_\Lambda$. □

For later use we state the explicit congruence formulae for ∂_M found in the proof as a corollary.

Corollary 7.3. *Let $B = \{b_1, \dots, b_m\}$ be a basis of the Frobenius lattice Λ in Theorem 7.2, $D := D_B(\Phi_q) \in \mathrm{GL}_m(\mathcal{O}_F)$ and $D_l := \phi_q^l(D)$ for $l \in \mathbb{N}$. Then the ∂_M-derivative of $x = B\boldsymbol{y} \in \Lambda$ has the property*

$$\partial_M(B\boldsymbol{y}) = BD_0 \cdots D_{l-1}\partial_F(D_{l-1}^{-1} \cdots D_0^{-1}\boldsymbol{y}) \pmod{r^l}. \tag{50}$$

The differential module (M, ∂_M) constructed in Theorem 7.2 is called the D-*module associated to* (M, Φ_q^F) and the triple $(M, \partial_M, \Phi_q^F)$ is a *p-adic* D*-module with Frobenius structure* or a *p-adic* DF*-module* for short.

7.4. From Theorem 7.2, we also can derive a criterion for the solutions of the D-module (M, ∂_M) associated to an integral p-adic Frobenius module. As in Section 3.3 for any extension D-field E of F the solution space of M over E is denoted by

$$\mathrm{Sol}_E(M) := \{a \otimes x \in E \otimes_F M \mid \partial_E(a) \otimes x + a \otimes \partial_M(x) = 0\}. \tag{51}$$

It is a vector space over the field of constants K_E of (E, ∂_E) with $\dim_{K_E}(\mathrm{Sol}_E(M)) \leq \dim_F(M)$. Correspondingly for the Frobenius lattice Λ in M we define

$$\mathrm{Sol}_E(\Lambda) := \{a \otimes x \in \mathcal{O}_E \otimes_{\mathcal{O}_F} \Lambda \mid \partial_E(a) \otimes x + a \otimes \partial_M(x) = 0\} \tag{52}$$

which is an $\mathcal{O}_{K_E}$-lattice of full rank in $\mathrm{Sol}_E(M)$. In the following, we use the notations M_E for $E \otimes_F M$ and Λ_E for $\mathcal{O}_E \otimes_{\mathcal{O}_F} \Lambda$.

Proposition 7.4. *Let $(M, \partial_M, \phi_q^F)$ be an integral p-adic* DF*-module with Frobenius lattice Λ as in Theorem 7.2 and Corollary 7.3. Then for any* D*-field extension E of F the following statements are equivalent.*

(a) $x = B\boldsymbol{y} \in \mathrm{Sol}_E(\Lambda)$,

(b) $\partial_E(\boldsymbol{y}) \equiv A_l\boldsymbol{y} \pmod{r^{l+1}}$ *for $l \in \mathbb{N}$ with*

$$A_l := \partial_F(D_0 \cdots D_l)(D_0 \cdots D_l)^{-1},$$

(c) $\partial_E(\boldsymbol{y}) = A\boldsymbol{y}$ *with* $A = \lim_{l \to \infty}(A_l) \in \mathcal{O}_F^{m \times m}$.

Proof. As in the proof of Theorem 7.2 let $B_l = BD_0 \cdots D_{l-1}$ be an $\mathcal{O}_l$-basis of Λ_l. Then any $x = B\boldsymbol{y} \in \Lambda_E$ with $\boldsymbol{y} \in \mathcal{O}_E^m$ can be written as

$x = B_l \boldsymbol{y}_l$ with $\boldsymbol{y}_l = (D_0 \cdots D_{l-1})^{-1}\boldsymbol{y} \in \mathcal{O}_E^m$. Now $x = B\boldsymbol{y}$ belongs to $\mathrm{Sol}_E(\Lambda)$ if and only if $\partial_E(\boldsymbol{y}_l) \equiv 0 \pmod{r^l}$ for all $l \in \mathbb{N}$. The last congruence is equivalent to

$$\partial_E(\boldsymbol{y}) = \partial_E(D_0 \cdots D_{l-1}\boldsymbol{y}_l) \equiv \partial_F(D_0 \cdots D_{l-1})\boldsymbol{y}_l = A_{l-1}\boldsymbol{y} \pmod{r^l}. \tag{53}$$

From $D_l \in \mathrm{GL}_m(\mathcal{O}_l)$ follows

$$A_l = \partial_F(D_0 \cdots D_l)(D_0 \cdots D_l)^{-1} \equiv A_{l-1} \pmod{r^l}. \tag{54}$$

Hence $A := \lim\limits_{l\to\infty}(A_l)$ exists in $\mathcal{O}_F^{m\times m}$ and the congruences in (b) are equivalent to the one equation $\partial_E(\boldsymbol{y}) = A\boldsymbol{y}$. □

Proposition 7.4 further implies the defining property of a (ϕ, ∂)-module (compare for example [17] or [4]).

Corollary 7.5. *For an integral p-adic DF-module $(M, \partial_M, \Phi_q^F)$, we have*

$$\partial_M \circ \Phi_q^F = z_F \Phi_q^F \circ \partial_M \quad \textit{with } z_F \textit{ as in (42)}. \tag{55}$$

Proof. The definition of A_l gives

$$\begin{aligned} A_l D_0 &= \partial_F(D_0 \cdots D_l)(D_0 \cdots D_l)^{-1} D_0 \\ &= \partial_F(D_0) + D_0 \partial_F(\phi_q^F(D_0 \cdots D_{l-1}))\phi_q^F(D_0 \cdots D_{l-1})^{-1} \\ &= \partial_F(D_0) + z_F D_0 \phi_q^F(\partial_F(D_0 \cdots D_{l-1})(D_o \cdots D_{l-1})^{-1}) \\ &= \partial_F(D_0) + z_F D_0 \phi_q^F(A_{l-1}) \end{aligned} \tag{56}$$

leading to the matrix identity

$$AD_0 = \partial_F(D_0) + z_F D_0 \phi_q^F(A). \tag{57}$$

But then for the basis B of M we obtain

$$\begin{aligned} \partial_M(\Phi_q^F(B)) &= \partial_M(BD_0) = -BAD_0 + B\partial_F(D_0) \\ &= -z_F BD_0\phi_q^F(A) = z_F\Phi_q^F(-BA) = z_F \Phi_q^F(\partial_M(B)). \end{aligned} \tag{58}$$

□

7.5. Now we want to apply Proposition 7.4 to two examples. First let $F = \widehat{\mathbb{Q}_p(t)}$ and $M = Fb$ the 1-dimensional Frobenius module with $\Phi_p^F(b) = tb$. Then from $D_B(\Phi_p^F) = (t)$ we obtain

$$A_l = \frac{1}{t}(1 + p + \cdots + p^l) \quad \text{and} \quad A = \frac{1}{(1-p)t}. \tag{59}$$

Thus the solution field E of the differential module M is generated by an element y with $\partial_E(y) = \frac{1}{(1-p)t}y$, hence by $y = t^{\frac{1}{1-p}} = \sqrt[p-1]{t^{-1}}$.

For the second example we use the same base field F and $M = Fb_1 \oplus Fb_2$ with $D_B(\Phi_p^F) = \begin{pmatrix} 1 & t \\ 0 & 1 \end{pmatrix}$. This leads to

$$A = \begin{pmatrix} 0 & g \\ 0 & 0 \end{pmatrix} \quad \text{with} \quad g = \sum_{j \in \mathbb{N}} p^j t^{p^j - 1} \in F. \tag{60}$$

Hence the fundamental solution matrix Y of M has the form

$$Y = \begin{pmatrix} 1 & f \\ 0 & 1 \end{pmatrix} \quad \text{with} \quad f = \sum_{j \in \mathbb{N}} t^{p^j} \notin F. \tag{61}$$

Since the differential Galois group of the solution field E/F is a subgroup of the additive group, f has to be transcendental over F.

8. The Solution Field of an Integral DF-Module

8.1. As before let F be a discrete valued p-adic DF-field with field of constants K and (M, Φ_q^F) an integral Frobenius module over F. By Theorem 7.2, the Frobenius endomorphism Φ_q^F induces a natural derivation ∂_M on M providing M with the structure of a D-module over F. Then differential Galois theory proves the existence of a *Picard-Vessiot extension* (PV-extension) $\overline{E}/\overline{F}$ for (M, ∂_M) over $\overline{F} := \overline{K} \otimes_K F$, where $\overline{K}$ denotes an algebraic closure of K (see for example [12]). By definition $\overline{E}$ is a solution field of (M, ∂_M) over $\overline{F}$. We will see that the PV-extension $\overline{E}/\overline{F}$ is already defined over a finite extension of constants of F.

Proposition 8.1. *Let (M, ∂_M) be a* D*-module over a* D*-field (F, ∂_F) with field of constants K. Then (M, ∂_M) has a solution field (E, ∂_E) with field of constants K_E of finite degree over K. Moreover, using $F_E := K_E \otimes_K F$, the field extension E/F_E is a* PV*-extension of (M, ∂_M) over F_E.*

Proof. The proof follows from the construction of a PV-ring R and a PV-field E as presented in [12] in the case of an algebraically closed field of constants. There the field of constants K_E of E (and R) is finitely generated and algebraic over K, hence finite. □

By Proposition 8.1 the D-module (M, ∂_M) associated to an integral p-adic Frobenius module (M, Φ_q^F) has a solution field E which is a PV-extension over a finite extension of constants. In contrast to the finite case, we cannot expect that its solution space $\mathrm{Sol}_E(M)$ always contains solutions of the original Frobenius module: We have already seen in Section 6.4 that the solution field of a "constant" Frobenius module over F may contain transcendental elements over F.

8.2. In this section we study the relation between the solution spaces of a DF-module with respect to its Frobenius and differential structure and prove the existence of Frobenius solution fields.

Theorem 8.2. *Let (M, Φ_q^F) be an integral p-adic Frobenius module over a discrete valued p-adic* DF*-field $(F, \partial_F, \phi_q^F)$ with field of constants $K \geq \mathbb{Q}_q$.*

(a) *Assume an* F*-field (E, ϕ_q^E) is a solution field of (M, Φ_q^F) over F with solution space $\mathrm{Sol}_E^\Phi(M)$. Then E is a* PV*-extension for the associated* D*-module (M, ∂_M) over some constant field extension $F_E := K_E \otimes_K F$, and the corresponding solution spaces are related by*

$$\mathrm{Sol}_E(M) = K_E \otimes_{K_E^\phi} \mathrm{Sol}_E^\Phi(M) \tag{62}$$

where K_E^ϕ is the fixed field of K_E with respect to ϕ_q^E.

(b) *Assume a PV-extension E/F_E with $F_E = K_E \otimes_K F$ is a solution field of (M, ∂_M). Then there exists an unramified extension $\tilde{K}/K_E$ inside $\mathbb{C}_p$ such that the Frobenius module (M, Φ_q) becomes trivial over $\tilde{E} := \tilde{K} \otimes_K E$. Moreover the solution spaces over $\tilde{F} := \tilde{K} \otimes_K F$ are related by*

$$\mathrm{Sol}_{\tilde{E}}^\Phi(M) = \mathrm{Sol}_{\tilde{E}}(M)^{\Phi_q}. \tag{63}$$

Proof. Let Λ be a Frobenius lattice of M, $B = \{b_1, \ldots, b_m\}$ an $\mathcal{O}_F$-basis of Λ and $D = D_B(\Phi_q^F) \in \mathrm{GL}_m(\mathcal{O}_F)$ the representing matrix of Φ_q^F with respect to B. Then by Proposition 6.1 there exists a fundamental solution matrix Y over E, i.e., $Y \in \mathrm{GL}_m(E)$ with $Y = D\phi_q^E(Y)$. By induction it follows that $Y = D_0 \cdots D_{l-1} Y_l$ with $D_k = \phi_q^k(D)$ and $Y_l = \phi_q^l(Y) \in \mathrm{Gl}_m(E_l)$. Hence the congruences

$$\partial_E(Y) \equiv \partial_F(D_0 \cdots D_{l-1}) Y_l = A_l Y \pmod{r^l} \tag{64}$$

with $A_l = \partial_F(D_0 \cdots D_{l-1})(D_0 \cdots D_{l-1})^{-1}$ define a derivation ∂_E on E extending ∂_F with $\partial_E(Y) = AY$ for $A = \lim_{l \to \infty}(A_l)$ as in Proposition 7.4. Thus Y is a fundamental solution matrix of (M, ∂_M) over F. By Proposition 6.1 the solution space of (M, ∂_M) is given by

$$\mathrm{Sol}_E(M) = K_E \otimes_{E^\phi} \mathrm{Sol}_E^\Phi(M) \tag{65}$$

where $E^{\phi} = K_E^{\phi}$ because of $\mathrm{Sol}_E^{\Phi}(M) \subseteq \mathrm{Sol}_E(M)$.

To prove (b) we first observe that equation (55) in Corollary 7.5 or its matrix version (57), respectively, implies that $\partial_M(\Phi_q(\mathrm{Sol}_E(M))) = 0$ from which we conclude that $\Phi_q(\mathrm{Sol}_E(M)) \subseteq \mathrm{Sol}_E(M)$. Thus the K_E-vector space $\mathrm{Sol}_E(M)$ is a Frobenius module over the field of constants K_E. By Theorem 6.2, $(\mathrm{Sol}_E(M), \Phi_q)$ possesses an unramified solution field $\tilde{K}/K_E$ inside $\mathbb{C}_p$. Hence (M, Φ_q) becomes trivial over $\tilde{E}$, and the solution spaces are related by formula (63). □

8.3. Next we are going to prove without using PV-theory that an integral p-adic Frobenius module (M, Φ_q^F) over a rational p-adic DF-field $(F, \partial_F, \phi_q^F)$, where the field of constants K has finite degree over $\mathbb{Q}_q$, has a solution field whose field of constants is unramified over K. In this case $(F, \partial_F, \phi_q^F)$ is the *field of analytic elements* $(\widehat{K(t)}, \widehat{\partial}_t, \phi_q^F)$ *with coefficients in* K (compare [3], Def. 2.1.3). Further let $\mathbb{M}_K \leq \widehat{K((t))}$ denote *the field of meromorphic functions on the open disk* $\mathcal{D}_K = \{a \in K \mid |a| < 1\}$ which becomes a DF-field by continuous extension of ∂_t and ϕ_q^F ([3], Ch. 2.5).

Proposition 8.3. *Let* $(M, \partial_M, \Phi_q^F)$ *be an integral p-adic* DF*-module over a rational p-adic* DF*-field* $(F, \partial_F, \phi_q^F)$ *with field of constants* K *of finite degree over* $\mathbb{Q}_q$. *Then the Frobenius module* (M, Φ_q^F) *has a solution* DF*-field* E *whose field of constants* K_E *is unramified over* K.

Proof. Let Λ be a Frobenius lattice in M, $B = \{b_1, \ldots, b_m\}$ an $\mathcal{O}_F$-basis of Λ and $D := D_B(\Phi_q^F) \in \mathrm{GL}_m(\mathcal{O}_F)$ the representing matrix of the Frobenius endomorphism with respect to B. By a linear transformation of t we can assume that $t = 0$ is regular for D, i.e., $D(0) \in \mathrm{GL}_m(\mathcal{O}_K)$. Then by Theorem 6.2 there exists an unramified extension L/K and a matrix $C \in \mathrm{GL}_m(\mathcal{O}_L)$ such that

$$D_0 := C^{-1} D \phi_q^L(C) \equiv I \pmod{t\mathcal{O}_L}. \tag{66}$$

Hence there exists a $G_0 \in \mathcal{O}_L^{m\times m}$ with $D_0 = I + tG_0$. This leads to

$$D_l := \phi_q^l(D_0) = I + \phi_q^l(t)\phi_q^l(G_0) = I + \phi_q^l(t)G_l \in \mathrm{GL}_m(\mathcal{O}_L). \tag{67}$$

Thus the sequence of matrices $Z_l := D_0 \cdots D_l \in \mathrm{GL}_m(\mathcal{O}_L)$ converges in $\mathrm{GL}_m(\mathbb{M}_L)$ with limit $Z := \lim_{l\to\infty}(Z_l)$. Defining E_L by $E_L := F_L(Z)$ where $F_L := F \otimes_K L$, the field E_L is a subfield of $\mathbb{M}_L$ with field of constants L and hence a DF-field.

By construction we obtain $Z = D_0\phi_q^{E_L}(Z)$. For $Y := CZ$ then follows

$$Y = CD_0\phi_q^{E_L}(Z) = CC^{-1}D\phi_q^L(C)\phi_q^{E_L}(Z) = D\phi_q^{E_L}(Y). \tag{68}$$

Hence $Y \in \mathrm{GL}_m(E)$ is a fundamental solution matrix of (M, Φ_q^F) over L. The subfield $E := F(Y)$ of E_L is a solution DF-field of (M, Φ_q^F) over F whose field of constants $K_E \leq L$ is unramified over K. □

8.4. Finally we want to generalize the solutions of (M, Φ_q^F) constructed in Proposition 8.3 to solutions convergent on the generic disc. For this purpose the completion of K_E in Proposition 8.3 is again denoted by K. Further let $\widehat{F(z)}$ be the completion of the field of rational functions over F with $|z| = 1$ with respect to the Gauss norm and $\mathbb{M}_F^* \leq F\widehat{((z-t))}$ the *field of meromorphic functions on the generic disc*

$$\mathcal{D}_F^* = \{u \in \widehat{F(z)} \,|\, |u - t| < 1\} \tag{69}$$

with coefficients in F. Then the map

$$\tau_F : F \to \mathbb{M}_F^*, f(t) \mapsto \sum_{k \in \mathbb{N}} \frac{1}{k!} \widehat{\partial}_t^k(f)(z-t)^k \tag{70}$$

identifies the valued D-field $(F, \partial_F) = (\widehat{K(t)}, \widehat{\partial}_t)$ with the subfield $(F^*, \partial_{F^*}) := (\tau_F(F), \widehat{\partial}_z)$ of $\mathbb{M}_F^*$, where the D-structure is translated by

$$\tau_F(\widehat{\partial}_t(f)) = \widehat{\partial}_z(\tau_F(f)) \tag{71}$$

(compare [3], Prop. 2.5.1). Now from Theorem 8.3 we can derive:

Corollary 8.4. *With assumptions as in Proposition 8.3 for $F := F_E$ the Frobenius module (M, Φ_q^F) has a solution field inside the field of meromorphic functions $\mathbb{M}_F^*$ on the generic disc with coefficients in F.*

Proof. By continuous extension we obtain an injection $\tau_E : E \to \mathbb{M}_F^*$ of the solution field $E \leq K((t))$ into $\mathbb{M}_F^*$ extending τ_F. This translates the equation $Y = D\phi_q^E(Y)$ into

$$Y^* := \tau_E(Y) = \tau_F(D)\tau_E(\phi_q^E(Y)) = \tau_F(D)\phi_q^{E^*}(Y^*) \tag{72}$$

with $E^* := \tau_E(E)$. Hence $E^* = F^*(Y^*)$ is a solution field in $\mathbb{M}_F^*$ of (M, Φ_q^F) over F^* (compare [3], Prop. 4.6.3). □

9. Frobenius Automorphism Groups

9.1. As at the beginning of the second part, let (F, ϕ_q) be a p-adic Frobenius field and (M, Φ_q^F) an integral p-adic Frobenius module over F with Frobenius lattice Λ and basis B of Λ. Assuming the existence of the solution field (E, ϕ_q^E) of M (as for example shown in Theorem 6.2) we define the *Frobenius automorphism group of* M (or of E/F respectively) by

$$\mathrm{Aut}^\Phi(M) := \{\gamma \in \mathrm{Aut}(E/F) \mid \gamma \circ \phi_q^E = \phi_q^E \circ \gamma\}. \tag{73}$$

This group acts in a natural manner on $M_E = M \otimes_F E$ and on $\mathrm{Sol}_E^\Phi(M)$ via $\gamma(x \otimes y) = x \otimes \gamma(y)$. In case the fundamental matrix Y belongs to $\mathrm{GL}_m(\mathcal{O}_E)$ (as in Theorem 6.2), the intersection of $\mathrm{Sol}_E^\Phi(M)$ and $\Lambda_E = \Lambda \otimes_{\mathcal{O}_F} \mathcal{O}_E$ is an $\mathcal{O}_E^\phi$-lattice with basis BY called the *lattice of integral solutions of* M in E:

$$\mathrm{Sol}_E^\Phi(\Lambda) := \mathrm{Sol}_E^\Phi(M) \cap \Lambda_E. \tag{74}$$

With these assumptions we define the *Frobenius automorphism group of* Λ to be the subgroup of $\mathrm{Aut}^\Phi(M)$ stabilizing Λ_E:

$$\mathrm{Aut}^\Phi(\Lambda) := \{\gamma \in \mathrm{Aut}^\Phi(M) \mid \gamma(\Lambda_E) \subseteq \Lambda_E\}. \tag{75}$$

Proposition 9.1. *Let (M, Φ_q^F) be an integral p-adic Frobenius module of dimension m with Frobenius lattice Λ over a p-adic Frobenius field (F, ϕ_q^F) with solution field (E, ϕ_q^E) and $E^\phi = F^\phi$.*

(a) $\mathrm{Aut}^\Phi(M)$ *is a reduced Zariski-closed subgroup of* $\mathrm{GL}_m(F^\phi)$.

(b) $\mathrm{Aut}^\Phi(\Lambda)$ *is a Zariski-closed subgroup of* $\mathrm{GL}_m(\mathcal{O}_F^\phi)$ *and is Zariski-dense in* $\mathrm{Aut}^\Phi(M)$ *in case the latter group is connected.*

Proof. For (a), let $Y \in \mathrm{GL}_m(E)$ be a fundamental solution matrix of M, i.e., $Y = D\phi_q^E(Y)$ with $D = D_B(\Phi_q^F)$. Then for any $\gamma \in \mathrm{Aut}^\Phi(M)$

$$\gamma(Y) = D \cdot \gamma(\phi_q^E(Y)) = D \cdot \phi_q^E(\gamma(Y)). \tag{76}$$

Thus $\gamma(Y)$ is again a fundamental solution matrix of M. By Proposition 6.1, there exists a $C_\gamma \in \mathrm{GL}_m(E^\phi) = \mathrm{GL}_m(F^\phi)$ with $\gamma(Y) = YC_\gamma$. Hence

$$\Gamma : \mathrm{Aut}^\Phi(M) \to \mathrm{GL}_m(F^\phi), \ \ \gamma \mapsto C_\gamma \tag{77}$$

defines an injective homomorphism because of $E = F(Y)$. As in the case of D-modules $\mathrm{Aut}^\Phi(M)$ coincides with

$$\mathrm{Aut}^\Phi(M) = \{C \in \mathrm{GL}_m(F^\phi) \mid f(YC) = 0 \quad \text{for all} \quad f \in P\} \tag{78}$$

where

$$P = \{f \in F^\phi[\mathrm{GL}_m] \mid f(Y) = 0\} \lhd F^\phi[\mathrm{GL}_m] \tag{79}$$

denotes the ideal of $Y = (y_{ij})_{i,j=1}^m$ inside the coordinate ring of GL_m over F^ϕ (compare [12], Observations 4.2). Therefore $\mathrm{Aut}^\Phi(M)$ as a subgroup of $\mathrm{GL}_m(F^\phi)$ is defined by algebraic equations and hence is Zariski-closed in $\mathrm{GL}_m(F^\phi)$ and reduced.

In (b) M has a lattice of integral solutions $\mathrm{Sol}_E^\Phi(\Lambda)$. We obtain

$$\mathrm{Aut}^\Phi(\Lambda) = \mathrm{Aut}^\Phi(M) \cap \mathrm{GL}_m(\mathcal{O}_E^\phi). \tag{80}$$

Thus $\mathrm{Aut}^\Phi(\Lambda)$ too is Zariski-closed in $\mathrm{GL}_m(\mathcal{O}_E^\phi)$. If moreover $\mathrm{Aut}^\Phi(M)$ is connected then $\mathrm{Aut}^\Phi(\Lambda)$ is Zariski-dense in $\mathrm{Aut}^\Phi(M)$ by [11], Lemma 3.2. □

From Proposition 9.1 follows that $\mathrm{Aut}^\Phi(M)$ in addition is a p-adic analytic group and $\mathrm{Aut}^\Phi(\Lambda)$ is an open compact subgroup of $\mathrm{Aut}^\Phi(M)$ (using [11], Ch. 3.1). It remains to answer the question under what conditions F already is the fixed field of $\mathrm{Aut}^\Phi(M)$ or $\mathrm{Aut}^\Phi(\Lambda)$, respectively.

9.2. Now we assume that (F, ϕ_q) is a p-adic DF-field as in §7 and $(M, \partial_M, \Phi_q^F)$ is an integral DF-module over F. Then by Proposition 8.1, a solution field E of M is a Picard-Vessiot extension over a constant field extension F_E/F. Hence without loss of generality we can assume that E/F is a PV-extension with differential Galois group $\mathrm{Gal}_\mathrm{D}(M) = \mathrm{Gal}_\mathrm{D}(E/F)$. This then is a group $\mathcal{G}(K)$ of K-rational points of a linear algebraic group $\mathcal{G}$ defined over the common field of constants K of F and E. Now in analogy to the last section we define the DF-*automorphism group of* M by

$$\mathrm{Aut}_\mathrm{D}^\Phi(M) := \{\gamma \in \mathrm{Gal}_\mathrm{D}(M) \mid \gamma \circ \phi_q^E = \phi_q^E \circ \gamma\} \tag{81}$$

and the DF-*automorphism group of* Λ by

$$\mathrm{Aut}_\mathrm{D}^\Phi(\Lambda) := \{\gamma \in \mathrm{Aut}_\mathrm{D}^\Phi(M) \mid \gamma(\Lambda_E) \subseteq \Lambda_E\} \tag{82}$$

in case a lattice of integral solutions exists.

Proposition 9.2. *Let $(M, \partial_M, \Phi_q^F)$ be an integral p-adic* DF*-module over the* DF*-field $(F, \partial_F, \phi_q^F)$ with field of constants K, lattice Λ, solution field (E, ϕ_q^E) and differential Galois group* $\mathrm{Gal}_\mathrm{D}(M) \cong \mathcal{G}(K)$.

(a) *Suppose $\mathcal{G}$ is defined over the fixed field K^ϕ of $\phi_q^F|_K$ then*

$$\mathrm{Aut}_\mathrm{D}^\Phi(M) \cong \mathcal{G}(K^\phi) \tag{83}$$

is a locally compact p-adic analytic group.

(b) *Suppose $\mathcal{G}$ is defined over the ring of integers $\mathcal{O}_K^\phi$ of K^ϕ then*

$$\mathrm{Aut}_\mathrm{D}^\Phi(\Lambda) \cong \mathcal{G}(\mathcal{O}_K^\phi) \tag{84}$$

is a compact subgroup of $\mathcal{G}(K^\phi)$.

(c) *In case* $\mathcal{G}$ *is connected, both groups* $\mathrm{Aut}_{\mathrm{D}}^{\Phi}(M)$ *and* $\mathrm{Aut}_{\mathrm{D}}^{\Phi}(\Lambda)$ *are Zariski-dense in* $\mathcal{G}(K)$.

Proof. The proofs of (a) and (b) immediately follow from Proposition 9.1 and the remark following it. It only remains to verify (c), namely that $\mathcal{G}(K^\phi)$ and $\mathcal{G}(\mathcal{O}_K^\phi)$ are Zariski-dense in $\mathcal{G}(K)$ in case $\mathcal{G}$ is connected. But this is a consequence of [11], Thm. 2.2 and Lemma 3.2, respectively. □

By Proposition 9.2 (c) in the case of connected groups the fixed fields of $\mathrm{Aut}_{\mathrm{D}}^{\Phi}(M)$ and $\mathrm{Aut}_{\mathrm{D}}^{\Phi}(\Lambda)$ equal the base field F. Hence it seems to be appropriate to call these groups DF-*Galois group of* M or *of* Λ, respectively, and to denote them by $\mathrm{Gal}_{\mathrm{D}}^{\Phi}(M)$ and $\mathrm{Gal}_{\mathrm{D}}^{\Phi}(\Lambda)$.

9.3. Finally as in the case of finite Frobenius modules for a DF-module M we want to give an upper bound for $\mathrm{Gal}_{\mathrm{D}}(M)$ (and hence for $\mathrm{Gal}_{\mathrm{D}}^{\Phi}(M)$ and $\mathrm{Gal}_{\mathrm{D}}^{\Phi}(\Lambda)$, too) only depending on the representing matrix of the Frobenius endomorphism of M.

Theorem 9.3. *Let* $(M, \partial_M, \Phi_q^F)$ *be an integral* p*-adic* DF*-module defining a* PV*-extension over a* p*-adic* DF*-field* $(F, \partial_F, \phi_q^F)$ *and* Λ *a Frobenius lattice in* M. *Let* $\mathcal{G}$ *be a reduced connected linear group defined over the field of constants* K *of* F. *Assume there exists a basis* B *of* Λ *such that* $D = D_B(\Phi_q^F)$ *belongs to the group* $\mathcal{G}(\mathcal{O}_F)$ *of* $\mathcal{O}_F$*-rational points of* $\mathcal{G}$. *Then*

$$\mathrm{Gal}_{\mathrm{D}}(M) \leq \mathcal{G}(K). \tag{85}$$

Proof. The matrices $A_l = \partial_F(D_0 \cdots D_l)(D_0 \cdots D_l)^{-1}$ in Proposition 7.4 belong to the Lie algebra $\mathrm{Lie}_F(\mathcal{G})$ of $\mathcal{G}$ over F since they are images of the logarithmic derivative

$$\lambda : \mathcal{G}(F) \to \mathrm{Lie}_F(\mathcal{G}), D \mapsto \partial_F(D)D^{-1}. \tag{86}$$

Now $\mathrm{Lie}_F(\mathcal{G})$ is complete with respect to the p-adic topology induced from F. Hence by the proof of the Proposition 7.4 the matrices A_l converge to a limit $A \in \mathrm{Lie}_F(\mathcal{G})$ defining ∂_M. But this implies $\mathrm{Gal}_{\mathrm{D}}(M) \leq \mathcal{G}(K)$ according to [12], Cor. 4.3. □

9.4. With Theorem 9.3 and Proposition 9.2 we obtain the following Frobenius Galois groups for the examples studied in Section 7.5. In the first case $M = Fb$ with $D_B(\Phi_p^F) = (t)$, the solution field E/F is cyclic of order $p-1$. Thus

$$\mathrm{Gal}(E/F) = \mathrm{Gal}_{\mathrm{D}}^{\Phi}(M) = \mathrm{Gal}_{\mathrm{D}}^{\Phi}(\Lambda) \cong Z_{p-1}. \tag{87}$$

In the second example $M = Fb_1 \oplus Fb_2$ with $D_B(\Phi_p^F) = \begin{pmatrix} 1 & t \\ 0 & 1 \end{pmatrix}$, the differential Galois group is the additive group. This leads to

$$\mathrm{Gal}_{\mathrm{D}}(E/F) = \mathrm{Gal}_{\mathrm{D}}^{\Phi}(M) \cong \mathbb{Q}_p \quad \text{and} \quad \mathrm{Gal}_{\mathrm{D}}^{\Phi}(\Lambda) \cong \mathbb{Z}_p. \tag{88}$$

10. Reduction of p-adic DF-Modules

10.1. As before let $(M, \partial_M, \Phi_q^F)$ be an integral p-adic DF-module with Frobenius lattice Λ over a DF-field $(F, \partial_F, \phi_q^F)$ and with field of constants $K \geq \mathbb{Q}_q$. Then the residue field $\widetilde{F} := \mathcal{O}_F/\mathcal{P}_F$ of the valuation ring $\mathcal{O}_F$ (with respect to the Gauss norm) is a field of characteristic p. Assuming $\mathcal{O}_F$ is an ID-ring, i.e.,

$$\frac{1}{k!}\partial_F^k(\mathcal{O}_F) \subseteq \mathcal{O}_F \quad \text{for } k \in \mathbb{N}, \tag{89}$$

$\widetilde{F}$ is an IDF-field with respect to the reduced Frobenius action $\widetilde{\phi}_q$ and the *induced iterative derivation* $\partial_{\widetilde{F}}^*$ defined by

$$\partial_{\widetilde{F}}^{(k)}(a + \mathcal{P}_F) := \frac{1}{k!}\partial_F^k(a) + \mathcal{P}_F \quad \text{for } k \in \mathbb{N}. \tag{90}$$

Moreover the field of iterative differential constants $\widetilde{K}$ of $\widetilde{F}$ coincides with the residue field $\mathcal{O}_K/\mathcal{P}_K$.

Now let $B = \{b_1, \ldots, b_m\}$ be a $\mathcal{O}_F$-basis of Λ and $D = D_B(\Phi_q^F) \in \mathrm{GL}_m(\mathcal{O}_F)$ the representing matrix of Φ_q^F. Then $\Lambda = \bigoplus_{i=1}^m \mathcal{O}_F b_i$ reduces to the $\widetilde{F}$-vector space

$$\widetilde{M} := \Lambda/\mathcal{P}_F\Lambda = \bigoplus_{i=1}^m \widetilde{F}\widetilde{b}_i \tag{91}$$

with the reduced basis $\widetilde{B} = \{\widetilde{b}_1, \ldots, \widetilde{b}_m\}$ and with a $\widetilde{\phi}_q$-semilinear Frobenius action given by

$$\widetilde{\Phi}_q : \widetilde{M} \to \widetilde{M}, \quad \widetilde{B} \mapsto \widetilde{B} \cdot \widetilde{D}(\Phi_q^F), \tag{92}$$

where $\widetilde{D}(\Phi_q^F) \in \mathrm{GL}_m(\widetilde{F})$ denotes the matrix obtained from $D(\Phi_q^F)$ by reducing modulo $\mathcal{P}_F$. Hence $(\widetilde{M}, \widetilde{\Phi}_q)$ is a Frobenius module over $\widetilde{F}$. According to Proposition 3.1, the Frobenius endomorphism $\widetilde{\Phi}_q$ defines a canonical iterative derivation $\partial_{\widetilde{M}}^*$ on $\widetilde{M}$ such that $(\widetilde{M}, \partial_{\widetilde{M}}^*, \widetilde{\Phi}_q)$ becomes an IDF-module over $\widetilde{F}$, called the *residue module of* $(M, \partial_M^*, \Phi_q^F)$. The next proposition shows that any finite Frobenius module can be obtained as a residue module of an integral p-adic Frobenius module, i.e., every finite Frobenius module can be lifted to an integral p-adic Frobenius module.

Proposition 10.1. *Let $(F, \partial_F, \phi_q^F)$ be a discrete valued p-adic* DF*-field with $F \geq \mathbb{Q}_q$, valuation ring $\mathcal{O}_F$ and valuation ideal $\mathcal{P}_F$ with respect to the Gauss norm. If $\mathcal{O}_F$ is an* ID*-ring then the residue field $\widetilde{F} = \mathcal{O}_F/\mathcal{P}_F$ is an* IDF*-field with the induced iterative derivation $\partial^*_{\widetilde{F}}$. Let $(\widetilde{M}, \partial^*_{\widetilde{M}}, \widetilde{\Phi}_q)$ be an* IDF*-module over $\widetilde{F}$. Then there exists an integral p-adic* DF*-module $(M, \partial_M, \Phi_q^F)$ over F and a Frobenius lattice $\Lambda \subseteq M$ such that the residue module $\Lambda/\mathcal{P}_F\Lambda$ is isomorphic to $\widetilde{M}$ as an* IDF*-module.*

Proof. Let $\widetilde{B}$ be a basis of $\widetilde{M}$ and $\widetilde{D} := D_{\widetilde{B}}(\widetilde{\Phi}_q) \in \mathrm{GL}_m(\widetilde{F})$ be the matrix of $\widetilde{\Phi}_q$ with respect to $\widetilde{B}$. Then there exists a matrix $D \in \mathrm{GL}_m(\mathcal{O}_F)$ with residue matrix $\widetilde{D}$. On a free $\mathcal{O}_F$-module Λ of rank m with basis B we can define a ϕ_q^F-semilinear Frobenius action by $\Phi_q^F(B) := BD$ and we can extend it ϕ_q^F-semilinearly to the F-vector space $M := F \otimes_{\mathcal{O}_F} \Lambda$. By Theorem 7.2, Φ_q^F defines a canonical derivation ∂_M on M providing M with the structure of a integral p-adic DF-module $(M, \partial_M, \Phi_q^F)$. Obviously the residue module of $(M, \partial_M, \Phi_q^F)$ equals $(\widetilde{M}, \partial^*_{\widetilde{M}}, \widetilde{\Phi}_q^F)$ since by Proposition 3.1 the iterative derivation $\partial^*_{\widetilde{M}}$ is uniquely determined by $\widetilde{\Phi}_q$. □

10.2. Next we want to study the behavior of the Frobenius automorphism groups under reduction. For simplicity we restrict ourselves to the case of a rational p-adic DF-field $F = \widehat{K(t)}$. Then the valuation ring $\mathcal{O}_F$ is automatically an ID-ring (for $\partial_F = \partial_t$). Further by Proposition 8.3 we can assume without loss of generality that the solution field E has no new constants, i.e., E/F is a PV-extension.

Theorem 10.2. *Let $(M, \partial_M, \Phi_q^F)$ be an integral p-adic* DF*-module with Frobenius lattice Λ over a discrete valued rational p-adic* DF*-field $(F, \partial_F, \phi_q^F)$ with field of constants K. Assume that the solution field E/F has no new constants and that the differential Galois group $\mathrm{Gal}_{\mathrm{D}}(M)$ is connected and defined over $\mathcal{O}_K^{\phi}$. Then the residue module $(\widetilde{M}, \partial^*_{\widetilde{M}}, \widetilde{\Phi}_q)$ of M is a finite Frobenius module over the residue field $\widetilde{F} = \mathcal{O}_F/\mathcal{P}_F$ whose Galois group is bounded by*

$$\mathrm{Gal}^{\widetilde{\Phi}}(\widetilde{M}) \leq \mathrm{Gal}^{\Phi}_{\mathrm{D}}(\Lambda)/\,\mathrm{Gal}^{\Phi}_{\mathrm{D}}(\Lambda)_1 \qquad (93)$$

where $\mathrm{Gal}_{\mathrm{D}}^{\Phi}(\Lambda)_1$ *denotes the principal congruence subgroup of the p-adic analytic group* $\mathrm{Gal}_{\mathrm{D}}^{\Phi}(\Lambda)$.

Proof. For the proof we follow the construction of the PV-ring of M over $\mathcal{O}_F$ instead of F. Let $B = \{b_1, \ldots, b_m\}$ be a basis of Λ and $D = D_B(\Phi_q^F) \in \mathrm{GL}_m(\mathcal{O}_F)$ the matrix of Φ_q^F with respect to B. Then we have

$$\Phi_q^F(B) = BD \quad \text{and} \quad \partial_M(B) = -BA \tag{94}$$

with the matrix $A \in \mathcal{O}_F^{m\times m}$ constructed in Proposition 7.4(c). The ring $U := \mathcal{O}_F[\mathrm{GL}_m] = \mathcal{O}_F[x_{ij}, \det(x_{ij})^{-1}]_{i,j=1}^m$ becomes a DF-ring with Frobenius endomorphism ϕ_q^U and derivation ∂_U defined by

$$\phi_q^U(X) = D^{-1}X \quad \text{and} \quad \partial_U(X) = AX \quad \text{for} \quad X = (x_{ij})_{i,j=1}^m. \tag{95}$$

Further U is an ID-ring by Corollary 7.5 because of

$$\partial_U^k(X) = \partial_U^k(D_0 \cdots D_{l-1}X_l) \equiv \partial_U^k(D_0 \cdots D_{l-1})X_l \pmod{r^l} \tag{96}$$

with $X_l = D_{l-1}^{-1} \cdots D_0^{-1}X \in (\phi_q^U)^l(U)$. Now let $Q \lhd U$ be a maximal ideal stable under the action of ϕ_q^U and of ∂_U and let $P \lhd U$ be a maximal differential ideal containing the "DF-ideal" Q. Then the $\mathcal{O}_F$-algebra $R := U/P$ is an "integral PV-ring" with quotient field E (up to differential isomorphism). By assumption E is the solution field of the Frobenius module (M, Φ_q^F), too. Hence there exists a fundamental solution matrix $Y \in \mathrm{GL}_m(R)$ with

$$\partial_E(Y) = AY \quad \text{and} \quad \phi_q^E(Y) = D^{-1}Y. \tag{97}$$

Thus the ring R of integral differentially finite elements in E is closed by the action of the induced Frobenius ϕ_q^E of ϕ_q^U (i.e., the maximal differential ideal P already is a DF-ideal) and an ID-ideal. By assumption $\mathrm{Gal}_{\mathrm{D}}(M) = \mathrm{Gal}_{\mathrm{D}}(E/F)$ is connected, hence by Proposition 9.2 $\mathrm{Gal}_{\mathrm{D}}^{\Phi}(\Lambda)$ is Zariski-dense in $\mathrm{Gal}_{\mathrm{D}}(M)$. Its matrix representation on the solution space is given by

$$\Gamma : \mathrm{Gal}_{\mathrm{D}}^{\Phi}(\Lambda) \to \mathrm{GL}_m(\mathcal{O}_K^{\phi}), \gamma \mapsto C_\gamma \tag{98}$$

where $\gamma(Y) = YC_\gamma$.

Now let $\widetilde{S} := \widetilde{F} \otimes_{\mathcal{O}_F} R$ be the residue ring of R over $\widetilde{F}$, i.e., $\widetilde{S} = \widetilde{F}[\widetilde{y}_{ij}, \det(\widetilde{y}_{ij})]_{i,j=1}^m$ with $\widetilde{y}_{ij} = 1 \otimes y_{ij} \in \widetilde{S}$. Because of $D \in \mathrm{GL}_m(\mathcal{O}_F)$, the residue matrix $\widetilde{D}$ belongs to $\mathrm{GL}_m(\widetilde{F})$ and defines the reduced Frobenius endomorphism $\widetilde{\phi}_q$ on $\widetilde{S}$ whose restriction to $\widetilde{F}$ coincides with the q-power Frobenius endomorphism. The derivation ∂_E restricted to R defines an iterative derivation $\partial^*_{\widetilde{S}}$ on $\widetilde{S}$ as in Section 10.1 compatible with the Frobenius structure, i.e., $(\widetilde{S}, \partial^*_{\widetilde{S}}, \widetilde{\phi}_q)$ is an IDF-ring. Hence $\widetilde{S}$ can be obtained as a quotient of $\widetilde{F}[\mathrm{GL}_m]$ by an IDF-ideal $\widetilde{Q}$:

$$\widetilde{S} = \widetilde{F}[\mathrm{GL}_m]/\widetilde{Q}. \tag{99}$$

The Frobenius automorphism group

$$\mathrm{Aut}^{\widetilde{\Phi}}(\widetilde{S}/\widetilde{F}) = \{\widetilde{\gamma} \in \mathrm{GL}_m(\widetilde{F}^{\phi}) \mid \widetilde{\gamma}(\widetilde{Q}) \subseteq \widetilde{Q}\} \tag{100}$$

is a finite group since $\widetilde{F}^{\phi} = \mathbb{F}_q$. Because $\widetilde{\gamma(Y)} = \widetilde{YC_\gamma} = \widetilde{Y}\widetilde{C}_\gamma$ for $\gamma \in \mathrm{Gal}^{\Phi}_{\mathrm{D}}(\Lambda)$, the restriction map

$$\widetilde{\ } : \mathrm{Gal}^{\Phi}_{\mathrm{D}}(\Lambda) \to \mathrm{Aut}^{\widetilde{\Phi}}(\widetilde{S}/\widetilde{F}), C_\gamma \mapsto \widetilde{C}_\gamma \tag{101}$$

is a group homomorphism with kernel $\mathrm{Gal}^{\Phi}_{\mathrm{D}}(\Lambda)_1$. It is surjective since the fixed field of $\widetilde{S}$ by $\mathrm{Aut}^{\widetilde{\Phi}}(\widetilde{S}/\widetilde{F})$ can also be obtained as the reduced fixed field of R by $\mathrm{Aut}^{\Phi}_{\mathrm{D}}(\Lambda)$ which equals F. This proves

$$\mathrm{Aut}^{\widetilde{\Phi}}(\widetilde{S}/\widetilde{F}) \cong \mathrm{Gal}^{\Phi}_{\mathrm{D}}(\Lambda)/\,\mathrm{Gal}^{\Phi}_{\mathrm{D}}(\Lambda)_1. \tag{102}$$

Now let $\widetilde{P} \lhd \widetilde{F}[\mathrm{GL}_m]$ be a maximal ID-ideal containing $\widetilde{Q}$. Then $\widetilde{R} := \widetilde{F}[\mathrm{GL}_m]/\widetilde{P}$ is an IPV-ring for $\widetilde{M}$ with Frobenius structure and with

$$\begin{aligned}\mathrm{Gal}^{\widetilde{\Phi}}(\widetilde{M}) = \mathrm{Gal}^{\widetilde{\Phi}}(\widetilde{R}/\widetilde{F}) &= \{\widetilde{\gamma} \in \mathrm{GL}_m(\mathbb{F}_q) \mid \widetilde{\gamma}(\widetilde{P}) \subseteq \widetilde{P}\} \\ &\le \{\widetilde{\gamma} \in \mathrm{GL}_m(\mathbb{F}_q) \mid \widetilde{\gamma}(\widetilde{Q}) \subseteq \widetilde{Q}\} = \mathrm{Aut}^{\widetilde{\Phi}}(\widetilde{S}/\widetilde{F}) \\ &\cong \mathrm{Gal}^{\Phi}_{\mathrm{D}}(\Lambda)/\,\mathrm{Gal}^{\Phi}_{\mathrm{D}}(\Lambda)_1.\end{aligned} \tag{103}$$

□

There remains the question under which conditions equality can be proved. This will be answered in the next section in some special cases.

10.3. Nori's examples in Theorem 5.2 are Galois extensions over $\mathbb{F}_q(t)$ unramified outside the denominator of (t) with unipotently generated Galois group $\mathcal{G}(\mathbb{F}_q)$. By construction these belong to effective Frobenius modules $(\widetilde{M}, \widetilde{\Phi}_q)$ over $\widetilde{F} = \mathbb{F}_q(t)$ with $D_{\widetilde{B}}(\widetilde{\Phi}_q) \in \mathcal{G}(\mathbb{F}_q[t])$ with respect to a suitable basis $\widetilde{B}$ of $\widetilde{M}$. (compare [9], Cor. 7.6 or [8], Cor. 7.16, respectively). For modules of this type the following holds.

Proposition 10.3. *Let $\mathcal{G}$ be a reduced connected linear group defined over $\mathbb{Z}_q$ with the property that $\mathcal{H}(\mathbb{F}_q)$ is a proper subgroup of $\mathcal{G}(\mathbb{F}_q)$ for all* *proper connected subgroups* $\mathcal{H}$ *of* $\mathcal{G}$*. Let* $(\widetilde{F}, \partial^*_{\widetilde{F}}) = (\widetilde{K}(t), \partial^*_t)$ *be a field with iterative derivation and finite field of constants* $\widetilde{K} \geq \mathbb{F}_q$*. Assume* $(\widetilde{M}, \widetilde{\Phi}_q)$ *is a Frobenius module over* $\widetilde{F}$ *with*

$$\mathrm{Gal}^{\widetilde{\Phi}}(\widetilde{M}) = \mathcal{G}(\mathbb{F}_q) \quad \text{and} \quad D_{\widetilde{B}}(\widetilde{\Phi}_q) \in \mathcal{G}(\widetilde{K}[t]) \tag{104}$$

for a suitable basis $\widetilde{B}$ *of* $\widetilde{M}$*. Then there exists an effective integral p-adic* DF*-module* $(M, \partial_M, \Phi^F_q)$ *over* $F = \widehat{K(t)}$*, where* $\mathrm{Quot}(\mathbb{W}(\widetilde{K})) \leq K \leq \widehat{\mathbb{T}}_p$ *with differential Galois group*

$$\mathrm{Gal}_{\mathrm{D}}(M) \cong \mathcal{G}(K) \quad \text{and} \quad \mathrm{Gal}^{\Phi}_{\mathrm{D}}(\Lambda) \cong \mathcal{G}(\mathbb{Z}_q) \tag{105}$$

for the corresponding Frobenius lattice Λ *in* M*. The residue module of* $(M, \partial_M, \Phi^F_q)$ *coincides with* $(\widetilde{M}, \partial^*_{\widetilde{M}}, \widetilde{\Phi}_q)$*, and the Galois groups are related by*

$$\mathrm{Gal}^{\widetilde{\Phi}}(\widetilde{M}) = \mathrm{Gal}^{\Phi}_{\mathrm{D}}(\Lambda) / \mathrm{Gal}^{\Phi}_{\mathrm{D}}(\Lambda)_1. \tag{106}$$

Proof. For the proof we almost only have to collect earlier results. First let K be a quotient field of $\mathbb{W}(\widetilde{K})$ and $F := \widehat{K(t)}$ the rational p-adic DF-field of analytic elements with $\partial_F = \partial_t$. Then we can find a matrix $D \in \mathcal{G}(\widehat{K[t]})$ with residue matrix $\widetilde{D}$. Using D as representing matrix of a Frobenius operator of a free $\mathcal{O}_F$-module Λ, we obtain a Frobenius lattice and by $M := F \otimes_{\mathcal{O}_F} \Lambda$ an integral p-adic Frobenius module over F together with a canonical derivation ∂_M by Theorem 7.2. By Proposition 8.3 M has a solution field E whose field of constants K_E is a subfield of $\widehat{\mathbb{T}}_p$. Now we substitute K by the completion $\widehat{K}_E$ of K_E and F by $\widehat{K(t)}$. Then E/F is a PV-extension with $\mathrm{Gal}_{\mathrm{D}}(E/F) \leq \mathcal{G}(K)$ by Theorem 9.3.

From $D \in \mathcal{G}(\widehat{K[t]})$ it follows by Proposition 7.4 that in the induced differential equation $\partial_E(\boldsymbol{y}) = A\boldsymbol{y}$ the matrix $A \in \mathrm{Lie}_F(\mathcal{G})$ has entries belonging to $\widehat{K(t)}$. Hence E/F has at most one singular point and therefore a connected differential Galois group $\mathrm{Gal}_{\mathrm{D}}(E/F)$, which is of the form $\mathcal{H}(K)$ with $\mathcal{H}$ an algebraic subgroup of $\mathcal{G}$. Then by Theorem 10.2 the Galois group of the residue module $\widetilde{M}$ is a subgroup of $\mathcal{H}(\mathbb{F}_q) = \mathcal{H}(K)/\mathcal{H}(K)_1$. This implies $\mathcal{H}(\mathbb{F}_q) = \mathcal{G}(\mathbb{F}_q)$ and thus $\mathcal{H}(K) = \mathcal{G}(K)$ by assumption. In particular, (M, Φ^F_q) is an effective Frobenius module over F. □

10.4. An elementary numerical example is given by the additive group already treated in Sections 7.5 and 9.4. Here the field $F = \widehat{\mathbb{Q}_p(t)}$ of analytic elements over $\mathbb{Q}_p$ reduces to the rational function field $\widetilde{F} =$

$\mathbb{F}_p(t)$, and the integral DF-module $(M, \Phi_p^F, \partial_M)$ over F with $D_B(\Phi_p^F) = \begin{pmatrix} 1 & t \\ 0 & 1 \end{pmatrix}$ and $\mathrm{Gal}_{\mathrm{D}}^{\Phi}(\Lambda) = \mathbb{Z}_p$ has the residue module $(\widetilde{M}, \widetilde{\Phi}_p, \partial_{\widetilde{M}}^*)$ with $D_{\widetilde{B}}(\widetilde{\Phi}_p) = \begin{pmatrix} 1 & t \\ 0 & 1 \end{pmatrix}$ and

$$\mathrm{Gal}^{\widetilde{\Phi}}(\widetilde{M}) \cong \mathbb{F}_p \cong \mathbb{Z}_p/p\mathbb{Z}_p \cong \mathrm{Gal}_{\mathrm{D}}^{\Phi}(\Lambda)/\,\mathrm{Gal}_{\mathrm{D}}^{\Phi}(\Lambda)_1. \tag{107}$$

References

[1] A. Borel and J.-P. Serre, *Théorèmes de finitude en cohomologie galoisienne*, Comment. Math. Helvet. **39** (1964), 111–164.

[2] J.-B. Bost et al., "Courbes Semi-stables et Groupe Fondamental en Géometrie Algébrique," Birkhäuser, Boston, 2000.

[3] G. Christol, "Modules différentiels et equations différentielles p-adiques," Queen's Paper in Pure and Applied Mathematics, 66. Queen's University, Kingston, 1983.

[4] P. Colmez, *Les conjectures de monodromie p-adiques*, Astérisque, (to appear).

[5] D. Goss, "Basic Structure of Function Field Arithmetic," Springer-Verlag, Berlin, 1996.

[6] M. Kreuzer and L. Robbiano, "Computational Commutative Algebra 1," Springer-Verlag, Berlin, 2000.

[7] G. Malle, *Explicit realization of the Dickson groups* $\mathrm{G}_2(q)$ *as Galois groups*, Pacific J. Math., (to appear).

[8] B. H. Matzat, "Differential Galois Theory in Positive Characteristic," IWR-Preprint 2001-35.

[9] B. H. Matzat and M. van der Put, *Iterative differential equations and the Abhyankar conjecture*, J. reine angew. Math. **257** (2003), 1–52.

[10] M. V. Nori, *Unramified coverings of the affine line in positive characteristic*, in "Algebraic Geometry and its Applications (C. L. Bajaj, ed.)," 209–212, Springer-Verlag, 1994.

[11] V. Platonov and A. Rapinchuk, "Algebraic Groups and Number Theory," Academic Press, Boston, 1994.

[12] M. van der Put, *Galois theory of differential equations, algebraic groups and Lie algebras*, J. Symb. Comput. **28** (1999), 441–472.

[13] M. van der Put, *Bounded p-adic differential equations*, in "Circumspice, Various Papers in and Around Mathematics in Honor of Arnoud van Rooij," Subfaculteit Wiskunde, Katholieke Universiteit Nijmegen, 2001.

[14] D. J. Saltman, *Generic Galois extensions and problems in field theory*, Adv. Math. **43** (1982), 250–283.

[15] T. A. Springer, "Linear Algebraic Groups," Birkhäuser, Boston, 1998.

[16] Y. Taguchi, *Regular singularity of Drinfeld modules*, Internat. J. of Math. **5** (1994), 595–608.

[17] N. Tsuzuki, *The overconvergence of morphisms of etale φ-∇-spaces on a local field*, Compositio math. **103** (1996), 227–239.

[18] B. A. F. Wehrfritz, "Infinite Linear Groups," Springer-Verlag, Berlin, 1973.

Part III

Algebraic number theory

Galois Theory and Modular Forms pp 271–286
K. Hashimoto, K. Miyake, and H. Nakamura, Editors
© 2003 Kluwer Academic Publishers

ON QUADRATIC NUMBER FIELDS EACH HAVING AN UNRAMIFIED EXTENSION WHICH PROPERLY CONTAINS THE HILBERT CLASS FIELD OF ITS GENUS FIELD*

Ken YAMAMURA
Department of Mathematics,
National Defense Academy,
Hashirimizu Yokosuka 239-8686, Japan
yamamura@cc.nda.ac.jp

1. Introduction

The purpose of this article is to describe simple ways to construct quadratic number fields each having an unramified extension which properly contains the Hilbert class field of its genus field (in the wide sense). The motivation of this study is the author's observation that under the Generalized Riemann Hypothesis (GRH), for most quadratic number fields of small conductors, their maximal unramified extensions coincide with the Hilbert class fields of their genus fields. More precisely, under GRH, among the 305 imaginary quadratic number fields with discriminants larger than -1000, at most 16 fields are exceptional [39], [40], and among the 1690 real quadratic number fields with discriminants less than or equal to 5565, only 4 fields are exceptional [41]. Thus, the following problem arises naturally:

Problem. *Characterize the quadratic number fields K having an unramified extension which properly contains the Hilbert class field of the genus field of K.*

*This article is based on the author's talk in the conference "Galois theory and modular forms" held at Saga University (December 14–17, 1999). It was long before the decision of the publication of the proceedings and some parts were already published. Therefore, we slightly change some parts and add many new numerical data.

[received: Jul. 03, 2002; accepted in revised form: Oct. 14, 2002]

Before describing our basic ideas, we introduce some notations. We denote by $K = \mathbb{Q}(\sqrt{d})$ a quadratic number field with discriminant d. K_{ur} denotes its maximal unramified extension and K_{gen} its genus field in the wide sense, that is, the maximal unramified abelian extension of K which is abelian over $\mathbb{Q}$. Note that we refer also to the unramifiedness of the infinite primes and therefore in the case where K is real, this is not the usual one which admits the ramification of the infinite primes. For a number field F, we denote by the sequence $F = F_0 \subseteqq F_1 \subseteqq F_2 \subseteqq \cdots$ the class field tower of F, that is, F_{n+1} is the Hilbert class field of F_n. The smallest number l with $F_{l+1} = F_l$ is called the length of the tower. In these notations, the above problem is to characterize K with $K_{\mathrm{ur}} \supsetneqq (K_{\mathrm{gen}})_1$.

We have the following inclusions:

$$K_{\mathrm{ur}} \supseteqq K_\infty \supseteqq K_3 \supseteqq K_2 \supseteqq (K_{\mathrm{gen}})_1 \supseteqq K_1 \supseteqq K_{\mathrm{gen}} \supseteqq K,$$

where $K_\infty = \cup_n K_n$ is the top of the class field tower of K, that is, the maximal unramified solvable Galois extension of K. These inclusions lead us to the following three basic ideas.

(1) If $K_{\mathrm{ur}} \supsetneqq K_\infty$, that is, if K has an unramified nonsolvable Galois extension, then $K_{\mathrm{ur}} \supsetneqq (K_{\mathrm{gen}})_1$.

(2) If $K_3 \neq K_2$, that is, if the length of the class field tower of K is at least three, then $K_{\mathrm{ur}} \supsetneqq (K_{\mathrm{gen}})_1$.

(3) If $K_2 \supsetneqq (K_{\mathrm{gen}})_1$, then $K_{\mathrm{ur}} \supsetneqq (K_{\mathrm{gen}})_1$.

On the other hand, the composition is also a basic idea. For some quadratic number fields K, there exists a Galois extension L of $\mathbb{Q}$ such that the compositum KL is an unramified extension of K not contained in $(K_{\mathrm{gen}})_1$. If F is a quadratic number field with $F_{\mathrm{ur}} \supsetneqq (F_{\mathrm{gen}})_1$, then any quadratic number field K with $K_{\mathrm{gen}} \supseteqq F$ has the same property $K_{\mathrm{ur}} \supsetneqq (K_{\mathrm{gen}})_1$. Therefore the existence of such one field implies that a positive rate of the quadratic number fields have the same property.

Now for simpler description, we introduce the new terminology "quadratic-ramified".

Definition. Let F be an algebraic number field of finite degree. We call an extension E of F *quadratic-ramified* (over F) if there exists a quadratic extension K/F such that the compositum KE/K is unramified at all finite primes.

Note that an extension E/F is quadratic-ramified if and only if L/F is so, where L is the normal closure of E over F. We will study quadratic-ramified normal extensions of the rational number field $\mathbb{Q}$.

The organization of this article is as follows. In Section 2, we describe the simplest way to construct quadratic number fields K with $K_{\mathrm{ur}} \supsetneqq K_{\mathrm{gen}}$. This way is the one to construct quadratic-ramified S_n-extensions of $\mathbb{Q}$. In Section 3, we review some facts on the structure of Galois groups of quadratic-ramified normal extensions of $\mathbb{Q}$. In Section 4, we describe quadratic number fields with class field towers of length $\geqq 3$ and give a simple way to obtain imaginary quadratic number fields with class field towers of length at least three.

2. Unramified A_n-extensions of quadratic number fields

In this section we will describe the simplest way to construct quadratic number fields K with $K_{\mathrm{ur}} \supsetneqq K_{\mathrm{gen}}$.

First, we review the following general fact.

Lemma 1 ([16]. See also [27], [37], [38].). *Let E be an algebraic number field of degree $n \geqq 3$. If the discriminant d_E of E is a fundamental quadratic discriminant, that is, a discriminant of a quadratic number field, then the normal closure M of E is an S_n-extension of $\mathbb{Q}$ and is unramified at all finite primes over its quadratic subfield $\mathbb{Q}(\sqrt{d_E})$.*

Sketch of Proof. If d_E is a fundamental quadratic discriminant, then for its any rational prime divisor p, the prime p has exactly one ramified divisor in E and its ramification index is two. Then for any prime ideal $\mathfrak{P}$ of M above p, its inertia group $I_{\mathfrak{P}}$ is generated by a transposition if we consider $\mathrm{Gal}(M/\mathbb{Q})$ as a (transitive) subgroup of S_n. By the Čebotarev monodromy theorem, $\mathrm{Gal}(M/\mathbb{Q})$ is generated by all $I_{\mathfrak{P}}$ for $\mathfrak{P} \mid d_E$, and therefore $\mathrm{Gal}(M/\mathbb{Q}) \cong S_n$, because any transitive subgroup of S_n generated by transpositions is S_n (see Proposition 1 below).

The unramifiedness follows from $I_{\mathfrak{P}} \cap A_n = \{1\}$. □

By this lemma from an algebraic number field of degree $n \geqq 3$ whose discriminant d_E is a fundamental quadratic discriminant, we obtain a quadratic number field $\mathbb{Q}(\sqrt{d_E})$ having an unramified (at all finite primes) A_n-extension. Therefore if E is totally real or $d_E < 0$, and $n \geqq 4$, then M is an unramified extension of $\mathbb{Q}(\sqrt{d_E})$ not contained in $(\mathbb{Q}(\sqrt{d_E})_{\mathrm{gen}})_1$. (Note that since $\mathbb{Q}(\sqrt{d_E})_{\mathrm{gen}}$ and M are linearly disjoint over $\mathbb{Q}(\sqrt{d_E})$, we have $\mathrm{Gal}((\mathbb{Q}(\sqrt{d_E})_{\mathrm{gen}})_1 M/\mathbb{Q}(\sqrt{d_E})_{\mathrm{gen}}) \cong \mathrm{Gal}(M/\mathbb{Q}(\sqrt{d_E})) \cong A_n$. Moreover, for any $K \neq \mathbb{Q}(\sqrt{d_E})$, we have $\mathrm{Gal}(KM/K) \cong \mathrm{Gal}(M/\mathbb{Q}) \cong S_n$ and if furthermore $K_{\mathrm{gen}} \supset \mathbb{Q}(\sqrt{d_E})$, then KM/K is unramified at all finite primes.) By constructing (infinitely many) number fields with squarefree discriminants †, Y. Yamamoto and K. Uchida independently proved the following [33], [34], [35].

Theorem 1. *For each natural number $n \geqq 3$, there exist infinitely many real and infinitely many imaginary quadratic number fields having an unramified (at all finite primes) A_n-extension (S_n-extension).*

Therefore, we conclude that to find such a field E is the simplest way to obtain K with desired property. (For $n \geqq 5$, this gives also the easiest way to obtain examples of unramified nonsolvable Galois extensions of number fields.) In fact, probabilistic argument leads the author to the conjecture stating that at least 60% of algebraic number fields of each degree $n \geqq 3$ have discriminants that are fundamental quadratic and the conjecture stating that for at least 40% of monic polynomials with rational integer coefficients of each degree $n \geqq 3$, their root generates an algebraic number field with squarefree discriminant [38]. Therefore we may expect that it is easy to find such fields unless we require that the fields are totally real. We exhibit below a list of negative and totally real squarefree discriminants of degree n for $4 \leqq n \leqq 16$ that are smallest in absolute value among known ones. (Here, for simplicity, we call the discriminant of a number field *totally real* if the field is totally real.) They are really known to be smallest for $n \leqq 7$ and for $n = 8, 9$ in the totally real case ([7], [8], [10], [25], [26], [6], [13], [32]). Most of the others are due to D. Simon [31] ‡.

Table 1

n	negative	totally real
4	−283	1957
5	−4511	24217
6	−14731	592661
7	−184607	20134393
8	−4296211	483345053
9	−109880167	9685993193
10	−215067767	513087549389
11	−5781612911	30733206894581
12	−172922533711	1272819520434029
13	−5894076408919	62557040066579369
14	−9977823237443	6013825446222379157
15	−336785064838151	363145034778499996657
16	−13627506010977623	75341762604432007766521

† They do not note the squarefreeness of discriminants, we can easily verify it.

‡ D. Simon kindly gave the author his large data [31] few of which can be found at his homepage. The author expresses gratitude to him.

3. Galois groups of quadratic-ramified normal extensions

In the previous section we describe a way to obtain quadratic-ramified S_n-extensions of $\mathbb{Q}$. In this section we review some facts on the structure of Galois groups of quadratic-ramified normal extensions of $\mathbb{Q}$ as primitive permutation groups, and give examples of such extensions.

Our study is based on the following.

Lemma 2. *Let F be an algebraic number field with no nontrivial extension unramified at all finite primes (e.g. the rational number field $\mathbb{Q}$, any imaginary quadratic number field with class number one, any cyclotomic field with class number one, etc.). Then for any quadratic-ramified normal extension L of F, its Galois group G is generated by elements of order two.*

Proof. Let H be the subgroup of G generated by the inertia groups of the prime ideals of L which are ramified over F and M the corresponding subextension of L/F. Then M/F is unramified at all finite primes and therefore M must coincide with F. Hence H coincides with G. Since L/F is quadratic-ramified, any inertia group of a ramified prime ideal is a cyclic group of order two. □

Let E/F be an extension of degree n and L its normal closure over F. Then the Galois group of L/F can be considered as a transitive subgroup of S_n, the symmetric group of degree n. Now, we consider primitive subgroups of S_n generated by elements of order two. An element of order two of S_n is a transposition or a product of disjoint transpositions. We know the following facts on such subgroups.

Proposition 1. *If a transitive subgroup of S_n is generated by transpositions, this subgroup coincides with S_n.*

Proposition 2 ([18, Theorem 3]). *Let G be a primitive subgroup of S_n other than S_n and A_n. If G contains a product of two disjoint transpositions, then one of the following holds:*

(1) $n = 5$ *and* $G \cong D_5$*, the dihedral group of order ten.*

(2) $n = 6$ *and* $G \cong \mathrm{PSL}(2,5) \cong A_5$*, the alternating group of degree five.*

(3) $n = 6$ *and* $G \cong \mathrm{PGL}(2,5) \cong S_5$*, the symmetric group of degree five.*

(4) $n = 7$ *and* $G \cong \mathrm{PSL}(2,7)$.

(5) $n = 8$ *and* $G \cong \mathrm{Hol}(C_2^3)$, *the holomorph of* C_2^3, *the elementary abelian 2-groups of order eight.*

Moreover, all these subgroups except for those of (3) *are generated by products of two disjoint transpositions. (The exceptional groups contain odd permutations*[§]*. The information for these groups is useless for our study.)*

As described in the previous section, Yamamoto and Uchida proved the existence of infinitely many quadratic-ramified S_n-extensions of $\mathbb{Q}$ for each $n \geqq 3$. All such extensions are unramified at all finite primes over their quadratic subfields. (We note that for $n \geqq 4$ there exist quadratic-ramified S_n-extensions of $\mathbb{Q}$ that are not unramified at all finite primes over their quadratic subfields.) Moreover, the author studied the distribution of algebraic number fields of degree $n \geqq 3$ whose normal closures are quadratic-ramified S_n-extensions of $\mathbb{Q}$ that are unramified at all finite primes over their quadratic subfields, and posed a conjecture for its density.

T. Kondo proved the following.

Theorem 2 ([18, Theorem 1]). *Let* E *be an algebraic number field of degree* $n \geqq 3$ *whose discriminant* d_E *is the square of an odd squarefree rational integer* δ *and* M *its normal closure. We put* $G = \mathrm{Gal}(M/\mathbb{Q})$ *and consider it as a transitive subgroup of* S_n. *If* G *is primitive, then one of the following holds:*

(a) $G \cong A_n$.

(b) $n = 5$ *and* $G \cong D_5$.

(c) $n = 6$ *and* $G \cong \mathrm{PSL}(2,5) \cong A_5$.

(d) $n = 7$ *and* $G \cong \mathrm{PSL}(2,7)$.

(e) $n = 8$ *and* $G \cong \mathrm{Hol}(C_2^3)$.

Moreover, in the cases (b)–(e), M *is quadratic-ramified and for any quadratic number field* K *whose discriminant is divisible by* δ, MK/K *is unramified at all finite primes and furthermore* MK/K *is a* G*-extension unless* $K \subset M$ *in the case* (b).

Note. The oddness of d_E is not essential for the quadratic-ramified-ness. However, we cannot weaken this to that $\sqrt{d_E}$ is fundamental quadratic.

[§]They are missing in [18, Theorem 3]. Since Kondo considers [18, Theorem 3] as a lemma for [18, Theorem 1] (Theorem 2 below), he forgets to list.

In fact, for some G in the cases (b)–(e), there exist G-extensions with d_E even and $\sqrt{d_E}$ fundamental that are not quadratic-ramified. Nevertheless, this restriction is meaningful, because once one knows the Galois groups in the cases (b)–(e), the quadratic-ramified-ness follows from the cubefreeness of d_E.

Sketch of Proof. By the assumption $d_E = \delta^2$, for any prime ideal of M dividing d_E, its inertia group is generated by a 3-cycle or a product of two disjoint transpositions. Any primitive group containing a 3-cycle is A_n. If G does not contain a 3-cycle, only the other cases can occur by Proposition 2, because $G \subsetneqq A_n$ from $d_E = \delta^2$. □

Thus, the problem is the existence of quadratic-ramified G-extensions of $\mathbb{Q}$.

Note that in the case (b) where $G \cong D_5$, M has exactly one quadratic subfield and is unramified over it. Therefore the class number of the unique quadratic subfield of M is divisible by five. Conversely, we obtain a quadratic-ramified D_5-extension of $\mathbb{Q}$ from a quadratic number field with class number divisible by five. More generally, Y. Yamamoto [35] proved the existence of infinitely many real and infinitely many imaginary quadratic-ramified D_m-extensions of $\mathbb{Q}$ for general $m \geqq 3$ by constructing real and imaginary quadratic number fields whose class groups contain an element of order m.

T. Kondo [1], [17], [18] verified that there exist many quadratic-ramified A_5-extensions of $\mathbb{Q}$. They are realized by specializations of sextic polynomial constructed by A. Brumer (as splitting fields). A. Brumer constructed the following family of curves of genus 2 with three parameters b, c, d whose Jacobian varieties have real multiplication:

$$\begin{aligned}
y^2 = f(x; b, c, d) :=& \\
x^6 + 2cx^5 + (c^2 + 2c + 2 - bd)x^4 &+ (2c^2 + 2c + 2 - 2bd + b - 4d)x^3 \\
&+ (c^2 + 4c + 5 - bd + 3b)x^2 + (2c + 6 + 3b)x + (b + 1).
\end{aligned}$$

(Until now his result seems not to have been published yet, though his result has been often cited. K. Hashimoto [9] reconstructed this family.) K. Anai and T. Kondo [1], [17] verified that the Galois group of $f(x; b, c, d)$ (over the field $\mathbb{Q}(b, c, d)$) is isomorphic to A_5. (It is expected that $f(x; b, c, d)$ is parametric for A_5-extensions of $\mathbb{Q}$.) T. Kondo [17], [18] observed that many specializations of b, c, d to rational integers yield polynomials whose splitting fields are quadratic-ramified A_5-extensions of $\mathbb{Q}$. We give below some examples of quadratic-ramified A_5-extensions of $\mathbb{Q}$ with small discriminant δ^2. (Imaginary ones in Table 2 and real ones in Table 3.) Each such an extension M is given by the coefficients of a sextic polynomial $X^6+a_1X^5+\cdots+a_5X+a_6$ whose minimal splitting field is M.

Table 2

a_1	a_2	a_3	a_4	a_5	a_6	δ
-2	1	-3	2	4	1	653 (prime)
-2	4	0	-7	4	1	1061 (prime)
-1	1	-7	-7	-1	1	1381 (prime)
-3	1	-1	3	-1	-1	$1507 = 11 \cdot 137$
-1	-3	0	4	0	1	1553 (prime)

Table 3

a_1	a_2	a_3	a_4	a_5	a_6	δ
-3	-7	16	10	-12	1	$5567 = 19 \cdot 293$
-3	-5	16	2	-8	1	8311 (prime)
-3	-11	15	51	29	1	10463 (prime)
-1	-19	-18	26	22	-3	10687 (prime)
-1	-20	21	17	-21	4	$10931 = 17 \cdot 643$

Now we give an important remark. When we search for quadratic-ramified A_5-extensions of $\mathbb{Q}$, it is much better to consult a list of sextic fields of type A_5 than to consult a list of quintic fields of type A_5. (Here, we mean by "of type A_5" that the normal closure is an A_5-extension of $\mathbb{Q}$.) In fact, quadratic-ramified fields are rare in the latter list, however, are often found in the former. (Compare the table in [2] and the database due to Klüners and Malle [14]. Among the 308 (non-isomorphic) quintic fields of signature $(1,2)$ and type A_5 with discriminants $\leqq 2083^2$, only 8 are quadratic-ramified. On the other hand, among (known) 211 (non-isomorphic) sextic fields of signature $(2,2)$ and type A_5 with discriminants $\leqq 21524^2$, at least 70 are quadratic-ramified. (Check is omitted for fields with even discriminants.)) Therefore Theorem 2 is meaningful for our study. Why does such a difference occur? Before explaining the reason, we first note that in most lists of algebraic number fields, fields are ordered by absolute values of discriminants and we briefly review the relationship between discriminants and ramifications. Let E be an algebraic number field with discriminant d_E. We know that for a prime ideal of $\mathfrak{p}$ of E, we have $\mathfrak{p} \mid d_E$ if and only if $\mathfrak{p}$ is ramified and that $N_{E/\mathbb{Q}}(\mathfrak{p})^{e-1} \mid d_E$, where e is the ramification index of $\mathfrak{p}$. Therefore the factorization of $|d_E|$ is closely related to the ramification in E. Let M be an A_5-extension of $\mathbb{Q}$ and E_5 (resp. E_6) its quintic (resp. sextic) subfield. If M is quadratic-ramified and p is an odd prime number ramified in M, then the factorization of p in E_5 is 1^21^21 or 2^21. (The former means that p has exactly three prime divisors, their degrees are all one, and their ramification indices are $2, 2, 1$, respectively.) In both cases, $p^2 \parallel d_{E_5}$, and the factorization of p in E_6 is $1^2 1^2 11$ and $1^2 1^2 2$, respectively and therefore $p^2 \parallel d_{E_6}$. On the other hand, if M is not quadratic-ramified but d_{E_5} is cubefree, any prime divisor q of d_{E_5} without factorization of the above types is neither 2 nor 3, and its factorization in E_5 is $1^3 11$ or $1^3 2$. (In fact, such factorizations can occur.) In these cases, the factorization of q in E_6 is $1^3 1^3$ and we have $q^4 \parallel d_{E_6}$. Similar argument works for the other possible types of factorization. (All possible types of factorization of ramified primes in quintic fields of type A_5 are tabulated in [3, p. 46, Table 3.1].) Thus, quadratic-ramified fields often appear in a list of sextic fields of type A_5. The above consideration leads us to the conjecture that among sextic fields of type A_5, quadratic-ramified fields have a density > 0.5.

The author verified that there exist many quadratic-ramified PSL(2, 7)-extensions of $\mathbb{Q}$. They are realized by specializations of the following septic polynomial with two parameters a and A whose Galois group over $\mathbb{Q}(a, A)$ is isomorphic to PSL(2, 7) constructed by S. E. LaMacchina [19]:

$$x^7 + (-6a+2)x^6 + (8a^2+4a-3)x^5 + (-14a^2+6a-2-A)x^4 + (-8a^3+6a^2-4a+2+A)x^3 + (8a^3+16a^2)x^2 + (8a^3-12a^2)x - 8a^3.$$

The author observed that many specializations of a and A to rational integers yield polynomials whose splitting fields are quadratic-ramified PSL(2, 7)-extensions of $\mathbb{Q}$. We give below some examples of quadratic-ramified PSL(2, 7)-extensions of $\mathbb{Q}$. (Imaginary ones in Table 4 and real ones in Table 5.) Each such an extension M is given by the coefficients of a septic polynomial $X^7 + a_1 X^6 + \cdots + a_6 X + a_7$ whose minimal splitting field is M.

Table 4

a_1	a_2	a_3	a_4	a_5	a_6	a_7	δ
−1	−3	0	4	−1	−2	1	$1417 = 13 \cdot 109$
−3	2	4	−3	−2	−1	−1	2741 (prime)
0	−2	−4	−1	5	5	1	3803 (prime)
−1	−3	−1	2	4	4	1	$3983 = 7 \cdot 569$
−2	0	4	−3	−4	2	1	4157 (prime)

Table 5

a_1	a_2	a_3	a_4	a_5	a_6	a_7	δ
0	−13	−8	36	27	−12	1	$46517 = 181 \cdot 257$
−3	−7	15	6	−11	−3	1	$62185 = 5 \cdot 12437$
−1	−11	13	24	−35	11	−1	$81109 = 7 \cdot 11587$
0	−22	0	112	−48	−82	1	110849 (prime)
−3	−8	25	8	−28	−15	−1	122693 (prime)

We have the following observation. Among the list of septic fields of type PSL(2, 7) in the database due to Klüners and Malle, quadratic-ramified ones are often found. More precisely, among known 222 (non-isomorphic but possibly arithmetically equivalent) septic fields of type PSL(2, 7) of signature (3, 2) with discriminants $\leqq 70549^2$, at least 103 fields are quadratic-ramified, and among known 136 totally real septic fields of type PSL(2, 7) with discriminants $\leqq 4531655^2$, at least 36 fields are quadratic-ramified.

By the database due to Klüners and Malle we verified the existence of quadratic-ramified $\mathrm{Hol}(C_2^3)$-extensions of $\mathbb{Q}$. We give below some examples of such extensions of $\mathbb{Q}$. (Imaginary ones in Table 6 and real ones in Table 7.) Each such an extension M is given by the coefficients of an octic polynomial $X^8 + a_1X^7 + \cdots + a_7X + a_8$ whose minimal splitting field is M.

Table 6

a_1	a_2	a_3	a_4	a_5	a_6	a_7	a_8	δ
−4	8	−11	12	−10	6	−3	2	5717 (prime)
−2	0	0	3	−2	0	−1	2	6037 (prime)
−3	5	−4	0	3	0	−3	2	6971 (prime)
−2	2	−3	5	−4	4	−2	1	8317 (prime)
0	0	−3	−1	2	6	2	1	$10055 = 5 \cdot 2011$

Table 7

a_1	a_2	a_3	a_4	a_5	a_6	a_7	a_8	δ
−2	−17	15	53	−1	−37	−15	−1	640663 (prime)
−1	−21	23	126	−113	−241	126	1	749971 (prime)
−2	−17	40	30	−78	−14	35	1	$855659 = 7 \cdot 251 \cdot 487$
3	−11	−34	31	108	−8	−96	−25	$1018757 = 953 \cdot 1069$
0	−12	5	31	−12	−14	8	−1	$1038325 = 5 \cdot 41 \cdot 1013$

We have the following observation. Among the list of octic fields of type $\mathrm{Hol}(C_2^3)$ in the database due to Klüners and Malle, quadratic-ramified ones are often found. More precisely, among known 234 (non-isomorphic) totally imaginary octic fields of type $\mathrm{Hol}(C_2^3)$ with discriminants $\leqq 420257044152^2$, at least 47 fields are quadratic-ramified, among known 134 (non-isomorphic) octic fields of type $\mathrm{Hol}(C_2^3)$ of signature (4, 2) with discriminants $\leqq 982007569^2$, at least 41 fields are quadratic-ramified, and among known 596 totally real octic fields of type $\mathrm{Hol}(C_2^3)$ with discriminants $\leqq 6650745841216^2$, at least 169 fields are quadratic-ramified. We note that there exist many quadratic-ramified octic fields of type $\mathrm{Hol}(C_2^3)$ whose discriminants are divisible by the fourth power of an odd prime. For example, the splitting field of the polynomial

$X^8 + 4X^7 - 408X^6 - 2158X^5 + 43577X^4 + 282422X^3 - 505558X^2 + 89217X - 10321$ is an imaginary quadratic-ramified $\mathrm{Hol}(C_2^3)$-extension of $\mathbb{Q}$, in which only the rational prime 5717 is ramified. (The factorization of the prime 5717 in any octic field defined by this polynomial is $1^2 1^2 2^2$.) By this example we can see that the primitive subgroups of S_8 of order 1344 (unique up to conjugacy and all isomorphic to $\mathrm{Hol}(C_2^3)$) are generated by products of four disjoint transpositions.

Thus, also for $G = A_5, \mathrm{PSL}(2,7), \mathrm{Hol}(C_2^3)$, we expect that there exist infinitely many real and infinitely many imaginary quadratic-ramified G-extensions of $\mathbb{Q}$, and its proof is our next subject. By our observation of the rate of quadratic-ramified G-extensions in the database, it is expected that for these groups G, among G-extensions of $\mathbb{Q}$, quadratic-ramified ones have a positive density. Since $\mathrm{Hol}(C_2^3)$ is an (group) extension of $\mathrm{PSL}(2,7)$ by C_2^3, for this group, we can treat the problem also as an embedding problem.

To conclude this section, we give examples of some other quadratic-ramified nonsolvable Galois extensions of $\mathbb{Q}$ found by consulting Klüner and Malle's database.

The splitting field of the polynomial $X^8 - X^7 - 29X^6 + 111X^5 - 139X^4 + 37X^3 + 32X^2 - 10X - 1$ is a real quadratic-ramified $\mathrm{PGL}(2,7)$-extension of $\mathbb{Q}$, in which only the rational prime 107509 is ramified. (The factorization of the prime 107509 in any octic field defined by this polynomial is $1^2 1^2 1^2 11$.) By this example we can see that the primitive subgroups of S_8 of order 336 (unique up to conjugacy and all isomorphic to $\mathrm{PGL}(2,7)$) are generated by products of three disjoint transpositions.

The splitting field of the polynomial $X^{10} - 4X^9 - 110X^8 + 660X^7 - 131X^6 - 2311X^5 + 30X^4 + 2453X^3 + 1401X^2 + 261X + 14$ is a real quadratic-ramified $\mathrm{PSL}(2,9)$-extension of $\mathbb{Q}$, in which only the rational prime 354317 is ramified. (The factorization of the prime 354317 in any field of degree 10defined by this polynomial is $112^2 2^2$.) Note that $\mathrm{PSL}(2,9) \cong A_6$. By this example we can see that the primitive subgroup of S_{10} of order 360 (unique up to conjugacy and all isomorphic to $\mathrm{PSL}(2,9)$) are generated by products of four disjoint transpositions. Like as search for quadratic-ramified A_5-extensions, in order to find quadratic-ramified A_6-extensions, it is much better to consult list of fields of degree 10 than to consult list of fields of degree 6.

The splitting field of the polynomial¶ $X^{11} - 4X^{10} - 25X^9 + 81X^8 + 237X^7 - 562X^6 - 1010X^5 + 1574X^4 + 1805X^3 - 1586X^2 - 847X + 579$ is a real quadratic-ramified $\mathrm{PSL}(2,11)$-extension of $\mathbb{Q}$, in which only the

¶This polynomial is found also in [12].

rational prime 74843 is ramified. (The factorization of the prime 74843 in any field of degree 11 defined by this polynomial is $1^2 1^2 1 2^2 2$.) By this example we can see that the primitive subgroups of S_{11} of order 660 (unique up to conjugacy and all isomorphic to PSL(2, 11)) are generated by products of four disjoint transpositions.

The splitting field of the polynomial $X^{15} + 4X^{14} + 17X^{13} + 59X^{12} + 148X^{11} + 295X^{10} + 406X^9 + 75X^8 - 132X^7 - 514X^6 - 436X^5 - 124X^4 + 145X^3 + 106X^2 + 20X - 1$ is an imaginary quadratic-ramified PSL(4, 2)-extension of $\mathbb{Q}$, in which only the rational prime 12037 is ramified. (The factorization of the prime 12037 in any field of degree 15 defined by this polynomial is $1 2 2^2 4^2$.) By this example we can see that the primitive subgroups of S_{15} of order 20160 (unique up to conjugacy and all isomorphic to PSL(4, 2)) are generated by products of six disjoint transpositions.

4. Class field towers of quadratic number fields

Here we discuss class field towers of quadratic number fields.

First we review sufficient conditions for the infiniteness. Let p be a prime number. For a number field F, we denote by the sequence $F = F_0^{(p)} \subsetneqq F_1^{(p)} \subsetneqq \cdots \subsetneqq F_n^{(p)} \subsetneqq F_{n+1}^{(p)} \cdots$ the p-class field tower of F, i.e., $F_{n+1}^{(p)}$ is the Hilbert p-class field of $F_n^{(p)}$. If $\mathrm{Cl}^{(p)}(K)$, the p-part of the class group $\mathrm{Cl}(K)$ of K, has sufficiently large p-rank, then $[K_\infty^{(p)} : K] = \infty$: For odd p, if the p-rank of $\mathrm{Cl}^{(p)}(K)$ is at least three, then $[K_\infty^{(p)} : K] = \infty$ (Venkov-Koch-Schoof [15], [28]). If K is imaginary (resp. real) and the 2-rank of $\mathrm{Cl}^{(2)}(K)$ is at least five (resp. six), then $[K_\infty^{(2)} : K] = \infty$. Since the 2-rank of $\mathrm{Cl}^{(2)}(K)$ is t or $t-1$, where t is the number of prime divisors of d, we can easily construct K with infinite 2-class field tower. However, for odd p, though it is conjectured that there exist infinitely many real and infinitely many imaginary quadratic number fields K whose p-class groups have rank $\geqq 3$, it is difficult to construct such K. The infiniteness of such fields has been proven only for $p = 3$. For $p \geqq 5$, only few examples of such imaginary fields are known for $p = 5, 7, 11$.

Next we consider smaller rank cases. For odd p, if the p-rank of $\mathrm{Cl}^{(p)}(K)$ is at least two, then the length of p-class field tower of K is at least two. In this case we have

$$K_2 \supseteqq K_2^{(p)}(K_{\mathrm{gen}})_1 \supsetneqq (K_{\mathrm{gen}})_1.$$

In fact, since K_{gen} and $K_2^{(p)}$ are linearly disjoint over K, $\mathrm{Gal}(K_2^{(p)} K_{\mathrm{gen}}/K_{\mathrm{gen}}) \cong \mathrm{Gal}(K_2^{(p)}/K)$ is not abelian. Therefore, if $K_2 =$

$(K_{\text{gen}})_1$, the odd part of $\text{Cl}(K)$ is cyclic. We note that the Cohen-Lenstra heuristics [5] state that the probability that an imaginary quadratic number field has class group whose odd part is cyclic is about 0.97757 and this conjecture matchs well numerical data. Thus, the following problems arise.

Problem 1. *Characterize the quadratic number fields K with*

$$K_2 = (K_{\text{gen}})_1 \neq K_1.$$

Problem 2. *Characterize the quadratic number fields K with*

$$K_2^{(2)} = (K_{\text{gen}})_1^{(2)} \neq K_1^{(2)}.$$

We conclude this article by giving a simple way to obtain imaginary quadratic number fields with class field towers of length at least three.

First, we repeat the idea described in [40]. Let F be a real real quadratic number field with $\text{Cl}^{(2)}(F) \cong C_{2^m}$ and $\text{Cl}_+^{(2)}(F) \cong C_{2^{m+1}}$ for some $m \geqq 1$, where $\text{Cl}_+^{(2)}(F)$ denotes the 2-part of the narrow class group $\text{Cl}_+(F)$ of F. Then the narrow Hilbert 2-class field $F_{1,nar}^{(2)}$ of F is a dihedral CM-field: $\text{Gal}(F_{1,nar}^{(2)}/\mathbb{Q}) \cong D_{2^{m+1}}$. For simplicity, we put $M = F_{1,nar}^{(2)}$ and $M_+ = F_1^{(2)}$. Now we assume that

(∗) the relative class number $h^-(M)$ of M is greater than one.

Then from F we can construct a family of infinite imaginary quadratic number fields whose class field towers are of length at least three. In fact, let d' be any negative fundamental discriminant prime to $d(F)$ and put $d = d' \cdot d(F)$. Then $K = \mathbb{Q}(\sqrt{d})$ is an imaginary quadratic number field and the length of the class field tower of K is at least three. Now the Hilbert class field M_1 of M is normal over $\mathbb{Q}$ and $H := \text{Gal}(M_1/\mathbb{Q})$ is a solvable group with $H'' \neq \{1\}$. This implies $K_3 \neq K_2$, because the compositum KM_1 is an unramified Galois extension of K with $\text{Gal}(KM_1/K) \cong H$. For since H' corresponds (by Galois theory) to the maximal abelian subfield of M_1 and since M_1/F is unramified at all finite primes, this field is the genus field F_g of F (in the narrow sense). By the assumption on F, we have $F \subsetneqq F_g \subseteqq M_+$. On the other hand, since $2 \nmid h(M)$ by [39, Lemma 9], we have $\text{Cl}_+(M_+) \cong C_2 \times \text{Cl}(M_+)$. Hence the condition (∗) $h^-(M) > 1$ implies that M_1/M_+ is not abelian, neither is M_1/F_g. Thus, $H'' \neq \{1\}$ and $K_3 \neq K_2$.

Now we characterize such real real quadratic number fields F in detail. For $\text{Cl}^{(2)}(F) \cong C_{2^m}$ and $\text{Cl}_+^{(2)}(F) \cong C_{2^{m+1}}$ for some $m \geqq 1$, it is necessary and sufficient that $d(F)$ is of the form $8p$ or pq, where p and

q are distinct prime numbers with $p \equiv q \equiv 1 \pmod 4$, and the norm of the fundamental unit ε of F is 1. We know necessary and sufficient conditions for $m = 1$, and those for $m = 2$ (see [36]).

We know that there exist only finitely many normal CM-fields with relative class number one [24] and all dihedral CM-fields with relative class number one have already determined by S. Louboutin et al [20], [21], [22].

For $m = 1$ the above is summarized in [40]. Here we summarize the above for $m = 2$ as follows.

Proposition 3. *Let p and q are distinct prime numbers satisfying the following conditions*:

(i) $p \equiv 1,\ q \not\equiv 3 \pmod 4$.

(ii) $(p/q)_4 = (q/p)_4 = 1$.

(iii) *The fundamental unit of* $\mathbb{Q}(\sqrt{pq})$ *has norm* 1.

(iv) *If* $q = 2,\ p \neq 257$. *Otherwise*, $\{p, q\} \neq \{5, 101\}$, $\{5, 181\}$, $\{13, 53\}$, $\{13, 61\}$.

Then for any negative discriminant d' prime to pq, the imaginary quadratic number field $\mathbb{Q}(\sqrt{pqd'})$ has class field tower of length at least three.

Remark. For the pairs $\{p, q\}$ excluded in (iii), $F_{1,nar}$ is a dihedral CM-field of degree 16 with relative class number one.

Example. Let $F = \mathbb{Q}(\sqrt{5 \cdot 401})$. Then $\mathrm{Cl}(F) \cong C_4$, and $\mathrm{Cl}_+(F) \cong C_8$. We take $d' = -3$, that is, let $K = \mathbb{Q}(\sqrt{(-3) \cdot 5 \cdot 401})$. Then $d(K) = -6015$ and $\mathrm{Cl}(K) \cong V_4 \times C_9$. By the results of Kisilevsky in [11], we have $K_\infty^{(2)} = K_2^{(2)}$ and $\mathrm{Gal}(K_2^{(2)}/K) \cong D_8$. Thus, the length of the 2-class field tower of K is two, the length of the 3-class field tower of K is one, the length of the l-class field tower of K is zero for any odd $l \neq 3$, but the length of the class field tower of K is at least three.

References

[1] H. Anai and T. Kondo, *A family of sextic polynomials with Galois group A_5 – computation of splitting fields and Galois groups*, in "Proceedings of AC95 (Tokyo, 1995)," 1996; electoronical; ftp://tnt.math.metro-u.ac.jp/pub/ac95/anai.tar.gz.

[2] J. Basmaji and I. Kiming, *A table of A_5-fields*, in "On Artin's conjecture for odd 2-dimensional representations (G. Frey, ed.)," Lecture Notes in Mathematics, 1585. 37–46, 122–141, Springer-Verlag, New York and Berlin, 1994.

[3] J. P. Buhler, "Icosahedral Galois representations," Lecture Notes in Mathematics, 654. Springer-Verlag, New York and Berlin, 1978.

[4] G. Butler and J. McKay, *The transitive groups of degree up to eleven*, Comm. Algebra **11** (1983), 863–911.

[5] H. Cohen and H. W. Lenstra, Jr., *Heuristics on class groups of number fields*, in "Number Theory, Noordwijkerhout, 1983 (H. Jager, ed.)," Lecture Notes in Mathematics, 1068. 33–62, Springer-Verlag, Berlin and New York, 1984.

[6] F. Diaz y Diaz, *Valuers minima du discriminant des corps de degré* 7 *ayant une seule place réelle*, C. R. Acad. Sci. Paris Sér. I Math. **296** (1983), 137–139.

[7] H. J. Godwin, *Real quartic fields with small discriminant*, J. London Math. Soc. **31** (1956), 478–485.

[8] H. J. Godwin, *On quartic fields of signature one with small discriminant*, Quart. J. Math. Oxford (2) **8** (1957), 214–222.

[9] K. Hashimoto, *On Brumer's family of RM-curves of genus two*, Tôhoku Math. J. (2) **52** (2000), 475–488.

[10] J. Hunter, *The minimum discriminants of quintic fields*, Proc. Glasgow Math. Assoc. **5** (1957), 57–67.

[11] H. Kisilevsky, *Number fields with class number congruent to* 4 *mod* 8 *and Hilbert's Theorem* 94, J. Number Theory **8** (1976), 271–279.

[12] J. Klüners, *A Polynomial with Galois Group* $\mathrm{SL}_2(11)$, J. Symbolic Comput. **30** (2000), 733-737.

[13] J. Klüners and G. Malle, *A Database for field extension of the rationals*, LMS J. Comput. Math. **4** (2001), 182–196.

[14] J. Klüners and G. Malle, *A Database for Polynomials over the Rationals*, available from http://www.iwr.uni-heidelberg.de/~Juergen.Klueners/datbase.ps.

[15] H. Koch and B. B. Venkov, *The p-tower of class fields for an imaginary quadratic field*, Modules and representations, Zap. Naučn. Sem. Leningrad. Otdel. Mat. Inst. Steklov. (LOMI) **46** (1974), 5–13, 140, (Russian); English translation, J. Soviet Math. **9** (1978), 291–199.

[16] T. Kondo, *Algebraic number fields with the discriminant equal to that of a quadratic number field*, J. Math. Soc. Japan **47** (1995), 31–36.

[17] T. Kondo, *Some family of sextic polynomials and their Galois groups*, in "Proc. twelfth symp. on algebraic combinatorial theory (M. Kitazume, ed.)," 165–176, 1995, (Japanese).

[18] T. Kondo, *Some examples of unramified extensions over quadratic number fields*, Sci. Rep. Tokyo Woman's Christian Univ., Nos. 121–124 (1997), 1399–1410.

[19] S. E. LaMacchina, *Polynomials with Galois group* $\mathrm{PSL}(2,7)$, Comm. Algebra **8** (1980), 983–992.

[20] Y. Lefeuvre, *Corps diédraux à multiplication comlexe principaux*, Ann. Inst. Fourier (Grenoble) **50** (2000), 67–103.

[21] Y. Lefeuvre and S. Louboutin, *The class number one problem for dihedral CM-fields*, in "Algebraic number theory and Diophantine analysis (Graz, 1998)," 249–275, de Gruyter, Berlin, 2000.

[22] S. Louboutin, *The class number one problem for the dihedral and dicyclic CM-fields*, Colloq. Math. **80** (1999), 259–265.

[23] J. Nakagawa, *On the Galois group of a number field with square free discriminant*, Comm. Math. Univ. Santi Pauli **37** (1988), 95–98.

[24] A. M. Odlyzko *Some analytic estimates of class numbers and discriminants*, Invent. Math. **29** (1975), 275–286.

[25] M. Olivier, *Corps sextiques primitifs*, Ann. Inst. Fourier (Grenoble) **40** (1990), 757–767.

[26] M. Pohst, *The minimum discriminant of seventh degree totally real algebraic number field*, in "Number Theory and Algebra (H. Zassenhaus, ed.)," 235–240, Academic Press, New York, 1977.

[27] A. Scholz, *Aufgabe* 208, Jahresber. Deutsch. Math.-Verein. **45** (1935), 110; *Lösung der Aufgabe* 208, ibid. **47** (1937), 47.

[28] R. Schoof, *Infinite class field towers of quadratic fields*, J. reine angew. Math. **372** (1987), 209–220.

[29] D. Simon, *Équations dans les corps de nombres et discriminants minimeaux*, Thèse, Univérsite Bordeaux I, 1998.

[30] D. Simon, *Petits Discriminants de Polynômes Irréductibles*, http://www.math.unicaen.fr/~simon/maths/TableSmallDisc.html.

[31] D. Simon, *davantagepolynomes.gp.gz*, On demand.

[32] K. Takeuchi, *Totaly real algebraic number fields of degree* 9 *with small discriminant*, Saitama Math. J. **17** (1999), 63–85.

[33] K. Uchida, *Unramified extensions of quadratic number fields, I*, Tôhoku Math. J. **23** (1970), 138–140.

[34] K. Uchida, *Unramified extensions of quadratic number fields, II*, Tôhoku Math. J. **23** (1970), 220–224.

[35] Y. Yamamoto, *On unramified Galois extensions of quadratic number fields*, Osaka J. Math. **7** (1970), 57–76.

[36] Y. Yamamoto, *Divisibility by* 16 *of class numbers of quadratic fields whose* 2-*class groups are cyclic*, Osaka J. Math. **21** (1984), 1–22.

[37] K. Yamamura, *On unramified Galois extensions of real quadratic number fields*, Osaka J. Math. **23** (1986), 471-486

[38] K. Yamamura, *Some analogue of Hilbert's irreducibility theorem and the distribution of algebraic number fields*, J. Fac. Sci. Univ. Tokyo Sect. IA **38** (1991), 99–135.

[39] K. Yamamura, *Maximal unramified extensions of imaginary quadratic number fields of small conductors*, J. Théor. Nombres Bordeaux **9** (1997), 405–448.

[40] K. Yamamura, *Maximal unramified extensions of imaginary quadratic number fields of small conductors. II*, J. Théor. Nombres Bordeaux **13** (2001), 633–649.

[41] K. Yamamura, *Maximal unramified extensions of real quadratic number fields of small conductors*, in preparation.

Galois Theory and Modular Forms pp 287–304
K. Hashimoto, K. Miyake, and H. Nakamura, Editors
© 2003 Kluwer Academic Publishers

DISTRIBUTION OF UNITS OF AN ALGEBRAIC NUMBER FIELD

Yoshiyuki KITAOKA*
Department of Mathematics
Meijo University,
Tenpaku, Nagoya,
468-8502, Japan
kitaoka@ccmfs.meijo-u.ac.jp

Let F be an algebraic number field with maximal order o_F and let $o_F^\times$ be the group of all units of F. We assume that $o_F^\times$ is an infinite group and we are interested in their distribution. For an integral ideal $\mathfrak{n}$ of F, we set $E(\mathfrak{n}) := \{u \bmod \mathfrak{n} \mid u \in o_F^\times\}$ $(\subset (o_F/\mathfrak{n})^\times)$ and $I(\mathfrak{n}) := [(o_F/\mathfrak{n})^\times : E(\mathfrak{n})]$, which is equal to the extension degree of the ramified part of the ray class field corresponding to the ideal $\mathfrak{n}$ over F. Let K be a Galois extension of the rational number field $\mathbb{Q}$, which contains the field F, and fix an element η of $\mathrm{Gal}(K/\mathbb{Q})$. We consider a primitive integral polynomial $g(x)$ such that $\{\epsilon^{g(\eta)} \mid \epsilon \in o_F^\times\}$ is a finite group with order δ_1. Here we choose such a polynomial $g(x)$ of minimal degree. Then $h(x) := (x^d - 1)/g(x)$ is in $\mathbb{Z}[x]$ for $d := [\langle\eta\rangle : \langle\eta\rangle \cap \mathrm{Gal}(K/F)]$. Next we take a maximal natural number δ_0 such that $\sqrt[\delta_0]{\epsilon^{\delta_1 g(\rho)}} = 1$ holds for every $\epsilon \in o_F^\times$ and any extension ρ of η. Let p be a prime number and $\mathfrak{p}$ a prime ideal of F lying above p such that there is a prime ideal $\mathfrak{P}$ of K with $\mathfrak{p} = \mathfrak{P} \cap F$ whose Frobenius automorphism is η. Then $\ell(p) := \delta_0 h(p)/\delta_1$ is an integer dividing $I(\mathfrak{p})$ and we conjecture that the set of such prime ideals with $I(\mathfrak{p}) = \ell(p)$ has a positive density $\kappa(\eta)$. We will give the conjectural density $\kappa(\eta)$ by the Euler product and study its positivity. Degrees of prime ideals in question are greater than or equal to 1 and so our density is a modified form of a natural density.

*Partially supported by Grant-in-Aid for Scientific Research (C), The Ministry of Education, Science, Sports and Culture.
[received: Feb. 21, 2002; accepted in revised form: Nov. 5, 2002]

Notations. For an algebraic number field L, $o_L, o_L^\times$ denote the ring of integers and the group of units in L, respectively. dL stands for the discriminant of L. For an automorphism ρ of L, we denote the action of ρ to $\alpha \in L$ by α^ρ and for a polynomial $f(x) = a_n x^n + a_{n-1}x^{n-1} + \cdots + a_0 \in \mathbb{Z}[x]$ we set $\alpha^{f(\rho)} := \prod_{k=0}^{n} (\alpha^{a_k})^{\rho^k}$. ζ_n denotes an nth primitive root of unity. The number of elements of a set S is denoted by $\#S$. $\langle a \rangle$ means the group generated by a.

Hereafter throughout the paper, we fix finite extension fields $F \subset K$ of the rational number field $\mathbb{Q}$ and assume that K is a Galois extension of $\mathbb{Q}$, and that $o_F^\times$ is an infinite group. Note that we do not assume that K is a Galois closure of F over $\mathbb{Q}$. We fix an arbitrary automorphism $\eta \in \mathrm{Gal}(K/\mathbb{Q})$ once and for all. K_n denotes the finite extension field of K adjoined by all nth roots of all units of K. For an integral ideal $\mathfrak{n}$ of F, we set

$$E(\mathfrak{n}) := \{u \bmod \mathfrak{n} \mid u \in o_F^\times\} \subset (o_F/\mathfrak{n})^\times,$$
$$I(\mathfrak{n}) := [(o_F/\mathfrak{n})^\times : E(\mathfrak{n})].$$

1.

In this section, we formulate the conjecture.

Lemma 1. *Let $g(x)$ be a polynomial in $\mathbb{Z}[x]$ such that*

$$W_1(g(x)) := \{\epsilon^{g(\eta)} \mid \epsilon \in o_F^\times\}$$

is a finite group. We fix a primitive polynomial $g(x)$ of minimal degree among them. Then it divides $x^d - 1$ in $\mathbb{Z}[x]$ for $d := [\langle\eta\rangle : \langle\eta\rangle \cap \mathrm{Gal}(K/F)]$.

Proof. By virtue of $\eta^d \in \mathrm{Gal}(K/F)$, $W_1(x^d - 1) = \{1\}$ is clear and let $g(x)$ be that as in the assertion. Take an integer a and integral polynomials $q(x), r(x)$ so that $a(x^d - 1) = q(x)g(x) + r(x)$ and $\deg r(x) < \deg g(x)$. The assumption $\#o_F^\times = \infty$ implies $\deg g(x) \geq 1$. For $\epsilon \in o_F^\times$, we have $\epsilon^{r(\eta)} = \epsilon^{a(\eta^d - 1) - q(\eta)g(\eta)} = (\epsilon^{g(\eta)})^{-q(\eta)}$ and hence $W_1(r(x))$ is a finite group. The minimality of $\deg g(x)$ implies $r(x) = 0$ and then the primitiveness of both $x^d - 1$ and $g(x)$ yields that a divides $q(x)$. Thus $g(x)$ divides $x^d - 1$. □

Hereafter the polynomial $g(x)$ means that defined in Lemma 1 and put, for $d := [\langle\eta\rangle : \langle\eta\rangle \cap \mathrm{Gal}(K/F)]$

$$h(x) := (x^d - 1)/g(x) \, (\in \mathbb{Z}[x]), \quad \delta_1 := \#W_1(g(x)).$$

Remark. Let F_0 be the Galois closure of F over $\mathbb{Q}$. The polynomial $g(x)$, $d = [\langle\eta\rangle : \langle\eta\rangle \cap \mathrm{Gal}(K/F)]$, δ_1 and hence $h(x)$ depend only on F_0 and $\eta_{|F_0}$.

The explicit form of $g(x)$ is complicated, but when $K = F$ is a Galois extension of $\mathbb{Q}$, it will be given at the end of this section.

Let $\mathfrak{P}$ be a prime ideal of K whose Frobenius automorphism $((K/\mathbb{Q})/\mathfrak{P})$ is η; then we say a prime number and a prime ideal of F lying below $\mathfrak{P}$ correspond to η if $p \nmid 2dK$. By ramification theory, we know $[\langle\eta\rangle : \langle\eta\rangle \cap \mathrm{Gal}(K/F)] = \deg(\mathfrak{P} \cap F)$.

Lemma 2. *If a prime number p ($\nmid 2dK$) corresponds to η, then δ_1 divides $h(p)$.*

Proof. Take a unit $\epsilon \in o_F^\times$ such that $\epsilon^{g(\eta)}$ is a primitive δ_1th root of unity, and let $\mathfrak{P}$ be a prime ideal of K lying above p such that $((K/\mathbb{Q})/\mathfrak{P}) = \eta$, and we put $\mathfrak{p} = \mathfrak{P} \cap F$. For a generator $\alpha \in o_F$ of $(o_F/\mathfrak{p})^\times$, we put $\epsilon \equiv \alpha^a \bmod \mathfrak{p}$ ($a \in \mathbb{Z}$). Then we have

$$1 = \epsilon^{\delta_1 g(\eta)} \equiv \epsilon^{\delta_1 g(p)} \bmod \mathfrak{P} \equiv \alpha^{a\delta_1 g(p)} \bmod \mathfrak{P},$$

which implies $1 \equiv \alpha^{a\delta_1 g(p)} \bmod \mathfrak{p}$. Since $d := [\langle\eta\rangle : \langle\eta\rangle \cap \mathrm{Gal}(K/F)] = \deg \mathfrak{p}$, there is an integer b such that $a\delta_1 g(p) = (p^d - 1)b$ and hence $a\delta_1 = h(p)b$. We have only to show $(\delta_1, b) = 1$. Suppose that q is a prime number dividing (δ_1, b); then we have

$$\epsilon^{\delta_1 g(p)/q} \equiv \alpha^{a\delta_1 g(p)/q} \equiv \alpha^{(p^d-1)b/q} \equiv 1 \bmod \mathfrak{p},$$

which implies $\epsilon^{g(\eta)\delta_1/q} \equiv 1 \bmod \mathfrak{P}$. Since $\epsilon^{g(\eta)\delta_1/q}$ is a qth root of unity in K, if it is not the unity, then $\mathfrak{P} \mid \zeta_q - 1$ holds and $p = q$ divides $2dK$, which is a contradiction. Hence we have $\epsilon^{g(\eta)\delta_1/q} = 1$, which contradicts the definition of δ_1. □

Lemma 3. *Let m be a natural number, and let p and $\mathfrak{p}$ ($\mid p$) be a prime number and a prime ideal of F corresponding to η. Then we have*

$$\frac{mh(p)}{\delta_1} \mid I(\mathfrak{p}) \iff \begin{cases} m \mid \delta_1 g(p) \text{ and} \\ \epsilon^{\delta_1 g(p)/m} \equiv 1 \bmod \mathfrak{p} \text{ for } {}^\forall\epsilon \in o_F^\times. \end{cases}$$

Proof. The assertion follows from

$$\begin{aligned} \frac{mh(p)}{\delta_1} \mid I(\mathfrak{p}) &\iff \frac{mh(p)}{\delta_1} n = \frac{p^d - 1}{\#E(\mathfrak{p})} \text{ for } {}^\exists n \in \mathbb{Z} \ \ (d := \deg \mathfrak{p}) \\ &\iff n\#E(\mathfrak{p}) = \delta_1 g(p)/m \text{ for } {}^\exists n \in \mathbb{Z} \\ &\iff m \mid \delta_1 g(p) \text{ and } \epsilon^{\delta_1 g(p)/m} \equiv 1 \bmod \mathfrak{p} \text{ for } {}^\forall\epsilon \in o_F^\times. \end{aligned}$$

□

Lemma 4. *Let m be a natural number and $\mathfrak{p}$ a prime ideal of F corresponding to η and we take a prime ideal $\mathfrak{P}_m$ of K_m lying above $\mathfrak{p}$ such that $\rho := ((K_m/\mathbb{Q})/\mathfrak{P}_m)$ is identical with η on K. If $\mathfrak{p} \nmid m$ moreover, then we have*

$$\frac{mh(p)}{\delta_1} \mid I(\mathfrak{p}) \iff \sqrt[m]{\epsilon}^{\delta_1 g(\rho)} = 1 \text{ for } {}^\forall \epsilon \in o_F^\times,$$

where p is a prime number lying below $\mathfrak{p}$.

Proof. Since $\sqrt[m]{\epsilon}^{\delta_1 g(\rho)}$ $(\epsilon \in o_F^\times)$ is an mth root of unity in K_m, the assumption $p \nmid 2mdK$ and the previous lemma yield

$$\frac{mh(p)}{\delta_1} \mid I(\mathfrak{p}) \iff \begin{cases} m \mid \delta_1 g(p) \text{ and} \\ \sqrt[m]{\epsilon}^{\delta_1 g(\rho)} = 1 \text{ for } {}^\forall \epsilon \in o_F^\times, \end{cases}$$

noting $\zeta_n \equiv 1 \bmod \mathfrak{P}_m$ implies $n \mid p^\infty$ for a natural number n. Here the condition $m \mid \delta_1 g(p)$ follows from the other condition with $\epsilon = 1$ and we complete the proof. □

We put for a natural number m

$$H_m(\eta) := \{\rho \in \mathrm{Gal}(K_m/\mathbb{Q}) \mid \rho_{|K} = \eta \text{ and } \sqrt[m]{\epsilon}^{\delta_1 g(\rho)} = 1 \text{ for } {}^\forall \epsilon \in o_F^\times\}.$$

Note that $\zeta_m^{\delta_1 g(\rho)} = 1$ for $\rho \in H_m(\eta)$. By what we have proved,

Proposition 1. *The following conditions* (i),(ii),(iii) *are equivalent for a natural number m.*

(i) $\#H_m(\eta) = [K_m : K]$.

(ii) *For every $\rho \in \mathrm{Gal}(K_m/\mathbb{Q})$ which satisfies $\rho_{|K} = \eta$ we have*

$$\sqrt[m]{\epsilon}^{\delta_1 g(\rho)} = 1 \text{ for } {}^\forall \epsilon \in o_F^\times.$$

(iii) *Let p and $\mathfrak{p}$ $(\mid p)$ be a prime number and a prime ideal of F corresponding to η; if $p \nmid 2mdK$, then we have*

$$mh(p)/\delta_1 \mid I(\mathfrak{p}).$$

Lemma 5. *If $\sigma \in \mathrm{Gal}(K_m/F)$ satisfies $\sigma_{|K}\eta = \eta\sigma_{|K}$, then we have $\sigma^{-1}H_m(\eta)\sigma = H_m(\eta)$.*

Proof. Let ρ be an automorphism in $H_m(\eta)$ and put $\rho' = \sigma^{-1}\rho\sigma$; then $\rho'_{|K} = (\sigma^{-1}\rho\sigma)_{|K} = \sigma_{|K}{}^{-1}\eta\sigma_{|K} = \eta$ is obvious. For $\epsilon \in o_F^\times$, we put

$\sqrt[m]{\epsilon}^{\sigma^{-1}} = \zeta \sqrt[m]{\epsilon}$, where ζ is an mth root of unity. Then we have $\sqrt[m]{\epsilon}^{\delta_1 g(\rho')} = \sqrt[m]{\epsilon}^{\sigma^{-1}\delta_1 g(\rho)\sigma} = \zeta^{\delta_1 g(\rho)\sigma} \sqrt[m]{\epsilon}^{\delta_1 g(\rho)\sigma} = (\zeta^{\delta_1 g(\rho)})^{\sigma}(\sqrt[m]{\epsilon}^{\delta_1 g(\rho)})^{\sigma} = 1$ and hence $\rho' \in H_m(\eta)$. □

Lemma 6. *Let K_η be the fixed subfield of K by $\langle\eta\rangle$. For $\sigma \in \mathrm{Gal}(K_m/K_\eta)$, we have $\sigma H_m(\eta)\sigma^{-1} = H_m(\eta)$.*

Proof. Let ρ be in $H_m(\eta)$; then $\rho_{|K} = \eta$, and the assumption yields $\sigma_{|K} = \eta^t = {\rho_{|K}}^t$ for some integer t. Hence $\sigma\rho\sigma^{-1}{}_{|K} = \eta$ is clear. For $\epsilon \in o_F^\times$, we put $\sqrt[m]{\epsilon}^{\sigma} = \zeta \sqrt[m]{\epsilon}^{\rho^t}$ with an mth root ζ of unity. Therefore we have $\sqrt[m]{\epsilon}^{\delta_1 g(\sigma\rho\sigma^{-1})} = \sqrt[m]{\epsilon}^{\delta_1 \sigma g(\rho)\sigma^{-1}} = (\zeta^{\delta_1 g(\rho)})^{\sigma^{-1}}(\sqrt[m]{\epsilon}^{\rho^t\delta_1 g(\rho)\sigma^{-1}}) = (\zeta^{\delta_1 g(\rho)})^{\sigma^{-1}}(\sqrt[m]{\epsilon}^{\delta_1 g(\rho)})^{\rho^t\sigma^{-1}} = 1$. Thus we have shown $\sigma\rho\sigma^{-1} \in H_m(\eta)$. □

Proposition 2. *Let m be a natural number. Denoting a prime ideal of F (resp. K_η) by $\mathfrak{p}$ (resp. $\mathfrak{p}_\eta$) and a prime number lying below them by p, we have*

$$\begin{aligned}
&\#\{\mathfrak{p} \mid p \le x,\ p \nmid 2mdK,\ mh(p)/\delta_1 \mid I(\mathfrak{p}),\ \textit{and } \mathfrak{p} \textit{ corresponds to } \eta\ \} \\
&= [\{\sigma \in \mathrm{Gal}(K/F) \mid \sigma\eta = \eta\sigma\} : \mathrm{Gal}(K/F) \cap \langle\eta\rangle]^{-1} \\
&\quad \times \#\left\{\mathfrak{p}_\eta \,\middle|\, \begin{array}{l} p \le x,\ p \nmid 2mdK,\ ((K_m/K_\eta)/\mathfrak{P}_m) \in H_m(\eta) \textit{ for some} \\ \textit{ideal } \mathfrak{P}_m \textit{ of } K_m \textit{ lying above } \mathfrak{p}_\eta,\ \textit{and } \deg \mathfrak{p}_\eta = 1 \end{array}\right\}.
\end{aligned}$$

Proof. By Lemma 4, the left-hand side is equal, denoting a prime ideal of K_m lying above $\mathfrak{p}$ by $\mathfrak{P}_m$ to

$$\#\{\mathfrak{p} \mid p \le x,\ p \nmid 2mdK \text{ and } ((K_m/\mathbb{Q})/\mathfrak{P}_m) \in H_m(\eta) \text{ for } {}^{\exists}\mathfrak{P}_m\}.$$

Suppose that for a prime ideal $\mathfrak{p}$ of F, there is a prime ideal $\mathfrak{P}_m$ of K_m such that $\mathfrak{P}_m \cap F = \mathfrak{p}$ and $((K_m/\mathbb{Q})/\mathfrak{P}_m) \in H_m(\eta)$; if a prime ideal $\mathfrak{P}'_m$ of K_m also satisfies $\mathfrak{P}'_m \cap F = \mathfrak{p}$ and $((K_m/\mathbb{Q})/\mathfrak{P}'_m) \in H_m(\eta)$, then we have $\mathfrak{P}'_m \cap K = (\mathfrak{P}_m \cap K)^\sigma$ for $\sigma \in \mathrm{Gal}(K/F)$ with $\sigma\eta = \eta\sigma$, since Frobenius automorphisms of $\mathfrak{P}_m \cap K$ and $\mathfrak{P}'_m \cap K$ are equal to η. If, conversely a prime ideal $\mathfrak{P}'_m$ of K_m satisfies $\mathfrak{P}'_m \cap K = (\mathfrak{P}_m \cap K)^\sigma$ for $\sigma \in \mathrm{Gal}(K/F)$ with $\sigma\eta = \eta\sigma$, then ${\mathfrak{P}'_m}^{\sigma^{-1}} = \mathfrak{P}_m^\kappa$ for some $\kappa \in \mathrm{Gal}(K_m/K)$ and then $((K_m/\mathbb{Q})/\mathfrak{P}'_m) = (\kappa\sigma)^{-1}((K_m/\mathbb{Q})/\mathfrak{P}_m)(\kappa\sigma) \in H_m(\eta)$ by Lemma 5,6. Hence possible prime ideals $\mathfrak{P}'_m \cap K$ of K lying above $\mathfrak{p}$ are transformed by $\sigma \in \mathrm{Gal}(K/F)$ with $\sigma\eta = \eta\sigma$ and therefore the above is equal to

$$\begin{aligned}
&[\{\sigma \in \mathrm{Gal}(K/F) \mid \sigma\eta = \eta\sigma\} : \mathrm{Gal}(K/F) \cap \langle\eta\rangle]^{-1} \\
&\quad \times \#\left\{\mathfrak{P} \,\middle|\, \begin{array}{l} p \le x,\ p \nmid 2mdK \text{ and } ((K_m/\mathbb{Q})/\mathfrak{P}_m) \in H_m(\eta) \\ \text{for } {}^{\exists}\mathfrak{P}_m \text{ lying above } \mathfrak{P} \end{array}\right\}.
\end{aligned}$$

Here $\mathfrak{P}$ denotes a prime ideal of K. Put $\mathfrak{p}_\eta := \mathfrak{P}_m \cap K_\eta = \mathfrak{P} \cap K_\eta$; then since the decomposition group of $\mathfrak{P}$ is $\langle \eta \rangle$, $\deg \mathfrak{p}_\eta = 1$ and $((K_m/\mathbb{Q})/\mathfrak{P}_m) = ((K_m/K_\eta)/\mathfrak{P}_m)$ hold and $\mathfrak{P}$ is the unique prime ideal of K lying above $\mathfrak{p}_\eta$. Therefore we have the assertion. □

Now Chebotarev's density theorem yields

Theorem 1. *Let m be a natural number. By denoting a prime ideal of F by $\mathfrak{p}$ and a prime number lying below it by p, the density of the set*

$$\{\mathfrak{p} \mid p \nmid 2mdK,\ mh(p)/\delta_1 \mid I(\mathfrak{p}),\ \text{and } \mathfrak{p} \text{ corresponds to } \eta\ \}$$

is equal to

$$[\{\sigma \in \mathrm{Gal}(K/F) \mid \sigma\eta = \eta\sigma\} : \mathrm{Gal}(K/F) \cap \langle \eta \rangle]^{-1} \frac{\#H_m(\eta)}{[K_m : K_\eta]}.$$

Hence we propose the following

Conjecture. *Setting, for a natural number δ*

$$\kappa(\eta;\delta) := [\{\sigma \in \mathrm{Gal}(K/F) \mid \sigma\eta = \eta\sigma\} : \mathrm{Gal}(K/F) \cap \langle \eta \rangle]^{-1} \times \sum_{m=1}^{\infty} \frac{\mu(m)\#H_{\delta m}(\eta)}{[K_{\delta m} : K_\eta]},$$

and denoting a prime number and a prime ideal of F by $p, \mathfrak{p}$ $(\mathfrak{p} \mid p)$, we have

$$\#\{\mathfrak{p} \mid p \leq x,\ p \nmid 2dK,\ I(\mathfrak{p}) = \delta|h(p)|/\delta_1,\ \text{and } \mathfrak{p} \text{ corresponds to } \eta\ \} \sim \kappa(\eta;\delta)\mathrm{Li}(x).$$

Here $\mu(m)$ denotes the Möbius function. We note that the condition $p < x$ is used instead of $N_{F/\mathbb{Q}}(\mathfrak{p}) < x$, and so this is a modification of the usual natural density. This conjecture is a generalization of [CKY], [K] and hence the conjecture for a real quadratic field $K = F$ is true under the Generalized Riemann Hypothesis with δ being δ_0 in the next section. When $\eta \in \mathrm{Gal}(K/F)$, we know $\deg \mathfrak{p} = [\langle \eta \rangle : \langle \eta \rangle \cap \mathrm{Gal}(K/F)] = 1$ and hence both $((K/\mathbb{Q})/\mathfrak{P}) = ((K/F)/\mathfrak{P})$ and $h(x) = 1$ hold. Then the conjecture is true under G.R.H. by a result of [L]. The situation of [L] is as follows: Let K/F be a finite Galois extension and let C be a union of conjugacy classes of $\mathrm{Gal}(K/F)$, and W is a finitely generated subgroup of $F^\times$ of finite rank (≥ 1) modulo its torsion subgroup, and k is an integer (> 0). In [L], under G.R.H. it is shown the density of the set $M(F, K, C, W, k)$ of prime ideals $\mathfrak{p}$ of F exists. Here $\mathfrak{p}$ is in $M(F, K, C, W, k)$ if and only if for a prime ideal $\mathfrak{P}$ of K lying above $\mathfrak{p}$, the Frobenius automorphism $((K/F)/\mathfrak{P})$ is in C, $\mathrm{ord}_{\mathfrak{p}}(w) = 0$ for all $w \in W$ and the index $[(o_F/\mathfrak{p})^\times : \{w \mod \mathfrak{p} \mid w \in W\}]$ divides k. So the index is bounded. On the contrary, our case allows that the index $I(\mathfrak{p})$ tends to the infinity.

Proposition 3. *Suppose that $K = F$ is a Galois extension of $\mathbb{Q}$; then the polynomial $g(x)$ is given as follows:*

(R) *The case where F is real.*

- **(R1)** $g(x) = x^{d-1} + x^{d-2} + \cdots + 1$ *if* $\mathrm{Gal}(F/\mathbb{Q}) = \langle\eta\rangle$.
- **(R2)** $g(x) = x^d - 1$ *otherwise.*

(I) *The case where F is imaginary.*

We denote the complex conjugation by J.

- **(I1)** $g(x) = x^{d-1} + x^{d-2} + \cdots + 1$ *if* $[\mathrm{Gal}(F/\mathbb{Q}) : \langle\eta\rangle] = 2$ *and* $J \notin \langle\eta\rangle$.
- **(I2)** $g(x) = x^d - 1$ *if* $[\mathrm{Gal}(F/\mathbb{Q}) : \langle\eta\rangle] > 2$ *and* $J \notin \langle\eta\rangle$.
- **(I3)** $g(x) = x^{d/2-1} + x^{d/2-2} + \cdots + 1$ *if* $\mathrm{Gal}(F/\mathbb{Q}) = \langle\eta\rangle$.
- **(I4)** *The case of* $J \in \langle\eta\rangle \neq \mathrm{Gal}(F/\mathbb{Q})$.
  - **(i)** *If there is an element* $u \in \mathrm{Gal}(F/\mathbb{Q})$ *such that* $JuJ^{-1}u^{-1} \notin \langle\eta\rangle$, *then* $g(x) = x^d - 1$.
  - **(ii)** *If* $JuJ^{-1}u^{-1} \in \langle\eta\rangle$ *holds for every element* $u \in \mathrm{Gal}(F/\mathbb{Q})$, *then* $g(x) = (x^{d/2}-1)(\sum_i x^{a_i} - \sum_i x^{b_i})$, *where integers* a_i, b_i *are defined by*

$$\langle JuJ^{-1}u^{-1} \mid u \in \mathrm{Gal}(F/\mathbb{Q})\rangle$$
$$= \{\eta^{a_i} \mid 0 \leq a_i < d/2\} \cup \{\eta^{b_i} J \mid 0 \leq b_i < d/2\}.$$

In case of (I4), $g(x) = x^{d/2} - 1$ *if the maximal real subfield of F is a Galois extension over $\mathbb{Q}$.*

Proof. Let us define the condition (A1) for an integral polynomial $G(x)$ by that $W_1(G(x)) := \{\epsilon^{G(\eta)} \mid \epsilon \in o_F^\times\}$ is a finite group. Let $\varphi_1, \varphi_2, \cdots$ be all embeddings of F into the field of complex numbers and as usual we suppose that F^{φ_i} is in the field of real numbers for $1 \leq i \leq r_1$ and $\varphi_{r_1+r_2+i} = \overline{\varphi_{r_1+i}}$. We remark that r_1 or $r_2 = 0$, and the relation

$$\sum_{1 \leq i \leq r_1} \log|\epsilon^{\varphi_i}| + 2 \sum_{r_1+1 \leq i \leq r_1+r_2} \log|\epsilon^{\varphi_i}| = 0 \tag{0}$$

is the unique one for $\epsilon \in o_F^\times$. We note that a unit ϵ is a root of unity if $|\epsilon^\sigma| = 1$ for all $\sigma \in \mathrm{Gal}(F/\mathbb{Q})$. Suppose that an integral polynomial

$G(x) = a(n)x^n + a(n-1)x^{n-1} + \cdots + a(0)$ satisfies the condition (A1). We assume that $(a(n), a(n-1), \cdots, a(0)) = 1$, and that $G(x)$ is of minimal degree. We show that $G(x)$ is equal to the polynomial $g(x)$ up to sign in the assertion. By the condition (A1), we have

$$(1) \qquad \sum_{i=0}^{n} a(i) \log |\epsilon^{\eta^i}| = 0 \text{ for } {}^\forall\epsilon \in o_F^\times.$$

We can see easily that in cases except (I4), the polynomial $g(x)$ in the assertion satisfies the condition (A1). The condition (1) is not valid for any integral polynomial of lower degree than $g(x)$ by the remark above. Hence the cases except (I4) are over. In the rest, we assume $J \in \langle \eta \rangle \neq \mathrm{Gal}(F/\mathbb{Q})$. Hence $d =$ the order of $\eta \le [F:\mathbb{Q}]/2$ is clear. We suppose that $n < d$ and put $T := \{\eta^i \mid a(i) \neq 0,\ 0 \le i \le n\} = T_1 \cup T_2 \cup T_2 J$. Here the union is disjoint and $T \cap T_1 J = \emptyset$. For simplicity, for $t \in T$ we let $a(t)$ stand for $a(i)$ if $t = \eta^i$ $(0 \le i \le n)$; then by the assumption on $G(x)$, we have

$$\log |\epsilon^{G(\eta)}| = \sum_{t_1 \in T_1} a(t_1) \log |\epsilon^{t_1}| + \sum_{t_2 \in T_2} (a(t_2) + a(t_2 J)) \log |\epsilon^{t_2}| = 0$$

for ${}^\forall\epsilon \in o_F^\times$. We note that $a(t) \neq 0$ if and only if $t \in T$. If T_1 is not empty or $a(t) + a(tJ) \neq 0$ for some $t \in T_2$, then this relation should be the relation (0) and hence $\#T_1 + \#T_2 = [F:\mathbb{Q}]/2$ holds. On the other hand, $\#T_1 + 2\#T_2 = \#T \le n+1 \le d \le [F:\mathbb{Q}]/2 = \#T_1 + \#T_2$ yields $\#T_2 = 0$, and $\#T_1 = d$. Then $T_1 \subset \langle \eta \rangle$ implies $T_1 = \langle \eta \rangle$. By virtue of $T \cap T_1 J = \emptyset$ and $J \in \langle \eta \rangle$, it is a contradiction. Hence we have $T_1 = \emptyset$ and $a(t) = -a(tJ)$ for $t \in T_2$. Exchanging t and tJ if necessary, we may assume $a(t) > 0$ for ${}^\forall t \in T_2 J$, and

$$\epsilon^{-G(\eta)} = \prod_{t \in T_2} (\epsilon^{tJ}/\epsilon^t)^{a(t)},$$

is a root of unity. Now, we have for $u \in \mathrm{Gal}(F/\mathbb{Q})$

$$\sum_{t \in T_2} a(t)(\log |\epsilon^{tJu}| - \log |\epsilon^{tu}|) = 0.$$

This is not the relation (0) and so it should be a trivial equation and we can make pairs $(t_1, t_2) \in T_2 \times T_2$ such that $a(t_1) = a(t_2)$ and $t_1 J u = t_2 u J$, and so we get $JuJ^{-1}u^{-1} = t_1^{-1} t_2 \in \langle \eta \rangle$, and the case (i) is over. Suppose $H := \langle JuJ^{-1}u^{-1} \mid u \in \mathrm{Gal}(F/\mathbb{Q}) \rangle \subset \langle \eta \rangle$; then the relation $t_2 = t_1 JuJ^{-1}u^{-1}$ shows that H acts on the set T_2, keeping $a(t)$. Hence there is a subset $A \subset \langle \eta \rangle$ such that

$$\epsilon^{-G(\eta)} = \prod_{t \in A} (\prod_{h \in H} \epsilon^{thJ}/\epsilon^{th})^{a(t)}.$$

It is easy to see that for $u \in \mathrm{Gal}(F/\mathbb{Q})$ and $\kappa := JuJ^{-1}u^{-1} \in H$

$$(\prod_{h \in H} \epsilon^{thJ}/\epsilon^{th})^u = \prod_{h \in H} (\epsilon^{th\kappa uJ}/\epsilon^{thu}) = \prod_{h \in H} (\epsilon^{thuJ}/\epsilon^{thu})$$

is of absolute value 1 and hence each factor $\prod_{h \in H}(\epsilon^{thJ}/\epsilon^{th})$ of $\epsilon^{-G(\eta)}$ is a root of unity. Hence the minimality of $\deg G(x)$ implies $\epsilon^{-G(\eta)} = \prod_{h \in H}(\epsilon^{thJ}/\epsilon^{th})$ for some $t \in A$. Hence, putting

$$tH = \{\eta^{A_i} \mid 0 \le A_i < d/2\} \cup \{\eta^{B_j} J \mid 0 \le B_i < d/2\}$$

we have $-G(x) = (x^{d/2} - 1)(\sum_i x^{A_i} - \sum_i x^{B_i})$. The minimality of $\deg G(x)$ implies A_i or $B_i = 0$ for some i, which implies t or $tJ \in H$ and hence $\epsilon^{\mp G(\eta)} = \prod_{h \in H}(\epsilon^{hJ}/\epsilon^{h})$ holds and we can replace tH by H in the above equation and $G(x)$ is equal to $\pm g(x)$ in the assertion. The last remark follows from $H = \{1\}$ if $\langle J \rangle$ is a normal subgroup. □

2.

First we show

Theorem 2. *Let R be an algebraic number field and E a subgroup of $\mathfrak{o}_R^\times$. Suppose that a primitive wth root ζ_w of unity and $\epsilon_1, \cdots, \epsilon_r$ generate E, and $\{\epsilon_1, \cdots, \epsilon_r\}$ is a free basis of $E/\langle \zeta_w \rangle$. Let A be a square-free integer such that a prime number p divides A if p divides $2dR$ or if for some $\alpha \in R \setminus E$ and some $\epsilon \in E$, $\alpha^p = \epsilon$ holds. Set $R_m = R(\{\sqrt[m]{\epsilon}\} \mid \epsilon \in E)$, where $\{\sqrt[m]{\epsilon}\}$ means all mth roots of ϵ. For $m = ab$ with $a \mid A^\infty$ and $(b, A) = 1$, we have*

$$[R_m : R] = [R_b : R][R_a : R] = b^r \varphi(b)[R_a : R] \text{ and } R_a \cap R_b = R,$$

where φ is Euler's function.

We need several lemmas.

Lemma 7. *Setting $L = R(\zeta_{mw})$, we have $[L(\sqrt[m]{\epsilon_1}) : L] = b[L(\sqrt[a]{\epsilon_1}) : L]$.*

Proof. Since $[L(\sqrt[m]{\epsilon_1}) : L] = [L(\sqrt[a]{\epsilon_1})(\sqrt[b]{\epsilon_1}) : L(\sqrt[a]{\epsilon_1})][L(\sqrt[a]{\epsilon_1}) : L]$, we have only to show that the polynomial $x^b - \epsilon_1$ is irreducible over $L(\sqrt[a]{\epsilon_1})$. Assume that it is reducible; then by virtue of $2 \nmid b$, there are a prime divisor p of b and $\alpha \in L(\sqrt[a]{\epsilon_1})$ satisfying $\alpha^p = \epsilon_1$. Since $L \subset L(\alpha) \subset L(\sqrt[a]{\epsilon_1})$ and $L(\sqrt[a]{\epsilon_1})/L$ is a Kummer extension, $L(\alpha)/L$ is abelian. If $\alpha \notin L$ holds then $x^p - \epsilon_1$ is irreducible and hence we have a contradiction $p = [L(\alpha) : L] \mid [L(\sqrt[a]{\epsilon_1}) : L] \mid a$. Thus $\alpha \in L$ holds and then $R \subset R(\alpha) \subset L = R(\zeta_{mw})$, and $p \nmid A$ implies that $x^p - \epsilon_1$ is irreducible over R and so $R(\alpha)/R$ is a Galois extension of degree p, because L/R is abelian. It implies that for a pth primitive root ζ_p, $\zeta_p\alpha \in R$ and so $\zeta_p\alpha/\alpha \in R$ holds. In the sequence $R \subset R(\zeta_p) \subset R(\alpha)$, we have $[R(\zeta_p) : R] \mid p-1$ and $[R(\alpha) : R] = p$. Thus $[R(\zeta_p) : R] = 1$, i.e., $\zeta_p \in R$ occurs. It contradicts $p \nmid dR$. □

Lemma 8. $[L(\sqrt[m]{\epsilon_1}, \sqrt[m]{\epsilon_2}) : L] = b^2[L(\sqrt[a]{\epsilon_1}, \sqrt[a]{\epsilon_2}) : L]$.

Proof. By virtue of

$$\begin{aligned}[L(\sqrt[m]{\epsilon_1}, \sqrt[m]{\epsilon_2}) : L] = {} & [L(\sqrt[m]{\epsilon_1}, \sqrt[a]{\epsilon_2})(\sqrt[b]{\epsilon_2}) : L(\sqrt[m]{\epsilon_1}, \sqrt[a]{\epsilon_2})] \\ & \times [L(\sqrt[m]{\epsilon_1}, \sqrt[a]{\epsilon_2}) : L(\sqrt[m]{\epsilon_1})][L(\sqrt[m]{\epsilon_1}) : L],\end{aligned}$$

we have only to show

(i) $x^b - \epsilon_2$ is irreducible over $L(\sqrt[m]{\epsilon_1}, \sqrt[a]{\epsilon_2})$,

(ii) $[L(\sqrt[m]{\epsilon_1}, \sqrt[a]{\epsilon_2}) : L(\sqrt[m]{\epsilon_1})] = [L(\sqrt[a]{\epsilon_1}, \sqrt[a]{\epsilon_2}) : L(\sqrt[a]{\epsilon_1})]$.

Suppose that $x^b - \epsilon_2$ is reducible over $L(\sqrt[m]{\epsilon_1}, \sqrt[a]{\epsilon_2})$; then there exist a prime factor p of b and $\alpha \in L(\sqrt[m]{\epsilon_1}, \sqrt[a]{\epsilon_2})$ satisfying $\alpha^p = \epsilon_2$. Therefore the inclusion $L(\sqrt[m]{\epsilon_1}) \subset L(\sqrt[m]{\epsilon_1}, \alpha) \subset L(\sqrt[m]{\epsilon_1}, \sqrt[a]{\epsilon_2})$ implies $[L(\sqrt[m]{\epsilon_1}, \alpha) : L(\sqrt[m]{\epsilon_1})] \mid a$. On the other hand, $[L(\sqrt[m]{\epsilon_1}, \alpha) : L(\sqrt[m]{\epsilon_1})] \mid p$ holds by $\alpha^p = \epsilon_2 \in L$ and $\zeta_p \in L$. The assumption $(a, b) = 1$ yields $[L(\sqrt[m]{\epsilon_1}, \alpha) : L(\sqrt[m]{\epsilon_1})] = 1$, and so $\alpha \in L(\sqrt[m]{\epsilon_1})$. From the property $\alpha^p (= \epsilon_2) \in L$ follows that there is an integer n satisfying $\alpha/\sqrt[m]{\epsilon_1}^{\,n} \in L$. Set $pn/m = d/c$ with $(c, d) = 1$; then we have

$$(\alpha/\sqrt[m]{\epsilon_1}^{\,n})^{cp} = \epsilon_2^c/\epsilon_1^d \in R$$

and from $p \nmid A$ follows that there is no element β in R satisfying $\beta^p = \epsilon_2^c/\epsilon_1^d$, in particular $(\alpha/\sqrt[m]{\epsilon_1}^{\,n})^c \notin R$. Thus we have

$$R \subset R((\alpha/\sqrt[m]{\epsilon_1}^{\,n})^c) \subset R(\alpha/\sqrt[m]{\epsilon_1}^{\,n}) \subset L = R(\zeta_{mw}),$$

which yields that $R((\alpha/\sqrt[m]{\epsilon_1}^{\,n})^c)$ is a Galois extension of R of degree p, since any pth root of $(\alpha/\sqrt[m]{\epsilon_1}^{\,n})^{cp} = \epsilon_2^c/\epsilon_1^d$ is not in R. Thus we have $\zeta_p \in R((\alpha/\sqrt[m]{\epsilon_1}^{\,n})^c)$, and then $[R(\zeta_p) : R] \mid p - 1$ implies $[R(\zeta_p) : R] = 1$. Since p is odd and $\zeta_p \in R$, p divides dR, which contradicts $p \mid b$. The assertion (i) has been proven. To show (ii), noting that $[L(\sqrt[m]{\epsilon_1}) : L(\sqrt[a]{\epsilon_1})] = b$, $[L(\sqrt[a]{\epsilon_1}, \sqrt[a]{\epsilon_2}) : L(\sqrt[a]{\epsilon_1})] \mid a$ and $(a, b) = 1$, we have $L(\sqrt[m]{\epsilon_1}) \cap L(\sqrt[a]{\epsilon_1}, \sqrt[a]{\epsilon_2}) = L(\sqrt[a]{\epsilon_1})$ and hence $L(\sqrt[a]{\epsilon_1}) \cap L(\sqrt[a]{\epsilon_2}) = (L(\sqrt[m]{\epsilon_1}) \cap L(\sqrt[a]{\epsilon_1}, \sqrt[a]{\epsilon_2})) \cap L(\sqrt[a]{\epsilon_2}) = L(\sqrt[m]{\epsilon_1}) \cap L(\sqrt[a]{\epsilon_2})$. Thus we obtain

$$\begin{aligned}&[L(\sqrt[m]{\epsilon_1}, \sqrt[a]{\epsilon_2}) : L(\sqrt[m]{\epsilon_1})] = [L(\sqrt[a]{\epsilon_2}) : L(\sqrt[a]{\epsilon_2}) \cap L(\sqrt[m]{\epsilon_1})] \\ &= [L(\sqrt[a]{\epsilon_2}) : L(\sqrt[a]{\epsilon_2}) \cap L(\sqrt[a]{\epsilon_1})] = [L(\sqrt[a]{\epsilon_1}, \sqrt[a]{\epsilon_2}) : L(\sqrt[a]{\epsilon_1})].\end{aligned}$$

□

Lemma 9. *Let $3 \le t \le r$, and suppose that*

$$[L(\sqrt[m]{\epsilon_1}, \cdots, \sqrt[m]{\epsilon_T}) : L] = b^T [L(\sqrt[a]{\epsilon_1}, \cdots, \sqrt[a]{\epsilon_T}) : L]$$

holds for $T = t-1$. Then it holds for $T = t$.

Proof. Set $M = L(\sqrt[m]{\epsilon_1}, \cdots, \sqrt[m]{\epsilon_{t-1}})$. Since

$$\begin{aligned}[L(\sqrt[m]{\epsilon_1}, \cdots, \sqrt[m]{\epsilon_t}) : L] &= [M(\sqrt[m]{\epsilon_t}) : M] \\ &\quad \times b^{t-1}[L(\sqrt[a]{\epsilon_1}, \cdots, \sqrt[a]{\epsilon_{t-1}}) : L] \\ &= [M(\sqrt[a]{\epsilon_t})(\sqrt[b]{\epsilon_t}) : M(\sqrt[a]{\epsilon_t})][M(\sqrt[a]{\epsilon_t}) : M] \\ &\quad \times b^{t-1}[L(\sqrt[a]{\epsilon_1}, \cdots, \sqrt[a]{\epsilon_{t-1}}) : L],\end{aligned}$$

we have only to show

(i) $x^b - \epsilon_t$ is irreducible over $M(\sqrt[a]{\epsilon_t})$,

(ii) $[M(\sqrt[a]{\epsilon_t}) : M] = [L(\sqrt[a]{\epsilon_1}, \cdots, \sqrt[a]{\epsilon_t}) : L(\sqrt[a]{\epsilon_1}, \cdots, \sqrt[a]{\epsilon_{t-1}})]$.

Suppose that $x^b - \epsilon_t$ is reducible over $M(\sqrt[a]{\epsilon_t})$; then there exist a prime factor p of b and an element $\alpha \in M(\sqrt[a]{\epsilon_t})$ satisfying $\alpha^p = \epsilon_t$. Then $M \subset M(\alpha) \subset M(\sqrt[a]{\epsilon_t})$ implies $[M(\alpha) : M] \mid a$ because $M(\sqrt[a]{\epsilon_t})/M$ is a Kummer extension. On the other hand, $[M(\alpha) : M] \mid p$ holds by $\zeta_p \in M$. Thus $\alpha \in M$ holds. Setting $N := L(\sqrt[m]{\epsilon_1}, \cdots, \sqrt[m]{\epsilon_{t-2}})$, we have $N \subset N(\alpha) \subset M = N(\sqrt[m]{\epsilon_{t-1}})$ and then there is an integer n satisfying $\alpha / \sqrt[m]{\epsilon_{t-1}}^n \in N$, since $\alpha^p = \epsilon_t \in N$. Writing $pn/m = d/c$ with $(c, d) = 1$, we have

$$(\alpha / \sqrt[m]{\epsilon_{t-1}}^n)^{cp} = \epsilon_t^c / \epsilon_{t-1}^d \in E \subset R \text{ and } (\alpha / \sqrt[m]{\epsilon_{t-1}^n})^c \notin R,$$

by $p \nmid A$. Setting $\epsilon'_{t-1} := \epsilon_t^c / \epsilon_{t-1}^d$, we can extend $\{\epsilon_1, \cdots, \epsilon_{t-2}, \epsilon'_{t-1}\}$ to a basis of $E/\langle \zeta_w \rangle$ and then the assumption of induction on t yields that $x^p - \epsilon'_{t-1}$ is irreducible over $L(\sqrt[m]{\epsilon_1}, \cdots, \sqrt[m]{\epsilon_{t-2}})$. Thus we have $[N(\sqrt[p]{\epsilon'_{t-1}}) : N] = p$. On the other hand, $\sqrt[p]{\epsilon'_{t-1}} = (\alpha / \sqrt[m]{\epsilon_{t-1}}^n)^c \in N$ contradicts it and so we have proved (i).

Secondly, since

$$[L(\sqrt[m]{\epsilon_1},\cdots,\sqrt[m]{\epsilon_{t-1}}):L(\sqrt[a]{\epsilon_1},\cdots,\sqrt[a]{\epsilon_{t-1}})]=b^{t-1}$$

and

$$[L(\sqrt[a]{\epsilon_1},\cdots,\sqrt[a]{\epsilon_t}):L(\sqrt[a]{\epsilon_1},\cdots,\sqrt[a]{\epsilon_{t-1}})]\mid a$$

imply

$$\begin{aligned}L(\sqrt[a]{\epsilon_1},\cdots,\sqrt[a]{\epsilon_t})\cap M&=L(\sqrt[a]{\epsilon_1},\cdots,\sqrt[a]{\epsilon_t})\cap L(\sqrt[m]{\epsilon_1},\cdots,\sqrt[m]{\epsilon_{t-1}})\\&=L(\sqrt[a]{\epsilon_1},\cdots,\sqrt[a]{\epsilon_{t-1}}),\end{aligned}$$

we have

$$\begin{aligned}&[M(\sqrt[a]{\epsilon_t}):M]=[L(\sqrt[a]{\epsilon_t}):L(\sqrt[a]{\epsilon_t})\cap M]\\&=[L(\sqrt[a]{\epsilon_t}):(L(\sqrt[a]{\epsilon_t})\cap L(\sqrt[a]{\epsilon_1},\cdots,\sqrt[a]{\epsilon_t}))\cap M]\\&=[L(\sqrt[a]{\epsilon_t}):L(\sqrt[a]{\epsilon_t})\cap L(\sqrt[a]{\epsilon_1},\cdots,\sqrt[a]{\epsilon_{t-1}})]\\&=[L(\sqrt[a]{\epsilon_1},\cdots,\sqrt[a]{\epsilon_{t-1}},\sqrt[a]{\epsilon_t}):L(\sqrt[a]{\epsilon_1},\cdots,\sqrt[a]{\epsilon_{t-1}})].\end{aligned}$$

Thus the assertion (ii) has been proved. □

Lemma 10. $[L(\sqrt[a]{\epsilon_1},\cdots,\sqrt[a]{\epsilon_r}):R]=\varphi(b)[R(\sqrt[a]{E}):R]$.

Proof. It is easy to see

$$\begin{aligned}&[L(\sqrt[a]{\epsilon_1},\cdots,\sqrt[a]{\epsilon_r}):R]\\&=[R(\zeta_{mw})(\sqrt[a]{\epsilon_1},\cdots,\sqrt[a]{\epsilon_r}):R(\sqrt[a]{E})][R(\sqrt[a]{E}):R]\\&=[R(\sqrt[a]{E})(\zeta_b):R(\sqrt[a]{E})][R(\sqrt[a]{E}):R]\\&=[\mathbb{Q}(\zeta_b):\mathbb{Q}(\zeta_b)\cap R(\sqrt[a]{E})][R(\sqrt[a]{E}):R].\end{aligned}$$

Since $dR(\sqrt[a]{E})\mid A^\infty$ and $(A,b)=1$, we have $\mathbb{Q}(\zeta_b)\cap R(\sqrt[a]{E})=\mathbb{Q}$, which completes the proof. □

Now we have

$$\begin{aligned}&[R_m:R]=[R(\sqrt[m]{E}):R]\\&=[L(\sqrt[m]{\epsilon_1},\cdots,\sqrt[m]{\epsilon_r}):R]\\&=b^r[L(\sqrt[a]{\epsilon_1},\cdots,\sqrt[a]{\epsilon_r}):R]\\&=\varphi(b)b^r[R_a:R]=[R_b:R][R_a:R],\end{aligned}$$

which implies $R_a\cap R_b=R$. Thus we have completed the proof of the theorem.

By virtue of theorem 2, we see for natural numbers $\delta = \gamma_1\gamma_2,\ a, b$ satisfying $\gamma_1 \mid (2dK)^\infty$, $(\gamma_2, 2dK) = 1$ and $(b\gamma_2, 2adK) = 1$,

$$\begin{aligned}
&\#H_{\delta ab}(\eta) \\
&= \#\{\rho \in \mathrm{Gal}(K_{\delta ab}/\mathbb{Q}) \mid \rho_{|K} = \eta,\ \sqrt[\delta ab]{\epsilon}^{-\delta_1 g(\rho)} = 1 \text{ for } {}^\forall\epsilon \in o_F^\times\} \\
&= \#\left\{\rho \in \mathrm{Gal}(K_{\delta ab}/\mathbb{Q}) \,\middle|\, \begin{array}{l} \rho_{|K} = \eta,\ \sqrt[\gamma_1 a]{\epsilon}^{-\delta_1 g(\rho)} = 1 \text{ and} \\ \sqrt[\gamma_2 b]{\epsilon}^{-\delta_1 g(\rho)} = 1 \text{ for } {}^\forall\epsilon \in o_F^\times \end{array}\right\} \\
&= \#\{\rho \in \mathrm{Gal}(K_{\gamma_1 a}/\mathbb{Q}) \mid \rho_{|K} = \eta,\ \sqrt[\gamma_1 a]{\epsilon}^{-\delta_1 g(\rho)} = 1 \text{ for } {}^\forall\epsilon \in o_F^\times\} \\
&\quad \times \#\{\rho \in \mathrm{Gal}(K_{\gamma_2 b}/\mathbb{Q}) \mid \rho_{|K} = \eta,\ \sqrt[\gamma_2 b]{\epsilon}^{-\delta_1 g(\rho)} = 1 \text{ for } {}^\forall\epsilon \in o_F^\times\} \\
&= \#H_{\gamma_1 a}(\eta)\#H_{\gamma_2 b}(\eta),
\end{aligned}$$

which yields

$$\begin{aligned}
&[\{\sigma \in \mathrm{Gal}(K/F) \mid \sigma\eta = \eta\sigma\} : \mathrm{Gal}(K/F) \cap \langle\eta\rangle]\kappa(\eta;\delta) \\
&= \sum_{a|2dK} \frac{\mu(a)\#H_{\gamma_1 a}(\eta)}{[K_{\gamma_1 a} : K_\eta]} \times \sum_{(b,2dK)=1} \frac{\mu(b)\#H_{\gamma_2 b}(\eta)}{[K_{\gamma_2 b} : K]} \\
&= \sum_{a|2dK} \frac{\mu(a)\#H_{\gamma_1 a}(\eta)}{[K_{\gamma_1 a} : K_\eta]} \times \prod_{p\nmid 2dK} (1 - \frac{\#H_{\hat{p}}(\eta)}{\hat{p}^{r(K)}\varphi(\hat{p})}),
\end{aligned}$$

where $r(K) := \dim_{\mathbb{Q}}(o_K^\times \otimes_{\mathbb{Z}} \mathbb{Q})$ and $\hat{p} := p^{\nu_p(\gamma_2)+1}$ for $\nu_p(\gamma_2)$ defined by $p^{\nu_p(\gamma_2)} \| \gamma_2$.

The evaluation of $\#H_{\hat{p}}(\eta)$ is not easy. However, the following assures the absolute convergence of $\kappa(\eta;\delta)$.

Proposition 4. *For almost all prime numbers p we have*

$$\frac{\#H_p(\eta)}{p^{r(K)}\varphi(p)} \le \frac{\deg g(x)}{p(p-1)}.$$

Proof. Let q be a natural number and U a subgroup of $o_K^\times$ generated by ϵ^{η^m} ($\epsilon \in o_F^\times$, $m \in \mathbb{Z}$). Denote by K' an extension field of K adjoined by all qth roots of all elements of U. Any extension ρ of η induces an automorphism of K', since $\eta(U) = U$. Let $\rho \in \mathrm{Aut}(K')$ be an extension of η. If $\sqrt[q]{u}^{-\delta_1 g(\rho)} = 1$ for ${}^\forall u \in U$ holds, then $\sqrt[q]{\epsilon}^{-\delta_1 g(\rho)} = 1$ for ${}^\forall\epsilon \in o_F^\times$ holds clearly. Conversely suppose $\sqrt[q]{\epsilon}^{-\delta_1 g(\rho)} = 1$ for ${}^\forall\epsilon \in o_F^\times$; then $1 = \sqrt[q]{\epsilon}^{-\delta_1 g(\rho)\rho^m} = \sqrt[q]{\epsilon}^{-\rho^m\delta_1 g(\rho)} = \sqrt[q]{\epsilon^{\eta^m}}^{-\delta_1 g(\rho)}$ holds, because of $\zeta_q^{-\delta_1 g(\rho)} = 1$. Hence $\sqrt[q]{\epsilon}^{-\delta_1 g(\rho)} = 1$ for ${}^\forall\epsilon \in o_F^\times$ holds if and only if $\sqrt[q]{u}^{-\delta_1 g(\rho)} = 1$ for

${}^\forall u \in U$ holds and then

$$\frac{\#H_q(\eta)}{[K_q:K]} = [K':K]^{-1}$$
$$\times \#\{\rho \in \mathrm{Aut}(K') \mid \rho_{|K} = \eta,\ \sqrt[q]{u}^{\delta_1 g(\rho)} = 1 \text{ for } {}^\forall u \in U\}.$$

Let $W = \langle \zeta_w \rangle$ be a set of all roots of unity in U and let $\{u_1, \cdots, u_s\} \subset U$ induce a basis of U/W, and define a unimodular matrix $A = (a_{ij})$ by

$$u_i^\eta = \zeta_w^{B_i} \prod_j u_j^{a_{ij}}.$$

Then it is easy to see $g(A) = 0$. Let $\rho \in \mathrm{Aut}(K')$ be an automorphism of K' with $\rho_{|K} = \eta$ and set

$$(*) \qquad \zeta_{qw}^\rho = \zeta_{qw}^c,\ \sqrt[q]{u_i}^\rho = \zeta_{qw}^{b_i} \prod_j \sqrt[q]{u_j}^{a_{ij}}.$$

We note that $c \bmod w$ is determined by $\zeta_w^\eta = \zeta_w^c$, and $b_i \equiv B_i \bmod w$ is clear. Putting $A^t = (a_{ij}(t))$, we can see easily by induction on $t\ (\geq 0)$

$$\sqrt[q]{u_i}^{\rho^t} = \zeta_{qw}^{\sum_{k=0}^{t-1} c^{t-k-1} \sum_{j=1}^s b_j a_{ij}(k)} \prod_\ell \sqrt[q]{u_\ell}^{a_{i\ell}(t)}.$$

Then setting $g(x) = g_n x^n + \cdots + g_0$, we have

$$\sqrt[q]{u_i}^{\delta_1 g(\rho)} = \zeta_{qw}^{\delta_1 \sum_{t=0}^n g_t \sum_{k=0}^{t-1} c^{t-k-1} \sum_{j=1}^s b_j a_{ij}(k)},$$

by virtue of $g(A) = 0$. Therefore $\sqrt[q]{u}^{\delta_1 g(\rho)} = 1$ holds for ${}^\forall u \in U$ if and only if

$$\delta_1 g(c) \equiv 0 \bmod qw \quad (\text{ in the case of } u = \zeta_w),$$

$$\delta_1 \sum_{t=0}^n g_t \sum_{k=0}^{t-1} c^{t-k-1} \sum_{j=1}^s b_j a_{ij}(k) \equiv 0 \bmod qw \quad (\text{ in the case of } u = u_i).$$

Put

$$g_A(c) := \sum_{t=0}^n g_t \sum_{k=0}^{t-1} c^{t-k-1} A^k \quad \text{and} \quad \boldsymbol{b} := (b_1, \cdots, b_s).$$

Then we have shown

$$\#\{\rho \in \mathrm{Aut}(K') \mid \rho_{|K} = \eta,\ \sqrt[q]{u}^{\delta_1 g(\rho)} = 1 \text{ for } {}^\forall u \in U\}$$
$$= \#\left\{ c \bmod qw, \boldsymbol{b} \bmod qw \;\middle|\; \begin{array}{l} \delta_1 g(c) \equiv 0 \bmod qw, \\ \delta_1 g_A(c)^t \boldsymbol{b} \equiv 0 \bmod qw \\ \text{and the extension condition} \end{array} \right\}$$

where the extension condition means that the extension automorphism ρ of η exists subject to the condition $(*)$. Next, let us show that the minimal polynomial of A is equal to $g(x)$. $g(A) = 0$ is known. Suppose that $G(x)$ is the minimal polynomial of A and we write $G(x) = \sum_i G_i(x)\omega_i$ where $G_i(x) \in \mathbb{Z}[x]$ and $\{\omega_i\}$ is a basis of $\mathbb{C}$ over $\mathbb{Q}$;then $G(A) = 0$ implies $G_i(A) = 0$ and hence $\{u^{G_i(\eta)} \mid u \in U\}$ is a finite group and the minimality of $\deg g(x)$ yields $\deg G_i(x) \geq \deg g(x)$ and so $\deg G(x) \geq \deg g(x)$. Thus $g(x)$ is the minimal polynomial of A. Under these let us prove the proposition. Assume that q is a prime number p which does not divide $2d \cdot dK$ and that if $u \in U$ is a pth power in $o_K^\times$, then u is a pth power in U. Then by virtue of theorem 2, we have $[K' : K] = p^s(p-1)$ and there exists an extension ρ of η defined by $(*)$ for any integers c, b_i which satisfy $\zeta_w^\eta = \zeta_w^c$, $b_i \equiv B_i \bmod w$. Suppose that ρ satisfies $\sqrt[p]{u}^{\delta_1 g(\rho)} = 1$ for ${}^\forall u \in U$ and let c, b_i correspond to ρ; then $g(c) \equiv 0 \bmod p$ and $g_A(c)^t\boldsymbol{b} \equiv 0 \bmod p$ hold, since $\delta_1 \mid (2dK)^\infty$ and $p \nmid 2dK$. We can take a non-zero vector $\boldsymbol{e} \in \mathbb{Z}^s$ such that $\boldsymbol{e}\, A \equiv c\boldsymbol{e} \bmod p$ because the minimal polynomial $g(x)$ of A divides the characteristic polynomial. Then $\boldsymbol{e}\, g_A(c) \equiv g'(c)\boldsymbol{e} \bmod p$ yields $\boldsymbol{e}\, g_A(c)^t\boldsymbol{b} \equiv g'(c)\boldsymbol{e}\, {}^t\boldsymbol{b} \equiv 0 \bmod p$. Since $g(x)$ divides $x^d - 1 = g(x)h(x)$, $g'(c) \not\equiv 0 \bmod p$ holds by virtue of $p \nmid d$. Thus the equation $g'(c)\boldsymbol{e}\, {}^t\boldsymbol{b} \equiv 0 \bmod p$ is non-trivial and then the number of extension ρ with a fixed c is less than or equal to p^{s-1}, which completes the proof. □

Lemma 11. *There exists the maximal natural number δ_0 such that*

$$\#H_{\delta_0}(\eta) = [K_{\delta_0} : K].$$

Proof. Since the condition $\#H_N(\eta) = [K_N : K]$ implies $\#H_a(\eta) = [K_a : K]$ for every divisor a of N, it is easy to see for $(a,b) = 1$ $\#H_{ab}(\eta) = [K_{ab} : K]$ holds if and only if $\#H_a(\eta) = [K_a : K]$ and $\#H_b(\eta) = [K_b : K]$ hold. First suppose $\#H_q(\eta) = [K_q : K]$ for a prime number q; then we have for $r = r(K)$, $q^r(q-1) = [K_q : K] = \#H_q(\eta) \leq (\deg g(x))q^{r-1}$ if q does not divide some integer by theorem 2 and proposition 4. Thus the condition $\#H_\delta(\eta) = [K_\delta : K]$ yields that prime divisors of δ are bounded. Next, we suppose that $\#H_{q^n}(\eta) = [K_{q^n} : K]$ holds for an integer n and a fixed prime number q. We apply Lemma 4 to $m = q^n$. Let $\mathfrak{P}$ be a prime ideal of K such that $\mathfrak{P} \nmid 2qdK$, $((K/\mathbb{Q})/\mathfrak{P}) = \eta$ and $g(p) \neq 0$ for a prime number p lying below $\mathfrak{P}$. $\mathfrak{P}$ is independent of n. For a prime ideal $\mathfrak{P}_{q^n}$ of K_{q^n} lying above $\mathfrak{P}$, we put $\rho = ((K_{q^n}, \mathbb{Q})/\mathfrak{P}_{q^n})$. Then Lemma 4 implies $q^n h(p)/\delta_1 \mid I(\mathfrak{p})$ for $\mathfrak{p} = \mathfrak{P} \cap F$ and hence by Lemma 3 we have $q^n \mid \delta_1 g(p)$ and so $n \log q \leq \log|\delta_1 g(p)|$. □

Remark. If a prime number p divides δ_0, then p divides $2d \cdot dK$ or $[L \cap \mathbb{Q}M : M]$, where L is a $\mathbb{Z}$-lattice $o_K^\times/\{\text{roots of unity in } K\}$ and M is a submodule of L corresponding to $\langle \epsilon^{\eta^t} \mid t \in \mathbb{Z}, \epsilon \in o_F^\times \rangle$.

To study the positivity of $\kappa(\eta; \delta_0)$, we need the following well-known fact.

Lemma 12. *Let S be a finite set and A_i $(i = 1, 2, \cdots, n)$ a subset of S. For a subset A of S, we set $P(A) = \#A/\#S$. Then we have*

$$P(\cup_{i=1}^n A_i) = \sum_{k=1}^n (-1)^{k+1} \sum_{1 \le i_1 < i_2 < \cdots < i_k \le n} P(A_{i_1} \cap \cdots \cap A_{i_k}).$$

Proposition 5. *Let $\delta = \gamma_1\gamma_2$ with $\gamma_1 \mid (2dK)^\infty$, $(\gamma_2, 2dK) = 1$. $\kappa(\eta; \delta)$ is positive if and only if both $\#H_{\hat{p}}(\eta) \neq \hat{p}^{r(K)}\varphi(\hat{p})$ holds for $\hat{p} := p^{\nu_p(\gamma_2)+1}$ in case of $p \nmid 2dK$ and there is an automorphism ρ such that $\rho_{|K_{\gamma_1}} \in H_{\gamma_1}(\eta)$ and $\rho_{|K_{\gamma_1 q}}$ is not in $H_{\gamma_1 q}(\eta)$ for every prime divisor q of $2dK$.*

Proof. By virtue of what we have shown, we have only to study

$$\sum_{a|2dK} \frac{\mu(a)\#H_{\gamma_1 a}(\eta)}{[K_{\gamma_1 a} : K_\eta]}.$$

Put $S := \mathrm{Gal}(K_{2\gamma_1 dK}/K_\eta)$ and $A(q) := \{\rho \in S \mid \rho_{|K_{\gamma_1 q}} \in H_{\gamma_1 q}(\eta)\}$ for a prime number q. Then for a square-free integer a, we see

$$\begin{aligned}\cap_{q|a} A(q) &= \{\rho \in S \mid \rho_{|K} = \eta,\ \sqrt[\gamma_1 q]{\epsilon}^{\,\delta_1 g(\rho)} = 1 \text{ for } {}^\forall\epsilon \in o_F^\times \text{ and } {}^\forall q \mid a\} \\ &= \{\rho \in S \mid \rho_{|K_{\gamma_1 a}} \in H_{\gamma_1 a}(\eta)\}.\end{aligned}$$

Therefore we have

$$\#(\cap_{q|a} A(q))/\#S = \frac{\#\{\rho \in S \mid \rho_{|K_{\gamma_1 a}} \in H_{\gamma_1 a}(\eta)\}}{[K_{2\gamma_1 dK} : K_\eta]} = \frac{\#H_{\gamma_1 a}(\eta)}{[K_{\gamma_1 a} : K_\eta]}$$

and using Lemma 12, we see

$$\begin{aligned}&\sum_{a|2dK} \frac{\mu(a)\#H_{\gamma_1 a}(\eta)}{[K_{\gamma_1 a} : K_\eta]} \\ &= \frac{\#H_{\gamma_1}(\eta)}{[K_{\gamma_1} : K_\eta]} - \#(\cup_{q|2dK} A(q))/\#S \\ &= [K_{2\gamma_1 dK} : K_\eta]^{-1} \Big\{\#\{\rho \in S \mid \rho_{|K} = \eta,\ \sqrt[\gamma_1]{\epsilon}^{\,\delta_1 g(\rho)} = 1 \text{ for } {}^\forall\epsilon \in o_F^\times\} \\ &\quad - \#\{\rho \in S \mid \rho_{|K} = \eta,\ \sqrt[\gamma_1 q]{\epsilon}^{\,\delta_1 g(\rho)} = 1 \text{ for } {}^\forall\epsilon \in o_F^\times \text{ holds for } {}^\exists q \mid 2dK\}\Big\}.\end{aligned}$$

Thus we have shown the assertion. □

Theorem 3. *$\kappa(\eta;\delta_0)$ is positive.*

Proof. We define integers n_p by $p^{n_p}\|\delta_0$ for prime divisors p of $2dK$. We have only to show that there exist a unit $u_p \in o_F^\times$ and an extension ρ_p of η such that $[K(\sqrt[p^{n_p+1}]{u_p}) : K]$ is a power of p and $\sqrt[p^{n_p+1}]{u_p}^{\delta_1 g(\rho_p)} \neq 1$, since fields $K(\sqrt[p^{n_p+1}]{u_p})$ are linearly disjoint over K and hence the composite ρ of ρ_p exists and $\rho \notin H_{\delta_0 p}(\eta)$ for any prime $p \mid 2dK$. By the maximality of δ_0, there exist a unit $u_p \in o_F^\times$ and an extension ρ_p of η such that $\sqrt[p^{n_p+1}]{u_p}^{\delta_1 g(\rho_p)} \neq 1$. If u_p is a pth power of $v \in F$, then $\sqrt[p^{n_p}]{v}^{\delta_1 g(\rho_p)} \neq 1$ and hence $\sqrt[p^{n_p+1}]{v}^{\delta_1 g(\rho_p)} \neq 1$. Repeating this, we may assume u_p is not a pth power in F. Suppose $u_p = v^p$ for $^\exists v \in K$; then $x^p - u_p$ is irreducible over F and it is reducible over K. Since K/F is a Galois extension, it yields $\zeta_p \in K$ and therefore $[K(\sqrt[p^{n_p+1}]{u_p}) : K]$ is a power of p. If u_p is not a pth power in K, then $x^{p^{n_p+1}} - u_p$ is irreducible over K in the case of $p \neq 2$ and hence we have $[K(\sqrt[p^{n_p+1}]{u_p}) : K]$ is a power of p. In case of $p = 2$, it is clear. Thus we have completed the proof. □

References

[CKY] Y-M. J. Chen, Y. Kitaoka and J. Yu, *Distribution of units of real quadratic number fields,* Nagoya Math. J. **158** (2000), 167–184

[H] C. Hooley, *On Artin's Conjecture,* J. reine angew. Math. **225** (1967), 209–220.

[IK] M. Ishikawa and Y. Kitaoka, *On the distribution of units modulo prime ideals in real quadratic fields,* J. reine angew. Math. **494** (1998), 65–72.

[K] Y. Kitaoka, *Distribution of units of a cubic field with negative discriminant,* J. of Number Theory **91** (2001), 318–355.

[L] H. W. Lenstra, Jr., *On Artin's conjecture and Euclid' algorithm in global fields,* Inventiones math. **42** (1977), 201–224.

[M] K. Masima, *On the distribution of units in the residue class field of real quadratic fields and Artin's conjecture* (in Japanese), RIMS *Kokyuroku* **1026** (1998), 156–166.

[R] H. Roskam, *A quadratic analogue of Artin's conjecture on primitive roots,* J. of Number Theory **81** (2000), 93–109.

Galois Theory and Modular Forms
K. Hashimoto, K. Miyake, and H. Nakamura, Editors
pp 305–314
© 2003 Kluwer Academic Publishers

ON CAPITULATION PROBLEM FOR 3-MANIFOLDS

Dedicated to Professor Yasuo Morita

Masanori MORISHITA
Department of Mathematics
Kanazawa University,
Kakuma-machi, Kanazawa, Ishikawa,
920-1192 Japan
morisita@kenroku.kanazawa-u.ac.jp

Introduction

The purpose of this paper is to investigate analogues for 3-manifolds after the model of capitulation theorems for number fields. This will give one of the foundational analogies between 3-dimensional topology and algebraic number theory.

The capitulation problem deals with the structure of ideal classes which capitulate in an extension of number fields, and it is formulated as a purely group-theoretic problem on the transfer by means of the Artin reciprocity map. In recent years, H. Suzuki ([Su1,2]) has established a fundamental theorem on this group-theoretic version which includes the three classical capitulation theorems, Hilbert's theorem 94, the principal ideal theorem and Tannaka-Terada's theorem.

In this paper, we shall consider a similar problem for homology classes in a covering of manifolds. Since the problem is again translated into a similar problem on the group-transfer by means of the Hurewicz isomorphism, we obtain a topological analog of Suzuki's theorem (Theorem 1.3 below). For a Galois covering of 3-manifolds, using the spectral sequence and Poincaré duality, we obtain more precise results on capitulation, analogous to the theorems by Iwasawa, Chevalley-Tate, and Suzuki for number fields (Theorem 2.3, Corollaries 2.4, 2.6 below). We note finally that some analogues of capitualtion theorems for coverings of topological spaces were also studied in the recent Master thesis of K. Fuluta ([Fu]).

[received: May 24, 2002; accepted in revised form: Oct. 18, 2002]

1. Capitulation problem

For a number field F,* we denote by H_F the ideal class group of F. For an extension K/k of number fields, the capitulation problem deals with the structure of the finite abelian group

$$C_{K/k} := \mathrm{Ker}(t_{K/k} : H_k \longrightarrow H_K)$$

where $t_{K/k}$ is the natural homomorphism induced by the extension of an ideal under the natural inclusion $k \to K$. For a useful survey on this topic, we refer to [Mi]. A fundamental theorem is due to H. Suzuki.

Theorem 1.1 ([Su1,2]). *Let k/k_0 be a finite cyclic extension of numer fields of Galois group G_0 and let K/k be an unramified extension so that K is abelian over k_0. Then the number of G_0-invariant ideal classes in $C_{K/k}$ is divisible by the degree $[K : k]$ of K over k.*

We note that Suzuki's theorem includes three classical theorems, Hilbert's theorem 94 ([Hi]) for $k_0 = k$ and cyclic K/k, the principal ideal theorem ([F]) for case that $k_0 = k$ and K is the Hilbert class field of k, and finally Tannaka-Terada's theorem ([T]) for case that K is the genus field of k/k_0

Since these capitulation theorems are translated into purely group theoretical statements by means of the Artin reciprocity map, we have their analogues for a wide class of coverings of topological spaces by means of the Hurewicz isomorphism. Let $p : N \to M$ be a finite abelian covering of path-connected topological manifolds and let $t_{N/M} : H_1(M, \mathbf{Z}) \to H_1(N, \mathbf{Z})$ be the transfer map defined by $t_{N/M}([c]) = [p^{-1}(c)]$. We set

$$C_{N/M} := \mathrm{Ker}(t_{N/M} : H_1(M, \mathbf{Z}) \to H_1(N, \mathbf{Z}))$$

and we say that a subgroup A of $H_1(M, \mathbf{Z})$ *capitulates in* N if $A \subset C_{N/M}$. Let M^a and N^a be the maximal abelian coverings of M and N respectively, and set $\Gamma = \mathrm{Gal}(N^a/M)$ and $H = \mathrm{Gal}(N^a/N)$. We write $\mathcal{G}'$ for the commutator subgroup of a group $\mathcal{G}$ throughout this paper. Consider the diagram

$$\begin{array}{ccc} H_1(M, \mathbf{Z}) & \xrightarrow{\sim} & \Gamma/\Gamma' \\ {\scriptstyle t_{N/M}} \downarrow & & \downarrow {\scriptstyle V_{\Gamma/H}} \\ H_1(N, \mathbf{Z}) & \xrightarrow{\sim} & H/H' = H \end{array}$$

*In this paper, we shall mean by a number field a finite extension over the rational number field $\mathbf{Q}$, although the capitulation problem makes sense for any algebraic extensions over $\mathbf{Q}$.

where the horizontal maps are Hurewicz isomorphisms and $V_{\Gamma/H}$ is the group-transfer. Basically, by applying the group-theoretical versions of capitulation theorems for number fields to the above diagram, we have analogous statements for $C_{N/M}$. However, for a topological setting above, the groups Γ and H/H' are possibly infinite (cf. Examples 1.5 below), while Γ is assumed to be finite in the group-theoretic version of Theorem 1.1 ([Su2], Th. 2) for the application to number fields. So, we give here a bit generalized form of Theorem 2 in [Su2], of which was informed by H. Suzuki, for our purpose.

Theorem 1.2. *Let Γ be a finitely generated group such that Γ/Γ' is finite. Let H be a normal subgroup of Γ with finite index in Γ and containing Γ'. Let σ be an automorphism of Γ such that $\sigma(H) \subset H$ and that σ induces the identity map on $G = \Gamma/H$. Then the order of the subgroup*

$$\{\gamma\Gamma' \in \mathrm{Ker}(V_{\Gamma/H}) \mid \sigma(\gamma)\gamma^{-1} \in \Gamma'\}$$

of the transfer kernel $\mathrm{Ker}(V_{\Gamma/H})$ *is divisible by* $\#G = [\Gamma : H]$.

Proof. (due to H. Suzuki). Since Γ/Γ' is finite, we have $V_{\Gamma/H}(\Gamma/\Gamma') \subset (H/H')_{tor}$ where A_{tor} denotes the torsion part of an abelian group A. Take a positive integer m so that m is a multiple of the exponents of Γ/Γ' and $(H/H')_{tor}$. Set $M := H^m H'$. Then we see that M is a normal subgroup of Γ with finite index and is contained in Γ'. Consider the following commutative diagram:

$$\begin{array}{lccc}
V_{\Gamma/H}: & \Gamma/\Gamma' & \longrightarrow & (H/H')_{tor} \\
 & \| & & \downarrow \\
V_{(\Gamma/M)/(H/M)}: & (\Gamma/M)(\Gamma/M)' & \longrightarrow & H/M.
\end{array}$$

Since $M/H' \cap (H/H')_{tor} = \{0\}$ by our choice of m, the right vertical map is injective and so we find $\mathrm{Ker}(V_{\Gamma/H}) = \mathrm{Ker}(V_{\Gamma/H} \bmod M)$. Noting that M is stable under the action of σ, we can reduce our assertion to the case Γ is finite. $\square$

Theorem 1.2 yields immediately the following topological counterpart of Suzuki's theorem 1.1.

Theorem 1.3. *Let $M \to M_0$ be a finite cyclic branched covering of path-connected compact manifolds with Galois group G_0 and let $N \to M$ be a finite covering so that the composite $N \to M_0$ is an abelian covering. Assume that $H_1(M, \mathbf{Z})$ is finite. Then the number of G_0-invariant 1st* *homology classes in $C_{N/M}$ is divisible by the degree $[N : M]$ of N over M.*

Remark 1.4. In [Mo5], we introduced the notion of the genus covering of an abelian covering of a link in the 3-sphere. However, this notion is purely group-theoretically defined and so we can define the genus covering of an abelian M/M_0 to be the maximal abelian subcovering of M^a over M_0. Then we have a topological Tannaka-Terada theorem: *any G_0-invariant 1st homology class of M becomes trivial in the genus covering of M/M_0.*

Example 1.5. Let M be a $3m$-fold cyclic covering of S^3 ramified over the figure eight knot, $m \geq 1$. It was shown in [He2] that $H_1(M, \mathbf{Z})$ is finite and M has an abelian covering N of degree 4 such that $H_1(N, \mathbf{Z}) \simeq H_1(M, \mathbf{Z}) \oplus \mathbf{Z}^3$ modulo 2-torsion. By Th. 1.3, $\#C_{N/M}$ is divisible by 4. Of course, if $H_1(M, \mathbf{Z})$ is finite and $H_1(N, \mathbf{Z})$ is free with rank ≥ 1, then the whole $H_1(M, \mathbf{Z})$ capitulates in N. J. Hempel informed me of the following such examples: Let M be the union of 2 copies of twisted I-bundles (cf. [He1]) along a common boundary which is an oriented compact surface Σ. Then we have a double covering $N = \Sigma \times S^1 \to M$ with finite $H_1(M, \mathbf{Z})$ and infinite and free $H_1(N, \mathbf{Z})$. A particular case is that $M = \mathbb{P}^1 \# \mathbb{P}^1$(connected sum) and $N = S^2 \times S^1$.

2. Capitulation theorems for 3-manifolds

For coverings of 3-manifolds, we can derive more precise information on the capitualtion problem using the Poincaré duality and the cohomology spectral sequence. We find that a similar cohomology spectral sequence for number fields leads a simple proof of a classical theorem by Iwasawa ([I1]) on $C_{K/k}$. This would deepen our underlying philosophy, namely, analogies between 3-manifolds and number fields. Here is a part of a basic dictionary of this analogy which we shall employ in the following. Hereafter, we shall mean by a 3-manifold an oriented, connected, closed 3-dimensional manifold, and we shall write $\mathcal{O}_k$ for the ring of integers of a number field k.

3-manifold $M \to S^3$	$\longleftrightarrow$	number field $k/\mathbf{Q}$
knot, link $\subset M$ end	$\longleftrightarrow$	prime ideal($\neq 0$), ideal $\subset \mathcal{O}_k$ infinite prime
surface $\Sigma \subset M$ $\Sigma \mapsto \partial\Sigma$	$\longleftrightarrow$	$a \in \mathcal{O}_k$ $a \mapsto a\mathcal{O}_k$

chain complex $C_*(M, \mathbf{Z})$	$\longleftrightarrow$	$k^\times \to \bigoplus_{\mathfrak{p}:\text{primes}} \mathbf{Z}$
(torsion part of) $H_1(M, \mathbf{Z})$	$\longleftrightarrow$	$H_k =$ (ideal class group of k)
Hurewicz isomorphism $H_1(M, \mathbf{Z}) \simeq \mathrm{Gal}(M^a/M)$ ($M^a =$ max. abel cover of M)	$\longleftrightarrow$	Unramified classfield theory $H_k \simeq \mathrm{Gal}(k^a/k)$ ($k^a =$ Hilbert classfield of k)
$H_2(M, \mathbf{Z})(\simeq H^1(M, \mathbf{Z}))$	$\longleftrightarrow$	(free part of) $\mathcal{O}_k^\times$

For more and precise analogies between knots and primes, 3-manifolds and number fields, we refer to [K], [Ma1,2], [Mo1~7], [Ra], [R1,2], [Si], [W].

Now, we shall firstly give a simple proof of the following well-known theorem by Iwasawa, which relates $C_{K/k}$ with the Galois cohomology of the unit group, by using cohomology spectral sequence.

Theorem 2.1 ([I1]). *For a finite unramified Galois extension K/k of number fields with Galois group G, we have*

$$C_{K/k} \simeq H^1(G, \mathcal{O}_K^\times)$$

where $\mathcal{O}_K^\times$ is the unit group of the ring of integers of K.

Proof. Consider the Hochschild-Serre spectral sequence associated to the Galois étale covering $\mathrm{Spec}(\mathcal{O}_K) \to \mathrm{Spec}(\mathcal{O}_k)$

$$E_2^{i,j} = H^i(G, H_{et}^j(\mathrm{Spec}(\mathcal{O}_K), \mathbf{G}_m)) \Longrightarrow H_{et}^{i+j}(\mathrm{Spec}(\mathcal{O}_k), \mathbf{G}_m)$$

with the associated differentials $d_r^{i,j} : E_r^{i,j} \to E_r^{i+r,j-r+1}$ and the filtration $H_k = H_{et}^1(\mathrm{Spec}(\mathcal{O}_k), \mathbf{G}_m) = F_0 \supset F_1 = E_\infty^{1,0}$. Then we see easily that

$$H_k/F_1 = E_\infty^{0,1} \simeq \cdots \simeq E_3^{0,1} = \mathrm{Ker}(d_2^{0,1}) \subset E_2^{0,1} = H^0(G, H_K) \subset H_k,$$

$$F_1 = E_\infty^{1,0} \simeq \cdots \simeq E_2^{1,0} = H^1(G, \mathcal{O}_K^\times).$$

Hence we have $C_{K/k} = F_1 \simeq H^1(G, \mathcal{O}_K^\times)$. □

When K/k is cyclic, the theorem by Chevalley and Tate on the Herbrand quotient of the unit group ([AT], Ch 5, 3) yields the following precise version of Hilbert's theorem 94.

Theorem 2.2 (Chevalley-Tate). *For a finite unramified cyclic extension K/k of number fields, we have*

$$\#C_{K/k} = [K:k]\#\hat{H}^0(G, \mathcal{O}_K^\times)$$

where $\hat{H}^0$ denotes the Tate cohomology.

Now, let $p: N \longrightarrow M$ be a finite Galois covering of 3-manifolds with Galois group G. By the Poincaré duality, we have

$$C_{N/M} \simeq \mathrm{Ker}(p^*: H^2(M, \mathbf{Z}) \longrightarrow H^2(N, \mathbf{Z})).$$

Then we can derive the following information on $C_{N/M} = \mathrm{Ker}(p^*)$ as in our proof of Iwasawa's theorem 2.1 from the Hochschild-Serre spectral sequence

$$E_2^{i,j} = H^i(G, H^j(N, \mathbf{Z})) \Longrightarrow H^{i+j}(M, \mathbf{Z}).$$

Theorem 2.3. *Notation being as above, we have the following exact sequence*

$$0 \longrightarrow \mathrm{Cok}(d_2^{0,1}) \longrightarrow C_{N/M} \longrightarrow \mathrm{Ker}(d_2^{1,1}) \longrightarrow 0$$

where $d_2^{0,1}: H^0(G, H^1(N, \mathbf{Z})) \rightarrow H^2(G, \mathbf{Z})$ and $d_2^{1,1}: H^1(G, H^1(N, \mathbf{Z})) \rightarrow H^3(G, \mathbf{Z})$ are the differentials attached to the spectral sequence.

In particular, $C_{N/M}$ is a finite abelian group.

Proof. We let

$$d_r^{i,j}: E_r^{i,j} \longrightarrow E_r^{i+r, j-r+1}$$

be the differentials associated to the spectral sequence and let

$$H^2(M, \mathbf{Z}) = F_0 \supset F_1 \supset F_2 \cdots$$

be the filtration of $H^2(M, \mathbf{Z})$ such that $F_i/F_{i+1} \simeq E_\infty^{i,2-i}$. Then we see that

(1) $H^2(M, \mathbf{Z})/F_1 = E_\infty^{0,2} \simeq \cdots \simeq E_4^{0,2} \simeq \mathrm{Ker}(d_3^{0,2}) \subset E_3^{0,2} \simeq \mathrm{Ker}(d_2^{0,2}) \subset E_2^{0,2} = H^0(G, H^2(N, \mathbf{Z}))$,

(2) $F_1/F_2 = E_\infty^{1,1} \simeq \cdots \simeq E_3^{1,1} = \mathrm{Ker}(d_2^{1,1}) \subset E_2^{1,1} = H^1(G, H^1(N, \mathbf{Z}))$,

(3) $F_2 = E_\infty^{2,0} \simeq \cdots \simeq E_3^{2,0} = \mathrm{Ker}(d_2^{2,0})/\mathrm{Im}(d_2^{0,1}) = H^2(G, \mathbf{Z})/\mathrm{Im}(d_2^{0,1})$.

By (1), we have $C_{N/M} \simeq F_1$. Hence, by (2), (3), we get the assertion. □

Recall that a 3-manifold M is called a *rational homology 3-sphere* if $H_i(M, \mathbf{Q}) = H_i(S^3, \mathbf{Q})$ for all i, which is equivalent to $H_1(M, \mathbf{Q}) = 0$

(or $H_1(M, \mathbf{Z})$ being finite) by the Poincaré duality. We see easily by the Hochschild-Serre spectral sequence that M is a rational homology 3-sphere if a finite Galois covering space N of M is so. The following is seen as a more precise analog of Suzuki's theorem for 3-manifolds.

Corollary 2.4. *If N is a rational homology 3-sphere, then we have*

$$C_{N/M} \simeq G/G' \ (\textit{non-canonical}).$$

Hence, $\#C_{N/M}$ is equal to the degree of the maximal abelian subcovering of N over M.

Proof. For a rational homology 3-sphere N, $H^1(N, \mathbf{Z}) = \mathrm{Hom}(H_1(N, \mathbf{Z}), \mathbf{Z}) = 0$ and so $\mathrm{Ker}(d_2^{1,1}) = 0$ in Th. 2.3. Hence by Th. 2.3 again, $C_{N/M} \simeq H^2(G, \mathbf{Z}) \simeq \mathrm{Hom}(G/G', \mathbf{Q}/\mathbf{Z})$. □

Remark 2.5. Suppose that M and its maximal abelian covering M^a are rational homology 3-spheres. Then Cor. 2.4 tells us that $H_1(M, \mathbf{Z})$ capitulates in M^a and cannot capitulate already in a proper subcovering N of M^a. In fact, if $H_1(M, \mathbf{Z})$ capitulates in a subcovering N of M^a, by Cor. 2.4, we have $[N : M] = \#C_{N/M} = \#H_1(M, \mathbf{Z}) = [M^a : M]$, hence $N = M^a$. On the other hand, it can happen that all ideal classes of a number field capitulate in a proper subextension of the Hilbert class field (e.g, [I2]).

Corollary 2.6. *If M is a rational homology 3-sphere and G is cyclic, we have the equality*

$$\#C_{N/M} = [N : M]h(H_2(N, \mathbf{Z}))$$

where $h(H_2(N, \mathbf{Z})) := \prod_{i=0}^{1} \#H^i(G, H_2(N, \mathbf{Z}))^{(-1)^{i+1}}$ is an integer.

Proof. Since G is cyclic, we have $H^3(G, \mathbf{Z}) = H^1(G, \mathbf{Z}) = 0$ and so $\mathrm{Ker}(d_2^{1,1}) = H^1(G, H^1(N, \mathbf{Z})) = H^1(G, H_2(N, \mathbf{Z}))$ by the Poincaré duality. Since $H^1(M, \mathbf{Z}) = \mathrm{Hom}(H_1(M, \mathbf{Z}), \mathbf{Z}) = 0$ by the Poincaré duality, a piece of the spectral sequence shows the injectivity of $d_2^{0,1}$. Hence we have the equality in our assertion by Th. 2.3. By Th. 1.3, $\#C_{N/M}$ is divisible by $[N : M]$ and so $h(H_2(N, \mathbf{Z}))$ is an integer. □

Since $H_2(N, \mathbf{Z})$ is an analog of the unit group $\mathcal{O}_K^\times$, Cor. 2.6 has a flavour analogous to Th. 2.2 by Chevalley and Tate.

Example 2.7. Let M be the lens space $L(m, n)$ which has the fundamental group $\mathbf{Z}/m\mathbf{Z}$. Hence, for any cyclic covering $N \to M$ of degree d, a divisor of m, we have $C_{N/M} \simeq \mathbf{Z}/d\mathbf{Z}$ by Cor. 2.4. Suppose m is even. Then M is the double cover of S^3 ramified over the 2-bridge link $B(m, n)$

([BZ], Ch.12). By [Mo5], the genus covering M^g of M over S^3 is the double cover $L(m/2, n)$ of M and $\#H_1(M, \mathbf{Z})^{G_0} = 2$, $G_0 = \mathrm{Gal}(M/S^3)$. Hence, by 1.4, we have $C_{M^g/M} = H_1(M, \mathbf{Z})^{G_0}$.

Acknowledgments

I thank M. Furuta for informing me of the method to compute $C_{N/M}$ by the Poincaré duality and spectral sequence. Many thanks go to J. Hempel and H. Suzuki for useful communications and to H. Sato for pointing out a mistake in the first manuscript. I thank K. Hashimoto, K. Miyake and H. Nakamura for giving me an opportunity to join the conference "Galois theory and Modular forms" held at Tokyo Metropolitan University, Sep. 25-29, 2001. This work was partially supported by the Grants-in-Aid for Scientific Research (C) (No.13640014), Ministry of Education, Culture, Sports, Science and Technology, Japan.

References

[AT] E. Artin, J. Tate, "Class field theory," 2nd edition. Advanced Book Classics. Addison-Wesley Publishing Company, Advanced Book Program, Redwood City, CA, 1990.

[BZ] G. Burde, H. Zieschang, "Knots," de Gruyter Studies in Math. 5, Walter de Gruyter and Co., Berlin, 1985.

[F] Ph. Furtwängler, *Beweis des Hauptidealsatzes für Klassenkörper algebraischer Zahlkörper*, Abh. Math. Sem. Univ. Hamburg **7** (1930), 14–36.

[Fu] K. Fuluta, *On Hilbert's theorem* 94 *for covering spaces*, Master thesis at Tohoku University (Japanese), 2002.

[Hi] D. Hilbert, *Bericht: Die Theorie der algebraischen Zahlkörper*, Jer. dt. Math.-Ver. **4** (1897), 175–546; Gesam, Abh. I., 63–363.

[He1] J. Hempel, "3-manifolds," Ann. of Math. Studies, No. 86, Princeton Univ. Press, 1976.

[He2] J. Hempel, *Homology of coverings*, Pacific J. Math. **112** (1984), 83–113.

[I1] K. Iwasawa, *A note on the group of units of an algebraic number field*, J. Math. pures et appl. (9) **35** (1956), 189–192.

[I2] K. Iwasawa, *A note on capitulation problem for number fields*, Proc. Japan Academy **65** (1989), 59-61; II, ibid, No.6, (1989), 183–186.

[K] M. Kapranov, *Analogies between number fields and* 3*-manifolds*, (unpublished note).

[Ma1] B. Mazur, unpublished mimeographed note, circa, 1965.

[Ma2] B. Mazur, *The theme of p-adic variation*, Mathematics: frontiers and perspectives, 433–459, Amer. Math. Soc., Providence, RI, (2000).

[Mi] K. Miyake, *Algebraic investigation of Hilbert's theorem* 94, *the principal ideal theorem and the capitulation problem*, Expo. Math. **7** (1989), 289–346.

[Mo1] M. Morishita, *Milnor's link invariants attached to certain Galois groups over* **Q**, Proc. Japan Academy **76** (2000), 18–21.

[Mo2] M. Morishita, *On certain analogies between knots and primes*, J. reine und angew. Math. **550** (2002), 141–167.

[Mo3] M. Morishita, *Knots and primes,* 3*-manifolds and number fields*, Algebraic number theory and related topics (Japanese) (Kyoto, 2000, Dec.), RIMS report, **1200**, (2001), 103–115.

[Mo4] M. Morishita, *Knots and primes,* 3*-manifolds and number fields*, A conference report "Art of Low Dimensional Topology VII," Kansai Seminar House, February, (2001), 99–109 .

[Mo5] M. Morishita, *A theory of genera for cyclic coverings of links*, Proc. Japan Academy **77** (2001), 115–118.

[Mo6] M. Morishita, *Analogies between knots and primes,* 3*-manifolds and number fields*, 2001 (submitted).

[Mo7] M. Morishita, *Milnor invariants and Massey products for prime numbers*, Compositio Math. (2003), (to appear).

[Ra] N. Ramachandran, *A note on arithmetic topology*, C. R. Math. Acad. Sci. Soc. R. Can. **23** (2001), 130–135.

[R1] A. Reznikov, *Three-manifolds class field theory* (*Homology of coverings for a nonvirtually b_1-positive manifold*), Sel. math. New ser. **3** (1997), 361–399.

[R2] A. Reznikov, *Embedded incompressible surfaces and homology of ramified coverings of three-manifolds*, Sel. math. New ser. **6** (2000), 1–39.

[Si] A. Sikora, *Analogies between group actions on* 3*-manifolds and number fields*, (2001), (to appear).

[Su1] H. Suzuki, *A generalization of Hilbert's theorem* 94, Nagoya Math. J. **121** (1991), 161–169.

[Su2] H. Suzuki, *On the capitulation problem*, Class field theory—its centenary and prospect (Tokyo, 1998), Adv. Stud. Pure Math., **30**, Math. Soc. Japan, Tokyo, (2001), 483-507.

[T] F. Terada, *On a generalization of the principal ideal theorem*, Tôhoku Math. J. (2) **1** (1950), 229–269.

[W] J.-L. Waldspurger, *Entrelacements sur Spec*(**Z**), Bull. Sc. math. **100** (1976), 113–139.

Galois Theory and Modular Forms pp 315–326
K. Hashimoto, K. Miyake, and H. Nakamura, Editors
© 2003 Kluwer Academic Publishers

ON THE IWASAWA μ-INVARIANT OF THE CYCLOTOMIC $\mathbb{Z}_p$-EXTENSION OF CERTAIN QUARTIC FIELDS

Tsuyoshi ITOH
Department of Mathematical Sciences,
School of Science and Engineering
Waseda University,
3-4-1 Okubo, Shinjuku, Tokyo,
169-8555, Japan
tsitoh@aoni.waseda.jp

Abstract We give a criterion of vanishing of the Iwasawa μ-invariant of the cyclotomic $\mathbb{Z}_3$-extension of certain quartic fields.

1. Introduction.

Let $\mathbb{Q}$ be the field of rational numbers and p a prime number. We denote by $\mathbb{Q}_\infty$ the cyclotomic $\mathbb{Z}_p$-extension of $\mathbb{Q}$, that is, the unique Galois extension of $\mathbb{Q}$ whose Galois group is isomorphic to the additive group of the p-adic integer ring $\mathbb{Z}_p$. Let $\mathbb{Q}_n$ be the unique subfield of $\mathbb{Q}_\infty$ whose degree over $\mathbb{Q}$ is p^n. Moreover, for any algebraic number field k, we put $k_\infty = k\mathbb{Q}_\infty$ and $k_n = k\mathbb{Q}_n$. The field k_∞ is called the cyclotomic $\mathbb{Z}_p$-extension of k.

Let $A(k_n)$ be the Sylow p-subgroup of the ideal class group of k_n. For a finite set S, we denote the order of S by $|S|$. According to the Iwasawa theory, there exist non-negative integers λ, μ and an integer ν which depend only on k and p such that

$$|A(k_n)| = p^{\lambda n + \mu p^n + \nu}$$

for all sufficiently large n.

[received: June 1, 2002; accepted in revised form: October, 29, 2002]

The determination of the invariants λ, μ and ν for any given number field k and any prime p is an important problem in the theory of $\mathbb{Z}_p$-extensions. In particular, the following conjecture is well-known:

Conjecture (Iwasawa [6]). *The μ-invariant of the cyclotomic $\mathbb{Z}_p$-extension vanishes for every number field k and every prime p.*

Concerning this conjecture, the following result is known.

Theorem (Ferrero-Washington [4]). *If k is an abelian extension field over $\mathbb{Q}$, then the μ-invariant of the cyclotomic $\mathbb{Z}_p$-extension of k vanishes for every prime p.*

But the above conjecture is not completely solved until now. We shall deal with the case where the base field is a certain non-Galois quartic extension over $\mathbb{Q}$. Let F be a real quadratic field which is embedded in the field of real numbers. Taking an element α of F which satisfies $\alpha > 0$ and $\alpha' < 0$, where α' is the conjugate of α, we put $M = F(\sqrt{\alpha})$. Then M is a non-Galois quartic field. Therefore it is to be questioned whether the μ-invariant of the cyclotomic $\mathbb{Z}_p$-extension of M vanishes or not.

But in the case where $p = 2$, we can show the following result:

Proposition 1. *The μ-invariant of the cyclotomic $\mathbb{Z}_2$-extension of M vanishes.*

Proof. It is easy to see that the Galois closure M^* of M over $\mathbb{Q}$ is a dihedral extension over $\mathbb{Q}$ of degree 8 and M^* contains an imaginary quadratic field k. By the theorem of Ferrero-Washington, the μ-invariant of the cyclotomic $\mathbb{Z}_2$-extension of k vanishes. Since M^*/k is a quartic extension, Iwasawa's theorem ([6, Theorem 3]) says that the μ-invariant of the cyclotomic $\mathbb{Z}_2$-extension of M^* vanishes. Thus the proposition follows. □

Therefore, the case $p > 2$ is to be treated. In this paper, we shall give a criterion for vanishing of the μ-invariant of the cyclotomic $\mathbb{Z}_3$-extension. Our method is based on the work of Kraft and Schoof [8] about Greenberg's conjecture for real quadratic fields.

Theorem 2. *Let F be a real quadratic number field and α an element of F which satisfies the condition that $\alpha > 0$ and $\alpha' < 0$ (α' denotes the conjugate of α). We put $M = F(\sqrt{\alpha})$. Moreover, let M_∞ be the cyclotomic $\mathbb{Z}_3$-extension of M and M_n the n-th layer of M_∞/M. We assume that every prime of F above 3 does not split in M. We fix a generator γ of $\mathrm{Gal}(M_1/M)$. Let $\eta_1 = X_{\mathfrak{f}_1}(1, G_1)$ be the Stark-Shintani unit of M_1 which will be defined later.*

If $\eta_1^{1-\gamma}$ is not a cube in M_1, the μ-invariant of the cyclotomic $\mathbb{Z}_3$-extension of M vanishes.

The Stark-Shintani units are special ones in a certain abelian extension field of a real quadratic field. The general existence of these units is not known (*Stark-Shintani conjecture*, see Stark [13] , Shintani [12]). But in the case where we treat in this paper, the Stark-Shintani units exist in every layer M_n of the cyclotomic $\mathbb{Z}_p$-extension of M and these units have behave themselves like the cyclotomic units in a real abelian field (see Nakagawa [9], [10]).

2. Proof of Theorem 2.

We begin by explaining our notation. For an algebraic number field k, we denote by E_k the unit group of k and by $A(k)$ the Sylow p-subgroup of the ideal class group of k. Let K be a finite Galois extension of k. We denote by $\mathrm{Gal}(K/k)$ the Galois group of K over k and by $N_{K/k}$ the norm mapping from K to k.

Let F and M be as in the previous section and p an odd prime. We fix a generator τ of the Galois group $\mathrm{Gal}(M_\infty/F_\infty)$. We know that τ has order 2 and we shall abuse the same letter τ as its restriction to M_n. Since p is an odd prime, we can decompose $A(M_n)$ by the action of τ. We put

$$A_n^+ = \{c \in A(M_n) \mid c^\tau = c\}$$
$$A_n^- = \{c \in A(M_n) \mid c^\tau = c^{-1}\}.$$

Then we have $A(M_n) = A_n^+ \oplus A_n^-$ and $A_n^+ \cong A(F_n)$.

For a $\mathbb{Z}_p$-module Y, the p-rank of Y denotes the dimension of Y/pY as a vector space over the finite field $\mathbb{F}_p$. It is known that the μ-invariant of the cyclotomic $\mathbb{Z}_p$-extension of M vanishes if and only if the p-rank of $A(M_n)$ is bounded as $n \to \infty$ (cf. Washington [14, Proposition 13.23]). Because F is a real quadratic field, the p-rank of $A(F_n)$ is bounded as $n \to \infty$ by the theorem of Ferrero-Washington. Thus we may restrict our attention to A_n^-.

Lemma 3. *Let p be an odd prime, K an algebraic number field and L a quadratic extension of K. We put $\mathrm{Gal}(L_\infty/K_\infty) = \langle\tau\rangle$ and*

$$A(L_n)^- = \{c \in A(L_n) \mid c^\tau = c^{-1}\}.$$

We assume that every prime of K above p does not split in L. If $|A(L_0)^-| = 1$, then $|A(L_n)^-| = 1$ for all n.

Proof. We fix a positive integer n and we put $G = \mathrm{Gal}(L_n/L)$. Since any element of G is commutative with τ, we have $A(L_n)^G \cong A(K_n)^G \oplus (A(L_n)^-)^G$. Then, by the genus formula and the assumption that every prime of K above p does not split in L, we obtain the equation

$$|(A(L_n)^-)^G| = \frac{|A(L_n)^G|}{|A(K_n)^G|} = |A(L_0)^-| \cdot \frac{(E_{K_0} : E_{K_0} \cap N_{K_n/K_0}(K_n^\times))}{(E_{L_0} : E_{L_0} \cap N_{L_n/L_0}(L_n^\times))}.$$

By using the method which is given in the proof of Lemma 1 of Kida [7], we can show that the following map:

$$E_{K_0}/(E_{K_0} \cap N_{K_n/K_0}(K_n^\times)) \to E_{L_0}/(E_{L_0} \cap N_{L_n/L_0}(L_n^\times))$$

induced from the natural embedding is injective.

Considering the above fact, if $|A(L_0)^-| = 1$, we have $|(A(L_n)^-)^G| = 1$. Therefore we have $|A(L_n)^-| = 1$, because both of $A(L_n)^-$ and G are p-groups. □

Next, we define the Stark-Shintani units (cf. Shintani [12], Nakagawa [9]). We fix a positive integer n. Let $\mathfrak{f}_n$ be the finite part of the conductor of M_n/F. We denote by $H(\mathfrak{f}_n)$ the *narrow* ray class group modulo $\mathfrak{f}_n$ of F. Take a totally positive integer ν of F with the property $\nu + 1 \in \mathfrak{f}_n$, and denote by the same letter ν the narrow ray class modulo $\mathfrak{f}_n$ represented by the principal ideal (ν). For each element c of $H(\mathfrak{f}_n)$, we define

$$\zeta_F(s, c) = \sum_{\mathfrak{a}} N(\mathfrak{a})^{-s},$$

where $\mathfrak{a}$ runs over all integral ideals of F belonging to the class c.

The Stark-Shintani ray class invariant $X_{\mathfrak{f}_n}(c)$ is now defined by

$$X_{\mathfrak{f}_n}(c) = \exp(\zeta_F'(0, c) - \zeta_F'(0, c\nu)).$$

Moreover, for the subgroup G_n of $H(\mathfrak{f}_n)$ corresponding to M_n, we put $X_{\mathfrak{f}_n}(c, G_n) = \prod_{g \in G_n} X_{\mathfrak{f}_n}(cg)$.

From now on, we assume that *every prime divisor of p in F does not split in M.*

By Theorem 2 of Shintani [12] and Theorem 1 of Nakagawa [10], the ray class invariant $X_{\mathfrak{f}_n}(c, G_n)$ is a unit of M_n, satisfying the reciprocity law

$$X_{\mathfrak{f}_n}(c, G_n)^{\sigma(c')} = \pm X_{\mathfrak{f}_n}(cc', G_n)$$

for $c, c' \in H(\mathfrak{f}_n)$, where σ denotes the Artin map from $H(\mathfrak{f}_n)$ to $\mathrm{Gal}(M_n/F)$. Then the group

$$C_n = \langle -1, X_{\mathfrak{f}_n}(c, G_n) | c \in H(\mathfrak{f}_n) \rangle$$

is a subgroup of E_{M_n} and it is stable under the action of $\mathrm{Gal}(M_n/F)$. If there is a prime divisor $\mathfrak{p}$ of p with the property $\mathfrak{p} \nmid \mathfrak{f}_0$, then we replace C_0 by the group generated by -1 and $X_{\mathfrak{f}_0}(c, G_0)^{2^e}$, where e is the number of such prime divisors $\mathfrak{p}$ of p.

By Theorem 1 of Nakagawa [10], it is known that C_n is a subgroup of

$$E_n^- = \{u \in E_{M_n} \mid u^\tau = u^{-1}\}$$

with finite index and the p-primary part of the index $(E_n^- : C_n)$ corresponds to the order of A_n^-. Let B_n be the Sylow p-subgroup of the quotient group E_n^-/C_n.

For any integers $m \geq n \geq 0$, we put $\Gamma_{m,n} = \mathrm{Gal}(M_m/M_n)$.

Lemma 4. *We have $B_m^{\Gamma_{m,n}} \cong B_n$ for all $m \geq n \geq 0$.*

Proof. By Proposition 1.2 of Nakagawa [10], both of the Tate cohomology groups $\hat{H}^0(\Gamma_{m,n}, C_m)$ and $\hat{H}^1(\Gamma_{m,n}, C_m)$ vanish. Then we get the following exact sequence:

$$0 \to C_m^{\Gamma_{m,n}} \to E_n^- \to (E_m^-/C_m)^{\Gamma_{m,n}} \to \hat{H}^1(\Gamma_{m,n}, C_m) = 0.$$

Considering the fact that $C_m^{\Gamma_{m,n}} = C_n$, we obtain the lemma. □

The above lemma says that the natural map $B_n \to B_m$ is injective for all $m \geq n \geq 0$. We put

$$\hat{B}_n = \mathrm{Hom}_{\mathbb{Z}_p}(B_n, \mathbb{Q}_p/\mathbb{Z}_p)$$

and

$$\hat{B} = \varprojlim \hat{B}_n,$$

where the projective limit is taken with respect to the dual map of the natural injection B_n to B_m for $m > n$.

We define the action of $\gamma \in \Gamma_{n,0}$ on $f \in \hat{B}_n$ by $\gamma(f)(u) = f(u^{\gamma^{-1}})$ for all $u \in B_n$. Then $\hat{B}$ is a module over the complete group ring $\mathbb{Z}_p[[\mathrm{Gal}(M_\infty/M)]] \cong \mathbb{Z}_p[[T]]$. ($1+T$ acts as a fixed topological generator of $\mathrm{Gal}(M_\infty/M)$.)

Proposition 5. *If the p-rank of B_0 is equal to the p-rank of B_1, then the μ-invariant of the cyclotomic $\mathbb{Z}_p$-extension of M vanishes.*

Proof. First, we show that $\hat{B}$ is a finitely generated $\mathbb{Z}_p$-module. The proof is similar to that of Theorem 1 of Fukuda [5].

From the assumption, we have $\hat{B}_0/p\hat{B}_0 \cong \hat{B}_1/p\hat{B}_1$. By Lemma 4, we have $\hat{B}/\omega_n\hat{B} \cong \hat{B}_n$ for all n, where $\omega_n = (1+T)^{p^n} - 1$. Then we see

$$\omega_0\hat{B} + p\hat{B} \cong \omega_1\hat{B} + p\hat{B}.$$

Since the element ω_1/ω_0 is contained in the unique maximal ideal of $\mathbb{Z}_p[[T]]$, we can show that $\omega_0\hat{B}$ is contained in $p\hat{B}$ by using Nakayama's lemma.

From the above argument, we have

$$\begin{aligned} p\text{-rank of } \hat{B}_0 &= p\text{-rank of } (\hat{B}/\omega_0\hat{B} + p\hat{B}) \\ &= p\text{-rank of } \hat{B}/p\hat{B} \\ &= p\text{-rank of } \hat{B}. \end{aligned}$$

Thus the p-rank of $\hat{B}$ is finite.

By the structure theorem of finitely generated $\mathbb{Z}_p[[T]]$-modules (cf. Washington [14, Proposition 13.19, Lemma 13.21]), there exist integers λ and ν such that

$$|\hat{B}_n| = |\hat{B}/\omega_n\hat{B}| = p^{\lambda n+\nu}$$

for all sufficiently large n. Since $|A_n^-| = |\hat{B}_n|$ and the theorem of Ferrero-Washington, the p-index of the order of A_n increases linearly with respect to n. Then the μ-invariant of the cyclotomic $\mathbb{Z}_p$-extension of M vanishes by Corollary 2.5 of Sands [11]. □

Now, we shall finish the proof of Theorem 2. Consider the exact sequence,

$$0 \to C_1(E_1^-)^3/(E_1^-)^3 \to E_1^-/(E_1^-)^3 \to E_1^-/C_1(E_1^-)^3 \to 0.$$

Because $E_1^-/(E_1^-)^3 \cong (\mathbb{Z}/3\mathbb{Z})^3$, if the 3-rank of $C_1(E_1^-)^3/(E_1^-)^3$ is at least 2, then the 3-rank of $E_1^-/C_1(E_1^-)^3$ is at most 1 and therefore B_1 is cyclic or trivial. Then we shall calculate the 3-rank of $C_1(E_1^-)^3/(E_1^-)^3 \cong C_1/(C_1 \cap (E_1^-)^3)$.

We put $\eta_1 = X_{\mathfrak{f}_1}(1, G_1)$ and fix a generator γ of $\mathrm{Gal}(M_1/M)$. Since it is known that $C_1/\{\pm 1\}$ is generated by $\eta_1, \eta_1^\gamma, \eta_1^{\gamma^2}$ (cf. Nakagawa [10]), we have only to examine the relations of these elements in $C_1/(C_1 \cap (E_1^-)^3)$. In fact, if $\eta_1^{1-\gamma}$ is not a cube in M_1, we can conclude that the 3-rank of $C_1(E_1^-)^3/(E_1^-)^3$ is at least 2. Then the group B_1 is cyclic or trivial. In this case, we conclude that the 3-rank of B_0 is equal to the 3-rank of B_1 from Lemma 3. Thus the theorem follows.

3. Computing examples.

In this section, we give examples of vanishing of the μ-invariant of the cyclotomic $\mathbb{Z}_3$-extension by computer. Let the notation be as in the previous sections.

First, we shall state an algorithm for computing of the minimal polynomial of the Stark-Shintani unit $X_{\mathfrak{f}_1}(c, G_1)$ in the first layer M_1. The method of computation is well-known (cf. Chapter 6 of Cohen [2], Shintani [12]).

We assume that approximate values of $X_{\mathfrak{f}_1}(c, G_1)$ are known for all $c \in H(\mathfrak{f}_1)$. Since $X_{\mathfrak{f}_1}(1, G_1)$ is a unit in M_1, we see that

$$P(T) = \prod_{c \in H(\mathfrak{f}_1)/G_1} (T - X_{\mathfrak{f}_1}(1, G_1)^{\sigma(c)})$$

is a polynomial whose coefficients are all algebraic integers in F. Moreover, it is known that

$$|\rho(X_{\mathfrak{f}_1}(1, G_1)^{\sigma(c)})| = 1$$

for every embedding ρ of M_1 into $\mathbb{C}$, which includes the non-trivial automorphism on F, and all $c \in H(\mathfrak{f}_1)$. We note that $[M_1 : F] = 6$ and we write $P(T) = \sum_{i=0}^{6} a_i T^i$. From the above fact, we have

$$|a_i'| \leq \binom{6}{i}.$$

for each i (Recall that a_i' is the conjugate of a_i).

Since a_i is an algebraic integer, if we compute an approximate value of a_i with sufficiently precision, we can determine the exact value of a_i as an element of F.

We know that $X_{\mathfrak{f}_1}(1, G_1)^{\sigma(c)} = \pm X_{\mathfrak{f}_1}(c, G_1)$. Therefore if we compute

$$\prod_{c \in H(\mathfrak{f}_1)/G_1} (T \pm X_{\mathfrak{f}_1}(c, G_1))$$

for all signs and find only one possibility of signs, we can determine the minimal polynomial $P(T)$ of $\eta_1 = X_{\mathfrak{f}_1}(1, G_1)$ over F.

Using this, the minimal polynomial $Q(T)$ of $\eta_1^{1-\gamma}$ over F is also determined. We use this polynomial for checking whether the element is a cube or not.

Now, we shall give some examples. By virtue of Lemma 3, we shall treat the case that the relative class number $|A(M)^-|$ is divisible by 3.

We shall treat the following case. Let l be a prime which satisfies $l \equiv 3 \pmod 4$. We put $F = \mathbb{Q}(\sqrt{3l})$. Assume that the class number of F is one. Let $\mathfrak{p}$ be the unique prime of F which divides 3. If the order of the narrow ray class group modulo $\mathfrak{p}$ of F is 4, there exists a unique quadratic extension M of F which is contained in the narrow ray class field modulo $\mathfrak{p}$ of F and satisfies the assumption of Theorem 2. We shall calculate the μ-invariant of the cyclotomic $\mathbb{Z}_3$-extension of M.

Now, we put $F = \mathbb{Q}(\sqrt{1857})$ and $\mathfrak{p}$ be the unique prime of F which divides 3. We can see that the order of the narrow ray class group modulo

$\mathfrak{p}$ of F is 4. Therefore, let M be the uniquely determined quadratic extension of F given by the above. In fact,

$$M = F\left(\sqrt{1847277637 + 87771412\left(\frac{1+\sqrt{1857}}{2}\right)}\right).$$

By using KASH [3], we know that $|A(M)^-| = 3$.

We computed approximate values of η_1^σ for all $\sigma \in \mathrm{Gal}(M_1/F)$ by using the function bnrL1 of PARI-GP [1] with a precision of 150 decimal digits. The values are the following:

$$\begin{aligned} \eta_1 &= 127829214079.2490851840966\cdots \\ \eta_1^\gamma &= 9.0180798486182840178379 40\cdots \\ \eta_1^{\gamma^2} &= 0.0001520176698600726988 46\cdots . \end{aligned}$$

By using the above mentioned method, we can compute the minimal polynomial of $\eta_1^{1-\gamma}$ over F, hence we can also obtain its minimal polynomial over $\mathbb{Q}$. The minimal polynomial $R(T)$ of $\eta_1^{1-\gamma}$ over $\mathbb{Q}$ is the following :

$$\begin{aligned} R(T) = T^{12} &- 840898094422485T^{11} \\ &+ 1191938663965301296915326 6T^{10} \\ &- 7071321187095864938820427234 18T^{9} \\ &+ 2749828994632785471910474645701T^{8} \\ &- 5454060080455494682796432087649T^{7} \\ &+ 6699174123340475549716147390896T^{6} \\ &- 5454060080455494682796432087649T^{5} \\ &+ 2749828994632785471910474645701T^{4} \\ &- 7071321187095864938820427234 18T^{3} \\ &+ 1191938663965301296915326 6T^{2} \\ &- 840898094422485T + 1. \end{aligned}$$

It is easy to see that the polynomial $R(T^3)$ is irreducible over $\mathbb{Q}$. Then $\eta_1^{1-\gamma}$ is not a cube in M_1. Therefore, the μ-invariant of the cyclotomic $\mathbb{Z}_3$-extension of M vanishes.

Moreover, applying Lemma 3 or Theorem 2, we have the following:

Corollary 6. *Let l be a prime which satisfies $l \equiv 3 \pmod 4$. We put $F = \mathbb{Q}(\sqrt{3l})$. Assume that the class number of F is one and the order of the narrow ray class group modulo (9) of F is 12. Let $\mathfrak{p}$ be the* *unique prime of F above 3 and $M(l)$ the unique quadratic extension of F which is contained in the narrow ray class field modulo $\mathfrak{p}$ of F and satisfies the assumption of Theorem 2. Then for all $l < 3000$ satisfying the above assumption, the μ-invariant of the cyclotomic $\mathbb{Z}_p$-extension of $M(l)$ vanishes.*

By using KASH, we can see that $|A(M(l))^-| = 1$ except for the case that $l = 619, 1051, 1447, 1627, 2203, 2851$ for all $l < 3000$ which satisfies the assumption of Corollary 6. We shall give the minimal polynomial of $\eta_1^{1-\gamma}$ for the excepted primes. (The case that $l = 619$ is already given.)

$l = 1051$
$T^{12} - 2542278225978963204433T^{11} + 116939363400084637512843294208267T^{10}$
$-6463178573633562178302908174331212589691O7T^{9}$
$-12791541530839596503320566556018092597069 9T^{8}$
$-17510886700629410408639168998702036906395 3T^{7}$
$-22362235366589168011778720575193656551859 2T^{6}$
$-17510886700629410408639168998702036906395 3T^{5}$
$-12791541530839596503320566556018092597069 9T^{4}$
$-646317857363356217830290817433121258969107T^{3}$
$+116939363400084637512843294208267T^{2} - 2542278225978963204433T + 1$

$l = 1447$
$T^{12} - 1136723853309632767T^{11} + 242459172756828236936890993935174 6T^{10}$
$-116900289004956057597551632340344607T^{9}$
$+195784464317284823502236350997565277T^{8}$
$-278923525526400852271212175103420407T^{7}$
$+344052257332171279804341160424219167T^{6}$
$-278923525526400852271212175103420407T^{5}$
$+195784464317284823502236350997565277T^{4}$
$-116900289004956057597551632340344607T^{3}$
$+2424591727568282369368909939351746T^{2}$
$-113672385330963276T + 1$

$l = 1627$
$T^{12} - 580818947397927845089087651692T^{11}$
$+113029613762306360676648702561906749056384157442T^{10}$
$-4578484134886744288477691702953216916870662606 32T^{9}$
$+10360174539750928921499459804753033047087372992 87T^{8}$
$-18313936539546977090020682598040804707683008739 16T^{7}$
$+18459756804255730652993145620149916580536480658 08T^{6}$
$-18313936539546977090020682598040804707683008739 16T^{5}$
$+10360174539750928921499459804753033047087372992 87T^{4}$
$-4578484134886744288477691702953216916870662606 32T^{3}$
$+113029613762306360676648702561906749056384157442T^{2}$
$-580818947397927845089087651692T + 1$

$l = 2203$
$T^{12} - 25437120536557149090797783433T^{11}$
$+512568729156253582510784888374639282630950566634T^{10}$
$-6470471023147510275828555588292666914668376541991412577718T^{9}$
$+27122483901710098062280641065599524194318108640895216817749T^{8}$
$-55683038145350424776754520413764187840788698850551591905545T^{7}$
$+69510031812744698030247789733057890287060677303905602679367T^{6}$
$-55683038145350424776754520413764187840788698850551591905545T^{5}$
$+27122483901710098062280641065599524194318108640895216817749T^{4}$
$-6470471023147510275828555588292666914668376541991412577718T^{3}$
$+512568729156253582510784888374639282630950566634T^{2}$
$-25437120536557149090797783433T + 1$

$l = 2851$
$T^{12} - 19355331325194992531499530674599657 3T^{11}$
$+1238827640600725711506236524591460158665268257635577810 6T^{10}$
$-37462885070807392959051492934251075271574908048541902516443664421583922T^{9}$
$-111398835362485827371779955278897436481955831146807532805279059108108539T^{8}$
$-219591672808453587575387446499053253583352535397425539746057326454167361T^{7}$
$-255802047289847856767973574463044610817826190920616524617543293974661472T^{6}$
$-219591672808453587575387446499053253583352535397425539746057326454167361T^{5}$
$-111398835362485827371779955278897436481955831146807532805279059108108539T^{4}$
$-37462885070807392959051492934251075271574908048541902516443664421583922T^{3}$
$+12388276406007257115062365245914601586652682576355577810 6T^{2}$
$-193553313251949925314995306745996573T + 1$

References

[1] C. Batut, K. Belabas, D. Bernardi, H. Cohen and H. Oliver, "User's guide to Pari-GP."

[2] H. Cohen, "Advanced topics in computational number theory," Graduate Texts in Mathematics, 193. Springer-Verlag, New York, 2000.

[3] M. Daberkow, C. Fieker, J. Klüners, M. Pohst, K. Roegner, M. Schörnig, and K. Wildanger, *KANT V4*, J. Symbolic Comput. **24** (1997), 267–283.

[4] B. Ferrero and L. Washigton, *The Iwasawa invariant μ_p vanishes for abelian number fields*, Ann. of Math. **109** (1979), 377–395.

[5] T. Fukuda, *Remarks on $\mathbb{Z}_p$-extensions of number fields*, Proc. Japan Acad. Ser. A **70** (1994), 264–266.

[6] K. Iwasawa, *On the μ-invariants of $\mathbb{Z}_l$-extensions*, in "Number theory, algebraic geometry and commutative algebra, in honor of Yasuo Akizuki," Kinokuniya, Tokyo, 1973, 1–11.

[7] Y. Kida, *l-extensions of CM-fields and cyclotomic invariants*, J. Number Theory **12** (1980), no. 4, 519–528.

[8] J. Kraft and R. Schoof, *Computing Iwasawa modules of real quadratic number fields*, Compositio Math. **97** (1995), no. 1-2, 135–155.

[9] J. Nakagawa, *On the Stark-Shintani conjecture and cyclotomic $\mathbb{Z}_p$-extensions of class fields over real quadratic fields*, J. Math. Soc. Japan **36** (1984), 577–588.

[10] J. Nakagawa, *On the Stark-Shintani conjecture and cyclotomic $\mathbb{Z}_p$-extensions of class fields over real quadratic fields. II*, Tohoku Math. J. **36** (1984), 439–452.

[11] J. Sands, *On small Iwasawa invariants and imaginary quadratic fields*, Proc. Amer. Math. Soc. **112** (1991), 671–684.

[12] T. Shintani, *On certain ray class invariants of real quadratic fields*, J. Math. Soc. Japan **30** (1978), 139–167.

[13] H. M. Stark, *L-functions at $s = 1$. III. Totally real fields and Hilbert's twelfth problem*, Advances in Math. **22** (1976), 64–84.

[14] L. Washington, "Introduction to cyclotomic fields. Second edition," Graduate Texts in Mathematics, 83. Springer-Verlag, New York, 1997.

Part IV

Modular forms and arithmetic functions

Galois Theory and Modular Forms
K. Hashimoto, K. Miyake, and H. Nakamura, Editors
© 2003 Kluwer Academic Publishers
pp 329–336

QUASIMODULAR SOLUTIONS OF A DIFFERENTIAL EQUATION OF HYPERGEOMETRIC TYPE

Masanobu KANEKO
Graduate School of Mathematics,
Kyushu University 33, Fukuoka,
812-8581, Japan
mkaneko@math.kyushu-u.ac.jp

Masao KOIKE
Graduate School of Mathematics,
Kyushu University, Ropponmatu, Fukuoka,
810-8560, Japan
koike@math.kyushu-u.ac.jp

1. Introduction and Main Theorem

In our previous paper [2], we studied further the solutions of the following differential equation in the upper half-plane $\mathfrak{H}$ which was originally found and studied in [4] in connection with the arithmetic of supersingular elliptic curves;

$$f''(\tau) - \frac{k+1}{6}E_2(\tau)f'(\tau) + \frac{k(k+1)}{12}E_2'(\tau)f(\tau) = 0.$$

Here, k is an integer or half an integer, the symbol $'$ denotes the differentiation $(2\pi i)^{-1}d/d\tau = q \cdot d/dq$ $(q = e^{2\pi i\tau})$, and $E_2(\tau)$ is the "quasimodular" Eisenstein series of weight 2 for the full modular group $\mathrm{SL}_2(\mathbb{Z})$:

$$E_2(\tau) = 1 - 24\sum_{n=1}^{\infty}(\sum_{d|n} d)q^n.$$

Let $p \geq 5$ be a prime number and $F_{p-1}(\tau)$ be the solution of the above differential equation for $k = p - 1$ which is modular on $\mathrm{SL}_2(\mathbb{Z})$

[received: June 4, 2002; accepted in revised form: July 31, 2002]

(such a solution exists and is unique up to a scalar multiple). For any zero τ_0 in $\mathfrak{H}$ of the form $F_{p-1}(\tau)$, the value of the j-function at τ_0 is algebraic and its reduction modulo (an extension of) p is a supersingular j-invariant of characteristic p, and conversely, all the supersingular j-invariants are obtained in this way from the single solution $F_{p-1}(\tau)$ with suitable choices of τ_0. This is the arithmetic connection that motivated our study of the differential equation.

Various modular forms on $\mathrm{SL}_2(\mathbb{Z})$ and its subgroups were obtained in [2] as solutions to this differential equation, the groups depending on the choice of k. Every modular solution is expressed in terms of a hypergeometric polynomial in a suitable modular function (hence the "hypergeometric type" in the title of the paper), also depending on the choice of k. For instance, if $k \equiv 0, 4 \bmod 12$, we have a modular solution

$$E_4(\tau)^{\frac{k}{4}} F(-\frac{k}{12}, -\frac{k-4}{12}, -\frac{k-5}{6}; \frac{1728}{j(\tau)}),$$

where

$$F(a,b,c;x) = \sum_{n=0}^{\infty} \frac{(a)_n (b)_n}{(c)_n} \frac{x^n}{n!}, \qquad (a)_n = a(a+1)\cdots(a+n-1)$$

is the Gauss hypergeometric series (which becomes a polynomial when a or b is a negative integer, which is the case here), $E_4(\tau)$ the Eisenstein series of weight 4 on $\mathrm{SL}_2(\mathbb{Z})$, and $j(\tau)$ the elliptic modular invariant.

In addition to the modular solutions, quite remarkable was an occurrence of a *quasimodular form*, not of weight k as in the modular case but of weight $k+1$. In the present paper, we give another supply of examples of quasimodular forms as solutions to an analogous differential equation attached to the group $\Gamma_0^*(2)$, which is *not* contained in $\mathrm{SL}_2(\mathbb{Z})$;

$$\Gamma_0^*(2) = \left\langle \Gamma_0(2), \begin{pmatrix} 0 & -1 \\ 2 & 0 \end{pmatrix} \right\rangle$$

where

$$\Gamma_0(2) = \left\{ \begin{pmatrix} a & b \\ c & d \end{pmatrix} \in \mathrm{SL}_2(\mathbb{Z}) \mid c \equiv 0 \mod 2 \right\}.$$

($\Gamma_0^*(2)$ is the triangular group "$2A$" in the notation of Conway-Norton [1].)

Let

$$E_{2A}(\tau) := (E_2(\tau) + 2E_2(2\tau))/3 = 1 - 8q - 40q^2 - 32q^3 - \cdots$$

be the quasimodular form of weight 2 on $\Gamma_0^*(2)$ which is the logarithmic derivative of the form

$$\Delta_{2A}(\tau) := \eta(\tau)^8 \eta(2\tau)^8 = q - 8q^2 + 12q^3 + 64q^4 - \cdots$$

of weight 8 on $\Gamma_0^*(2)$; $E_{2A}(\tau) = \Delta'_{2A}(\tau)/\Delta_{2A}(\tau)$, an analogous situation in the $\mathrm{SL}_2(\mathbb{Z})$ case where $E_2(\tau)$ is the logarithmic derivative of the Ramanujan $\Delta(\tau)$. Consider the following differential equation;

$$(\#)_k \qquad f''(\tau) - \frac{k+1}{4} E_{2A}(\tau) f'(\tau) + \frac{k(k+1)}{8} E'_{2A}(\tau) f(\tau) = 0.$$

Solutions which are modular on the group $\Gamma_0^*(2)$ and its subgroups were studied in [6, 7]. In particular, when k is a non-negative integer congruent to 0 or 6 modulo 8, the equation $(\#)_k$ has a one dimensional space of solutions which are modular on the group $\Gamma_0^*(2)$ itself. We note here that the equation $(\#)_k$ has a characterization by the invariance of the space of solutions under the action of $\Gamma_0^*(2)$, similar to the previous case for $\mathrm{SL}_2(\mathbb{Z})$, owing to the fact that there is no holomorphic modular form of weight 2 on $\Gamma_0^*(2)$ (see [5] and [2, §5]). By a general theory of ordinary differential equations, we see that the equation $(\#)_k$ has a quasimodular solution (which, since its transformation under $\tau \to -1/2\tau$ is also a solution, inevitably gives a solution having $\log q$ term in the expansion at $q = 0$) only when k is a positive integer congruent to 3 modulo 4.

In the following, we show there indeed exists a quasimodular solution in this case and describe explicitly the solution in terms of a certain orthogonal polynomials. First we need to develop some notations. Put

$$\begin{aligned} C(\tau) &:= 2E_2(2\tau) - E_2(\tau) \\ &= 1 + 24 \sum_{n=1}^{\infty} \Big(\sum_{\substack{d|n \\ d:\mathrm{odd}}} d \Big) q^n = 1 + 24q + 24q^2 + 96q^3 + \cdots, \end{aligned}$$

$$D(\tau) := \frac{\eta(2\tau)^{16}}{\eta(\tau)^8} = \sum_{n=1}^{\infty} \Big(\sum_{\substack{d|n \\ d:\mathrm{odd}}} (n/d)^3 \Big) q^n = q + 8q^2 + 28q^3 + 64q^4 + \cdots,$$

where

$$\eta(\tau) = q^{\frac{1}{24}} \prod_{n=1}^{\infty} (1 - q^n) = q^{\frac{1}{24}} - q^{\frac{25}{24}} - q^{\frac{49}{24}} + q^{\frac{121}{24}} + \cdots$$

is the Dedekind eta function. The functions $C(\tau)$ and $D(\tau)$ are modular forms of respective weights 2 and 4 on the group $\Gamma_0(2)$ (="$2B$") and the graded ring of modular forms of integral weights on $\Gamma_0(2)$ is generated by these $C(\tau)$ and $D(\tau)$. Recall that (see [3]) an element of degree k in the graded ring $\mathbb{C}[E_2(\tau), C(\tau), D(\tau)]$, where the generators $E_2(\tau), C(\tau), D(\tau)$ have degrees 2, 2, and 4 respectively, is referred to as a quasimodular form of weight k (on $\Gamma_0(2)$). Incidentally, the graded ring of modular forms of integral weights on $\Gamma_0^*(2)$ is generated by three elements $C(\tau)^2 = (E_4(\tau)+4E_4(2\tau))/5$, $C(\tau)^3-128C(\tau)D(\tau) = (E_6(\tau)+8E_6(2\tau))/9$, and $\Delta_{2A}(\tau)$ of respective weights $4,6,8$, of which $C(\tau)^2$ and $\Delta_{2A}(\tau)$ generate freely the subring consisting forms of weight being multiple of 4, and the whole space as a graded module is generated over this ring by $C(\tau)^3 - 128C(\tau)D(\tau)$.

Now define a sequence of polynomials $P_n(x)$ $(n = 0, 1, 2, \dots)$ by

$$P_0(x) = 1,\ P_1(x) = x,\quad P_{n+1}(x) = xP_n(x) + \lambda_n P_{n-1}(x) \quad (n = 1, 2, \dots)$$

where

$$\lambda_n = 4\frac{(4n+1)(4n+3)}{n(n+1)}.$$

First few examples are

$$P_2(x) = x^2 + 70,\ P_3(x) = x^3 + 136x,\ P_4(x) = x^4 + 201x^2 + 4550, \dots.$$

The $P_n(x)$ is even or odd polynomial according as n is even or odd. We also define a second series of polynomials $Q_n(x)$ by the same recursion (with different initial values):

$$Q_0(x) = 0,\ Q_1(x) = 1,\quad Q_{n+1}(x) = xQ_n(x)+\lambda_n Q_{n-1}(x) \quad (n = 1, 2, \dots),$$

a couple of examples being

$$Q_2(x) = x,\ Q_3(x) = x^2 + 66,\ Q_4(x) = x^3 + 131x, \dots.$$

The $Q_n(x)$ has opposite parity: It is even if n is odd and odd if n is even.

Put $G(\tau) = C(\tau)^2 - 128D(\tau)$ $(= (4E_4(2\tau) - E_4(\tau))/3)$.

Theorem. *Let $k = 4n+3$ $(n = 0, 1, 2, \dots)$. The following quasimodular form of weight $k + 1$ on $\Gamma_0(2)$ is a solution of $(\#)_k$:*

$$\sqrt{\Delta_{2A}(\tau)}^{\,n} P_n\Big(\frac{G(\tau)}{\sqrt{\Delta_{2A}(\tau)}}\Big)\frac{C'(\tau)}{24} - \sqrt{\Delta_{2A}(\tau)}^{\,n+1} Q_n\Big(\frac{G(\tau)}{\sqrt{\Delta_{2A}(\tau)}}\Big).$$

Remark. The appearance of the square root $\sqrt{\Delta_{2A}(\tau)}$ in the formula is superficial because of the parities of $P_n(x)$ and $Q_n(x)$, that is, the form is actually an element in $\mathbb{Q}[E_2(\tau), C(\tau), D(\tau)]$, by noting $\Delta_{2A}(\tau) = D(\tau)(C(\tau)^2 - 64D(\tau))$ and $C'(\tau) = (E_2(\tau)C(\tau) - C(\tau)^2)/6 + 32D(\tau)$. The form does not belong to $\Gamma_0^*(2)$.

2. Proof of Theorem

It is convenient to introduce the operator ϑ_k defined by

$$\vartheta_k(f)(\tau) = f'(\tau) - \frac{k}{8}E_{2A}(\tau)f(\tau).$$

By the quasimodular property of $E_2(\tau)$ or the fact that $E_{2A}(\tau)$ is the logarithmic derivative of $\Delta_{2A}(\tau)$, we have the transformation formulas

$$E_{2A}\left(\frac{a\tau+b}{c\tau+d}\right) = (c\tau+d)^2E_{2A}(\tau) + \frac{4}{\pi i}c(c\tau+d) \quad \left(\begin{pmatrix} a & b \\ c & d \end{pmatrix} \in \Gamma_0(2)\right)$$

and

$$E_{2A}\left(-\frac{1}{2\tau}\right) = 2\tau^2E_{2A}(\tau) + \frac{8}{\pi i}\tau.$$

From these we see that if f is modular of weight k on a subgroup of $\Gamma_0^*(2)$, then $\vartheta_k(f)$ is modular of weight $k+2$ on the same group. If f and g have weights k and l, the Leibniz rule

$$\vartheta_{k+l}(fg) = \vartheta_k(f)g + f\vartheta_l(g)$$

holds. We sometimes drop the suffix of the operator ϑ_k when the weights of modular forms we consider are clear. With this operator, the equation $(\#)_k$ can be rewritten as

$$(\#')_k \qquad \vartheta_{k+2}\vartheta_k(f)(\tau) = \frac{k(k+2)}{64}C(\tau)^2f(\tau),$$

(use $E'_{2A}(\tau) = (E_{2A}(\tau)^2 - C(\tau)^2)/8$).

Denote the form in the theorem by $F_k(\tau)$. We first establish the recurrence relation (note $n = (k-3)/4$):

$$F_{k+4}(\tau) = G(\tau)F_k(\tau) + \lambda_n\Delta_{2A}(\tau)F_{k-4}(\tau). \tag{1}$$

This is a consequence of the recursion of P_n and Q_n, namely (we often omit the variable τ hereafter)

$$\begin{aligned}
&GF_k + \lambda_n \Delta_{2A} F_{k-4} \\
&= G\Big(\sqrt{\Delta_{2A}}^{\,n} P_n\big(\frac{G}{\sqrt{\Delta_{2A}}}\big)\frac{C'}{24} - \sqrt{\Delta_{2A}}^{\,n+1} Q_n\big(\frac{G}{\sqrt{\Delta_{2A}}}\big)\Big) \\
&\quad + \lambda_n \Delta_{2A}\Big(\sqrt{\Delta_{2A}}^{\,n-1} P_{n-1}\big(\frac{G}{\sqrt{\Delta_{2A}}}\big)\frac{C'}{24} - \sqrt{\Delta_{2A}}^{\,n} Q_{n-1}\big(\frac{G}{\sqrt{\Delta_{2A}}}\big)\Big) \\
&= \sqrt{\Delta_{2A}}^{\,n+1}\Big(\frac{G}{\sqrt{\Delta_{2A}}} P_n\big(\frac{G}{\sqrt{\Delta_{2A}}}\big) + \lambda_n P_{n-1}\big(\frac{G}{\sqrt{\Delta_{2A}}}\big)\Big)\frac{C'}{24} \\
&\quad - \sqrt{\Delta_{2A}}^{\,n+2}\Big(\frac{G}{\sqrt{\Delta_{2A}}} Q_n\big(\frac{G}{\sqrt{\Delta_{2A}}}\big) + \lambda_n Q_{n-1}\big(\frac{G}{\sqrt{\Delta_{2A}}}\big)\Big) \\
&= \sqrt{\Delta_{2A}}^{\,n+1} P_{n+1}\big(\frac{G}{\sqrt{\Delta_{2A}}}\big)\frac{C'}{24} - \sqrt{\Delta_{2A}}^{\,n+2} Q_{n+1}\big(\frac{G}{\sqrt{\Delta_{2A}}}\big) \\
&= F_{k+4}.
\end{aligned}$$

Now we prove by induction that the $F_k(\tau)$ satisfies the equation $(\#')_k$. We can check the cases $k = 3$ and 7 directly. Assume F_{k-4} and F_k satisfy $(\#')_{k-4}$ and $(\#')_k$ respectively. Then by using (1) and the formulas

$$\vartheta(C) = -\frac{1}{4}G, \quad \vartheta(G) = -\frac{1}{2}C^3, \quad \vartheta(\Delta_{2A}) = 0$$

we have

$$\begin{aligned}
&\vartheta^2(F_{k+4}) \\
&= \vartheta\big(\vartheta(F_k)G - \frac{1}{2}C^3 F_k\big) + \lambda_n \Delta_{2A}\vartheta^2(F_{k-4}) \\
&= \vartheta^2(F_k)G - \frac{1}{2}\vartheta(F_k)C^3 + \frac{3}{8}C^2 G F_k - \frac{1}{2}C^3\vartheta(F_k) + \lambda_n \Delta_{2A}\vartheta^2(F_{k-4}) \\
&= \frac{k(k+2)}{64}C^2 G F_k - C^3\vartheta(F_k) + \frac{3}{8}C^2 G F_k \\
&\quad + \frac{(k-4)(k-2)}{64}\lambda_n \Delta_{2A} C^2 F_{k-4} \\
&= \frac{k^2+2k+24}{64}C^2 G F_k + \frac{(k-4)(k-2)}{64}\lambda_n \Delta_{2A} C^2 F_{k-4} - C^3\vartheta(F_k).
\end{aligned}$$

Hence we find

$$\begin{aligned}
&\vartheta^2(F_{k+4}) - \frac{(k+4)(k+6)}{64}C^2F_{k+4} \\
&= \left(\frac{k^2+2k+24}{64} - \frac{(k+4)(k+6)}{64}\right)C^2GF_k \\
&\quad + \left(\frac{(k-4)(k-2)}{64} - \frac{(k+4)(k+6)}{64}\right)\lambda_n\Delta_{2A}C^2F_{k-4} \\
&= -C^2\left(\frac{k}{8}GF_k + C\vartheta(F_k) + \frac{k+1}{4}\lambda_n\Delta_{2A}F_{k-4}\right).
\end{aligned}$$

The proof of the theorem therefore boils down to show the equation

$$\frac{k}{8}GF_k + C\vartheta(F_k) = -\frac{k+1}{4}\lambda_n\Delta_{2A}F_{k-4}.$$

For this we also proceed by induction. For $k = 7$ the equation is checked directly. Assuming that this is valid for k, we have

$$F_{k+4} = GF_k + \lambda_n\Delta_{2A}F_{k-4} = \frac{1}{2(k+1)}\big((k+2)GF_k - 8C\vartheta(F_k)\big)$$

and

$$\begin{aligned}
\frac{k+4}{8}GF_{k+4} + C\vartheta(F_{k+4}) &= \frac{k+4}{16(k+1)}G\big((k+2)GF_k - 8C\vartheta(F_k)\big) \\
&\quad + \frac{1}{2(k+1)}C\Big(-\frac{1}{2}(k+2)C^3F_k \\
&\quad + (k+2)G\vartheta(F_k) + 2G\vartheta(F_k) - 8C\vartheta^2(F_k)\Big) \\
&= \frac{(k+2)(k+4)}{16(k+1)}(G^2 - C^4)F_k \\
&= -\frac{k+5}{4}\lambda_{n+1}\Delta_{2A}F_k.
\end{aligned}$$

Here we have used the (previous) induction assumption that F_k satisfies $(\#')_k$ and the relation $G^2 - C^4 = -256\Delta_{2A}$. This completes our proof.

References

[1] J. Conway and S. Norton, *Monstrous Moonshine*, Bull. London Math. Soc. **11** (1979), 308–339.

[2] M. Kaneko and M. Koike, *On modular forms arising from a differential equation of hypergeometric type*, to appear in Ramanujan J.

[3] M. Kaneko and D. Zagier, *A generalized Jacobi theta function and quasimodular forms*, in "The Moduli Space of Curves (Texel Island, 1994)," 165–172, Progress in Math., 129. Birkhäuser Boston, Boston, MA, 1995.

[4] M. Kaneko and D. Zagier, *Supersingular j-invariants, Hypergeometric series, and Atkin's orthogonal polynomials*, in " Computational perspectives on number theory (Chicago, IL, 1995)," 97–126, AMS/IP Studies in Advanced Mathematics, 7. Amer. Math. Soc., Providence, RI, 1998.

[5] M. Koike, *Modular forms on non-compact arithmetic triangle groups*, preprint.

[6] M. Koike and T. Saijo, in preparation.

[7] T. Saijo, *Modular forms and differential equations*, Kyushu University Master's thesis, 1999.

Galois Theory and Modular Forms pp 337–356
K. Hashimoto, K. Miyake, and H. Nakamura, Editors
© 2003 Kluwer Academic Publishers

SPECIAL VALUES OF THE STANDARD ZETA FUNCTIONS

Hidenori KATSURADA
Muroran Institute of Technology,
27-1 Mizumoto, Muroran,
050-8585, Japan
hidenori@mmm.muroran-it.ac.jp

1. Introduction

Let M and k be positive integers, and ϕ a Dirichlet character modulo M such that $\phi(-1) = (-1)^k$. Let

$$f(z) = \sum_{m=1}^{\infty} a(m) \exp(2\pi i m z)$$

be a normalized cuspidal Hecke eigenform of weight k and Nebentypus ϕ belonging to $\Gamma_0(M)$. Then, for a Dirichlet character χ modulo N, we define the standard zeta function $L(f, s, \chi)$ twisted by χ as

$$L(f,s,\chi) = \prod_p \{(1 - \chi(p)\alpha_p\beta_p p^{-s-k+1})(1 - \chi(p)\alpha_p^2 p^{-s-k+1}) \times (1 - \chi(p)\beta_p^2 p^{-s-k+1})\}^{-1},$$

where α_p, β_p are complex numbers such that $\alpha_p + \beta_p = a(p)$ and $\alpha_p\beta_p = \phi(p)p^{k-1}$ for each prime number p. The standard zeta function of a modular form is an important subject in number theory. To explain this briefly, from now on we assume that k is even and $\chi(-1) = \phi(-1) = 1$. We then put

$$L^*(f,m,\chi) = \frac{L(f,m,\chi)}{\pi^{k+2m} < f,f >}$$

for an odd positive integer $m \leq k-1$, where $< *, * >$ is the normalized Petersson product. As is well known, these values are algebraic numbers and their qualitative natures have been fully investigated by many people

[received: June 6, 2002; accepted in revised form: November, 01, 2002] (cf. [Stur], [Sh2], [B-S]). Furthermore, these values are related to many other areas, especially to the Galois representations. As for this, see, for examples, [D-H-I] and [D]. To investigate various problems related to these values, it is important to compute these values exactly. Several people have considered algorithms for computing these values and have carried out the computations. Zagier [Z] gave an explicit formula expressing $L^*(f, m, \chi)$ in the case where M is a squarefree positive integer congruent to 1 modulo 4, ϕ is the Kronecker character $(\frac{M}{*})$ corresponding to the extension $\mathbf{Q}(\sqrt{M})/\mathbf{Q}$, and χ is trivial. Sturm [Stur] gave an algorithm for computing these values for a general χ. However, it seems difficult to give exact values by a direct use of his method. Stopple [Sto] gave an explicit formula expressing $L^*(f, m, \chi)$ in case $M = 1$ and χ is a quadratic character.

In this note, we give some formulas which seem useful for the computation of $L^*(f, m, \chi)$ in the case where $M = 1$ or a prime number congruent to 1 modulo 4, $\phi = (\frac{M}{*})$, and χ is not necessarily quadratic character of prime conductor p. To state our main result more precisely, let M, ϕ, χ and p be as above. For simplicity assume that $M \neq p$. Let k and l be even positive integers such that $l \leq k-2$. Then put $s = -1/2$ or 0 according as $l = 2$ and $\chi^2 = 1$, or not. Then for two positive integers m_1, m_2 put

$$\begin{aligned}\epsilon(m_1, m_2; l, s) \\ = \sum_{r^2 \leq 4m_1m_2} \tilde{c}_{2,l}\left(\begin{pmatrix} m_1 & r/2 \\ r/2 & m_2 \end{pmatrix}, s\right) G_l^{k-l}(m_1m_2, r)\chi(r)\tau(\bar{\chi}).\end{aligned}$$

Here, for a semi-positive definite half-integral matrix A of degree 2, $\tilde{c}_{2,l}(A, s)$ denotes (up to constant multiple) the A-th Fourier coefficient of a certain Eisenstein series $E^*_{2,l}(Z, Mp^2, \phi\bar{\chi}, s)$ (cf. Section 2), $G_l^{k-l}(u, v)$ is the polynomial introduced by Zagier [Z], and $\tau(\bar{\chi})$ is the Gauss sum (cf. Section 3). Furthermore, put

$$t(m; l, s) = \epsilon(p, p^2m; l, s) - p^{k-2}\epsilon(p, m; l, s).$$

Take a basis $\{f_i\}_{i=1}^{d_1}$ of $S_k(\Gamma_0(M), \phi)$ consisting of primitive forms. Write

$$f_i(z) = \sum_{m=1}^{\infty} a_i(m) \exp(2\pi imz)$$

with $a_i(1) = 1$. Then for any positive integer m prime to pM we have

$$t(m; l, s) = d_{k,l,p,M} \sum_{i=1}^{d_1} L^*(f_i, l + 2s - 1, \chi)\bar{c}_i a_i(M) a_i(p) a_i(m),$$

where $d_{k,l,p,M}$ is a rational number explicitly determined by k, l, p, M, and c_i is a certain complex number with absolute value 1. In case $M = p$ we have a similar formula (cf. Theorem 4.1.) We note that an explicit form of $\tilde{c}_{2,l}(A, s)$ is given (cf. Section 2.) Thus, by the above formula combined with the trace formula of Hecke operators, we can compute the norm $N_{K_{f,\chi}}(L^*(f, m, \chi))$ for a primitive form $f \in S_k(\Gamma_0(M), \phi)$ in the case where m is an odd integer $3 \le m \le k-3$, or $m = 1$ and $\chi^2 \ne 1$. Here $K_{f,\chi}$ is the field generated over $\mathbf{Q}$ by all the eigenvalues of Hecke operators relative to f and all the values of χ (cf. Theorem 4.2.) We can also compute $N_{K_{f,\chi}}(L^*(f, 0, \chi))$ in case $\chi^2 = 1, \chi \ne \phi$ by the same method, and by the functional equation due to Li [L], we can finally compute $N_{K_{f,\chi}}(L^*(f, 1, \chi))$ in this case. We note that $L^*(f, 1, \phi)$ has been obtained by a different method (cf. Proposition 4.4.)

The main tool of the proof of the above formula is the pull-back formula of the Siegel Eisenstein series of degree 2 due to Böcherer and Schmidt [B-S], Shimura [Sh2]. Such a formula has been used to study a qualitative nature of the special values of the standard zeta function. However, as far as the author knows, no one has used the formula to give its exact values. In this note, we carry out such a computation.

This note is organized as follows. In Section 2, we review the Siegel Eisenstein series of degree 2. In Section 3, we review the pull-back formula of Eisenstein series following [B-S]. In Section 4, by using the pull-back formula we give an explicit formula for $L^*(f, m, \chi)$. In Section 5, we give some numerical examples, and discuss some related topics.

The author is grateful to the referee for helpful comments on the original manuscript.

2. Eisenstein series

Let $GSp_n^+(\mathbf{R})$ be the group of proper symplectic similitudes of degree n, and $\mathbf{H}_n$ Siegel's upper half space of degree n. As usual we write $\gamma(Z) = (AZ + B)(CZ + D)^{-1}$ and $j(\gamma, Z) = \det(CZ + D)$ for $\gamma = \begin{pmatrix} A & B \\ C & D \end{pmatrix} \in GSp_n^+(\mathbf{R})$. We write $f|_k\gamma(z) = (\det\gamma)^{k/2} j(\gamma, z)^{-k} f(\gamma(z))$ for $\gamma \in GSp_n^+(\mathbf{R})$ and $f \in C^\infty(\mathbf{H}_n)$. We simply write $f|\gamma$ as $f|_k\gamma$ if there is no confusion. Let $Sp_n(\mathbf{Z})$ be the Siegel modular group of degree n. For a positive integer M, we denote by $\Gamma_0^{(n)}(M)$ (resp. ${\Gamma'_0}^{(n)}(M)$) the subgroup of $Sp_n(\mathbf{Z})$ consisting of matrices whose lower left $n \times n$ block (resp. upper right $n \times n$ block) is congruent to O modulo M. For a Dirichlet character modulo M, we denote by $\tilde{\phi}$ (resp. $\tilde{\phi}'$) the character of $\Gamma_0^{(n)}(M)$ (resp. $\Gamma_0'^{(n)}(M)$) defined by $\tilde{\phi}(\gamma) = \phi(\det D)$ (resp. $\tilde{\phi}'(\gamma) = \phi(\det A)$) for $\gamma = \begin{pmatrix} A & B \\ C & D \end{pmatrix}$.

For a Dirichlet character ϕ modulo M, we denote by $M_k(\Gamma_0^{(n)}(M), \phi)$ (resp. $M_k^\infty(\Gamma_0^{(n)}(M), \phi)$) the space of holomorphic (resp. C^∞-) modular forms of weight k and Nebentypus ϕ belonging to $\Gamma_0^{(n)}(M)$, and by $S_k(\Gamma_0^{(n)}(M), \phi)$ the subspace of $M_k(\Gamma_0^{(n)}(M), \phi)$ consisting of cusp forms. Furthermore, for a subgroup Γ of $Sp_n(\mathbf{Z})$ we denote by Γ_∞ the subgroup of Γ consisting of matrices whose lower left $n \times n$ block is O. For a function f on $\mathbf{H}_n$ we write $f^c(Z) = \overline{f(-\bar{Z})}$. Let dv denote the invariant volume element on $\mathbf{H}_n$ given by $dv = \det(\mathrm{Im}(Z))^{-n-1} \wedge_{1 \le j \le l \le n} (dx_{jl} \wedge dy_{jl})$. Here for $Z \in \mathbf{H}_n$ we write $Z = (x_{jl}) + i(y_{jl})$ with real matrices (x_{jl}) and (y_{jl}). For two C^∞-modular forms f and g of weight k and Nebentypus ϕ belonging $\Gamma_0^{(n)}(M)$, we define the Petersson scalar product $< f, g >_{\Gamma_0^{(n)}(M)}$ by

$$< f, g >_{\Gamma_0^{(n)}(M)} = \int_{\Gamma_0^{(n)}(M) \backslash \mathbf{H}_n} f(Z)\bar{g}(Z) \det(\mathrm{Im}(Z))^k dv,$$

provided the integral converges. Furthermore, we define the normalized Petersson scalar product

$$< f, g > = m(\Phi_{\Gamma_0^{(n)}(M)})^{-1} < f, g >_{\Gamma_0^{(n)}(M)},$$

where $\Phi_{\Gamma_0^{(n)}(M)}$ is the fundamental domain for $\mathbf{H}_n$ modulo $\Gamma_0^{(n)}(M)$, and $m(\Phi_{\Gamma_0^{(n)}(M)}) = \int_{\Gamma_0^{(n)}(M) \backslash \mathbf{H}_n} dv$. For a Dirichlet character ψ we denote by $L(s, \psi)$ the Dirichlet L-function associated to ψ. Let n, l and M be positive integers. For a Dirichlet character ϕ modulo M such that $\phi(-1) = (-1)^l$, we define the Eisenstein series $E'_{n,l}(Z; M, \phi, s)$ by

$$E'_{n,l}(Z; M, \phi, s) = \det \mathrm{Im}(Z)^s L(l + 2s, \phi) \prod_{i=1}^{[n/2]} L(2l + 4s - 2i, \phi^2) \times \sum_{\gamma \in \Gamma_0'^{(n)}(M)_\infty \backslash \Gamma_0'^{(n)}(M)} \tilde{\phi}'(\gamma) j(\gamma, Z)^{-l} |j(\gamma, Z)|^{-2s}.$$

We then define $E^*_{n,l}(Z; M, \phi, s)$ by

$$E^*_{n,l}(Z; M, \phi, s) = j(\iota, Z)^{-l} E'_{n,l}(\iota(Z); M, \phi, s),$$

where $\iota = \begin{pmatrix} 0_n & -1_n \\ 1_n & 0_n \end{pmatrix}$. Let $\mathcal{H}_n(\mathbf{Z})$ denote the set of half-integral matrices of degree n over $\mathbf{Z}$, and we denote by $\mathcal{H}_n(\mathbf{Z})_{>0}$ (resp. $\mathcal{H}_n(\mathbf{Z})_{\geq 0}$) the subset of $\mathcal{H}_n(\mathbf{Z})$ consisting of positive definite (resp. semi-positive definite) matrices. Then it is well known that $E^*_{n,l}(Z; M, \phi, s)$ belongs to $M_l^\infty(\Gamma_0^{(n)}(M), \phi)$, and has the following Fourier expansion:

$$E^*_{n,l}(X + iY; M, \phi, s) = \sum_{A \in \mathcal{H}_n(\mathbf{Z})} c_{n,l,s}(A, Y, M, \phi) \exp(2\pi i \mathrm{tr}(AX)),$$

where $\mathrm{tr}(*)$ denotes the trace of a matrix. In particular, if $E^*_{n,l}(Z; M, \phi, s)$ belongs to $M_l(\Gamma_0^{(n)}(M), \phi)$, it has the following Fourier expansion:

$$E^*_{n,l}(Z; M, \phi, s) = \sum_{A \in \mathcal{H}_n(\mathbf{Z})_{\geq 0}} c_{n,l}(A, M, \phi, s) \exp(2\pi i \mathrm{tr}(AZ)).$$

From now on we exclusively consider the case $n = 2$. Let l be an even positive integer. Let $M > 1$ be an integer, and let ϕ be a Dirichlet character modulo M such that $\phi(-1) = 1$. Then $E^*_{2,l}(Z; M, \phi, 0)$ belongs to $M_l(\Gamma_0^{(2)}(M), \phi)$ in case $l \geq 4$. Furthermore $E^*_{2,2}(Z; M, \phi, 0)$ belongs to $M_2(\Gamma_0^{(2)}(M), \phi)$ if $\phi^2 \neq 1$. We remark that $E^*_{2,2}(Z; M, \phi, 0)$ is neither holomorphic nor nearly holomorphic in the sense of [Sh2] if $\phi^2 = 1$. However, $E^*_{2,l}(Z; M, \phi, -1/2)$ belongs to $M_2(\Gamma_0^{(2)}(M), \phi)$ in this case.

Now to see the Fourier coefficient of the Eisenstein series, for an element $A = \begin{pmatrix} a_{11} & a_{12}/2 \\ a_{12}/2 & a_{22} \end{pmatrix} \in \mathcal{H}_2(\mathbf{Z})$, put $e = e_A = \mathrm{GCD}(a_{11}, a_{12}, a_{22})$. For an element $A \in \mathcal{H}_2(\mathbf{Z})$ such that rank $A = 1$ and for each prime number p define a polynomial $F_p(A, X)$ as

$$F_p(A, X) = \sum_{i=1}^{ord_p(e_A)} (pX)^i.$$

For an element $A = \begin{pmatrix} a_{11} & a_{12}/2 \\ a_{12}/2 & a_{22} \end{pmatrix} \in \mathcal{H}_2(\mathbf{Z})_{>0}$ $-4\det A = d_A f_A^2$ with d_A the fundamental discriminant of $\mathbf{Q}(\sqrt{-\det A})$ and f_A a positive integer. Furthermore, let $\chi_A = (\frac{d_A}{*})$ be the Kronecker character corresponding to $\mathbf{Q}(\sqrt{-\det A})/\mathbf{Q}$. For a prime number p define a polynomial $F_p(A,X)$ as

$$F_p(A,X) = \sum_{i=0}^{ord_p(e_A)} (p^2X)^i \sum_{j=0}^{ord_p(f_A)-i} (p^3X^2)^j - \chi_A(p)pX \sum_{i=0}^{ord_p(e_A)} (p^2X)^i \sum_{j=0}^{ord_p(f_A)-i-1} (p^3X^2)^j.$$

For a Dirichlet character ψ, let m_ψ denote its conductor, and $\psi^{(0)}$ the associated primitive character. Furthermore, let $B_{m,\psi}$ be the m-th generalized Bernoulli number associated with ψ, and let $\tau(\psi)$ be the Gauss sum defined by

$$\tau(\psi) = \sum_{X \bmod L} \psi(X)\exp(2\pi i X/L).$$

Let l be an even positive integer, and $s=0$ or $-1/2$. Let ϕ be a Dirichlet character such that $\phi(-1)=1$. Now assume that the triple (l,s,ϕ) satisfies one of the following conditions (h-1),(h-2),(h-3):

(h-1) $l \geq 4$ and $s=0$,

(h-2) $l=2, s=0$ and $\phi^2 \neq 1$,

(h-3) $l=2$ and $s=-1/2$.

First assume that (l,s,ϕ) satisfies either the condition (h-1) or (h-2). Then for $A \in \mathcal{H}_2(\mathbf{Z})_{\geq 0}$ put

$$\tilde{c}_{2,l}(A,0) = \tilde{c}_{2,l}(A;M,\phi,0) = \begin{cases} \prod_{p|f_A} F_p(A,\phi(p)p^{-l})|d_A|^{l-3/2} \\ \times f_A^{2l-3}(-i)\tau((\phi\chi_A)^{(0)})m_{(\phi\chi_A)^{(0)}}^{1-l} \\ \times B_{l-1,\overline{(\phi\chi_A)^{(0)}}} \prod_{p|M}(1-p^{1-l}(\phi\chi_A)^{(0)}(p)) & \text{if } A>0 \\ 0 & \text{otherwise.} \end{cases}$$

Next assume that (l,s,ϕ) satisfies the condition (h-3). Then for $A \in \mathcal{H}_2(\mathbf{Z})_{\geq 0}$ put

$$\tilde{c}_{2,2}(A,-1/2) = \tilde{c}_{2,2}(A;M,\phi,-1/2) = \begin{cases} \prod_{p|f_A} F_p(A,\phi(p)p^{-1})L(0,\phi\chi_A) & \text{if } A>0 \\ L(-1,\phi^2)\prod_{p|e_A} F_p(A,\phi(p)p^{-1}) & \text{if rank } A=1 \\ 1/2L(-1,\phi^2)L(-1,\phi) & \text{if } A=O. \end{cases}$$

Remark. Let ϕ be a primitive character with conductor m such that $\phi(-1) = 1$. If d_A is prime to m, we have

$$\tilde{c}_{2,l}(A,0) = \phi(d_A)\chi_A(m)\tau(\phi)\prod_{p|f_A} F_p(A,\phi(p)p^{-l}) f_A^{2l-3} m^{1-l} B_{l-1,\overline{(\phi\chi_A)^{(0)}}}$$

for $l \geq 4$. On the other hand, if $\phi^2 = 1$, we have

$$\begin{aligned}\tilde{c}_{2,l}(A,0) = &\prod_{p|f_A} F_p(A,\phi(p)p^{-l})|d_A|^{l-3/2} f_A^{2l-3} m_{(\phi\chi_A)^{(0)}}^{3/2-l} \\ &\times B_{l-1,\overline{(\phi\chi_A)^{(0)}}} \prod_{p|M}(1 - p^{1-l}(\phi\chi_A)^{(0)}(p)).\end{aligned}$$

Thus if $\phi^2 = 1$ and d_A is prime to m, we have

$$\tilde{c}_{2,l}(A,0) = \prod_{p|f_A} F_p(A,\phi(p)p^{-l}) f_A^{2l-3} m^{3/2-l} B_{l-1,\overline{(\phi\chi_A)^{(0)}}}.$$

Theorem 2.1 (cf. [Ka1], [Sh2]). *Let $M > 1$ be an integer, and ϕ a Dirichlet character modulo M such that $\phi(-1) = 1$. Let l be an even positive integer, and $s = 0$ or $-1/2$. Assume that the triple (l,s,ϕ) satisfies one of the conditions (h-1),(h-2),(h-3). Let*

$$A(l,s) = \frac{8\pi^{5/2}}{\Gamma(3/2)} \text{ or } \frac{(-1)^{l/2}2^l\pi^{3l-3/2}}{\Gamma(l)^2\Gamma(l-1/2)}$$

according as $l = 2$ and $s = -1/2$, or $l \geq 4$ and $s = 0$, where $\Gamma(*)$ is Gamma function. Then we have

$$c_{2,l}(A;M,\phi,s) = A(l,s)\tilde{c}_{2,l}(A;M,\phi,s)$$

for $s = 0$ or $-1/2$.

3. Pull-back formula

From now on put $\Gamma_0(N) = \Gamma_0^{(1)}(N)$, and in particular put $\Gamma = \Gamma^{(1)}$. For an even positive integer l and non-negative integer ν we define a polynomial $G_l^{2\nu}(u,v)$ in u,v by

$$G_l^{2\nu}(u,v) = \sum_{\mu=0}^{\nu}(-1)^{\mu}\frac{(l+2\nu-\mu-2)!}{(2\nu-2\mu)!\mu!}u^{\mu}v^{2\nu-2\mu}.$$

This type of polynomial was introduced by Zagier (cf. [Z], [I].) Let l be an even positive integer, and ϕ a Dirichlet character modulo M

such that $\phi(-1) = 1$. For an even positive integer $k \geq l$ and a primitive Dirichlet character χ modulo N, we define a function $\mathcal{E}(z_1, z_2) = \mathcal{E}_{2,k}(z_1, z_2; l, M, \phi, s)$ on $\mathbf{H}_1 \times \mathbf{H}_1$ by

$$\begin{aligned}
&\mathcal{E}_{2,k}(z_1, z_2; l, M, \phi, \chi, s) \\
&= \sum_{m_1=0}^{\infty} \sum_{m_2=0}^{\infty} \sum_{r^2 \leq 4m_1m_2} c_{2,l,s}(\begin{pmatrix} m_1 & r/2 \\ r/2 & m_2 \end{pmatrix}, \begin{pmatrix} y_1 & 0 \\ 0 & y_2 \end{pmatrix}, M, \phi\bar{\chi}) \\
&\qquad \times G_l^{k-l}(m_1m_2, r)\chi(r)\tau(\bar{\chi}) \exp(2\pi i m_1 x_1) \exp(2\pi i m_2 x_2),
\end{aligned}$$

where we write $z_1 = x_1 + iy_1, z_2 = x_1 + iy_2$. Assume that N^2 divides M. Then, $\mathcal{E}_{2,k}(z_1, z_2; l, M, \phi, \chi, s)$ defines an element of $M_k^{\infty}(\Gamma_0(M), \phi)$ when considered as a function of z_1 or z_2, and hence it belongs to $M_k^{\infty}(\Gamma_0(M), \phi) \otimes M_k^{\infty}(\Gamma_0(M), \phi)$ (cf. [B-S, Sections 2 and 6].) Thus for an element $f(z_1) \in S_k(\Gamma_0(M), \phi)$, we can define

$$< f, \mathcal{E}_{2,k}(*, -\bar{z}_2; l, M, \phi, \chi, \bar{s})) >_{\Gamma_0(M)}$$

when we consider $\mathcal{E}_{2,k}(*, -\bar{z}_2; l, M, \phi, \chi, \bar{s}))$ as a function of z_1. Then by [B-S, Theorem 3.1], we have

Theorem 3.1. *In addition to the notation and the assumption as above, assume that $M > 1$, N^2 divides $M, \phi^2 = 1$ and $\chi(-1) = 1$. Let $f \in S_k(\Gamma_0(M), \phi)$ be a common eigenfunction of all Hecke operators. Then we have*

$$\begin{aligned}
&< f, \mathcal{E}_{2,k}(*, -\bar{z}; l, M, \phi, \chi, \bar{s})) >_{\Gamma_0(M)} \\
&\qquad = \kappa_{l,k}(s) N^{k+l+2s-2} M^{1-k/2} L(f|W_M, l+2s-1, \chi) \\
&\qquad\qquad \times f|W_M|T(M/N^2)(z),
\end{aligned}$$

where

$$\begin{aligned}
\kappa_{l,k}(s) = &\frac{(-1)^{l/2}}{2^{-3+2k-l+2s}\pi^{k-l-1}} \frac{\Gamma(k+s-1/2)\Gamma(k+s-1)}{\Gamma(l+s)\Gamma(l+s-1/2)} \\
&\times \frac{\Gamma(k-1)}{\prod_{\mu=1}^{k-l}(\mu/2)(k-1-s-\mu/2)},
\end{aligned}$$

$T(M/N^2)$ is the Hecke operator, and $W_M = \begin{pmatrix} 0 & -1 \\ M & 0 \end{pmatrix}$.

Now let ϕ be as in Theorem 3.1, and assume that the triple $(l, s, \phi\bar{\chi})$ satisfies one of the conditions (h-1),(h-2),(h-3) in Section 2. Then we define a function $\tilde{\mathcal{E}}(z_1, z_2) = \tilde{\mathcal{E}}_{2,k}(z_1, z_2; l, M, \phi, \chi, s)$ on $\mathbf{H}_1 \times \mathbf{H}_1$ by

$$\tilde{\mathcal{E}}_{2,k}(z_1, z_2; l, M, \phi, \chi, s) = \sum_{m_1=0}^{\infty} \sum_{m_2=0}^{\infty} \sum_{r^2 \le 4m_1m_2} \tilde{c}_{2,l}(\begin{pmatrix} m_1 & r/2 \\ r/2 & m_2 \end{pmatrix}, M, \phi\bar{\chi}, s) \times G_l^{k-l}(m_1m_2, r)\chi(r)\tau(\bar{\chi})q_1^{m_1}q_2^{m_2},$$

where $q_1 = \exp(2\pi i z_1), q_2 = \exp(2\pi i z_2)$. Then $\tilde{\mathcal{E}}_{2,k}(z_1, z_2; l, M, \phi, \chi, s)$ belongs to $M_k(\Gamma_0(M), \phi) \otimes M_k(\Gamma_0(M), \phi)$, and, in particular if $k > l$, then $\tilde{\mathcal{E}}_{2,k}(z_1, z_2; l, M, \phi, \chi, s)$ belongs to $S_k(\Gamma_0(M), \phi) \otimes S_k(\Gamma_0(M), \phi)$. Furthermore, by Theorem 2.1 we have

$$\mathcal{E}_{2,k}(z_1, z_2; l, M, \phi, \chi, s) = A(l, s)\tilde{\mathcal{E}}_{2,k}(z_1, z_2; l, M, \phi, \chi, s)$$

From now on, for a Dirichlet character ψ modulo M_0 we use the same symbol ψ to denote the character modulo M induced from ψ if M_0 divides M. For a positive integer r put $\delta_r = \begin{pmatrix} r & 0 \\ 0 & 1 \end{pmatrix}$, and let $S_k(\Gamma_0(M), \phi)^{(r)} = \{f|\delta_r; f \in S_k(\Gamma_0(M), \phi)\}$, and $S_k(\Gamma_0(M), \phi)^{new}$ the space of new forms in $S_k(\Gamma_0(M), \phi)$. Furthermore for a primitive form f in $S_k(\Gamma_0(M), \phi)$ let c_f be the complex number such that $f|W_M = c_f f^c$. Let $\lambda_f(m)$ be the eigenvalue of the Hecke operator $T(m)$ for a positive integer m. For an odd positive integer $m \le k-1$, let

$$\Lambda(f, m, \chi) = \frac{\Gamma(k-1)\Gamma(k+m-1)\Gamma(m+1)}{\Gamma(k-m)} \frac{L(f, m, \chi)}{2^{2k+2m-4}\pi^{k+2m} < f, f >},$$

and

$$\Lambda(f, 0, \chi) = \Gamma(k-1)\frac{L(f, 0, \chi)}{2^{2k-3}\pi^k < f, f >}.$$

We note that $m(\Phi_{\Gamma_0(N)})) = \pi/3[\Gamma : \Gamma_0(N)]$. Thus by Theorem 3.1 we have

Theorem 3.2. *Under the above notation and the assumption, we have the following:*

(1) Let ϕ be a primitive quadratic character modulo p^r with $r \ge 1$. Let f be a primitive form in $S_k(\Gamma_0(p^r), \phi)^{new}$. Then

$$< f, \tilde{\mathcal{E}}_{2,k}(*, -\bar{z}; l, p^r, \phi, 1, s)) > = 3[\Gamma : \Gamma_0(p^r)]^{-1} p^{(1-k/2)r} \Lambda(f^c, l+2s-1, 1) < f, f > \bar{\lambda}_f(p^r) c_f f^c(z).$$

(2) Let ϕ be a primitive quadratic character modulo p, and f a primitive form in $S_k(\Gamma_0(p),\phi)^{new}$. Then we have

$$< f, \tilde{\mathcal{E}}_{2,k}(*,-\bar{z};l,p^2,\phi,\phi,s)) >$$
$$= 3[\Gamma:\Gamma_0(p^2)]^{-1}p^{l+2s}\Lambda(f^c|\delta_p,l+2s-1,\phi) < f|\delta_p, f|\delta_p > c_f f^c|\delta_p(z),$$

and

$$< f|\delta_p, \tilde{\mathcal{E}}_{2,k}(*,-\bar{z};l,p^2,\phi,\phi,s)) >$$
$$= 3[\Gamma:\Gamma_0(p^2)]^{-1}p^{l+2s}\Lambda(f^c,l+2s-1,\phi) < f,f > c_f f^c(z).$$

(3) Let $p_0 = 1$ or a prime number different from p. Let ϕ be a primitive character modulo p_0 such that $\phi^2 = 1$, and f a primitive form in $S_k(\Gamma_0(p_0),\phi)^{new}$. Then for a primitive character χ modulo p, we have

$$< f, \tilde{\mathcal{E}}_{2,k}(*,-\bar{z};l,p^2p_0,\phi,\chi,s)) >$$
$$= 3[\Gamma:\Gamma_0(p^2p_0)]^{-1}p^{l+2s}p_0^{1-k/2}\Lambda(f^c|\delta_{p^2},l+2s-1,\chi)$$
$$\times < f|\delta_{p^2}, f|\delta_{p^2} > c_f\bar{\lambda}_f(p_0)f^c|\delta_{p^2}(z),$$

$$< f|\delta_p, \tilde{\mathcal{E}}_{2,k}(*,-\bar{z};l,p^2p_0,\phi,\chi,s)) >$$
$$= 3[\Gamma:\Gamma_0(p^2p_0)]^{-1}p^{l+2s}p_0^{1-k/2}\Lambda(f^c|\delta_p,l+2s-1,\chi)$$
$$\times < f|\delta_p, f|\delta_p > c_f\bar{\lambda}_f(p_0)f^c|\delta_p(z),$$

and

$$< f|\delta_{p^2}, \tilde{\mathcal{E}}_{2,k}(*,-\bar{z};l,p^2p_0,\phi,\chi,s)) >$$
$$= 3[\Gamma:\Gamma_0(p^2p_0)]^{-1}p^{l+2s}p_0^{1-k/2}\Lambda(f^c,l+2s-1,\chi)$$
$$\times < f,f > c_f\bar{\lambda}_f(p_0)f^c(z).$$

4. Computation of $L(f,l,\chi)$

Let p be prime a number such that $p \equiv 1 \bmod 4$. Let $p_0 = 1$, or a prime number such that $(\frac{p_0}{p}) = 1$ and $p_0 \equiv 1 \bmod 4$. Here $(\frac{p_0}{p})$ denotes the Legendre symbol. In this section we give some formulas to compute $L(f,m,\chi)$ for a primitive form $f \in S_k(\Gamma_0(N),\psi)$ in the following three cases:

(1) $N = p$, $\psi = \left(\frac{p}{*}\right)$, and χ is trivial,

(2) $N = p$, $\psi = \left(\frac{p}{*}\right)$, and χ is a primitive character modulo p.

(3) $N = p_0$, $\psi = \left(\frac{p_0}{*}\right)$, and χ is a primitive character modulo p.

Put $M = p, p^2$ or $p_0 p^2$ according as the case (1), (2), or (3). Let $k \geq 4$ be an even positive integer, and l a positive integer such that $l \leq k-2$. Assume that the triple $(l, s, \psi\bar{\chi})$ satisfies one of the conditions (h-1),(h-2),(h-3) in Section 2. For two positive integers m_1, m_2 put

$$\begin{aligned}\epsilon(m_1, m_2; l, s) &= \epsilon(m_1, m_2; l, M, \psi, \chi, s)\\ &= \sum_{r^2 \leq 4m_1m_2} \tilde{c}_{2,l}(\begin{pmatrix} m_1 & r/2 \\ r/2 & m_2 \end{pmatrix}, M, \psi\bar{\chi}, s)\\ &\quad \times G_l^{k-l}(m_1m_2, r)\chi(r)\tau(\bar{\chi}).\end{aligned}$$

Furthermore, put

$$t(m; l, s) = \begin{cases} 3^{-1}(p+1)p^{k/2-1}\epsilon(1, m; l, s) & \text{case (1)} \\ 3^{-1}(p+1)p^{1/2-l-2s}\epsilon(p, m; l, s) & \text{case (2)} \\ 3^{-1}t_{p_0}p_0^{k/2-1}p^{-k-l-2s+2} & \\ \times(\epsilon(p, p^2m; l, s) - p^{k-2}\epsilon(p, m; l, s)) & \text{case (3),} \end{cases}$$

where $t_{p_0} = p_0 + 1$ or 1 according as p_0 is prime number or 1. Take a basis $\{f_i\}_{i=1}^{d_1}$ of $S_k(\Gamma_0(N), \psi)^{new}$ consisting of primitive forms. Let $f_i | W_N = c_i f_i^c$ with constant c_i, and write

$$f_i(z) = \sum_{m=1}^{\infty} a_i(m) \exp(2\pi i m z)$$

with $a_i(1) = 1$. Using Theorem 3.2 we shall show

Theorem 4.1. *Let the notation and the assumption be as above.*

(1) In case (1), for any positive integer m we have

$$t(m; l, s) = \sum_{i=1}^{d_1} \Lambda(f_i, l + 2s - 1, 1)\bar{c}_i a_i(p) a_i(m).$$

(2) In case (2), for any positive integer m we have

$$t(m; l, s) = \sum_{i=1}^{d_1} \Lambda(f_i, l + 2s - 1, \psi) a_i(p) a_i(m).$$

(3) In case (3), for any positive integer m *prime to* pp_0 *we have*

$$t(m;l,s)=\sum_{i=1}^{d_1}\Lambda(f_i,l+2s-1,\chi)\bar{c}_i a_i(p_0)a_i(p)a_i(m).$$

Sketch of the proof. Here we treat only the case (3), and show how to derive the formula from Theorem 3.2. As for the details, see [Kat2]. As is well known,

$$\begin{aligned}&S_k(\Gamma_0(p^2p_0),\psi)\\&=S_k(\Gamma_0(p_0),\psi)\oplus S_k(\Gamma_0(p_0),\psi)^{(p)}\oplus S_k(\Gamma_0(p_0),\psi)^{(p^2)}\\&\quad\perp S_k(\Gamma_0(pp_0),\psi)^{new}\perp S_k(\Gamma_0(pp_0),\psi)^{new\,(p)}\perp S_k(\Gamma_0(p^2p_0),\psi)^{new}.\end{aligned}$$

Take bases $\{g_i\}_{i=1}^{d_2}$ of $S_k(\Gamma_0(pp_0),\psi)^{new}$, and $\{h_i\}_{i=1}^{d_3}$ of $S_k(\Gamma_0(p^2p_0),\psi)^{new}$ consisting of primitive forms. Then $\{f_i\ (i=1,2,\ldots,d_1),\ f_i|\delta_p\ (i=1,2,\ldots,d_1),\ f_i|\delta_{p^2}\ (i=1,2,\ldots,d_1),\ g_i\ (i=1,2,\ldots,d_2),\ g_i|\delta_p\ (i=1,2,\ldots,d_2),\ h_i\ (i=1,2,\ldots,d_3)\}$ forms a basis of $S_k(\Gamma_0(p^2p_0),\psi)$. Thus we have

$$\begin{aligned}&\tilde{\mathcal{E}}_{2,k}(z_1,z_2;l,p^2p_0,\psi,\chi,s)\\&=\sum_{i,j=1}^{d_3}c_{ij}^{(0,0)}h_i(z_1)h_j(z_2)+\sum_{\alpha,\beta=0}^{1}\sum_{i,j=1}^{d_2}b_{ij}^{(\alpha,\beta)}g_i|\delta_{p^\alpha}(z_1)g_j|\delta_{p^\beta}(z_2)\\&\quad+\sum_{\alpha,\beta=0}^{2}\sum_{i,j=1}^{d_1}a_{ij}^{(\alpha,\beta)}f_i|\delta_{p^\alpha}(z_1)f_j|\delta_{p^\beta}(z_2)\end{aligned}$$

with $c_{ij}^{(0,0)},\ b_{ij}^{(\alpha,\beta)},\ c_{ij}^{(\alpha,\beta)}\in\mathbf{C}$. Now we have $f_i|W_{p^2p_0}=c_if_i^c|\delta_{p^2}$, $f_i|\delta_p|W_{p^2p_0}=c_if_i^c|\delta_p$. Similarly, $g_i|W_{p_0p^2}=c_i'g_i^c|\delta_p$ with constant c_i', and $h_i|W_{p_0p^2}=c_i''h_i^c$ with constant c_i''. From this we have $g_i|\delta_p|W_{p^2p_0}=\bar{c}_i'g_i^c$. For a positive integer write $\lambda_i(m)=\lambda_{f_i}(m)$, $\lambda_i(m)'=\lambda_{g_i}(m)$, and $\lambda_i(m)''=\lambda_{h_i}(m)$. By assumption, we have $\psi(p)=1$. Thus $\lambda_i(p)$ is a real number. Now by using the formula [Sh1, (2.5)], we can show that we have

$$<f_i|\delta_p,f_i>=\frac{p^{-k/2}\lambda_i(p)}{1+p^{-1}}<f_i,f_i>,$$

$$<f_i|\delta_{p^2},f_i>=\frac{p^{-k}\lambda_i(p)^2-p^{-1}(1+p^{-1})}{1+p^{-1}}<f_i,f_i>,$$

and

$$<g_i|\delta_p,g_i>=p^{-k/2}\overline{\lambda_i(p)'}<g_i,g_i>.$$

Then by a direct computation we have

$$\tilde{\mathcal{E}}_{2,k}(z_1,z_2;l,p^2p_0,\psi,\chi,s)$$
$$=\sum_{i}^{d_3} c_{ii}h_i(z_1)h_i(z_2)+\sum_{i=1}^{d_2} b_{ii}\{-p^{-k/2}\lambda_i(p)g_i(z_1)g_i(z_2)$$
$$+g_i(z_1)g_i|\delta_p(z_2)+g_i|\delta_p(z_1)g_i(z_2)-p^{-k/2}\lambda_i(p)g_i|\delta_p(z_1)g_i|\delta_p(z_2)\}$$
$$+\sum_{i=1}^{d_1} a_{ii}\{p^{-1}f_i|\delta_{p^2}(z_1)f_i|\delta_{p^2}(z_2)-p^{-k/2}\lambda_i(p)f_i|\delta_p(z_1)f_i|\delta_{p^2}(z_2)$$
$$+f_i(z_1)f_i|\delta_{p^2}(z_2)-p^{-k/2}\lambda_i(p)f_i|\delta_{p^2}(z_1)f_i|\delta_p(z_2)$$
$$+(1+\lambda_i(p)^2p^{-k}-p^{-2})f_i|\delta_p(z_1)f_i|\delta_p(z_2)-\lambda_i(p)p^{-k/2}f_i(z_1)f_i|\delta_p(z_2)$$
$$+f_i|\delta_{p^2}(z_1)f_i(z_2)-p^{-k/2}\lambda_i(p)f_i|\delta_p(z_1)f_i(z_2)+p^{-1}f_i(z_1)f_i(z_2)\},$$

where

$$c_{ii}=3p^{-1}(p+1)^{-1}t_{p_0}^{-1}p^{l+2s}p_0^{1-k/2}\Lambda(h_i,l+2s-1,\chi)\bar{c}_i''\lambda_i(p_0)'',$$

$$b_{ii}=3p^{-1}(p+1)^{-1}t_{p_0}^{-1}\frac{p^{l+2s}p_0^{1-k/2}\Lambda(g_i,l+2s-1,\chi)\bar{c}_i'\lambda_i(p_0)'}{1-p^{-2}},$$

and

$$a_{ii}=3p^{-1}(p+1)^{-1}t_{p_0}^{-1}\frac{(1+p^{-1})p^{l+2s}p_0^{1-k/2}\Lambda(f_i,l+2s-1,\chi)\bar{c}_i\lambda_i(p_0)}{(1-p^{-1})((1+p^{-1})^2-p^{-k}\lambda_i(p)^2)},$$

where $d_q=p_0+1$ or 1 according as p_0 is prime number or 1. Now let

$$g_i(z)=\sum_{m=1}^{\infty} b_i(m)\exp(2\pi imz)\ (i=1,2,...,d_2)$$

and

$$h_i(z)=\sum_{m=1}^{\infty} c_i(m)\exp(2\pi imz)\ (i=1,2,...,d_3)$$

with $b_i(1)=c_i(1)=1$. Put

$$\mathcal{E}_p(z_2)=\sum_{m_2=1}^{\infty}\epsilon(p,m_2;l,s)\exp(2\pi im_2z_2).$$

We note that $c_i(pm)=0$ and $b_i(pm)=\lambda_i(p)'b_i(m)$ for any positive integer m. Thus we have

$$\mathcal{E}_p(z_2)=\sum_{i=1}^{d_1} a_{ii}\{p^{k/2}(1-p^{-2})f_i|\delta_p(z_2)-(1-p^{-1})\lambda_i(p)f_i(z_2)\}$$

$$+\sum_{i=1}^{d_2} p^{k/2}(1-p^{-2})b_{ii}g_i(z_2).$$

We note that $b_i(p^2m) = p^{k-2}b_i(m)$ for any positive integer m prime to p (cf. [Miy, Theorem 4.6.17]). Thus the assertion follows by a direct computation. □

For a prime number q not dividing pp_0 put

$$\beta_{i+1} = \beta(i+1, q; l, s) = \sum_{r=0}^{[i/2]} ({}_iC_r - {}_iC_{r-1})q^{r(k-1)}t(q^{i-2r}; l, s)$$

for $i = 0, 1, ..., d_1 - 1$, where ${}_iC_r = \frac{i!}{r!(i-r)!}$. We understand that ${}_iC_{-1} = 0$. For a Hecke eigenform f let K_f be the field generated over $\mathbf{Q}$ by all the eigenvalues of the Hecke operators relative to f. Furthermore, for a character χ let $K_{f,\chi}$ be the field generated over K_f by all the values of χ. Put $e_f = [K_f : \mathbf{Q}]$, and we denote by $N_{K_f}(\alpha)$ the norm of α over $\mathbf{Q}$ for $\alpha \in K_f$. Similarly we define $e_{f,\chi}$ and $N_{K_{f,\chi}}(\alpha)$ for $\alpha \in K_{f,\chi}$. Let $\{f_i\}_{i=1}^d$ be the basis of $S_k(\Gamma_0(N), \psi)$ as above, and write $K_i = K_{f_i,\chi}$ and $e_i = e_{f_i,\chi}$. Let $\Phi(X)$ be the characteristic polynomial of $T(q)$ on $S_k(\Gamma_0(N), \psi)$. We note that $N_{K_i}(c_i) = 1$ in cases (1),(2),(3). Furthermore we have $N_{K_i}(a_i(p)) = p^{(k-1)e_i/2}$ in cases (1) and (2),and $N_{K_i}(a_i(p_0)) = p_0^{(k-1)e_i/2}$. Thus by [G, Lemma 2.2] we obtain

Theorem 4.2. *Let the notation and the assumption be as above. Let f be a primitive form in $S_k(\Gamma_0(N), \psi)$, and $a(q)$ be the q-th Fourier coefficient of f. Assume that $\Phi'(a(q)) \neq 0$. Write $\Phi(X) = \sum_{i=0}^{d_1} b_{d_1-i}X^i, K = K_{f,\chi}$, and $e = e_{f,\chi}$.*

(1) In case (1) and (2), we have

$$N_K(\Lambda(f, l+2s-1, \chi)) = p^{(1-k)e/2}N_K\left(\frac{\sum_{i=0}^{d_1-1}\sum_{j=i}^{d_1-1}\beta_{d_1-j}b_{j-i}a(q)^i}{\Phi'(a(q))}\right).$$

(2) In case (3), if $a(p) \neq 0$, we have

$$N_K(\Lambda(f, l+2s-1, \chi)) = p_0^{(1-k)e/2}N_K\left(\frac{\sum_{i=0}^{d_1-1}\sum_{j=i}^{d_1-1}\beta_{d_1-j}b_{j-i}a(q)^i}{a(p)\Phi'(a(q))}\right).$$

Remark 1. For a Hecke eigenform $f \in S_k(\Gamma_0(M), \psi)$,

$$\tilde{L}(f, s, 1) = \frac{L(f, s, 1)}{\prod_{p|M}(1 - \bar{a}(p)^2p^{-s-k+1})},$$

and

$$\tilde{L}(f,s,\psi)=\frac{L(f,s,\psi)}{\prod_{p|M}(1-p^{-s})}.$$

Put for an odd positive integer $l \leq k-1$

$$\tilde{\Lambda}(f,l,\chi)=\frac{\Gamma(k-1)\Gamma(k+l-1)\Gamma(l+1)}{\Gamma(k-l)}\frac{\tilde{L}(f,l,\chi)}{2^{2k+2l-4}\pi^{k+2l}<f,f>},$$

and

$$\tilde{\Lambda}(f,0,\chi)=\Gamma(k-1)\frac{\tilde{L}(f,0,\chi)}{2^{2k-3}\pi^{k}<f,f>}$$

for $\chi=1$ or ψ. Then Zagier [Z] gave a useful algorithm to compute the exact value of $\tilde{\Lambda}(f,l,1)$. Sturm [Stur] gives a general algorithm to compute $\Lambda(f,l,\chi)$ for an eigenform $f \in S_k(\Gamma_0(N),\psi)$ and a Dirichlet character χ. However, even in the case $N=1$ and χ is a quadratic character modulo a prime number p, one needs precise information about the Fourier coefficients of primitive forms in $S_k(\Gamma_0(p^2))$ to compute $\Lambda(f,l,\chi)$ by using his method. Thus it seems difficult to carry out his method directly for the computation of such a value. Such Fourier coefficients appear also in the course of our algorithms as seen above. However, they eventually disappear in the final formula. We note that Stopple [Sto] gave a formula similar to (3) in case $f \in S_k(SL_2(\mathbf{Z}))$ and χ is quadratic. His formula involves some non-elementary character sums. On the other hand, our formula does not contain such a non-elementary term.

Remark 2. By using the same method, we can give an exact value of $\Lambda(f,l,1)$ for $l \geq 3$ for a cuspidal Hecke eigen form of level 1.

Remark 3. The function $\tilde{\mathcal{E}}_{2,k}(z_1,z_2;k,M,\psi,\chi,0)$ does not belong to $S_k(\Gamma_0(M),\psi)\otimes S_k(\Gamma_0(M),\psi)$, but it belongs to $M_k(\Gamma_0(M),\psi)\otimes M_k(\Gamma_0(M),\psi)$. Thus, modifying the above method, we can compute the value $\Lambda(f,k-1,\chi)$. We remark that the numerator of $\Lambda(f,k-1,1)$ is related to the congruence between the Klingen-Eisenstein lifting of f and another Siegel modular form . As for this, see [Ku] and [Miz].

Remark 4. As seen above, if $\chi^2 \neq 1$, $E^*_{2,2}(Z;M,\psi\bar{\chi},0)$ belongs to $M_2(\Gamma_0(M),\psi)$. Thus we can use the same method to compute the value $\Lambda(f,1,\chi)$. On the other hand, if $\chi^2=1, E^*_{2,2}(Z;M,\psi\chi,0)$ does not belong to $M_2(\Gamma_0(M),\psi\bar{\chi})$, and thus we cannot give an exact value of $\Lambda(f,1,\chi)$ by a direct use of the above method. However, we can relate the value $\Lambda(f,1,\chi)$ to $\Lambda(f,0,\chi)$ by using the functional equation. We explain this in case (1), (3). First in case (1), put

$$\tilde{R}(f,s,1)=M^{(s+k-1)/2}\pi^{-3/2(s+k-1)} \times\Gamma((s+k-1)/2)\Gamma((s+k)/2)\Gamma((s+1)/2)\tilde{L}(f,s,1).$$

Next in case (3), for $f \in S_k(SL_2(\mathbf{Z}))$, and a quadratic character χ modulo M, put

$$R(f,s,\chi) = M^{3(s+k-1)/2}\pi^{-3/2(s+k-1)} \\ \times \Gamma((s+k-1)/2)\Gamma((s+k)/2)\Gamma((s+1)/2)L(f,s,\chi).$$

Then by Li [L], we have the following functional equation:

Proposition 4.3. *(1) In case (1),*

$$\tilde{R}(f,1-s,1) = \tilde{R}(f,s,1).$$

In particular

$$\tilde{\Lambda}(f,1,1) = p^{-1/2}\tilde{\Lambda}(f,0,1).$$

(2) In case (3), under the above notation and the assumption, we have

$$R(f,1-s,\chi) = R(f,s,\chi).$$

In particular, in case (3), we have

$$\Lambda(f,1,\chi) = p^{-3/2}\Lambda(f,0,\chi).$$

In case (2), the value $\Lambda(f,1,\phi)$ can be given by a different method (cf. [Z].)

Proposition 4.4. *In case (2), we have*

$$\Lambda(f,1,\phi) = 2/3(1-p^{-2}).$$

5. Numerical examples and comments

By Theorem 4.2 combined with Proposition 4.3 we can compute the values $\Lambda(f,m,\chi)$. A subspace S of $S_k(\Gamma_0(N),\psi)$ is called non-splitting if it is spanned by all Galois conjugates of a primitive form in S. Take a primitive form f of $S_k(SL_2(\mathbf{Z}))$. Assume that $S_k(SL_2(\mathbf{Z}))$ is non-splitting. Then, $N_{K_f}(\Lambda(f,l,(\frac{p}{*})))$ is independent of f. Thus, in this case, we write this value as $L(k,l,p)$. Similarly, in case $S_k(\Gamma_0(p);(\frac{p}{*}))$ is non-splitting, we define $\tilde{L}(k,l,p)$ as $N_{K_f}(\tilde{\Lambda}(f,l,1))$ for a primitive form f of $S_k(\Gamma_0(p),(\frac{p}{*}))$. It is conjectured by Maeda that $S_k(SL_2(\mathbf{Z}))$ is non-splitting, and so far, this conjecture has been verified at least for $k \leq 2000$ (cf. [H-M], [F-J]). We show some examples of $L(k,l,p)$ for various k,l,p. We have computed the values by using *Mathematica*.

(1) We compute $L(12,l,p)$ for $1 \leq l \leq k-3$ and some prime numbers p. The results are as follows:

l	$L(12,l,5)$
1	$2^{14}\cdot 3^5\cdot 7/5^{10}$
3	$2^{14}\cdot 3^5\cdot 7\cdot 2851/5^{13}$
5	$2^{19}\cdot 3^5\cdot 7\cdot 1511599/5^{16}$
7	$2^{19}\cdot 3^8\cdot 7^3\cdot 521\cdot 295387/5^{20}$
9	$2^{26}\cdot 3^{10}\cdot 7^2\cdot 110308273279/5^{24}$

l	$L(12,l,13)$
1	$2^{14}\cdot 3^8\cdot 5^3\cdot 7\cdot 563/13^{12}$
3	$2^{14}\cdot 3^5\cdot 5^2\cdot 7\cdot 41177\cdot 1445419/13^{16}$
5	$2^{19}\cdot 3^5\cdot 5^4\cdot 7\cdot 299696968678699/13^{20}$
7	$2^{19}\cdot 3^7\cdot 5^3\cdot 7^3\cdot 31^2\cdot 5479\cdot 306945156059/13^{23}$
9	$2^{26}\cdot 3^{10}\cdot 5^5\cdot 7^2\cdot 547\cdot 10267\cdot 1634679978646831/13^{26}$

l	$L(12,l,17)$
1	$2^{21}\cdot 3^5\cdot 5^3\cdot 7\cdot 2389/17^{12}$
3	$2^{17}\cdot 3^5\cdot 5^2\cdot 7\cdot 477200018623/17^{14}$
5	$2^{20}\cdot 3^5\cdot 5^4\cdot 7\cdot 23\cdot 29\cdot 997\cdot 46316422211/17^{20}$
7	$2^{22}\cdot 3^8\cdot 5^3\cdot 7^3\cdot 167\cdot 11003\cdot 322079\cdot 915248119/17^{24}$
9	$2^{30}\cdot 3^{10}\cdot 5^5\cdot 7^2\cdot 43\cdot 892028959\cdot 1604767911433/17^{27}$

We note that the value $L(12,l,p)$ has been obtained by Stopple [Sto] in the case where $p=5$ and $l=1,3,5,7$, or $p=13,17,29,37,41$ and $l=1$. We note that relatively large prime numbers appear in the numerator of $L(k,l,5)$ contrary to the untwisted case in [D]. At present, the author does not know whether these primes relate to any arithmetic algebraic geometry. We note that the numerator of $\Lambda(f,l,1)$ is related to the order of the Shafarevich-Tate group (cf. [D].)

(2) The values of the standard zeta functions at $s=1$ are particularly important. To explain this, let q be a prime number congruent to 1 modulo 4, and let $O_{\mathbf{Q}(\sqrt{q})}$ be the ring of integers in $\mathbf{Q}(\sqrt{q})$. Let $S_{k,k}(SL_2(O_{\mathbf{Q}(\sqrt{q})}))$ be the space of cusp forms of weight (k,k) with respect to $SL_2(O_{\mathbf{Q}(\sqrt{q})})$. Then $S_{k,k}(SL_2(O_{\mathbf{Q}(\sqrt{q})}))$ has the following decomposition:

$$S_{k,k}(SL_2(O_{\mathbf{Q}(\sqrt{q})})) = \hat{S}_k(SL_2(\mathbf{Z}))\perp\hat{S}_k(\Gamma_0(q),\left(\frac{q}{*}\right))\perp S^0_{k,k},$$

where $\hat{S}_k(SL_2(\mathbf{Z}))$ (resp. $\hat{S}_k(\Gamma_0(q),(\frac{q}{*})))$ is the image of $S_k(SL_2(\mathbf{Z}))$ (resp. $S_k(\Gamma_0(q),(\frac{q}{*})))$ under the Doi-Naganuma map, and $S^0_{k,k}$ the orthogonal complement of $\hat{S}_k(SL_2(\mathbf{Z}))\perp\hat{S}_k(\Gamma_0(q),(\frac{q}{*}))$ in $S_{k,k}(SL_2(O_{\mathbf{Q}(\sqrt{q})}))$ with respect to the Petersson product. Take a primitive form $g\in S^0_{k,k}$, and for an integral ideal $\mathfrak{A}$ in $\mathbf{Q}(\sqrt{q})$ let $c(\mathfrak{A};g)$ be the $\mathfrak{A}$-th Fourier coefficient of g. Let K_g be the field generated over $\mathbf{Q}$ by all $c(\mathfrak{A};g)'s$ and K_g^+ the subfield of K_g generated by $c((p);g)$ for all rational primes p. We denote by $D(K_g/K_g^+)$ the relative discriminant of K_g/K_g^+. Assume that $S_{k,k}^0$ is non-splitting, that is, it is spanned by all Galois conjugates of a primitive form in $S_{k,k}^0$. Then $N_{K_g^+}D(K_g/K_g^+)$ does not depend on g. Hence we write this as $D_{k,q}$. Then in [D-H-I] Doi, Hida, and Ishii, among others, conjectured the following:

> "Any prime factor of $D_{k,q}$ divides either the numerator of $N_{K_f}(\Lambda(f,1,(\frac{q}{*})))$ for some primitive form f in $S_k(SL_2(\mathbf{Z}))$ or the numerator of $N_{K_f}(\tilde{\Lambda}(f,1,1))$ for some primitive form f in $S_k(\Gamma_0(q),(\frac{q}{*}))$."

Doi, Hida and Ishii [D-H-I] computed an exact value of $\tilde{L}(k,1,p)$, and verified the above conjecture in some cases. Goto [G] computed the value $L(20,1,5)$, and Hiraoka [Hir] computed the values $L(22,1,5)$ and $L(24,1,5)$. Then, combining the result of [D-H-I], they verified the conjecture for $(k,q) = (20,5),(22,5)$, and $(24,5)$. Now let $k = 12$ and $q = 13$. In this case, $S_{12,12}^0$ and $S_k(\Gamma_0(13),(\frac{13}{*}))$ are non-splitting. Furthermore, we have $D_{12,13} = 13 \cdot 563 \cdot 6205151$ and the numerator of $\tilde{L}(12,1,13)$ is $5 \cdot 7 \cdot 13^{29} \cdot 6205151$. Thus it is expected that 563 appears in the numerator of $L(12,1,13)$. The example in (1) shows that this is true.

Now we compute $L(k,1,5)$ for $16 \leq k \leq 38$. As for the other numerical examples, see [Sto].

First let $k = 16,18,20,22,26$. Then we have $\dim S_k(SL_2(\mathbf{Z})) = 1$ and we have the following table:

k	$L(k,1,5)$
16	$2^{13}\cdot 3^4 \cdot 7^3 \cdot 11/5^{14}$
18	$2^{13}\cdot 3^4 \cdot 7 \cdot 11 \cdot 13/5^{13}$
20	$2^{15}\cdot 3^4 \cdot 7 \cdot 11 \cdot 13 \cdot 977/5^{18}$
22	$2^{15}\cdot 3^5 \cdot 7^2 \cdot 13 \cdot 17 \cdot 71/5^{18}$
26	$2^{13}\cdot 3^4 \cdot 7 \cdot 11 \cdot 17 \cdot 19 \cdot 337 \cdot 1409/5^{22}$

Next let $k = 24,28,30,32,34,38$. Then we have $\dim S_k(SL_2(\mathbf{Z})) = 2$. Furthermore, $K_{f_1} = K_{f_2}$ is a real quadratic field, and thus we write this as K. Then we have the following table:

k	K	$L(k,1,5)$
24	$\mathbf{Q}(\sqrt{144169})$	$2^{29}\cdot 3^9\cdot 7^4\cdot 11^4\cdot 13\cdot 17\cdot 19\cdot 109\cdot 54449/5^{39}\cdot 144169$
28	$\mathbf{Q}(\sqrt{131\cdot 139})$	$2^{26}\cdot 3^8\cdot 7^3\cdot 11^2\cdot 13^3\cdot 17\cdot 19\cdot 23\cdot 66876860429/5^{44}\cdot 131\cdot 139$
30	$\mathbf{Q}(\sqrt{51349})$	$2^{26}\cdot 3^{10}\cdot 7^4\cdot 11^2\cdot 13\cdot 17\cdot 19\cdot 23\cdot 3253\cdot 20017939/5^{43}\cdot 51349$
32	$\mathbf{Q}(\sqrt{67\cdot 273067})$	$2^{26}\cdot 3^{10}\cdot 7^4\cdot 11\cdot 13^2\cdot 17\cdot 19\cdot 23\cdot 37\cdot 157\cdot p_1/5^{54}\cdot 67\cdot 273067$
34	$\mathbf{Q}(\sqrt{479\cdot 4919})$	$2^{26}\cdot 3^{10}\cdot 7^3\cdot 11^3\cdot 13^2\cdot 19\cdot 23\cdot 29\cdot 191\cdot 3191\cdot p_2/5^{51}\cdot 479\cdot 4919$
38	$\mathbf{Q}(\sqrt{181\cdot 349\cdot 1009})$	$2^{28}\cdot 3^8\cdot 7^4\cdot 11^2\cdot 13\cdot 17\cdot 23\cdot 29\cdot 31\cdot p_3/5^{60}\cdot 181\cdot 349\cdot 1009$

$p_1 = 222142617427425679,$

$$p_2 = 121120620073,$$
$$p_3 = 245396303520197996115221087.$$

Remark. By using the pull-back formula of Eisenstein series of degree 4, Ibukiyama's differential operators in [I], and an explicit formula for the Siegel series in [Kat1], we can obtain exact values of standard zeta function of cuspidal Hecke eigenform of degree 2. As for this, see [Kat3].

References

[B-S] S. Böcherer and C. G. Schmidt, *p-adic measures attached to Siegel modular forms*, Ann. Inst. Fourier **50** (2000), 1375–1443.

[D] N. Dummigan, *Symmetric square L-functions and Shafarevich-Tate groups*, Exp. Math. **10** (2001), 383–400.

[D-H-I] K. Doi, H. Hida, and H. Ishii, *Discriminant of Hecke fields and twisted adjoint L-values for GL(2)*, Invent. Math. **134** (1998), 547–577.

[F-J] D. W. Farmer and K. James, *The irreducibility of some level-1 Hecke polynomials*, Math. Comp. **71** (2002), 1263–1270

[G] K. Goto, *A twisted adjoint L-values of an elliptic modular form*, J. Number Theory **73** (1998), 34–46

[H-M] H. Hida and Y. Maeda, *Non abelian base change for totally real fields*, Pacific J. Math. Olga Taussky-Todd Memorial Issue (1997), 189–218.

[Hir] Y. Hiraoka, *Numerical calculation of twisted adjoint L-values attached to modular forms*, Exp. Math. **9** (2000), 67–73.

[I] T. Ibukiyama, *On Differential operators on automorphic forms and invariant pluri-harmonic polynomials*, Comm. Math. Univ. St. Pauli **48** (1999), 103–118.

[Kat1] H. Katsurada, *An explicit formula for Siegel series*, Amer. J. Math. **121** (1999), 415–452.

[Kat2] H. Katsurada, *Special values of the twisted standard zeta function of an elliptic modular form*, preprint (2001).

[Kat3] H. Katsurada, *Exact values of the standard zeta function of a Siegel modular form of degree two*, preprint (2001).

[Ku] N. Kurokawa, *Congruence between Siegel modular forms of degree two*, Proc. Japan Acad. **55** (1979), 417–422.

[L] W. Li, *L series of Rankin type and their functional equations*, Math. Ann. **244** (1979), 135–166.

[Miy] T. Miyake, "Modular forms," Springer-Verlag, Berlin, 1989.

[Miz] S. Mizumoto, *Congruences for eigenvalues of Hecke operators on Siegel modular forms of degree two*, Math. Ann. **275** (1986), 149–161.

[Sh1] G. Shimura, *The special values of the zeta functions associated with cusp forms*, Comm. pure appl. Math. **29** (1976), 783–804.

[Sh2] G. Shimura, "Arithmeticity in the theory of automorphic forms," Mathematical Surveys and Monographs, 82. American Mathematical Society, Providence, 2000,

[Sto] J. Stopple, *Special values of twisted symmetric square L-functions and the trace formula*, Compositio Math. **104** (1996), 65–76.

[Stur] J. Sturm, *Special values of zeta functions and Eisenstein series of half integral weight*, Amer. J. Math. **102** (1980), 219–240.

[Z] D. Zagier, *Modular forms whose Fourier coefficients involve zeta functions of quadratic fields*, in "Modular functions of one variable, VI (Proc. Second Internat. Conf., Univ. Bonn, Bonn, 1976)," Lecture Notes in Mathematics, 627. Springer, Berlin, 105–169, 1977.

Galois Theory and Modular Forms pp 357–366
K. Hashimoto, K. Miyake, and H. Nakamura, Editors
© 2003 Kluwer Academic Publishers

p-ADIC PROPERTIES OF VALUES OF THE MODULAR j-FUNCTION*

Ken ONO
Department of Mathematics,
University of Wisconsin, Madison,
Wisconsin 53706 USA
ono@math.wisc.edu

Matthew A. PAPANIKOLAS
Department of Mathematics,
Brown University, Providence,
Rhode Island 02912 USA
map@math.brown.edu

1. Introduction and Statement of Results

As usual, let $q := e^{2\pi iz}$ and let $j(z)$ be the classical modular function

$$j(z) = \sum_{n=-1}^{\infty} c(n)q^n = q^{-1} + 744 + 196884q + \cdots .$$

The values of $j(z)$ and its coefficients play many important roles in mathematics. For example, its values generate class fields and its coefficients appear as dimensions of a graded representation of the Monster via the Moonshine phenomenon. In a recent paper, Kaneko [K] produced an interesting connection between the values of $j(z)$ at Heegner points, the so-called singular moduli, and its coefficients. Using recent formulas of Zagier, he systematically expresses the coefficients $c(n)$ in terms of singular moduli. In this paper we want to illustrate some peculiar p-adic

*The first author thanks the National Science Foundation, the Alfred P. Sloan Foundation, the David and Lucile Packard Foundation, and an H. I. Romnes Fellowship for their generous research support. Both authors thank the referee for his/her helpful suggestions and comments. The first author thanks Professors K. Hashimoto, K. Miyake and H. Nakamura for their kind hospitality during the 1999 and 2001 meetings. These were two wonderful conferences!
[received: February 6, 2002; accepted in revised from: May 16, 2002]

properties of certain values of the j-function in connection with class numbers and the integer 720.

To motivate our first result, we begin by recalling some classical formulas for values of the Riemann zeta-function at negative odd integers. To state these formulas, we begin by recalling some standard facts and notation. If $k \geq 4$ is even, then let $E_k(z)$ denote the usual weight k Eisenstein series for the full modular group $SL_2(\mathbb{Z})$

$$E_k(z) = 1 - \frac{2k}{B_k} \sum_{n=1}^{\infty} \sigma_{k-1}(n) q^n.$$

As usual, B_k is the k-th Bernoulli number and $\sigma_{k-1}(n) = \sum_{d|n} d^{k-1}$, and H/Γ denotes the usual fundamental domain of the action of the modular group $\Gamma = SL_2(\mathbb{Z})$ on the upper half of the complex plane. If $k \geq 4$ is even, then

$$\frac{2}{\zeta(1-k)} = 60k - \sum_{\tau \in H/\Gamma} e_\tau \mathrm{ord}_\tau(E_k(z)) j(\tau). \tag{1}$$

If $\tau \in H/\Gamma$, then e_τ is defined by

$$e_\tau := \begin{cases} 1/2 & \text{if } \tau = i, \\ 1/3 & \text{if } \tau = e^{2\pi i/3}, \\ 1 & \text{otherwise.} \end{cases} \tag{2}$$

These formulas follow from Euler's formula for $\zeta(1-k)$ and the Fourier expansion for $E_k(z)$. For example, consider the case where $k = 10$. The divisor of $E_{10}(z)$ is supported by simple zeros at $\tau = i$ and $e^{2\pi i/3}$, and so the fact that $j(i) = 1728$ and $j(e^{2\pi i/3}) = 0$ confirms (1):

$$\frac{2}{\zeta(-9)} = -264 = 600 - \frac{j(e^{2\pi i/3})}{3} - \frac{j(i)}{2}.$$

Here we prove a p-adic analog of (1), where the values of the Kubota-Leopoldt zeta-function at $s = -1$ are given by a p-adic limit of certain precise expressions in the values of the j-function.

Theorem 1. *If $\tau \in H/\Gamma$ is a point for which $j(\tau) \in \mathbb{Q}$, and $p \leq 7$ is a prime for which*

$$j(\tau) \equiv \begin{cases} 0 \pmod{p} & \text{if } p \leq 5, \\ 6 \pmod{7} & \text{if } p = 7, \end{cases}$$

then as p-adic numbers we have

$$\frac{2}{\zeta_p^*(-1)} = -\frac{744}{p-1} - \lim_{n\to+\infty}\left(\sum_{a=0}^{n}\sum_{b=0}^{p^a-1} j\left(\frac{p^{n-a}\tau+b}{p^a}\right)\right).$$

Note the similarity between (1) and the p-adic formulas appearing in Theorem 1. On the left hand side we have a zeta-value, and on the right hand side we have an expression involving values of the j-function. In (1) these expressions are traces of j-values over the divisor of an Eisenstein series, and in Theorem 1 these expressions are p-adic limits (as $n \to +\infty$) of the 'traces' of j-values over the images of τ under the upper triangular Möbius transformations with determinant p^n.

There are some important differences. First of all, the formulas in Theorem 1 only pertain to those zeta values at $s = 1-k = -1$ (i.e. where $k = 2$). This is a byproduct of the fact that Theorem 1 is a statement about the arithmetic of p-adic modular forms of weight $k = 2$. Furthermore, Theorem 1 does not appear to be a statement related to the arithmetic of Eisenstein series. However, this is not true. The proof of Theorem 1 requires the arithmetic of the Eisenstein series (at least E_4 and E_6) in a vital way.

Using the fact that $\zeta_p^*(-1) = -(p-1)/12$, Theorem 1 may be interpreted as a curious collection of p-adic formulas for the integer 720:

$$0 = 720 + (p-1)\lim_{n\to+\infty}\left(\sum_{a=0}^{n}\sum_{b=0}^{p^a-1} j\left(\frac{p^{n-a}\tau+b}{p^a}\right)\right). \tag{3}$$

Example. To illustrate Theorem 1, let $\tau = i$. Since $j(i) = 1728 \equiv 0 \pmod 2$, let $p = 2$ in Theorem 1. Formula (3), which is equivalent to Theorem 1, asserts that if H_n is defined by

$$H_n := 720 + \sum_{a=0}^{n}\sum_{b=0}^{2^a-1} j\left(\frac{2^{n-a}i+b}{2^a}\right),$$

then H_n tends to zero 2-adically. If $n = 0, 1, \ldots, 5$, then

$$\begin{aligned}
H_0 &= 720 + j(i) = 2448 = 2^4\cdot 3^2\cdot 17,\\
H_1 &= 2^5\cdot 3^2\cdot 5\cdot 401,\\
H_2 &= 2^6\cdot 3^2\cdot 5\cdot 167\cdot 341927,\\
H_3 &= 2^7\cdot 3^2\cdot 131\cdot 89604100069763687,\\
H_4 &= 2^8\cdot 3^2\cdot 11\cdots,\\
H_5 &= 2^9\cdot 3^2\cdot 124909\cdots.
\end{aligned}$$

Combining similar arguments with recent work of the first author and Bruinier and Kohnen [Th. 9, B-K-O], one obtains the following p-adic formulas for class numbers $H(-D)$ of quadratic orders with discriminant $-D$. First we recall the notion of a Heegner point. A complex number τ of the form $\tau = \frac{-b+\sqrt{b^2-4ac}}{2a}$ with $a, b, c \in \mathbb{Z}$, $\gcd(a, b, c) = 1$ and $b^2 - 4ac < 0$ is known as a *Heegner point*, and its discriminant is the integer $d_\tau := b^2 - 4ac$.

Theorem 2. *Suppose that $-D < -4$ is a fundamental discriminant of an imaginary quadratic field, and let $\tau \in H/\Gamma$ be a Heegner point of discriminant $-D$. If $K = \mathbb{Q}(j(\tau))$, then the following are true:*

(1) *If $D \equiv 3 \pmod 8$, then as 2-adic numbers we have*

$$H(-D) = -\frac{1}{720} \lim_{n\to+\infty} \mathrm{Tr}_{K/\mathbb{Q}} \left(\sum_{a=0}^{n} \sum_{b=0}^{2^a-1} j\left(\frac{2^{n-a}\tau + b}{2^a}\right)\right).$$

(2) *If $D \equiv 1 \pmod 3$, then as 3-adic numbers we have*

$$H(-D) = -\frac{1}{360} \lim_{n\to+\infty} \mathrm{Tr}_{K/\mathbb{Q}} \left(\sum_{a=0}^{n} \sum_{b=0}^{3^a-1} j\left(\frac{3^{n-a}\tau + b}{3^a}\right)\right).$$

(3) *If $D \equiv 2, 3 \pmod 5$, then as 5-adic numbers we have*

$$H(-D) = -\frac{1}{180} \lim_{n\to+\infty} \mathrm{Tr}_{K/\mathbb{Q}} \left(\sum_{a=0}^{n} \sum_{b=0}^{5^a-1} j\left(\frac{5^{n-a}\tau + b}{5^a}\right)\right).$$

(4) *If $D \equiv 1, 2, 4 \pmod 7$, then as 7-adic numbers we have*

$$H(-D) = -\frac{1}{120} \lim_{n\to+\infty} \mathrm{Tr}_{K/\mathbb{Q}} \left(\sum_{a=0}^{n} \sum_{b=0}^{7^a-1} j\left(\frac{7^{n-a}\tau + b}{7^a}\right)\right).$$

Example. Here we illustrate Theorem 2 using the Heegner point $\tau = (3+\sqrt{-7})/2$, using the classical evaluation $j(\tau) = -15^3$. Since $H(-7) = 1$, Theorem 2 (3) implies that

$$I_n := 180 + \sum_{a=0}^{n} \sum_{b=0}^{5^a-1} j\left(\frac{5^{n-a}\tau + b}{5^a}\right)$$

tends to zero 5-adically as $n \to +\infty$. For $n = 0, 1, 2$ and 3 we find that

$$\begin{aligned} I_0 &= -3195 = -3^2 \cdot 5 \cdot 71, \\ I_1 &= -3^2 \cdot 5^2 \cdot 11^2 \cdot 41118261707401, \\ I_2 &= -3^2 \cdot 5^3 \cdot 1123 \cdots, \\ I_3 &= -3^2 \cdot 5^4 \cdot 109 \cdots. \end{aligned}$$

Example. For another example, we consider the Heegner point $\tau = (1+\sqrt{-15})/2$, using the evaluation (see [§II.6, Si])

$$j(\tau) = \frac{-191025 - 85995\sqrt{5}}{2}.$$

Now $H(-15) = 2$, and so Theorem 2 (4) implies that

$$J_n := 240 + \mathrm{Tr}_{\mathbb{Q}(\sqrt{5})/\mathbb{Q}} \left(\sum_{a=0}^{n} \sum_{b=0}^{7^a-1} j\left(\frac{7^{n-a}\tau + b}{7^a} \right) \right)$$

tends to zero 7-adically as $n \to +\infty$. For $n = 0, 1$ and 2 we find that

$$\begin{aligned} J_0 &= -190785 = -3 \cdot 5 \cdot 7 \cdot 23 \cdot 79, \\ J_1 &= -3 \cdot 5 \cdot 7^2 \cdot 3678217 \cdots, \\ J_2 &= -2^2 \cdot 3 \cdot 5 \cdot 7^3 \cdots. \end{aligned}$$

In the next section we prove Theorem 1, and we sketch the proof of Theorem 2.

2. Proof of Theorems 1 and 2

We begin this section with two essential facts. First of all, we construct holomorphic modular forms with prescribed divisor whose Fourier expansions are trivial modulo primes $p \leq 7$. For our purposes, it suffices to observe that

$$E_4(z) = 1 + 240 \sum_{n=1}^{\infty} \sigma_3(n) q^n, \tag{4}$$

$$E_6(z) = 1 - 504 \sum_{n=1}^{\infty} \sigma_5(n) q^n. \tag{5}$$

Proposition 2.1. *Suppose that $\tau \in H/\Gamma$ is a point for which $j(\tau) \in \mathbb{Q}$. If $p \leq 7$ is a prime for which*

$$j(\tau) \equiv \begin{cases} 0 \pmod{p} & \text{if } p \leq 5, \\ 6 \pmod{7} & \text{if } p = 7, \end{cases}$$

then there is a holomorphic integer weight modular form with p-integral coefficients, say $\mathcal{E}_{\tau,p}(z)$, for which $\mathcal{E}_{\tau,p}(\tau) = 0$ and

$$\mathcal{E}_{\tau,p}(z) \equiv 1 \pmod{p}.$$

Proof. Define $\mathcal{E}_{\tau,p}(z)$ by

$$\mathcal{E}_{\tau,p}(z) := \Delta(z)(j(z) - j(\tau)). \tag{6}$$

Here $\Delta(z)$ is the classical cusp form

$$\Delta(z) = q\prod_{n=1}^{\infty}(1-q^n)^{24} = \frac{E_4(z)^3}{j(z)} = \frac{E_6(z)^2}{j(z)-1728}. \tag{7}$$

It follows that $\mathcal{E}_{\tau,p}(z)$ is a holomorphic modular form of weight 12 for which $\mathcal{E}_{\tau,p}(\tau) = 0$.

If $p \leq 5$ is prime, then the desired congruence follows from (4), (6) and (7). If $p = 7$, then (5), (6), (7) and the fact that $1728 \equiv 6 \pmod 7$ implies the desired congruence. □

Now we recall some important facts about Ramanujan's Theta-operator in connection with the values of the j-function. Ramanujan's Theta-operator is the differential operator defined by

$$\Theta\left(\sum_{n=h}^{\infty} a(n)q^n\right) := \sum_{n=h}^{\infty} na(n)q^n. \tag{8}$$

It is a classical fact that if $f(z) = \sum_{n=h}^{\infty} a(n)q^n$ is a weight k meromorphic modular form on $SL_2(\mathbb{Z})$, then

$$\Theta(f) = (\tilde{f} + kfE_2)/12, \tag{9}$$

where $\tilde{f}$ is a meromorphic modular form of weight $k+2$ on $SL_2(\mathbb{Z})$. Here E_2 denotes the weight 2 quasi-modular form

$$E_2(z) = 1 - 24\sum_{n=1}^{\infty}\sigma_1(n)q^n.$$

Although Θ is simple to define, its arithmetic nature is deeper and is dictated by the $\tilde{f}$ appearing in (9). There is an explicit formula for $\Theta(f)$ in terms of a natural sequence of modular functions $j_m(z)$. Let $j_0(z) := 1$, and for every positive integer m let $j_m(z)$ be the unique modular function whose Fourier expansion is of the form

$$j_m(z) = q^{-m} + \sum_{n=1}^{\infty} c_m(n)q^n,$$

which is also holomorphic on the upper half of the complex plane. Notice that if m is a positive integer, then a convenient description of these functions is given by using the normalized weight 0 Hecke operators $T(m)$:

$$j_m(z) := (j(z) - 744 \mid T(m)) = \sum_{\substack{ad=m,\, d>0 \\ 0\le b\le d-1}} \left(j\left(\frac{az+b}{d}\right) - 744 \right). \quad (10)$$

The first few j_m are:

$$\begin{aligned} j_0(z) &= 1, \\ j_1(z) &= j(z) - 744 \\ &= q^{-1} + 196884q + \cdots, \\ j_2(z) &= j(z)^2 - 1488j(z) + 159768 \\ &= q^{-2} + 42987520q + \cdots, \\ j_3(z) &= j(z)^3 - 2232j(z)^2 + 1069956j(z) - 36866976 \\ &= q^{-3} + 2592899910q + \cdots. \end{aligned}$$

Each j_m is a monic degree m polynomial in j with integer coefficients. The conclusions of the next theorem may be found in [A-K-N, B-K-O]:

Theorem 2.2. *For every point $\tau \in H/\Gamma$, define $H_\tau(z)$ by*

$$H_\tau(z) := \sum_{n=0}^{\infty} j_n(\tau) q^n.$$

(1) *For every point $\tau \in H/\Gamma$, we have*

$$H_\tau(z) = \frac{E_4^2(z)E_6(z)}{\Delta(z)} \cdot \frac{1}{j(z) - j(\tau)}.$$

In particular, H_τ is a weight 2 meromorphic modular form.

(2) *If $f = \sum_{n=h}^{\infty} a_f(n)q^n$ is a nonzero weight k meromorphic modular form on $SL_2(\mathbb{Z})$ for which $a_f(h) = 1$, then*

$$\Theta(f) = \frac{kE_2 f}{12} - f f_\Theta,$$

where f_Θ is defined by

$$f_\Theta := \sum_{\tau \in H/\Gamma} e_\tau \operatorname{ord}_\tau(f) H_\tau(z).$$

Corollary 2.3. *If $\tau \in H/\Gamma$, then*

$$\frac{\Theta(j(z)-j(\tau))}{j(z)-j(\tau)} = -H_\tau(z) = -1 - \sum_{n=1}^{\infty} j_n(\tau) q^n.$$

Therefore, the series appearing in Corollary 2.3 is a meromorphic modular form of weight 2. The following theorem of Serre [Th. 7, Se] is vital for all of the results in this paper.

Theorem 2.4. *If $p \leq 7$ is prime and*

$$f = \sum_{n=0}^{\infty} a(n) q^n$$

is a p-adic modular form of weight $k \neq 0$, then

$$a(0) = \frac{\zeta_p^*(1-k)}{2} \cdot \lim_{n \to +\infty} a(p^n).$$

Proof of Theorem 1. If s is a non-negative integer, then consider the modular form

$$P_{\tau,s}(z) := \mathcal{E}_{\tau,p}(z)^{p^s} \cdot \frac{\Theta(j(z)-j(\tau))}{j(z)-j(\tau)}. \tag{11}$$

By Proposition 2.1 and Corollary 2.3, $P_{\tau,s}(z)$ is a holomorphic modular form for which

$$P_{\tau,s}(z) \equiv -H_\tau(z) \pmod{p^{s+1}}. \tag{12}$$

Furthermore, $P_{\tau,s}(z)$ has weight $2 + 12 \cdot p^s$. By letting $s \to +\infty$, we find that $H_\tau(z)$ is a p-adic modular form of weight 2. The conclusion of Theorem 1 now follows from (10) and Theorem 2.4. □

Sketch of the Proof of Theorem 2. Let $\tau_1, \tau_2, \ldots \tau_{H(-D)} \in H/\Gamma$ be the Heegner points of discriminant $-D$. Then let $F_D(z)$ be the modular function defined by

$$F_D(z) := \prod_{t=1}^{H(-D)} (j(z) - j(\tau_t)). \tag{13}$$

Gross and Zagier proved that if [Cor. 2.5, G-Z] τ is a Heegner point with discriminant d_τ, then

$$\begin{aligned}
|d_\tau| \equiv 3 \pmod 8 \quad &\Longrightarrow \quad j(\tau) \equiv 0 \pmod{2^{15}},\\
|d_\tau| \equiv 1 \pmod 3 \quad &\Longrightarrow \quad j(\tau) \equiv 1728 \pmod{3^6},\\
|d_\tau| \equiv 2,3 \pmod 5 \quad &\Longrightarrow \quad j(\tau) \equiv 0 \pmod{5^3},\\
|d_\tau| \equiv 1,2,4 \pmod 7 \quad &\Longrightarrow \quad j(\tau) \equiv 1728 \pmod{7^2}.
\end{aligned}$$

Arguing as in the proof of Theorem 1, for every integer s the holomorphic modular form $Q_{D,s}(z)$ defined by

$$Q_{D,s}(z) := \frac{\Theta(F_D(z))}{F_D(z)} \cdot \prod_{t=1}^{H(-D)} \mathcal{E}_{\tau_t,p}(z)^{p^s}$$

satisfies the congruence

$$Q_{D,s}(z) \equiv \frac{\Theta(F_D(z))}{F_D(z)} = -\sum_{t=1}^{H(-D)} H_{\tau_t}(z) \pmod{p^{s+1}}.$$

By letting $s \to +\infty$, we find that $\frac{\Theta(F_D(z))}{F_D(z)}$ is a weight 2 p-adic modular form. Since the numbers $j(\tau_t)$ are algebraic integers which form a complete set of Galois conjugates over $\mathbb{Q}$, the result follows from Theorem 2.4. □

Remark. It would be interesting to see a direct and elementary proof of the p-adic convergence of the limits appearing in Theorems 1 and 2.

References

[A-K-N] T. Asai, M. Kaneko and H. Ninomiya, *Zeros of certain modular functions and an application*, Comm. Math. Univ. St. Pauli **46** (1997) 93-101.

[B-K-O] J. Bruinier, W. Kohnen and K. Ono, *The arithmetic of the values of modular functions and the divisors of modular forms*, Compositio Mathematica, accepted for publication.

[G-Z] B. Gross and D. Zagier, *On singular moduli*, J. reine angew. math. **355** (1985) 191-220.

[K] M. Kaneko, *Traces of singular moduli and the Fourier coefficients of the elliptic modular function $j(\tau)$*, in "Number theory (Ottawa, ON, 1996)," 173-176, CRM Proc. and Lecture Notes, 19. Amer. Math. Soc., Providence, RI, 1999.

[Se] J.-P. Serre, *Formes modulaires et fonctions zêta p-adiques*, in "Modular functions of one variable. III (Willem Kuyk and J.-P. Serre, ed.)," 191-268, Lecture Notes in Mathematics, 350. Springer-Verlag, Berlin-New York, 1973.

[Si] J. H. Silverman, "Advanced Topics in the Arithmetic of Elliptic Curves," Springer-Verlag, New York, 1994.

Galois Theory and Modular Forms pp 367–374
K. Hashimoto, K. Miyake, and H. Nakamura, Editors
© 2003 Kluwer Academic Publishers

THOMPSON SERIES AND RAMANUJAN'S IDENTITIES

Masao KOIKE
Graduate School of Mathematics
Kyushu University,
Fukuoka, 812-8581 Japan
koike@math.kyushu-u.ac.jp

Introduction

B. J. Birch published some manuscripts of Ramanujan which contained, among other things, a list of identities involving Rogers-Ramanujan functions

$$G(x) = \prod_{n=0}^{\infty} \frac{1}{(1-x^{5n+1})(1-x^{5n+4})},$$

$$H(x) = \prod_{n=0}^{\infty} \frac{1}{(1-x^{5n+2})(1-x^{5n+3})},$$

and

$$q(m) = \prod_{n=1}^{\infty} (1-x^{mn})$$

which is related to Dedekind η function

$$\eta(\tau) = e^{\frac{2\pi i\tau}{24}} \prod_{n=1}^{\infty} (1-e^{2n\pi i\tau}).$$

A typical one, the identity (2) in [1], reads as follows:

$$G(x)G(x^4) + xH(x)H(x^4) = \left(\frac{q(2)^2}{q(1)q(4)}\right)^2.$$

[received: October 25, 2001; accepted in revised form: September 9, 2002]

These identities involve the functions $G(x)$ and $H(x)$ on one side and the products of $q(m)$ for various m on the other side. All of them except one are already proved. (For the history of proofs, see [1], [2]). In [1], Birch wrote that these identities of Ramanujan seemed too complicated to guess, even for one with Ramanujan's incredible instinct for formulae. In this paper, we shall propose an idea which gives us an easier way to find, but not to prove, some of such complicated identities, moreover new identities which Ramanujan missed. The idea is to connect these results with Thompson series. Thompson series T_g is the normalized generator of a genus zero function field parametrized by the Monster M, the largest sporadic finite simple group. In [3], Conway and Norton gave all products for T_g+constant expressible in terms of $\eta(k\tau)$ for various k. We observed that many products of $\eta(k\tau)$ for various k which are involved are common in the both identities. Therefore we try to find many identities between Thompson series and Rogers-Ramanujan functions directly by using computers.

1. Preliminaries

$G(x)$ and $H(x)$ can be expressed as theta functions by using the Jacobi triple product identity. Put

$$g(\tau) = x^{-\frac{1}{60}} G(x), \quad h(\tau) = x^{\frac{11}{60}} H(x), \qquad x = e^{2\pi i \tau}.$$

$g(\tau)$ and $h(\tau)$ are modular funtions on $\Gamma_1(5)$ with some character. We define

$$F^{+}_{r,s} = g(r\tau)g(s\tau) + h(r\tau)h(s\tau), \quad \text{for } r + s \equiv 0 \pmod 5$$

and

$$F^{-}_{r,s} = g(r\tau)h(s\tau) - h(r\tau)g(s\tau), \quad \text{for } r - s \equiv 0 \pmod 5).$$

Then the leading term of Fourier expansion of $F^{+}_{r,s}, F^{-}_{r,s}$ is $x^{\frac{r+s}{60}}$, $x^{\frac{11s-r}{60}}$ respectively. Put $N = rs$. N is called the level of $F_{r,s}$. Since r and s satisfies $r + s \equiv 0 \pmod 5$ or $r - s \equiv 0 \pmod 5$, the level N satisfy $N \equiv 1$ or $4 \pmod 5$. Moreover we suppose that N is an order of any element of the Monster. Then N is in the set

$$S = \left\{ \begin{matrix} 4, 6, 9, 11, 14, 16, 19, 21, 24, 26, 29, 31, 34, 36, 39, 41, \\ 44, 46, 51, 54, 56, 59, 66, 69, 71, 84, 94, 104, 119 \end{matrix} \right\}.$$

By observing the list of Ramanujan's identities in [1] and Table 3 and Table 4 in [3], we could make a guess that if N is in the set S, then there exists an element g of order N in the Monster such that Thompson series T_g+constant can be expressed as a product of some Rogers-Ramanujan functions of level N.

We recall several notation from Conway and Norton [3]. For every $g \in M$, Thompson series T_g:

$$T_g = x^{-1} + 0 + H_1(g)x + H_2(g)x^2 + \cdots ,$$

where H_n are representations of M and $H_n(g)$ are character values for every $g \in M$.

Let N be a positive integer. Let e be a Hall divisor of N, namely, $(e, N/e) = 1$. Then the Atkin-Lehner involution $W_e = \begin{pmatrix} ae & b \\ cN & de \end{pmatrix}$, where a, b, c and d are integers and $ade^2 - bcN = e$, is in the normalizer of $\Gamma_0(N)$. We denote by $N+e,\ f, \cdots$ the group obtained from $\Gamma_0(N)$ by adjoining Atkin-Lehner involutions W_e, $W_f, \cdots$. Especially we denote by $N+$ (resp. $N+N$) the group obtained from $\Gamma_0(N)$ by adjoining all its Atkin-Lehner involutions (resp. by adjoining the particular Atkin-Lehner involution W_N, which is also called the Fricke involution). If Thompson series T_g is the normalized generator of the modular function field with respect to the Fuchsian group $N+e,\ f, \cdots$, T_g is also denoted by $T_{N+e,f,\cdots}$

Then we notice that the above guess is not true for $N = 11$, 16, 84, 94 and 104. When $N = 11$, there exists only one Rogers-Ramanujan's function $F^-_{11,1}$ of level 11 which is equal to 1. When $N = 16$, there exists two Roger-Ramanujan's functions $F^-_{16,1}$, $F^+_{8,2}$ of level 16. Then it holds that $(F^-_{16,1})^2 = F^+_{8,2}$ and $(F^+_{8,2})^6 = T_{8B}$. When $N = 84$, 104 the Fuchsian group associated with T_{NA} is not of the above form $N+e,\ f, \cdots$. When $N = 94$, there exist two Rogers-Ramanujan functions $F^+_{94,1}$ and $F^-_{47,2}$. The leading term of Fourier expansion of $F^+_{94,1}$ (resp. $F^-_{47,2}$) is $x^{-\frac{19}{12}}$ (resp. $x^{-\frac{5}{12}}$) , so there is no simple linear combination $19n+5m = 12$,which is different from others.

So our conjecture is

Conjecture 1.1. *If N is in the set*

$$\tilde{S} = \left\{ \begin{matrix} 4,\ 6,\ 9,\ 14,\ 19,\ 21,\ 24,\ 29,\ 26,\ 31,\ 34,\ 36,\ 39, \\ 41,\ 44,\ 46,\ 51,\ 54,\ 56,\ 59,\ 66,\ 69,\ 71,\ 119 \end{matrix} \right\},$$

then there exists an element g of order N in the Monster M such that Thompson series T_g+constant can be expressed as a product of some Rogers-Ramanujan functions of level N.

2. Results

The more precise statement of Conjecture 1.1 is given by the following list of formulae. We checked that the first fifty powers of x in the following formulae are correct. The first fifty coefficients of all Thompson series are given in [4]. The symbol $a^\alpha b^\beta \cdots / c^\gamma d^\delta \cdots$ means the function $\eta(a\tau)^\alpha \eta(b\tau)^\beta \cdots / \eta(c\tau)^\gamma \eta(d\tau)^\delta \cdots$.

Level	Formulae
4	$(F^+_{4,1})^{12} = T_{4+} + 24 = \frac{2^{48}}{1^{24}4^{24}}$
6	$(F^+_{3,2})^{12} = T_{6+6} + 12 = \frac{2^{12}3^{12}}{1^{12}6^{12}}$
9	$(F^+_{9,1})^6 = T_{9+} + 6 = \frac{1^3}{9^3} + 27\frac{9^3}{1^3} + 9$
14	$(F^+_{14,1})^4 = T_{14+14} + 4 = \frac{2^4 7^4}{1^4 14^4}$
19	$(F^+_{19,1})^3 = T_{19+} + 3$
21	$(F^+_{7,3})^6 = T_{21+} = \frac{3^2 7^2}{1^2 21^2} + \frac{1^2 21^2}{3^2 7^2} - 2$
24	$(F^+_{24,1})^2 F^+_{6,4} = T_{24+24} + 2 = \frac{2.3^2 8^2 12}{1^2 4.6.24^2}$
26	$(F^+_{13,2})^4 = T_{26+} = \frac{2^2 13^2}{1^2 26^2} + \frac{1^2 26^2}{2^2 13^2} - 2$
29	$(F^+_{29,1})^2 = T_{29+} + 2$
31	$(F^-_{31,1})^3 = T_{31+}$
34	$(F^+_{34,1} F^-_{17,2})^2 = T_{34+} + 2$
36	$\frac{F^-_{36,1}}{F^-_{9,4}} = T_{36+36} + 1 = \frac{4.9}{1.36}$

Level	Formulae
39	$\frac{F^{+}_{39,1}}{F^{-}_{13,3}} = T_{39+39} + 1 = \frac{3.13}{1.39}$
41	$(F^{-}_{41,1})^2 = T_{41+}$
44	$F^{+}_{44,1}F^{+}_{11,4} = T_{44+} + 1 = \frac{2^4 22^4}{1^2 4^2 11^2 44^2} - 1$
46	$F^{-}_{46,1}F^{+}_{23,2} = T_{46+} = \frac{1.23}{2.46} + 2\frac{2.46}{1.23} + 1$
51	$F^{-}_{51,1}F^{+}_{17,3} = T_{51+}$
54	$F^{+}_{54,1}F^{-}_{27,2} = T_{54+} + 1$
56	$F^{-}_{56,1}F^{+}_{8,7} = T_{56+} = \frac{2.4.14.28}{1.7.8.56} - 1$
59	$F^{+}_{59,1} = T_{59+} + 1$
66	$\frac{F^{+}_{22,3}F^{+}_{33,2}}{F^{-}_{11,6}F^{-}_{66,1}} = T_{66+6,11,66} + 1 = \frac{2.3.22.33}{1.6.11.66}$
69	$F^{+}_{69,1}F^{-}_{23,3} = T_{69+} + 1$
71	$F^{-}_{71,1} = T_{71+}$
119	$F^{+}_{119,1}F^{-}_{17,7} = T_{119+} + 1$

In the above list, Formulae for $N = 4, 6, 14, 39, 44, 46, 66$ follow from Ramanujan's identities in [1]. We prove the case $N = 4$ for example. The identity (2) in [1] which is given in the Introduction is equivalent to

$$x^{\frac{1}{12}}(g(\tau)g(4\tau) + h(\tau)h(4\tau)) = x^{\frac{1}{12}}\left(\frac{\eta(2\tau)^2}{\eta(\tau)\eta(4\tau)}\right)^2.$$

Dividing both sides by $x^{\frac{1}{12}}$ and taking 12-th power of both sides, we get the proof.

There exist relations between Rogers-Ramanujan functions of the same level as shown in the next list. Formulae for $N = 6, 14, 21, 24$ are contained in the list of Ramanujan's identities [1]. But there exist more relations. Even in the case that the level N is not contained in the set $\tilde{S}$ in Conjecture 1.1, relations between Rogers-Ramanujan functions of the same level exist. For example, let $N = 94$. The leading term of Fourier expansion of the product $F^{+}_{94,1}F^{-}_{47,2}$ is x^{-2}, so this may be equal to the polynomial of T_{94+} of degree 2, namely, $T^2_{94+} + T_{94+} - 1$. We checked that the first fifty powers of x are correct.

Level	Formulae
6	$F^{+}_{3,2}F^{-}_{6,1} = 1$
14	$F^{+}_{14,1}F^{-}_{7,2} = 1$
16	$(F^{-}_{16,1})^2 = F_{8,2+}$
21	$F^{+}_{7,3} = F^{-}_{21,1}$
24	$F^{-}_{8,3}F^{+}_{24,1} = 1$
26	$F^{+}_{13,2} = F^{-}_{26,1}$
34	$\frac{(F^{+}_{34,1})^3}{(F^{-}_{17,2})^3} + \frac{(F^{-}_{17,2})^3}{(F^{+}_{34,1})^3} = (T_{34+})^2 + 3T_{34+}$
36	$F^{+}_{12,3}F^{+}_{18,2}F^{-}_{36,1} = T_{36+36}$
39	$(F^{+}_{39,1})^3 = (T_{39+39})^2 + 3T_{39+39} + 2$
44	$(F^{+}_{44,1})^4 + (F^{+}_{11,4})^4 = (T_{44+})^3 + 4(T_{44+})^2 + 8T_{44+} + 6$
51	$(F^{-}_{51,1})^3 - (F^{+}_{17,3})^3 = (T_{51+})^2 - T_{51+} + 1$

Level	Formulae
66	$F^+_{33,2}F^+_{22,3} = T_{66+6,11,66}$
84	$F^+_{84,1} = \frac{F^-_{42,2}}{F^-_{12,7}}$ $\frac{F^+_{84,1}}{F^+_{21,4}} = \frac{F^-_{28,3}}{F^-_{12,7}}$
94	$F^+_{94,1}F^-_{47,2} = (T_{94+})^2 + T_{94+} - 1$
96	$F^-_{96,1} = F^+_{48,2}F^+_{32,3}$
104	$F^+_{104,1}F^-_{13,8} = F^-_{52,2}$

References

[1] B. J. Birch, *A look back at Ramanujan's Notebooks*, Math. Proc. Camb. Phil. Soc. **78** (1975), 73–79.

[2] A. J. F. Biagioli, *A proof of some identities of Ramanujan using modular forms*, Glasgow Math. J. **31** (1989), 271–295.

[3] J. H. Conway and S. P. Norton, *Monstrous Moonshine*, Bull. London Math. Soc. **11** (1979), 308–309.

[4] J. Mckay and H. Strauss, *The q-series of monstrous moonshine and the decomposition of the head characters*, Comm. in Algebra **18** (1990), 253–278.

Galois Theory and Modular Forms pp 375–394
K. Hashimoto, K. Miyake, and H. Nakamura, Editors
© 2003 Kluwer Academic Publishers

GENERALIZED RADEMACHER FUNCTIONS AND SOME CONGRUENCE PROPERTIES

Hiroaki NAKAMURA
Department of Mathematics
Faculty of Science,
Okayama University, Okayama
700-8530, JAPAN
h-naka@math.okayama-u.ac.jp

Abstract In §1, we introduce the classical Dedekind sum and the Rademacher function. Then, in §2, certain generalized Rademacher functions are introduced as the Eichler-Shimura type period integrals of Eisenstein series. In §3, we present a version of continued fraction algorithm which computes efficiently the generalized Rademacher functions. In §4, we show a congruence formula connecting values of the generalized Rademacher functions of weight 2 and weight $k > 2$. This formula will be applied in a forthcoming paper [N4].

1. The classical Dedekind sums

The classical Dedekind eta function

$$\eta(\tau) = q^{1/24} \prod_{n=1}^{\infty} (1 - q^n) \qquad (q = e^{2\pi i \tau},\ \tau \in \mathfrak{H}) \tag{1.1}$$

on the upper half plane $\mathfrak{H}$ is one of the most beautiful objects in number theory. Since $\Delta(\tau) = (2\pi i)^{12}\eta^{24}(\tau)$ is a cusp form of weight 12 for $\mathrm{SL}_2(\mathbb{Z})$, it is clear that, for any $A = \left(\begin{smallmatrix} a & b \\ c & d \end{smallmatrix}\right) \in \mathrm{SL}_2(\mathbb{Z})$, the ratio $\eta(A\tau)/\sqrt{\frac{c\tau+d}{i}}\eta(\tau)$ is a 24-th root of unity. The behavior of this little ratio with respect to $A \in \mathrm{SL}_2(\mathbb{Z})$ looks delicate and mysteri-

[received: June 18, 2002; accepted in revised form: October 18, 2002]

ous. R. Dedekind, in his note [D] included in *Gesammelte Mathematische Werke* of B. Riemann, studied the behavior with Riemann's method. Outstanding is that he found an explicit *integer valued* function $\varphi : \mathrm{SL}_2(\mathbb{Z}) \to \mathbb{Z}$ controlling this ratio in the form

$$\eta(A\tau) = \begin{cases} e^{\frac{2\pi i}{24}\varphi(A)}\eta(\tau), & (c=0); \\ e^{\frac{2\pi i}{24}\varphi(A)}\sqrt{\frac{c\tau+d}{i}}\eta(\tau), & (c>0). \end{cases} \tag{1.2}$$

Dedekind's formula for the function φ (which, by definition, factors through $\mathrm{PSL}_2(\mathbb{Z}) \to \mathbb{Z}$) reads as follows:

$$\varphi(A) = \begin{cases} \frac{b}{d}, & (c=0), \\ \frac{a+d}{c} - 12s(a,c), & (c>0). \end{cases} \tag{1.3}$$

Here $s(a,c)$, called now the Dedekind sum, is defined by

$$s(a,c) = \sum_{i=0}^{c-1} P_1(\frac{i}{c})P_1(\frac{ai}{c})$$

with $P_1(*)$ the "sawtooth" function:

$$P_1(x) = \begin{cases} 0 & (x \in \mathbb{Z}); \\ x - [x] - \frac{1}{2} & (x \notin \mathbb{Z}). \end{cases}$$

($[x]$: the greatest integer not exceeding $x \in \mathbb{R}$.)

In [R], H. Rademacher intensively studied algebraic properties of the function φ. He derived, for example, the composition formula

$$\varphi(AB) = \varphi(A) + \varphi(B) - 3\mathrm{sgn}(c_A c_B c_{AB}) \qquad (A, B \in \mathrm{SL}_2(\mathbb{Z})), \tag{1.4}$$

where, for any matrix $S \in \mathrm{SL}_2(\mathbb{Z})$, c_S denotes the lower left entry of S. Following [KM], we shall call φ the Rademacher (φ-)function.

One of the key ingredients that enabled Dedekind to find the above result was to consider the modular transformation of not only $\eta(\tau)$ but also of $\log \eta(\tau)$. In fact, Dedekind's key formula (for the case $c > 0$) reads:

$$\log \eta(A\tau) - \log \eta(\tau) = -\frac{1}{2}\log \frac{i}{c\tau+d} + \pi i\Big(\frac{a+d}{12c} - s(a,c)\Big). \tag{1.5}$$

This left hand side may also be written in the form:

$$\log \eta(A\tau) - \log \eta(\tau) = -\pi i \int_{\tau}^{A\tau} E_2(z)dz, \tag{1.6}$$

where

$$E_2(\tau) = \frac{1}{(2\pi i)^2} \sum_{m,n}{}' \frac{1}{(m\tau + n)^2}$$

is the Eisenstein series of weight 2 (with specified conditional convergence). In other words, the Rademacher function φ can be viewed almost as the periods of the Eisenstein series. Unfortunately, $E_2(\tau)$ is only a quasi-modular form and not a usual modular form. This causes the need of the adjustment term $\frac{1}{2}\log\frac{i}{c\tau+d}$ of (1.5) which connects $\varphi(A)$ and the period. To gain a more satisfactory viewpoint, we would need a more subtle treatment by using a non-holomorphic modification of E_2. However, we do not enter this subtlety any more here, because this sort of adjustment is rather exceptional phenomenon occurring in this original case among generalizations of E_2 discussed in the next section.

2. Generalized Rademacher functions

We wish to consider a generalization of the Rademacher function φ introduced in §1 by considering φ as the case of weight 2 and level 1. With regard to the appearance of E_2 in (1.6), we shall begin by introducing the generalized Eisenstein series $E_k^{(x_1,x_2)}$'s ($k \geq 2$, $(x_1, x_2) \in (\frac{1}{N}\mathbb{Z}/\mathbb{Z})^2$) given by

$$\begin{aligned} &E_k^{(x_1,x_2)}(\tau) \\ &:= \frac{(k-1)!}{(2\pi i)^k} \sum_{\mathbf{a}\in(\mathbb{Z}/N\mathbb{Z})^2} e^{2\pi i(x_1a_2 - x_2a_1)} \sum_{\mathbf{m}\equiv\mathbf{a}\,(N)}{}' \frac{1}{(m_1\tau + m_2)^k} \\ &= -\frac{P_k(x_1)}{k} + \sum_{\substack{s\in x_1+\mathbb{Z}\\ s>0}} \sum_{l=1}^{\infty} s^{k-1} e^{2\pi i l(x_2+s\tau)} + \sum_{\substack{s\in -x_1+\mathbb{Z}\\ s>0}} \sum_{l=1}^{\infty} s^{k-1} e^{2\pi i l(-x_2+s\tau)}, \end{aligned} \tag{2.1}$$

where $P_k(x) : \mathbb{R} \to \mathbb{R}$ is the k-th periodic Bernoulli function (whose definition will be recalled in (3.4)). It is well known that $E_k^{(\mathbf{x})}$ is a holomorphic modular form of level N and weight k, except for the original case $E_2 = E_2^{(\mathbf{0})}$ (cf. [St]).

Let $\Gamma(N) \subset \mathrm{SL}_2(\mathbb{Z})$ be the principal congruence subgroup of level $N \geq 1$. For any pair of $k \geq 2$ and $\mathbf{x} = (x_1, x_2) \in (\frac{1}{N}\mathbb{Z}/\mathbb{Z})^2$ with

$(k,\mathbf{x}) \neq (2,\mathbf{0})$, we then consider the Eichler integral of $E_k^{(\mathbf{x})}$:

$$F_k^{(\mathbf{x})}(\tau) = -\frac{1}{(k-2)!}\int_\tau^{i\infty}\left(E_k^{(\mathbf{x})}(u)+\frac{P_k(x_1)}{k}\right)(\tau-u)^{k-2}du \\ -\frac{P_k(x_1)}{k}\frac{\tau^{k-1}}{(k-1)!}. \tag{2.2}$$

This (indefinite) integral satisfies $(\frac{d}{d\tau})^{k-1}F_k^{(\mathbf{x})} = E_k^{(\mathbf{x})}$. Writing $j(A,\tau) := (c\tau+d)$ for $A = \left(\begin{smallmatrix} a & b \\ c & d \end{smallmatrix}\right) \in \Gamma(N)$, we then see that the difference

$$\phi_k^{(\mathbf{x})}(A)(\tau) := j(A,\tau)^{k-2}F_k^{(\mathbf{x})}(A\tau) - F_k^{(\mathbf{x})}(\tau) \tag{2.3}$$

is a polynomial in τ of degree $k-2$. What we wish to look closely at here is its "real part" $Re\phi_k^{(\mathbf{x})}(A)(\tau)$ which is, by definition, the polynomial in the "complex variable" τ whose coefficients consist of the real parts of the corresponding coefficients of $\phi_k^{(\mathbf{x})}(A)(\tau)$. It is known that $Re\phi_k^{(\mathbf{x})}(A)(\tau)$ has only rational coefficients.

The assignment $A \mapsto Re\phi_k^{(\mathbf{x})}(A)(\tau)$ enjoys the 1-cocycle property of the form:

$$Re\phi_k^{(\mathbf{x})}(AB)(\tau) = j(B,\tau)^{k-2}Re\phi_k^{(\mathbf{x})}(A)(B\tau) + Re\phi_k^{(\mathbf{x})}(B)(\tau) \tag{2.4}$$

on $A, B \in \Gamma(N)$.

To switch $Re\phi_k^{(\mathbf{x})}(A)(\tau)$ into the form of 1-cocycle for a left module, let us introduce the space $\mathrm{Sym}^{k-2}(\mathbb{Q}^2) = \mathbb{Q}[X,Y]_{deg=k-2}$ of the homogeneous polynomials in X, Y of degree $k-2$ on which $\mathrm{SL}_2(\mathbb{Z})$ acts on the left (written ρ) by

$$\rho\begin{pmatrix} a & b \\ c & d \end{pmatrix} \cdot \phi(X,Y) := \phi(aX+cY, bX+dY).$$

Definition (2.5). Let $(k,\mathbf{x}) \in \mathbb{Z}_{\geq 2} \times (\frac{1}{N}\mathbb{Z}/\mathbb{Z})^2$ with $(k,\mathbf{x}) \neq (2,\mathbf{0})$. We define the generalized Rademacher function

$$\Phi_{\mathbf{x}}^{(k)} : \Gamma(N) \longrightarrow \mathrm{Sym}^{k-2}(\mathbb{Q}^2) = \mathbb{Q}[X,Y]_{\mathrm{deg}=k-2}$$
$$\left(A = \begin{pmatrix} a & b \\ c & d \end{pmatrix} \longmapsto \sum_{r=0}^{k-2} \Phi_{\mathbf{x}}^{(r+1,k-1-r)}(A)X^rY^{k-2-r}\right)$$

by

$$\Phi_{\mathbf{x}}^{(k)}(A) := (-1)^{k-1}(k-2)!\,X^{k-2} \cdot Re\phi_k^{(\mathbf{x})}(A^{-1})\left(-\frac{Y}{X}\right).$$

As expected from the above definition, (2.4) can be converted into the following 1-cocycle property of $\Phi_{\mathbf{x}}^{(k)}$:

$$\Phi_{\mathbf{x}}^{(k)}(AB) = \Phi_{\mathbf{x}}^{(k)}(A) + \rho(A).\Phi_{\mathbf{x}}^{(k)}(B) \tag{2.6}$$

on $A, B \in \Gamma(N)$. But one can get a nicer result. In fact, G. Stevens [St] showed how to interpret $\Phi_{\mathbf{x}}^{(k)}$ as the periods of a certain real differential form extended to the Borel-Serre compactification $\bar{\mathfrak{H}}$ of the upper half plane $\mathfrak{H}$. With his method, we can extend $\Phi_{\mathbf{x}}^{(k)}$ canonically to a function from $\mathrm{SL}_2(\mathbb{Z})$ (or even from $\mathrm{GL}_2(\mathbb{Q})^+$) to $\mathrm{Sym}^{k-2}(\mathbb{Q}^2)$. Moreover the case of $(k, \mathbf{x}) = (2, \mathbf{0})$ may naturally be included in this unified construction of $\Phi_{\mathbf{x}}^{(k)}$ with regarding $\Phi_{\mathbf{0}}^{(2)} = -\varphi(A)/12$. Then, the above properties (1.4) and (2.6) can be generalized to

$$\Phi_{\mathbf{x}}^{(k)}(AB) = \Phi_{\mathbf{x}}^{(k)}(A) + \rho(A).\Phi_{\mathbf{x}A}^{(k)}(B) + \frac{1}{4}\delta_{\mathbf{x}}^{k=2}\mathrm{sgn}(c_A c_B c_{AB}). \tag{2.7}$$

Here, $\mathrm{sgn}(*) \in \{\pm 1, 0\}$ denotes the signature of $*$, and the Kronecker symbol $\delta_{\mathbf{x}}^{k=2}$ is defined by $\delta_{\mathbf{x}}^{k=2} = 1$ when $(k, \mathbf{x}) = (2, \mathbf{0})$ and $\delta_{\mathbf{x}}^{k=2} = 0$ otherwise.

It will also be nice to have generalizations of the beautiful formula (1.3) to the cases of our generalized Rademacher functions $\Phi_{\mathbf{x}}^{(k)}$. Although the author could not find a literature describing the explicit form, by careful calculations one obtains the following formula of $\Phi_{\mathbf{x}}^{(k)}$: $\mathrm{PSL}_2(\mathbb{Z}) \to \mathrm{Sym}^{k-2}(\mathbb{Q}^2) = \mathbb{Q}[X, Y]_{\deg = k-2}$;

$$\Phi_{(x_1,x_2)}^{(k)}\left(\begin{pmatrix} a & b \\ c & d \end{pmatrix}\right) =$$

$$\begin{cases} -\frac{P_k(x_1)}{k}\int_0^{\frac{b}{d}}(tX+Y)^{k-2}dt, & (c=0); \\ -\frac{P_k(x_1)}{k}\int_0^{\frac{a}{c}}(tX+Y)^{k-2}dt \\ \quad -\frac{P_k(ax_1+cx_2)}{k}\int_{-\frac{d}{c}}^{0}(t(aX+cY)+bX+dY)^{k-2}dt \\ \quad +\sum_{r=0}^{k-2}(-1)^r\binom{k-2}{r}X^r(aX+cY)^{k-2-r}s_{(x_1,x_2)}^{(k-1-r,r+1)}(a,c), & (c>0). \end{cases} \tag{2.8}$$

where the last factor (called the generalized Dedekind sum) is defined by

$$s_{(x_1,x_2)}^{(k-1-r,r+1)}(a,c) = \sum_{i=0}^{c-1}\frac{P_{k-1-r}(\frac{x_1+i}{c})}{k-1-r}\frac{P_{r+1}(x_2+a\frac{x_1+i}{c})}{r+1}. \tag{2.9}$$

The following distribution relation is an important property which connects values of $\Phi_{\mathbf{x}}^{(k)}(A)$ of different $\mathbf{x}$'s:

$$\Phi_{\mathbf{x}}^{(k)}(A) = n^{k-2} \sum_{\mathbf{y} \in \frac{1}{n}\mathbf{x}} \Phi_{\mathbf{y}}^{(k)}(A) \qquad (n \geq 1). \tag{2.10}$$

3. Continued fraction algorithm for $\Phi_{\mathbf{x}}^{(k)}$

The calculation of $\Phi_{\mathbf{x}}^{(k)}(\left(\begin{smallmatrix} a & b \\ c & d \end{smallmatrix}\right))$ by means of (2.8) involve those of generalized Dedekind sums $s_{\mathbf{x}}^{(k-1-r,r+1)}(a,c)$ of (2.9) which need lots of times when c is big. Variants of continued fraction algorithms have been known to improve this sort of inefficiency on computers. We shall exploit a version of it well suited to calculations of our generalized Rademacher functions. The implementation and various numerical tests have been given by Y. Morimoto [Mo] by using PARI-GP and Maple softwares.

Given a matrix $A = \left(\begin{smallmatrix} a & b \\ c & d \end{smallmatrix}\right) \in \mathrm{SL}_2(\mathbb{Z})$, choose an expression of a/c by a continued fraction of the form:

$$\begin{aligned} \frac{a}{c} &= k_0 - \frac{1}{k_1 -} \frac{1}{k_2 -} \cdots \frac{1}{-k_n} \\ &= k_0 - \cfrac{1}{k_1 - \cfrac{1}{k_2 - \cfrac{1}{\ddots \, \cfrac{1}{k_{n-1} - \frac{1}{k_n}}}}} \end{aligned} \tag{3.1}$$

with an integer sequence $(k_0, k_1, \ldots, k_n)$. This expression may be obtained by first putting $r_0 = a$, $r_1 = c > 0$ and then defining integers $\{k_i\}_i$ with the Euclidean method:

$$\begin{cases} r_0 = k_0 r_1 - r_2, & |r_1| > |r_2| > 0, \\ r_1 = k_1 r_2 - r_3, & |r_2| > |r_3| > 0, \\ \cdots & \cdots \\ r_{n-1} = k_{n-1} r_n - r_{n+1}, & |r_n| > |r_{n+1}| = 1, \\ r_n = k_n r_{n+1} - 0, & |r_{n+1}| = 1, \; r_{n+2} = 0, \end{cases} \tag{3.2}$$

where each k_i $(i = 0, \ldots, n)$ is chosen to be either $[r_i/r_{i+1}]$ (the greatest integer not exceeding r_i/r_{i+1}) or $\lceil r_i/r_{i+1} \rceil$ (the least integer greater than or equal to r_i/r_{i+1}). Let us write $\varepsilon = r_{n+1} \in \{\pm 1\}$.

Define also a sequence of matrices $\{A_i\}_i$ in $\mathrm{SL}_2(\mathbb{Z})$ by

$$A_{-1} = \begin{pmatrix} 1 & 0 \\ 0 & 1 \end{pmatrix}, \ A_0 = \begin{pmatrix} k_0 & -1 \\ 1 & 0 \end{pmatrix}, \ A_1 = \begin{pmatrix} k_0 & -1 \\ 1 & 0 \end{pmatrix} \begin{pmatrix} k_1 & -1 \\ 1 & 0 \end{pmatrix},$$

$$\dots$$

$$A_n = \begin{pmatrix} k_0 & -1 \\ 1 & 0 \end{pmatrix} \begin{pmatrix} k_1 & -1 \\ 1 & 0 \end{pmatrix} \cdots \begin{pmatrix} k_n & -1 \\ 1 & 0 \end{pmatrix}.$$

and put

$$A_i = \begin{pmatrix} -p_i & p_{i-1} \\ -q_i & q_{i-1} \end{pmatrix} \qquad (-1 \le i \le n).$$

Then,

Lemma (3.3). *Notations being as above, we have*

(i) $a = -\varepsilon p_n, \ c = -\varepsilon q_n$;

(ii) $\delta := (d - \varepsilon q_{n-1})/c \in \mathbb{Z}$;

(iii) $\begin{pmatrix} a & b \\ c & d \end{pmatrix} = \varepsilon \begin{pmatrix} k_0 & -1 \\ 1 & 0 \end{pmatrix} \cdots \begin{pmatrix} k_n & -1 \\ 1 & 0 \end{pmatrix} \begin{pmatrix} 1 & \delta \\ 0 & 1 \end{pmatrix}.$

Proof. The Euclidean sequence (3.2) may be rephrased as

$$\begin{pmatrix} a \\ c \end{pmatrix} = \begin{pmatrix} k_0 & -1 \\ 1 & 0 \end{pmatrix} \begin{pmatrix} k_1 & -1 \\ 1 & 0 \end{pmatrix} \cdots \begin{pmatrix} k_n & -1 \\ 1 & 0 \end{pmatrix} \begin{pmatrix} r_{n+1} \\ 0 \end{pmatrix}.$$

The claim (i) follows immediately from this. Both A and εA_n are in $\mathrm{SL}_2(\mathbb{Z})$ and have the same left column. Therefore, there exists $\delta \in \mathbb{Z}$ with $A = \varepsilon A_n \left(\begin{smallmatrix} 1 & \delta \\ 0 & 1 \end{smallmatrix}\right)$, i.e.,

$$\begin{pmatrix} a & b \\ c & d \end{pmatrix} = \varepsilon \begin{pmatrix} -p_n & p_{n-1} \\ -q_n & q_{n-1} \end{pmatrix} \begin{pmatrix} 1 & \delta \\ 0 & 1 \end{pmatrix}.$$

The claims (ii) and (iii) follow from this simultaneously. □

Definition (3.4). The Bernoulli polynomials $B_k(X)$ $(k = 0, 1, 2, \dots)$ are defined by the generating function

$$\frac{te^{tX}}{e^t - 1} = \sum_{k=0}^{\infty} B_k(X) \frac{t^k}{k!},$$

and the Bernoulli number B_k is by definition $B_k(1)$. We define the periodic Bernoulli function $P_k : \mathbb{R} \to \mathbb{R}$ by $P_k(x) = B_k(x - [x])$ except for the case $k=1$ and $x \in \mathbb{Z}$ where we set $P_1(n)=0$ ($n\in\mathbb{Z}$). For $\mathbf{x}=(x_1,x_2)\in\mathbb{Q}^2$ and $(a,b)\in\mathbb{Z}^2_{\geq 0}$, define

$$\boldsymbol{\beta}_{a,b}(\mathbf{x}) := \frac{P_a(x_1)}{\max\{a,1\}}\cdot\frac{P_b(x_2)}{\max\{b,1\}}.$$

Proposition (3.5) (Continued fraction algorithm). *Notations being as above, put* $k_{n+1}:=\delta$. *Then, for* $\mathbf{x}=(x_1,x_2)\in\mathbb{Q}^2$ *and* $A=\left(\begin{smallmatrix} a & b\\ c & d\end{smallmatrix}\right)\in \mathrm{SL}_2(\mathbb{Z})$, *we have*

$$\begin{aligned}\Phi^{(k)}_{\mathbf{x}}(A) = &- \sum_{i=0}^{n}\sum_{j=0}^{k-2}\boldsymbol{\beta}_{j+1,k-1-j}(\mathbf{x}A_i)\binom{k-2}{j}\\ &\quad\times (p_{i-1}X+q_{i-1}Y)^j(p_iX+q_iY)^{k-2-j}\\ &-\sum_{i=0}^{n+1}\boldsymbol{\beta}_{k,0}(\mathbf{x}A_{i-1})\\ &\quad\times\int_0^{k_i}\big((-p_{i-1}t+p_{i-2})X+(-q_{i-1}t+q_{i-2})Y\big)^{k-2}dt\\ &+\frac{\delta_{\mathbf{x}}^{k=2}}{4}(\mathrm{sgn}(q_{-1}q_0)+\mathrm{sgn}(q_0q_1)+\cdots+\mathrm{sgn}(q_{n-1}q_n)).\end{aligned}$$

Proof. Applying (2.7) iteratedly to (3.3)(iii), we obtain

$$\begin{aligned}\Phi^{(k)}_{\mathbf{x}} =& \sum_{i=0}^{n}\rho(A_{i-1}).\Phi^{(k)}_{\mathbf{x}A_{i-1}}(\begin{pmatrix} k_i & -1\\ 1 & 0\end{pmatrix})\\ &+\rho(A_n).\Phi^{(k)}_{\mathbf{x}A_n}(\begin{pmatrix} 1 & \delta\\ 0 & 1\end{pmatrix})+\frac{\delta_{\mathbf{x}}^{k=2}}{4}\sum_{i=0}^{n}\mathrm{sgn}(q_{i-1}q_i).\end{aligned}$$

Then, the formula (2.8) gives

$$\begin{aligned}\Phi^{(k)}_{\mathbf{x}A_n}(\begin{pmatrix} 1 & \delta\\ 0 & 1\end{pmatrix}) &= -\boldsymbol{\beta}_{k,0}(\mathbf{x}A_n)\int_0^{k_{n+1}}(tX+Y)^{k-2}dt,\\ \Phi^{(k)}_{\mathbf{x}A_{i-1}}(\begin{pmatrix} k_i & -1\\ 1 & 0\end{pmatrix}) &= -\boldsymbol{\beta}_{k,0}(\mathbf{x}A_{i-1})\int_0^{k_i}(tX+Y)^{k-2}dt\\ &\quad+\sum_{j=0}^{k-2}(-1)^j\binom{k-2}{j}\\ &\quad\times X^j(k_iX+Y)^{k-2-j}s^{(k-1-j,j+1)}_{\mathbf{x}A_{i-1}}(k_i,1).\end{aligned}$$

Writing here $(y_i, z_i) := \mathbf{x}A_{i-1}$ $(i = 0, \dots, n)$ so that

$$(y_i, z_i)\begin{pmatrix} k_i & -1 \\ 1 & 0 \end{pmatrix} = (y_{i+1}, z_{i+1}) \ (0 \le i < n),$$

we see that

$$\begin{aligned} s_{\mathbf{x}A_{i-1}}^{(k-1-j,j+1)}(k_i, 1) &= \frac{P_{k-1-j}(y_i)}{k-1-j}\frac{P_{j+1}(z_i + k_i y_i)}{j+1} \\ &= (-1)^{k-1-j}\boldsymbol{\beta}_{j+1,k-1-j}(\mathbf{x}A_i). \end{aligned}$$

Therefore,

$$\begin{aligned} \Phi_{\mathbf{x}}^{(k)} = &-\sum_{i=0}^{n+1} \rho(A_{i-1}).\boldsymbol{\beta}_{k,0}(\mathbf{x}A_{i-1}) \int_0^{k_i} (tX+Y)^{k-2} dt \\ &+ (-1)^{k-1} \sum_{i=0}^{n}\sum_{j=0}^{k-2} \rho(A_{i-1}).\binom{k-2}{j} \\ &\times X^j (k_i X + Y)^{k-2-j} \boldsymbol{\beta}_{j+1,k-1-j}(\mathbf{x}A_i) \\ &+ \frac{\delta_{\mathbf{x}}^{k=2}}{4} \sum_{i=0}^{n} \operatorname{sgn}(q_{i-1}q_i). \end{aligned}$$

Noticing that $\rho(A_{i-1})$ maps

$$\begin{cases} X & \longmapsto & -p_{i-1}X - q_{i-1}Y, \\ Y & \longmapsto & p_{i-2}X + q_{i-2}Y, \\ k_i X + Y & \longmapsto & -p_i X - q_i Y, \end{cases}$$

we conclude the proof of Proposition (3.5). □

4. Some congruence properties

In this section, we shall present a congruence formula (4.3) which connects N^2 values of $\Phi_{\mathbf{x}}^{(2)}(A)$ $(\mathbf{x} \in (\frac{1}{N}\mathbb{Z}/\mathbb{Z})^2)$ and the coefficients of $\Phi_{\mathbf{0}}^{(k)}(A)$. The congruence formula will be applied to the study of a certain measure function on the congruence kernel of SL_2 which governs the meta-abelian monodromy representation in the fundamental group of the universal elliptic curve minus the origin section ([N1-4]). We refer to the forthcoming paper [N4] as a main reference for details of such an application.

Let $\mathbb{Z}'_N \subset \mathbb{Q}$ denote the ring of fractional numbers with denominators prime to N. For $n \ge 1$, write $B_n^*(X) := B_n(X) - B_n(0)$, and define

d_n (resp. d_n^*) to be the least common multiple of the denominators of coefficients of the polynomial $\frac{1}{n}B_n(X)$ (resp. $\frac{1}{n}B_n^*(X)$). In the proof of Proposition (4.3) below, we shall frequently make use of the following well known properties:

$$\sum_{x=0}^{N-1} x^n = \frac{1}{n+1}B_{n+1}^*(N) \equiv 0 \pmod{\frac{N}{d_{n+1}^*}\mathbb{Z}'_N}, \tag{4.1}$$

$$\frac{N^{n-1}}{n}\left(P_n(\frac{x}{N}+u) - P_n(\frac{x}{N})\right) \equiv x^{n-1}\left(P_1(\frac{x}{N}+u) - P_1(\frac{x}{N})\right) \pmod{\frac{N}{d_n}\mathbb{Z}'_N} \tag{4.2}$$

for $x \in \mathbb{Z}$, $N \in \mathbb{N}$, $u \in \mathbb{Z}'_N$.

Proof. The formula (4.1) is quite popular, and its proof may be left to readers. To prove (4.2), let $\{t\} := t - [t]$ denote the fractional part of $t \in \mathbb{R}$, and put $v := \{\frac{x}{N}+u\} - \{\frac{x}{N}\}$ so that $\{u\} = \{v\}$. Then, $v \in \mathbb{Z}'_N$ and we compute

$$\begin{aligned}
&\frac{N^{n-1}}{n}(P_n(\frac{x}{N}+u) - P_n(\frac{x}{N})) \\
&= \frac{N^{n-1}}{n}(B_n(\{\frac{x}{N}\}+v) - B_n(\{\frac{x}{N}\})) \\
&\equiv \frac{N^{n-1}}{n}\left[\left((\{\frac{x}{N}\}+v)^n - \frac{n}{2}(\{\frac{x}{N}\}+v)^{n-1}\right) - \left(\{\frac{x}{N}\}^n - \frac{n}{2}\{\frac{x}{N}\}^{n-1}\right)\right] \\
&= \frac{N^{n-1}}{n}\left[n\{\frac{x}{N}\}^{n-1}v + P_{n-2}(\{\frac{x}{N}\}) + \frac{n}{2}Q_{n-2}(\{\frac{x}{N}\})\right].
\end{aligned}$$

Here, the congruence is taken modulo $\frac{N}{d_n}\mathbb{Z}'_N$, and P_{n-2}, Q_{n-2} are polynomials of degree $n-2$ with coefficients in $\mathbb{Z}$. Since $\frac{1}{n}B_n(X)$ is a polynomial of the form $\frac{1}{n}x^n - \frac{1}{2}x^{n-1} + \ldots$, it follows that d_n is a multiple of $LCM\{2, n\}$. Therefore, the second and third terms of the above last side vanish modulo $\frac{N}{d_n}\mathbb{Z}'_N$. Moreover the integer $x' := \{x/N\}N$ is congruent to $x \pmod N$. Thus, the above last side continues to

$$\equiv (x')^{n-1}v \equiv x^{n-1}v = x^{n-1}\left(P_1(\frac{x}{N}+u) - P_1(\frac{x}{N})\right).$$

□

Proposition (4.3). *Let N, r, k be integers satisfying $N \geq 1$, $k \geq 2$ and $0 \leq r \leq k-2$, and, for $s = 1, \ldots, k-2$, let e_s be the denominator of $\frac{B_k}{k(k-s)}$. Define $D_{k,r}$ to be the least common multiple of the set*

$$\left\{\begin{matrix} d_{j+1}d_{k-1-j}\ (r \leq j \leq k-2),\ e_1, \ldots, e_{k-2} \\ d_{r+1}^*d_{k-r}^*,\ d_{r+2}^*d_{k-1-r}^*,\ d_{r+1}^*d_{k-r+1}^*,\ d_{r+2}^*d_{k-r}^*,\ d_{r+3}^*d_{k-1-r}^* \end{matrix}\right\}.$$

Then, for $A \in \Gamma(N)$, we have the following congruence:

$$12 \sum_{x=0}^{N-1} \sum_{y=0}^{N-1} \binom{k-2}{r} x^{k-2-r}(-y)^r \Phi^{(2)}_{(\frac{x}{N},\frac{y}{N})}(A)$$
$$\equiv 12\Phi_{\mathbf{0}}^{(r+1,k-1-r)}(A) \ (\mathrm{mod} \ \binom{k-2}{r} \frac{GCD\{6,N\}N}{D_{k,r}} \mathbb{Z}'_N).$$

This congruence formula was first proved by the author in the special case of $r = 0$. The general case of the formula was then conjectured by Y. Morimoto after numerical computations and evidences [Mo].

(4.4) Proof of Proposition 4.3. Let $A = \left(\begin{smallmatrix} a & b \\ c & d \end{smallmatrix}\right) \in \Gamma(N)$. Since $\left(\begin{smallmatrix} 1 & gN^3 \\ 0 & 1 \end{smallmatrix}\right)\left(\begin{smallmatrix} a & b \\ c & d \end{smallmatrix}\right) = \left(\begin{smallmatrix} a+cgN^3 & b+dgN^3 \\ c & d \end{smallmatrix}\right)$, it is sufficient to prove the formula when $a \gg c > 0$. In what follows, we shall assume this. Applying (2.7) to the equation $\left(\begin{smallmatrix} 0 & -1 \\ 1 & 0 \end{smallmatrix}\right)\left(\begin{smallmatrix} c & d \\ -a & -b \end{smallmatrix}\right) = \left(\begin{smallmatrix} a & b \\ c & d \end{smallmatrix}\right)$, we find

$$\begin{aligned}
\Phi^{(k)}_{(x_1,x_2)}(A) = &- \frac{P_k(x_2)}{k} \int_0^{-\frac{c}{a}} (tY - X)^{k-2} dt \\
&- \frac{P_k(ax_1 + cx_2)}{k} \int_{-\frac{b}{a}}^{0} ((aX + cY)t + bX + dY)^{k-2} dt \\
&+ \sum_{r=0}^{k-2} X^r Y^{k-2-r} \binom{k-2}{r} (-1)^r \frac{P_{k-1-r}(x_1)}{k-1-r} \frac{P_{r+1}(x_2)}{r+1} \\
&- \sum_{r=0}^{k-2} (aX + cY)^r Y^{k-2-r} \binom{k-2}{r} \\
&\quad \times (-1)^r \sum_{i=0}^{a-1} \frac{P_{k-1-r}(x_1 + c\frac{x_2+i}{a})}{k-1-r} \frac{P_{r+1}(\frac{x_2+i}{a})}{r+1}. \qquad (4.4.0)
\end{aligned}$$

Let us first evaluate the left hand side of the congruence formula (4.3). When $k = 2$, $r = 0$, the above (4.4.0) gives

$$\begin{aligned}
\Phi^{(2)}_{\mathbf{x}}(A) = &\frac{c}{2a} P_2(x_2) - \frac{b}{2a} P_2(ax_1 + cx_2) \\
&+ \left\{ P_1(x_1)P_1(x_2) - \sum_{i=0}^{a-1} P_1(x_1 + c\frac{x_2+i}{a}) P_1(\frac{x_2+i}{a}) \right\}.
\end{aligned}$$

Since $a \equiv 1$, $b \equiv c \equiv 0 \pmod N$, for any integers x, y, $P_2(\frac{ax+cy}{N}) = P_2(\frac{x}{N})$. Thus,

$$12 \sum_{x=0}^{N-1} \sum_{y=0}^{N-1} \binom{k-2}{r} x^{k-2-r} (-y)^r \Phi^{(2)}_{(\frac{x}{N}, \frac{y}{N})}(A) \\ = \binom{k-2}{r} (-1)^r (S_0 + S_1 + S_2), \quad (4.4.1)$$

where

$$S_0 := 12 \sum_{x=0}^{N-1} \sum_{y=0}^{N-1} x^{k-2-r} y^r \left(\frac{c}{2a} P_2(\frac{y}{N}) - \frac{b}{2a} P_2(\frac{x}{N})\right),$$

$$S_1 := 12 \sum_{x=0}^{N-1} \sum_{y=0}^{N-1} x^{k-2-r} y^r P_1(\frac{x}{N}) P_1(\frac{y}{N}),$$

$$S_2 := -12 \sum_{x=0}^{N-1} \sum_{y=0}^{N-1} \sum_{i=0}^{a-1} x^{k-2-r} y^r P_1(\frac{x}{N} + c\frac{\frac{y}{N}+i}{a}) P_1(\frac{\frac{y}{N}+i}{a}).$$

Recalling the notation $B_s^*(X) := B_s(X) - B_s(0)$, by (4.1), we immediately see

$$\begin{aligned} S_0 = {} & \frac{6c}{a} \frac{B^*_{k-r-1}(N)}{k-r-1} \left(\frac{B^*_{r+3}(N)}{N^2(r+3)} - \frac{B^*_{r+2}(N)}{N(r+2)} + \frac{B^*_{r+1}(N)}{6(r+1)} \right) \\ & - \frac{6b}{a} \left(\frac{B^*_{k-r+1}(N)}{N^2(k-r+1)} - \frac{B^*_{k-r}(N)}{N(k-r)} + \frac{B^*_{k-r-1}(N)}{6(k-r-1)} \right) \frac{B^*_{r+1}(N)}{r+1} \\ \equiv {} & 0 \pmod{\frac{GCD\{6,N\}N}{D_{k,r}} \mathbb{Z}'_N}. \end{aligned}$$

(Note also here that $b \equiv c \equiv 0 \pmod N$.) Let us rewrite S_2. According to the decomposition of $P_1(\frac{\frac{y}{N}+i}{a})$ as $\frac{y}{aN} + (\frac{i}{a} - \frac{1}{2})$, we may write

$$S_2 = S_2' + S_2'', \quad (4.4.2)$$

where

$$S_2' = -12\sum_{x=0}^{N-1}\sum_{y=0}^{N-1}\sum_{i=0}^{a-1} x^{k-2-r}y^r P_1(\frac{x}{N}+c\frac{y+Ni}{aN})(\frac{y}{aN})$$
$$= -12\sum_{x=0}^{N-1}\sum_{y=0}^{N-1} x^{k-2-r}y^r P_1(\frac{ax+cy}{N})(\frac{y}{aN})$$
$$= -12\sum_{x=0}^{N-1}\sum_{y=0}^{N-1} x^{k-2-r}y^r P_1(\frac{x}{N})(\frac{y}{aN}),$$
$$S_2'' = -12\sum_{x=0}^{N-1}\sum_{y=0}^{N-1}\sum_{i=0}^{a-1} x^{k-2-r}y^r P_1(\frac{x}{N}+c\frac{y+Ni}{aN})(\frac{i}{a}-\frac{1}{2}).$$

Noticing that $\frac{i}{a}-\frac{1}{2}\in\frac{1}{2}\mathbb{Z}_N'$ and that $c\frac{y+iN}{aN}\in\mathbb{Z}_N'$ as $a\equiv 1$, $c\equiv 0 \pmod N$, we may apply (4.2) to the part "$x^{k-2-r}P_1(\frac{x}{N}+c\frac{y+iN}{aN})$" of S_2'' to get the congruence

$$S_2''\equiv T_1+T_2 \pmod{\frac{12N}{2d_{k-r-1}}\mathbb{Z}_N'} \tag{4.4.3}$$

where

$$T_1 := -12\sum_{x=0}^{N-1}\sum_{y=0}^{N-1}\sum_{i=0}^{a-1}\frac{N^{k-2-r}}{k-1-r}$$
$$\times\left(P_{k-1-r}(\frac{x}{N}+c\frac{y+Ni}{aN})-P_{k-1-r}(\frac{x}{N})\right)y^r(\frac{i}{a}-\frac{1}{2})$$
$$= -12\sum_{y=0}^{N-1}\sum_{i=0}^{a-1}\frac{1}{k-1-r}$$
$$\times\left(P_{k-1-r}(c\frac{y+Ni}{a})-P_{k-1-r}(0)\right)y^r(\frac{i}{a}-\frac{1}{2}),$$
$$T_2 := -12\sum_{x=0}^{N-1}\sum_{y=0}^{N-1}\sum_{i=0}^{a-1} x^{k-2-r}P_1(\frac{x}{N})y^r(\frac{i}{a}-\frac{1}{2})$$
$$= -12\sum_{x=0}^{N-1}\sum_{y=0}^{N-1} x^{k-2-r}P_1(\frac{x}{N})y^r(-\frac{1}{2}).$$

Combining the above S_1, S_2', T_2, we obtain

$$\begin{aligned}
&S_1 + S_2' + T_2 \\
&= 12\Big(\sum_x x^{k-2-r} P_1\big(\frac{x}{N}\big)\Big) \cdot \Big(\sum_y y^{r+1}\Big)\Big(\frac{1}{N} - \frac{1}{aN}\Big) \\
&= 12\frac{a-1}{aN}\frac{B^*_{r+2}(N)}{r+2}\Big(\frac{B^*_{k-r}(N)}{N(k-r)} - \frac{B^*_{k-r-1}(N)}{2(k-r-1)}\Big) \equiv 0 \ (\mathrm{mod}\ \frac{6N}{D_{k,r}}\mathbb{Z}'_N).
\end{aligned}$$

Meanwhile, since

$$\begin{aligned}
&-12\sum_{y=0}^{N-1}\sum_{i=0}^{a-1}\frac{1}{k-1-r}P_{k-1-r}\Big(c\frac{y+Ni}{a}\Big)y^r\Big(-\frac{1}{2}\Big) \\
&\qquad\qquad\qquad\qquad = \frac{6}{a^{k-r}}\frac{P_{k-1-r}(0)}{k-1-r}\frac{B^*_{r+1}(N)}{r+1}
\end{aligned}$$

and

$$-12\sum_{y=0}^{N-1}\sum_{i=0}^{a-1}\frac{P_{k-1-r}(0)}{k-1-r}y^r\Big(\frac{i}{a}-\frac{1}{2}\Big) = -6\frac{P_{k-1-r}(0)}{k-1-r}\frac{B^*_{r+1}(N)}{r+1},$$

it follows that

$$T_1 \equiv -12\sum_{y=0}^{N-1}\sum_{i=0}^{a-1}\frac{1}{k-1-r}P_{k-1-r}\Big(c\frac{y+Ni}{a}\Big)y^r\frac{i}{a} \ (\mathrm{mod}\ \frac{6N}{D_{k,r}}\mathbb{Z}'_N).$$

Summing up the above discussions with (4.4.2-3), we may rewrite (4.4.1) as

$$\begin{aligned}
&12\sum_{x=0}^{N-1}\sum_{y=0}^{N-1}\binom{k-2}{r}x^{k-2-r}(-y)^r\Phi^{(2)}_{(\frac{x}{N},\frac{y}{N})}(A) \\
&\equiv -12\binom{k-2}{r}(-1)^r\sum_{y=0}^{N-1}\sum_{i=0}^{a-1}\frac{1}{k-1-r}P_{k-1-r}\Big(c\frac{y+Ni}{a}\Big)y^r\frac{i}{a} \qquad (4.4.4)
\end{aligned}$$

modulo $\binom{k-2}{r}\frac{GCD\{6,N\}N}{D_{k,r}}\mathbb{Z}'_N$. Next, we shall consider the right hand side of the statement formula. Reading the coefficients of X^rY^{k-2-r} of (4.4.0) for $(x_1,x_2)=(0,0)$, we get

$$12\Phi_{\mathbf{0}}^{(r+1,k-1-r)}(A) = U_1 + U_2 + U_3,$$

where

$$U_1 = -12\binom{k-2}{r}\frac{P_k(0)}{k} \times \left\{ \int_0^{-c/a} (-1)^r t^{k-2-r} dt + \int_{-b/a}^{0} (at+b)^r (ct+d)^{k-2-r} dt \right\},$$

$$U_2 = \frac{12\binom{k-2}{r}(-1)^r}{(k-1-r)(r+1)} \times \left\{ P_{k-1-r}(0)P_{r+1}(0) - a^r \sum_{i=0}^{a-1} P_{k-1-r}(c\frac{i}{a})P_{r+1}(\frac{i}{a}) \right\},$$

$$U_3 = -12 \sum_{j=r+1}^{k-2} a^r c^{j-r} \binom{j}{r}\binom{k-2}{j}(-1)^j \sum_{i=0}^{a-1} \frac{P_{k-1-j}(ci/a)P_{j+1}(i/a)}{(k-1-j)(j+1)}.$$

As $b \equiv c \equiv 0 \pmod N$, it is easy to see $U_1 \equiv 0 \ (\mathrm{mod}\ \frac{12\binom{k-2}{r}N}{LCM\{e_1,\ldots,e_{k-2}\}}\mathbb{Z}'_N)$. Moreover, taking the equality $\binom{j}{r}\binom{k-2}{j} = \binom{k-2}{r}\binom{k-2-r}{j-r}$ into accounts, we easily see that U_3 vanishes modulo $\frac{12\binom{k-2}{r}N}{LCM\{d_{j+1}d_{k-1-j}|r+1\le j\le k-2\}}\mathbb{Z}'_N$ (again as $c \equiv 0 \pmod N$). Now, comparing the distribution relations

$$\begin{cases} \Phi_{\mathbf{0}}^{(r+1,k-1-r)}(A) = N^{k-2} \sum_{x=0}^{N-1}\sum_{y=0}^{N-1} \Phi_{(\frac{x}{N},\frac{y}{N})}^{(r+1,k-1-r)}(A), \\ P_k(0) = N^{k-2} \sum_{x=0}^{N-1}\sum_{y=0}^{N-1} P_k(\frac{y}{N}) = N^{k-2} \sum_{x=0}^{N-1}\sum_{y=0}^{N-1} P_k(\frac{ax+cy}{N}), \end{cases}$$

we obtain

$$\begin{aligned} U_2 = {} & \frac{12N^{k-2}\binom{k-2}{r}(-1)^r}{(k-1-r)(r+1)} \left\{ \sum_{x=0}^{N-1}\sum_{y=0}^{N-1} P_{k-1-r}(\frac{x}{N})P_{r+1}(\frac{y}{N}) \right. \\ & \left. - \sum_{x=0}^{N-1}\sum_{y=0}^{N-1}\sum_{i=0}^{a-1} a^r P_{k-1-r}(\frac{x}{N} + \frac{cy}{aN} + \frac{ci}{a})P_{r+1}(\frac{i+\frac{y}{N}}{a}) \right\} \\ = {} & 12\binom{k-2}{r}(-1)^r \left\{ \frac{P_{k-1-r}(0)}{k-1-r}\frac{P_{r+1}(0)}{r+1} \right. \\ & \left. - \sum_{y=0}^{N-1}\sum_{i=0}^{a-1} a^r \frac{P_{k-1-r}(\frac{cy}{a} + \frac{ciN}{a})}{k-1-r} N^r \frac{P_{r+1}(\frac{i+\frac{y}{N}}{a})}{r+1} \right\}. \end{aligned} \tag{4.4.5}$$

We shall decompose the second term of the above last side according to

$$\frac{(aN)^r}{r+1}P_{r+1}(\frac{y}{aN}+\frac{i}{a}) \equiv \frac{(aN)^r}{r+1}P_{r+1}(\frac{y}{aN}) + y^r\frac{i}{a} \ (\mathrm{mod}\ \frac{N}{d_{r+1}}\mathbb{Z}'_N),$$

which follows from (4.2) and $0 \le \frac{y}{aN} + \frac{i}{a} < 1$. We compute:

$$\begin{aligned}
&\sum_{i=0}^{a-1}\sum_{y=0}^{N-1}\frac{P_{k-1-r}(\frac{y+iN}{a}c)}{k-1-r}(aN)^r\frac{P_{r+1}(\frac{y}{aN})}{r+1}\\
&\quad = \sum_{y=0}^{N-1} a^{-(k-2-r)}\frac{P_{k-1-r}(0)}{k-1-r}(aN)^r\frac{P_{r+1}(\frac{y}{aN})}{r+1}\\
&\quad \equiv a^{-(k-2-r)}\frac{P_{k-1-r}(0)}{(k-1-r)(r+1)}\sum_{y=0}^{N-1}(\frac{y^{r+1}}{aN}-\frac{r+1}{2}y^r)\\
&\qquad (\mathrm{mod}\ \frac{N}{d_{r+1}d_{k-r-1}}\mathbb{Z}'_N).
\end{aligned}$$

Noticing that the coefficient of X in $B_{r+2}(X)$ is $(-1)^{r+1}(r+2)B_{r+1}$ and that $P_{r+1}(0) = 0$ if $r+1$ is odd, the above last side continues to

$$\begin{aligned}
&= a^{-(k-2-r)}\frac{P_{k-1-r}(0)}{k-1-r}(\frac{B^*_{r+2}(N)}{aN(r+1)(r+2)} - \frac{B^*_{r+1}(N)}{2(r+1)})\\
&\equiv \frac{P_{k-1-r}(0)}{k-1-r}\frac{(-1)^{r+1}}{r+1}P_{r+1}(0) \ (\mathrm{mod}\ \frac{N}{2D_{k,r}}\mathbb{Z}'_N)\\
&= \frac{P_{k-1-r}(0)}{k-1-r}\frac{P_{r+1}(0)}{r+1}.
\end{aligned}$$

Here we also used $a \equiv 1 \ (\mathrm{mod}\ N)$. Therefore, we conclude

$$U_2 \equiv -\frac{12\binom{k-2}{r}(-1)^r}{k-1-r}\sum_{i=0}^{a-1}\sum_{y=0}^{N-1}P_{k-1-r}(\frac{y+iN}{a}c)y^r\frac{i}{a} \ (\mathrm{mod}\ \frac{6\binom{k-2}{r}N}{D_{k,r}}\mathbb{Z}'_N),$$

which coincides with (4.4.4). This completes the proof of Proposition (4.3). □

Corollary (4.5). *Let $k \ge 2$. Then, there exists an integer $D(k)$ (depending only on k) such that for every $N \ge 1$,*

$$12\sum_{x=0}^{N-1}\sum_{y=0}^{N-1}(xY-yX)^{k-2}\Phi^{(2)}_{(\frac{x}{N},\frac{y}{N})}(A) \equiv 12\Phi_0^{(k)}(A) \ (\mathrm{mod}\ \frac{N}{D(k)}\mathbb{Z}'_N).$$

Proof. This is an immediate consequence of Proposition (4.2). We may take $D(k)$ so that

$$\frac{N}{D(k)}\mathbb{Z}'_N = \sum_{r=0}^{k-2}\binom{k-2}{r}\frac{GCD\{6,N\}N}{D_{k,r}}\mathbb{Z}'_N \subset \mathbb{Q}.$$

□

Example (4.6). Take a matrix $A = \begin{pmatrix} 122 & -4961 \\ 363 & -14761 \end{pmatrix} \in \Gamma(11^2)$, and let us examine our congruence formula (4.5) for $k = 6$, $N = 11$. In this case, for any $0 \le r \le 4$, $\binom{4}{r}\frac{GCD\{6,N\}11}{12D_{6,r}}\mathbb{Z}'_{11} = 11\mathbb{Z}'_{11}$, hence $\frac{N}{12D(6)}\mathbb{Z}'_{11}$ may be taken as $11\mathbb{Z}'_{11}$ Computation on RHS shows

$$\begin{aligned}
&\Phi_{\mathbf{0}}^{(6)}(A)(X,Y)\\
&= \frac{6157810527168637}{315}X^4 + \frac{117260782677249595}{504}YX^3\\
&\quad + \frac{37381997569467617}{36}Y^2X^2 + \frac{519057868253062293 7}{2520}Y^3X\\
&\quad + \frac{1930511018334372017}{1260}Y^4,
\end{aligned}$$

while that on LHS shows

$$\begin{aligned}
&\sum_{x,y=0}^{10}(xY-yX)^4\Phi^{(2)}_{(\frac{x}{11},\frac{y}{11})}(A)\\
&\qquad = \frac{157339}{6}X^4 + \frac{557102}{3}YX^3 - 102157Y^2X^2\\
&\qquad\quad - \frac{230146}{3}Y^3X + \frac{717205}{6}Y^4.
\end{aligned}$$

In both sides, each coefficient except for that of X^2Y^2 is prime to 11. Now, their difference may be computed as:

$$\begin{aligned}
&\Phi_{\mathbf{0}}^{(6)}(A)(X,Y) - \sum_{x,y=0}^{10}(xY-yX)^4\Phi^{(2)}_{(\frac{x}{11},\frac{y}{11})}(A)\\
&= \frac{12315621037816679}{630}X^4 + \frac{117260782583656459}{504}YX^3\\
&\quad + \frac{37381997573145269}{36}Y^2X^2 + \frac{5190578682723945577}{2520}Y^3X\\
&\quad + \frac{1930511018183758967}{1260}Y^4\\
&\equiv 0 \pmod{11\mathbb{Z}'_{11}}.
\end{aligned}$$

Example (4.7). This example was suggested by Y. Morimoto. Let $A = \begin{pmatrix} 6643 & 56295 \\ 1539 & 13042 \end{pmatrix} \in \Gamma(81)$, and consider the case of weight 10. Then, for $0 \leq r \leq 8$, $\binom{8}{r}\frac{GCD\{6,81\}81}{12D_{10,r}}\mathbb{Z}_3' = 3\mathbb{Z}_3'$. The coefficient $\Phi_{\mathbf{0}}^{(5,5)}(A)$ of X^4Y^4 in $\Phi_{\mathbf{0}}^{(10)}(A)$ is computed as

$$\Phi_{\mathbf{0}}^{(5,5)}(A) = -\frac{116073541903991309357774956489289951 9}{8}$$

which is prime to 3, while

$$\sum_{x,y=0}^{80} \binom{8}{4} x^4(-y)^4 \Phi^{(2)}_{(\frac{x}{81},\frac{y}{81})}(A) = -8836456074579123550$$

which is also prime to 3. Now the difference of these two values is

$$\Phi_{\mathbf{0}}^{(5,5)}(A) - \sum_{x,y} \binom{8}{4} x^4(-y)^4 \Phi^{(2)}_{(\frac{x}{81},\frac{y}{81})}(A)$$
$$= -\frac{1160735419039913022886100968259911119}{8}$$

which is divisible by 3 (but not divisible by 9).

Example (4.8). Let us consider the case of weight 4 for $A = \begin{pmatrix} 385 & 15616 \\ 15744 & 638593 \end{pmatrix} \in \Gamma(128)$. Then, $\binom{2}{r}\frac{GCD\{6,128\}128}{12D_{6,r}}\mathbb{Z}_2'$ is $4\mathbb{Z}_2'$ for $r = 0, 2$, and $8\mathbb{Z}_2'$ for $r = 1$. The computation shows

$$\Phi_{\mathbf{0}}^{(4)}(A) = \frac{1236394022}{45}X^2 + \frac{33706991732}{15}XY + \frac{229732849464}{5}Y^2,$$

and

$$\sum_{x,y=0}^{127} (xY - yX)^2 \Phi^{(2)}_{(\frac{x}{128},\frac{y}{128})}(A)$$
$$= \frac{80854186}{3}X^2 - \frac{10808380}{3}XY - \frac{80165720}{3}Y^2.$$

Both of them do not belong to $4\mathbb{Z}_2'[X, Y]$ (because of the coefficients of X^2). Now the difference of these two polynomials is computed as

$$\Phi_{\mathbf{0}}^{(4)}(A) - \sum_{x,y=0}^{127} (xY - yX)^2 \Phi^{(2)}_{(\frac{x}{128},\frac{y}{128})}(A)$$
$$= \frac{23581232}{45}X^2 + \frac{33761033632}{15}XY + \frac{689599376992}{15}Y^2$$

which belongs to $4\mathbb{Z}_2'[X,Y]$ (actually to $16\mathbb{Z}_2'[X,Y]$).

At the time of writing this paper, the author does not have an example assuring whether our estimate of the modulus is best possible with respect to 2-powers.

Acknowledgements

The author would like to thank Yasuhiko Morimoto for valuable discussions on the topic and useful calculations of numerical examples. He is also very grateful to the referee for crucial comments which were very helpful to improve the submitted version of this paper.

References

[D] R. Dedekind, *Erläuterungen zu zwei Fragmenten von Riemann*, in "Riemann's Gesammelte Math. Werke, 2nd edition," 466–472, 1892.

[KM] R. Kirby and P. Melvin, *Dedekind sums, μ-invariants and the signature cocycle*, Math. Ann. **299** (1994), 231–267.

[Mo] Y. Morimoto, *Arithmetic behavior of functions related with generalized Dedekind sums* (Japanese), Master Thesis, Tokyo Metropolitan University, January 2002.

[N1] H. Nakamura, *On exterior Galois representations associated with open elliptic curves*, J. Math. Sci. Univ. Tokyo **2** (1995), 197–231.

[N2] H. Nakamura, *Galois representations in the pro-l fundamental groups of punctured elliptic curve* (Japanese), in "RIMS Technical Report Series 884," 46–53, 1994.

[N3] H. Nakamura, *Tangential base points and Eisenstein power series*, in "Aspects of Galois Theory (H.Voelkein, D.Harbater, P.Mueller, J.G.Thompson, eds.)," 202–217, London Math. Soc. Lecture Note Ser., 256. Cambridge Univ. Press, Cambridge, 1999.

[N4] H. Nakamura, *On exterior monodromy representations associated with affine elliptic curves*, in preparation.

[R] H. Rademacher, *Zur Theorie der Modulfunktionen*, J. Reine Angew. Math. **167** (1931), 312–336.

[RG] H. Rademacher and E. Grosswald, "Dedekind sums," Carus Math. Monographs, 16. The Mathematical Association of America, Washington, D.C., 1972.

[Scz] R. Sczech, *Eisenstein cocycles for $GL_2\mathbb{Q}$ and values of L-functions in real quadratic fields*, Comment. Math. Helvetici **67** (1992), 363–382.

[St] G. Stevens, *The Eisenstein measure and real quadratic fields*, in "Proceedings of the International Conference on Number Theory (Quebec, PQ, 1987)," 887–927, de Gruyter, Berlin, 1987.